**NORTH-HOLLAND
PERSONAL LIBRARY**

ELLIPSOMETRY AND POLARIZED LIGHT

ELLIPSOMETRY AND POLARIZED LIGHT

R. M. A. AZZAM
Department of Electrical Engineering
University of New Orleans
Lakefront, New Orleans, Louisiana, USA

and

N. M. BASHARA
Electrical Materials Laboratory, Engineering Research Center,
College of Engineering, University of Nebraska, Lincoln, USA

NORTH-HOLLAND
AMSTERDAM · OXFORD · NEW YORK · TOKYO

© Elsevier Science Publishers B.V., 1977, 1987

All rights reserved. No part of this publication may be reproduced, stored in a retrieval system, or transmitted, in any form or by any means, electronic, mechanical, photocopying, recording or otherwise, without the prior permission of the publisher, Elsevier Science Publishers B.V. (North-Holland Physics Publishing Division), P.O. Box 103, 1000 AC Amsterdam, The Netherlands.
Special regulations for readers in the USA: This publication has been registered with the Copyright Clearance Center Inc. (CCC), Salem, Massachusetts. Information can be obtained from the CCC about conditions under which photocopies of parts of this publication may be made in the USA.
All other copyright questions, including photocopying outside of the USA, should be referred to the publisher.

ISBN: 0 444 87016 4

Published by:
North-Holland Physics Publishing
a division of
Elsevier Science Publishers B.V.
P.O. Box 103
1000 AC Amsterdam
The Netherlands

Sole distributors for the USA and Canada:
Elsevier Science Publishing Company, Inc.
52 Vanderbilt Avenue
New York, NY 10017
USA

First edition 1977
Reprinted 1979
Paperback edition 1987

Library of Congress Cataloging in Publication Data

The Library of Congress has Cataloged the first printing of this title as follows:

Azzam, R. M. A.
   Ellipsometry and polarized light.

   Bibliography: p.
   Includes index.
   1. Ellipsometry. 2. Polarization (Light)
I. Bashara, N. M., joint author. II. Title.
QC443.A96   535.5'2   76-55761
ISBN 0-7204-0694-3
      0-444-87016-4 (Paperback)

PRINTED IN THE NETHERLANDS

Preface to the Paperback Edition

Corrections have been made to the first edition and we thank the readers who brought several of them to our attention.

Progress continues to be made in all aspects of ellipsometry, theoretical and experimental, accompanied by increased sophistication in instrumentation and applications. Two international conferences dealing principally with ellipsometry have been held since this book was published (Berkeley, 1979 and Paris, 1983). The contents of their proceedings (along with those of earlier conferences) are listed at the end of this book. For research activities in the Soviet Union, we refer the reader to the book (in Russian) *Principles of Ellipsometry*, by A.V. Rzhanov et al. (NAUKA, Novosibirsk, 1979) and to the extensive bibliography at its end.

We are pleased that this edition makes our book affordable to individual students and researchers.

October, 1986

R.M.A. Azzam
N.M. Bashara

Preface

Ellipsometry is an optical technique for the characterization of, and observation of events at, an interface or film between two media and is based on exploiting the polarization transformation that occurs as a beam of polarized light is reflected from or transmitted through the interface or film. Two factors make ellipsometry particularly attractive: (1) its essentially non-perturbing character (when the wavelength and intensity of the light beam are properly chosen), hence its suitability for *in-situ* measurements, and (2) its remarkable sensitivity to minute interfacial effects, such as the formation of a sparsely distributed sub-monolayer of atoms or molecules.

The great diversity of situations in nature and man-made systems where interfaces and films play an important role has led to the application of ellipsometry in a wide spectrum of fields such as physics, chemistry, materials and photographic science, biology, as well as optical, electronic, mechanical, metallurgical and biomedical engineering.

In spite of the steadily growing interest in ellipsometry, no English monograph is available that is primarily devoted to this subject.* *Ellipsometry and Polarized Light* represents our attempt to remedy this deficiency; we trust that it provides adequate coverage of the theory, instrumentation and applications of ellipsometry. We hope it will be useful to both the newcomer to the field and the specialist.

The book proceeds according to the following plan. In chapter 1 we deal with the various mathematical representations that are employed to describe polarized light, and in chapter 2 we use these representations to discuss the

---

*There are several excellent reviews on ellipsometry, in addition to the monograph in Romanian by Moisil and Moisil (ref. [315] in the bibliography at the end of the book).

interaction of polarized light with the optical components that may compose a polarizing optical system, *e.g.*, an ellipsometer. We place special emphasis on the complex-plane representation of polarization, because this concise and useful representation has been grossly neglected in optics. The potential utility of this part of the book transcends the domain of ellipsometry and should be of value to anyone interested in polarized vector waves in general and polarized light in particular. In chapter 3, the mathematical tools developed earlier in chapters 1 and 2 are applied to analyze the theory of measurement in ellipsometers, including a comprehensive study of the effect of component imperfections. Chapter 4 on the reflection and transmission of polarized light by stratified planar structures is intended to provide results and techniques that are essential for the interpretation of ellipsometric data in terms of the macroscopic properties of the particular sample under measurement. Having developed the theoretical basis for ellipsometry in chapters 1–4, we devote the remainder of the book to the practical side of the subject. In chapter 5 we consider the instrumentation of ellipsometry, and in chapter 6 we present many of its interesting applications.

The book is structured so as to give an integrated treatment of the various aspects of ellipsometry. Readers who are primarily interested in applying ellipsometry will be mostly interested in chapters 4–6. On the other hand, instrument designers will find the methods and techniques of chapters 1–3 particularly useful.

In writing this book we have generally opted for detail, an approach that will certainly be helpful to the beginner. Only basic knowledge of matrix algebra and complex variables is required for understanding much of the theoretical parts of this book.

In addition to reliance on our own contributions to the field, which will be evident in certain parts of the book (particularly in chapters 2 and 3), we have liberally quoted from the publications of many other investigators (especially in chapters 5 and 6), with appropriate reference given in each case. While the bibliography at the end may not be complete, it is fairly representative of the field. Unavoidably, some interesting work has been left out.

R.M.A. Azzam gratefully acknowledges an extended leave of absence from Cairo University and the kind hospitality of the Hematology Division and Department of Internal Medicine of the University of Nebraska College of Medicine, Omaha. The Metallurgy and Materials Section of the National Science Foundation provided generous support to the Electrical Materials Laboratory that was essential for carrying out some of the work reported in this monograph. Support from the Metallurgy Program of the U.S. Office of Naval Research was also helpful. The secretarial assistance of Mrs. P. Barrows and Mrs. D. VanSchooten is kindly appreciated. Finally, we wish to dedicate this effort to our parents and to our wives.

R.M.A. Azzam, N.M. Bashara

Acknowledgements

We are indebted to the authors and publishers for permission to reproduce the following figures appearing in this book:

5.21, Appl. Opt. **5**, 759 (1966) H. Takasaki. 6.67, Appl. Phys. Lett. **24**, 200 (1974) N.J. Chou, Y.J. van der Meulen, R. Hammer and J. Cahill. 6.16, Can. J. Chem. **49**, 1115 (1971) M.J. Dignam, B. Rao, M. Moscovitz and R.W. Stobie (Reproduced by permission of the National Research Council of Canada). 6.56, 6.57, 6.58, Cytobios, **3**, 5 (1971) G. Poste and L.W. Greenham. 5.24, 5.25, 5.26, 5.27, IBM J. Res. Devel. **17**, 472 (1973) P.S. Hauge and F.H. Dill (Copyright © 1973 by International Business Machines Corporation). 6.72, Jap. J. App. Phys. **11**, 1205 (1972) K. Nakamura and M. Kondo. 6.42, 643, J. Chem. Phys. **48**, 671 (1968) A.K.N. Reddy, M.A. Genshaw and J. O'M. Bockris. 6.32, J. Electron. Mater. **2**, 115 (1973) N.J. Chou, J.M. Eldridge, R. Hammer and D.W. Dong. 6.28, J. Electrochem. Soc. **104**, 619 (1957) R.J. Archer (Reprinted by permission of the publisher, The Electrochemical Society, Inc.). 6.35, 6.36, J. Electrochem. Soc. **108**, 503 (1961) J. Kruger (Reprinted by permission of the publisher, The Electrochemical Society, Inc.). 6.40, 6.41, J. Electrochem. Soc. **113**, 277 (1966) L. Young and F.G.R. Zobel (Reprinted by permission of the publisher, The Electrochemical Society, Inc.). 6.44, 6.45, 6.46, J. Electrochem. Soc. **120**, 1225 (1973) C.J. Dell'Oca (Reprinted by permission of the publisher, The Electrochemical Society, Inc.). 6.47, 6.48, J. Electrochem. Soc. **113**, 1133 (1966) J. O'M. Bockris, A.K.N. Reddy and M.G.B. Rao (Reprinted by permission of the publisher, The Electrochemical Society, Inc.). 6.49, 6.50, J. Electrochem. Soc. **114**, 43 (1967) J. Kruger and J.P. Calvert (Reprinted by permission of the publisher, The Electrochemical Society, Inc.). 6.53, 6.54, J. Electrochem. Soc. **117**, 733 (1970) M. Novak, A.K.N. Reddy and H. Wroblowa (Reprinted by permission of the publisher, The Electrochemical Society, Inc.). 3.21, J. Opt. Soc. Am. **26**, 122 (1936) H.M. O'Bryan. 6.2, J. Opt. Soc. Am. **61**, 608 (1971) L.G. Holcomb and N.M. Bashara. 6.4, J. Opt. Soc. Am. **58**, 102 (1968) T.E. Faber and N.V. Smith. 6.7, J. Opt. Soc. Am. **64**, 804 (1974) Y.J. van der Meulen and N.C. Hien. 6.8, J. Opt. Soc. Am. **63**, 529 (1972) F. Meyer, E.E. Kluizenaar and D. den Engelsen. 6.9, J. Opt. Soc. Am. **57**, 751 (1967) R.T. Jacobsen and M. Kerker. 6.15, J. Opt. Soc. Am. **56**, 1320 (1966) N.M. Bashara and D.W. Peterson. 6.22, J. Opt. Soc. Am. **58**, 1069 (1968) T. Smith. 6.34, J. Opt. Soc. Am. **62**, 1524 (1972) R.J. Lederich. 6.70, J. Opt. Soc. Am. **65**, 914 (1975) A.B. Buckman, N.H. Hong and D. Wilson. 6.12, 6.13, 6.14, J. Phys. Chem. **65**, 2242 (1961) L.S. Bartell and D. Churchill (Copyright by the American Chemical Society). 6.38, 6.39, J. Phys. Chem. **74**, 4266 (1970) W.K. Paik, M.A. Genshaw, and J. O'M. Bockris (Copyright by the American Chemical Society). 5.28, 6.3, Optics Commun. **8**, 222 (1973) D.E. Aspnes. 6.10, 6.11, Phys. Rev. **174**, 719 (1968) A.B. Buckman and N.M. Bashara. 6.68, 6.69, Proc. R. Soc. Lond. **A. 335**, 39 (1973) W.D. Cornish and L. Young. 5.22, Rev. Sci. Instrum. **45**, 798 (1974) H.J. Mathieu, D.E. McClure and R.H. Muller. 5.29, Rev. Sci. Instrum. **40**, 761 (1969) S.N. Jasperson and S.E. Schnatterly. 5.23, Sci. Light **16**, 64 (1967) T. Yamaguchi and H. Hasunuma. 6.66, Solid State Commun. **12**, 615 (1973) K. Knorr and J.D. Leslie. 5.18, Surface Sci. **16**, 34 (1969) P.H. Smith. 5.30, Surface Sci. **37**, 548 (1973) S.N. Jasperson, D.K. Burge and R.C. O'Handley. 6.17, Surface Sci. **46**, 308 (1974) M.J. Dignam, B. Rao and R.W. Stobie. 6.25, 6.26, Surface Sci. **16**, 234 (1969) R.H. Muller, R.F. Steiger, G.A. Somorjai and J.M. Morabito. 6.29, Surface Sci. **30**, 91 (1972) F. Lukes. 6.30, 6.31, Surface Sci. **38**, 292 (1973) T. Smith. 6.43, Surface Sci. **46**, 1 (1974) J. Horkans, B.D. Cahan and E. Yeager. 6.61, 6.62, Surface Sci. **30**, 632 (1972) M.M. Ibrahim and N.M. Bashara. 6.64, 6.65, Surface Sci. **49**, 441 (1975) J.R. Adams and N.M. Bashara. 6.71, Surface Sci. **49**, 236 (1975) J.-F. Reber and R. Steiger. 6.63, Surface Sci. **30**, 680 (1972) M.M. Ibrahim and N.M. Bashara. 6.51, 6.52, *Advances in Electrochemistry and Electrochemical Engineering*, Vol. 9, *Optical Techniques in Electrochemistry*, ed. R.H. Muller (Wiley, New York, 1973) J. Kruger. 6.27, *Ellipsometry in the Measurement of Surfaces and Thin Films*, ed. E. Passaglia, R.R. Stromberg and J. Kruger (National Bureau of Standards, Washington, 1964) E. Passaglia, R.R. Stromberg and D.J. Tutas. 6.33, *Ellipsometry in the Measurement of surfaces and Thin Films*, ed. E. Passaglia, R.R. Stromberg and J. Kruger (National Bureau of Standards, Washington, 1964) P.C.S. Hayfield and G.W.T. White. 6.18, 6.19, 6.20, and 6.21, *Ellipsometry in the Measurement of Surfaces and Thin Films*, ed. E. Passaglia, R.R. Stromberg and J. Kruger (National Bureau of Standards, Washington, 1964) R.J. Archer. 6.5, 6.6, *Manual on Ellipsometry* (Gaertner Scientific Corporation, Chicago, 1968) R.J. Archer. 6.1, *Optical Properties of Thin Solid Films* (Dover, New York, 1965) O.S. Heavens. 6.23, 6.24, *Proceedings of the Symposium on Recent Developments in Ellipsometry*, Eds. – N.M. Bashara, A.B. Buckman, and A.C. Hall (North-Holland, Amsterdam, 1969) F. Meyer and G.A. Bootsma. 6.37, *Proceedings of the Symposium on Recent Developments in Ellipsometry*, Eds. – N.M. Bashara, A.B. Buckman and A.C. Hall (North-Holland, Amsterdam, 1969) D.K. Burge, J.M. Bennett, R.L. Peck and H.E. Bennett. 6.59, 6.60, *Progress in Surface Science*, Vol. 2 (Part 3) Ed. – S.G. Davison (North-Holland, Amsterdam, 1972) G. Poste and C. Moss. 2.16–2.21, 2.24, *Appl. Opt.* **12**, 764 (1973) R.M.A. Azzam, B.E. Merrill and N.M. Bashara. 4.29–4.35, *Appl. Opt.* **14**, 1652 (1975) R.M.A. Azzam, M. Elshazly-Zaghloul and N.M. Bashara. 5.12–5.14, *Appl. Opt.* **15**, 2568 (1976) D.L. Confer, R.M.A. Azzam and N.M. Bashara. 5.16, 5.17, *Appl. Opt.* **13**, 1938 (1974) J.R. Zeidler, R.B. Kohles and N.M. Bashara. 4.15–4.18, 4.20–4.25, *J. Opt. Soc. Am.* **65**, 252 (1975) R.M.A. Azzam, A.-R.M. Zaghloul and N.M. Bashara. 4.36, 4.37, *J. Opt. Soc. Am.* **61**, 1622 (1971) M.M. Ibrahim and N.M. Bashara. 5.10, *J. Res. Natl. Bur. Std.* **67A**, 363 (1963) F.L. McCrackin, E. Passaglia, R.R. Stromberg and H.L. Steinberg. 3.12–3.15, 3.17, 3.18, *Optics Commun.* **7**, 110 (1973) R.M.A. Azzam, T.L. Bundy and N.M. Bashara. 4.43, 4.44, *Surface Sci.* **16**, 85 (1969) C.A. Fenstermaker and F.L. McCrackin.

We are also indebted for permission to reproduce the following tables:

6.1, J. Opt. Soc. Am. **49**, 1195 (1959) J. Kruger and W.J. Ambs. 6.2, J. Opt. Soc. Am. **58**, 1069 (1968) T. Smith. 6.3, *Proceedings of the Symposium on Recent Developments in Ellipsometry*, Eds. – N.M. Bashara, A.B. Buckman and A.C. Hall (North-Holland, Amsterdam, 1969) F. Meyer and G.A. Bootsma. 6.4, Surface Sci. **30**, 632 (1972) M.M. Ibrahim and N.M. Bashara.

Contents

PREFACE TO THE PAPERBACK EDITION	v
PREFACE	vi
ACKNOWLEDGEMENTS	viii
CONTENTS	xi

CHAPTER 1.	THE POLARIZATION OF LIGHT WAVES	1
1.1.	The concept of polarization	1
1.2.	Polychromatic and monochromatic waves of light	2
1.3.	Polarization of monochromatic waves of arbitrary spatial structure . .	2
	1.3.1. Elliptical polarization	6
	1.3.2. Linear and circular polarizations	9
1.4.	The phasor (complex) representation of time-harmonic fields	10
1.5.	Uniform TE plane waves of light	11
1.6.	The Jones vector of a uniform TE plane wave	13
	1.6.1. Wave intensity	16
	1.6.2. Jones vectors of some states of polarization	17
	1.6.3. Orthogonal and orthonormal pairs of Jones vectors	19
	1.6.4. Transformation of the Jones vector under the effect of a coordinate rotation	19
	1.6.5. Basis Jones vectors and basis states of polarization	22
	1.6.5.1. Cartesian basis vectors	22
	1.6.5.2. Circular basis vectors	22
	1.6.5.3. Arbitrary basis vectors	25
	1.6.6. The Cartesian and circular Jones vectors of a given elliptical polarization state	26
1.7.	Representation of the polarization states of light by complex numbers	28
	1.7.1. The Cartesian complex-plane representation of polarized light	28
	1.7.1.1. Orthogonal states	30
	1.7.1.2. Equi-azimuth and equi-ellipticity contours	33

	1.7.2. The circular complex-plane representation of polarized light	39
	1.7.3. The generalized complex-plane representation of polarized light	43
1.8.	The Poincaré-sphere representation of polarized light	47
1.9.	The polarization of quasi-monochromatic waves	52
	1.9.1. The Jones vector of a quasi-monochromatic wave	53
	1.9.2. The Stokes parameters and the Stokes vector	55
	1.9.3. The coherency-matrix representation	60

CHAPTER 2. PROPAGATION OF POLARIZED LIGHT THROUGH POLARIZING OPTICAL SYSTEMS 66

2.1.	Polarizing optical systems	66
2.2.	The Jones-matrix formulation	67
	2.2.1. The significance of the elements T_{ij} of the Jones matrix	69
	2.2.2. Cascade of optical systems	70
	2.2.3. Jones matrices of basic optical devices	72
	2.2.3.1. Transmission-type devices	72
	2.2.3.2. Reflection-type devices	77
	2.2.4. Transformation of the Cartesian Jones matrix under the effect of a coordinate rotation	79
	2.2.5. Circular and generalized Jones matrices	81
2.3.	Separability of information on the ellipse of polarization in the Jones-matrix formulation – The polarization transfer function	84
2.4.	Transformation of the ellipse of polarization by an optical system in the complex-plane representation	85
2.5.	Transformation of polarization by an optical system in the Poincaré-sphere representation	92
2.6.	Separability of information on amplitude and phase in the Jones-matrix formulation – The amplitude and phase transfer functions	94
	2.6.1. The eigenvalue problem – General classification of optical devices	97
	2.6.1.1. Elliptic retarders	98
	2.6.1.2. Elliptic partial polarizers	100
	2.6.1.3. Elliptic ideal polarizers	101
	2.6.1.4. Other devices	102
2.7.	Polarization-dependent intensity transmittance of optical systems	103
	2.7.1. Loci of polarization states of equal attenuation or amplification	103
	2.7.2. Simplified expressions for the intensity transmittance	107
	2.7.3. Generalization of the Law of Malus	110
	2.7.4. Special cases of optical systems	111
2.8.	Resolution of an arbitrary polarization state into two given orthogonal states in the complex-plane representation	113
	2.8.1. Nearness of two polarization states in the complex-plane representation	113

2.8.2. Resolution of an arbitrary polarization state into two given orthogonal states in the Poincaré-sphere representation 116
2.9. Polarization-dependent phase shift introduced by an optical system .. 117
2.10. Propagation of polarized light in anisotropic media 119
    2.10.1. Evolution of the ellipse of polarization 122
        2.10.1.1. The case of homogeneous anisotropic media 123
        2.10.1.2. The case of inhomogeneous anisotropic media 131
    2.10.2. Evolution of the complex amplitude 137
2.11. Propagation of partially polarized quasi-monochromatic light through non-depolarizing optical systems – The coherency-matrix formulation . 141
    2.11.1. Intensity transmittance of a linear non-depolarizing optical system for partially polarized incident light 144
2.12. Propagation of partially polarized quasi-monochromatic light through depolarizing optical systems – The Mueller-matrix formulation 148

CHAPTER 3. THEORY AND ANALYSIS OF MEASUREMENTS IN ELLIPSOMETER SYSTEMS 153

3.1. Introduction 153
3.2. Ellipsometric measurement of the Jones matrix of an optical system – The basis of generalized ellipsometry 156
3.3. Ellipsometry on optical systems with known eigenpolarizations 158
3.4. Ellipsometric measurement of the ratio of eigenvalues of an optical system with orthogonal linear eigenpolarizations 159
    3.4.1. Null ellipsometry 166
    3.4.2. The fixed-compensator-azimuth nulling scheme with $C = \pm\frac{1}{4}\pi$, $\delta_C = -\frac{1}{2}\pi$ and $T_C = 1$ 167
    3.4.3. Alternate ellipsometer arrangement with the compensator placed after the optical system 169
    3.4.4. The ratio of eigenvalues ρ_S in reflection and transmission ellipsometry 173
    3.4.5. Unified treatment of the PCSA and PSCA ellipsometer arrangements – Example of the application of the bilinear polarization transfer function 175
3.5. Ellipsometric measurement of the ratio of eigenvalues of an optical system with orthogonal circular eigenpolarizations 180
3.6. Ellipsometric measurement of the ratio of eigenvalues of an optical system with elliptic eigenpolarizations 184
3.7. Effect of azimuth-angle errors and component imperfections on the ellipsometric measurement of ρ_S 186
    3.7.1. The imperfection plate of a non-ideal optical component 189
    3.7.2. Transposition of two optical components 191
    3.7.3. The PCWSW'A ellipsometer arrangement 192

		3.7.3.1. Azimuth-angle errors	193
		3.7.3.2. Component imperfections	195
		3.7.3.3. Errors in the ellipsometric angles ψ and Δ	202
		3.7.3.4. Two-zone averages	208
		3.7.3.5. Four-zone averages	210
		3.7.3.6. Choice of compensator azimuth and position	211
3.8.	Ellipsometry with imperfect components including incoherent effects		213
	3.8.1. Application to the PCWSW′A ellipsometer arrangement		216
		3.8.1.1. Polarizer and analyzer imperfections	219
		3.8.1.2. Compensator imperfection	224
		3.8.1.3. Entrance-window, system, and exit-window imperfections	225
	3.8.2. Zone averaging		226
	3.8.3. Component depolarization		229
	3.8.4. Discussion		231
3.9.	Generalized ellipsometry		233
	3.9.1. The polarizer-compensator combination as a controlled polarization filter		234
	3.9.2. Nulling schemes and conditions of compensation		238
		3.9.2.1. The fixed-analyzer nulling scheme	239
		3.9.2.2. The fixed-compensator nulling scheme	240
		3.9.2.3. The fixed-polarizer nulling scheme	242
	3.9.3. Multiple-null measurements		245
	3.9.4. Alternate PSCA ellipsometer arrangement		246
3.10.	Other ellipsometer arrangements		247
	3.10.1. Null ellipsometers that do not employ compensators		248
		3.10.1.1. An ellipsometer based on the detection of the azimuth of the polarization ellipse only, with no measurement of ellipticity	248
		3.10.1.2. Self-compensated ellipsometers based on the angle-of-incidence tunability of the reflection phase difference Δ	251
	3.10.2. Photometric ellipsometers		255
		3.10.2.1. Static photometric ellipsometers	256
		3.10.2.2. Dynamic photometric ellipsometers	257
	3.10.3. Interferometric Ellipsometers (IE)		262
3.11.	Modulated ellipsometry		265
	3.11.1. PSA arrangement		266
	3.11.2. PCSA arrangement		267

CHAPTER 4. REFLECTION AND TRANSMISSION OF POLARIZED LIGHT BY STRATIFIED PLANAR STRUCTURES 269

4.1. Introduction 269

4.2.	Reflection and refraction at the planar interface between two isotropic media	270
4.3.	Reflection and transmission by an ambient-film-substrate system	283
4.4.	The equations of reflection and transmission ellipsometry for ambient-film-substrate systems	288
	4.4.1. Reflection ellipsometry	288
	4.4.1.1. Constant-Angle-of-Incidence Contours (CAIC) of the ellipsometric function $\rho(\phi, d)$	289
	4.4.1.2. Constant-Thickness Contours (CTC) of the ellipsometric function $\rho(\phi, d)$	293
	4.4.1.3. Cartesian curves	298
	4.4.2. Transmission ellipsometry	299
	4.4.3. Linear approximation of the equations of reflection and transmission ellipsometry for ambient-film-substrate systems	301
	4.4.3.1. Reflection	301
	4.4.3.2. Transmission	304
	4.4.3.3. ψ and Δ sensitivity factors	305
	4.4.3.4. Validity of the linear approximations	312
4.5.	Numerical inversion of the exact equation of reflection ellipsometry	315
	4.5.1. Cases that allow analytical inversion	315
	4.5.2. General formulation of the numerical-inversion problem and choice of error function	317
	4.5.3. Multiple-Angle-of-Incidence (MAI) ellipsometry	320
	4.5.3.1. Independence (or interdependence) of the MAI system of equations: The parameter correlation test	321
	4.5.3.2. The Hessian matrix and rate of convergence	324
	4.5.3.3. The air–SiO_2–Si system – An example	326
4.6.	Reflection and transmission by isotropic stratified planar structures	332
4.7.	Reflection and transmission by anisotropic stratified planar structures	340
	4.7.1. Reflection and transmission by a finite anisotropic layer between semi-infinite isotropic ambient and substrate media	348
	4.7.2. Reflection and transmission by a semi-infinite anisotropic substrate in an isotropic ambient	352
	4.7.3. Explicit expressions for the reflection matrix in simple special cases	354
	4.7.3.1. Isotropic ambient and uniaxially anisotropic substrate	354
	4.7.3.2. Isotropic ambient and biaxially anisotropic substrate	355
	4.7.3.3. Uniaxially anisotropic film on an isotropic substrate in an isotropic ambient	356
	4.7.3.4. Biaxially anisotropic film on an isotropic substrate in an isotropic ambient	357
	4.7.3.5. Uniaxial film on a uniaxial substrate in an isotropic ambient	358
4.8.	Ellipsometry on surfaces covered with discontinuous films and on surfaces with rough boundaries	359

CHAPTER 5. INSTRUMENTATION AND TECHNIQUES OF ELLIPSOMETRY — 364

- 5.1. Introduction — 364
- 5.2. The basic instrument – The null ellipsometer — 364
  - 5.2.1. The optical devices — 366
    - 5.2.1.1. Linear polarizers — 367
    - 5.2.1.2. Retarders — 370
    - 5.2.1.3. Other polarizing optical devices — 372
  - 5.2.2. Optical components of an ellipsometer that perform no polarizing function — 373
  - 5.2.3. Light sources and detectors — 375
- 5.3. Operation of the null ellipsometer — 376
  - 5.3.1. Alignment of the telescopes — 376
  - 5.3.2. Calibration of the ellipsometer divided circles — 380
  - 5.3.3. Nulling schemes — 385
- 5.4. Sources of error and their correction — 386
  - 5.4.1. Azimuth-angle errors, component and cell-window imperfections — 388
  - 5.4.2. Measurement of ellipsometer imperfection parameters — 389
  - 5.4.3. Beam deviation errors — 392
  - 5.4.4. Effect of auto-collimated parasitic beams from multiple reflections — 397
  - 5.4.5. Errors that arise from certain assumptions concerning the ellipsometer light beam — 398
    - 5.4.5.1. Bandwidth — 398
    - 5.4.5.2. Source polarization and component depolarization — 398
    - 5.4.5.3. Collimation and non-uniform polarization — 400
  - 5.4.6. Polarization-dependent photodetector sensitivity — 401
  - 5.4.7. Residual mechanical imperfections — 402
  - 5.4.8. Other errors — 402
  - 5.4.9. Model errors — 402
- 5.5. Precision of null ellipsometers — 403
- 5.6. Automation of null ellipsometers — 405
  - 5.6.1. Motor-driven self-nulled ellipsometers — 406
  - 5.6.2. Electro-optic self-nulled ellipsometers — 408
- 5.7. Automatic photometric ellipsometers — 411
  - 5.7.1. The rotating-analyzer ellipsometer — 411
  - 5.7.2. The polarization-modulated ellipsometer — 415

CHAPTER 6. APPLICATIONS OF ELLIPSOMETRY — 417

- 6.1. Introduction — 417
- 6.2. Optical properties of materials and spectroscopic ellipsometry — 417
  - 6.2.1. Optical properties of materials in bulk phase — 417
  - 6.2.2. Optical properties of materials in thin-film phase — 428

6.3.	Physical and chemical adsorption	433
6.4.	Oxidation of semiconductor and metal surfaces	443
6.5.	Electrochemistry	452
6.6.	Applications of ellipsometry in biology and medicine	464
	6.6.1. Interaction of blood with foreign surfaces – Blood coagulation	465
	6.6.2. Antigen-antibody immunological reactions in thin films	466
	6.6.3. The immunoelectroadsorption test	468
	6.6.4. Measurement of cell-surface (coat) materials	469
6.7.	Other applications of ellipsometry	473

APPENDIX	487
REFERENCES	493
LISTINGS OF THE CONTENTS OF THE INTERNATIONAL ELLIPSOMETRY CONFERENCES	501
AUTHOR INDEX	517
SUBJECT INDEX	522

CHAPTER 1

The Polarization of Light Waves

1.1. The concept of polarization

Polarization is a property that is common to all types of vector waves. Electromagnetic waves possess this property and so do elastic and spin waves in solids, for example. *For all types of vector waves polarization refers to the behaviour with time of one of the field vectors appropriate to that wave* (electric field, strain or spin), *observed at a fixed point in space.*

Light waves are electromagnetic in nature and require four basic field vectors for their complete description: the electric-field strength **E**, the electric-displacement density **D**, the magnetic-field strength **H** and the magnetic-flux density **B**. Of these four vectors the *electric-field strength* **E** is chosen to define the state of polarization of light waves. This choice is based on the fact that when light interacts with matter, the force exerted on the electrons by the electric field of the light wave is much greater than the force exerted on these electrons by the magnetic field of the wave.[1] In general, once the polarization of **E** has been determined, the polarization of the three remaining field vectors **D**, **H** and **B** can be found. This is because **E**, **D**, **H** and **B** are interrelated by Maxwell's field equations and the associated constitutive (material) relations. In the following, we will focus our attention on the polarization of light as defined by the behaviour of the electric vector $E(r, t)$ observed at a fixed point in space, r, with time, t.

---

[1]The total force exerted by the electromagnetic field on a particle of charge q moving with a velocity v consists of two terms: the electric force $q\boldsymbol{E}$ and the Lorentz force $q\boldsymbol{v} \times \boldsymbol{B}$. The ratio of magnitudes of the latter to the former cannot exceed vB/E or v/c where c is the velocity of light. Because $v/c \ll 1$ (for all cases of interest), the Lorentz force can be neglected.

1.2. Polychromatic and monochromatic waves of light

Fourier analysis of the time variation of $E(r, t)$ for a light wave yields spectral (Fourier) components with frequencies anywhere in the range from 10^{12} Hertz in the far infrared (IR) to 10^{16} Hertz in the far ultraviolet (UV). This span of four decades of frequency defines the optical part of the electromagnetic spectrum of which only one octave ($\sim 4\text{-}8 \times 10^{14}$ Hertz) represents visible light. The optical region joins smoothly with the microwave region at its low-frequency end and with the X-ray region at its high-frequency end. A light wave is called polychromatic if the Fourier analysis produces a composite spectrum of spectral frequencies in a continuous or discrete distribution. On the other hand, the wave is monochromatic when it consists of a single *discrete* frequency component of zero spectral width. Between the two cases is the quasi-monochromatic wave which is described by a narrow spectral line of very small but non-zero width. In this book we are interested only in monochromatic and quasi-monochromatic light waves.

1.3. Polarization of monochromatic waves of arbitrary spatial structure

For a monochromatic wave the time variation of the electric vector E is exactly sinusoidal. The spatial structure of this *time-harmonic* optical field is assumed arbitrary, so that the discussion may apply to light waves of different types.[2]

At a fixed point in space the most general vibration of the electric vector E can be resolved into three *independent*, linear, simple-harmonic vibrations E_x, E_y and E_z along three mutually orthogonal directions x, y and z, respectively

$$E = E_x\hat{x} + E_y\hat{y} + E_z\hat{z}, \tag{1.1}$$

$$E_i = \tilde{E}_i \cos(\omega t + \delta_i), \quad i = x, y, z. \tag{1.2}$$

$\hat{x}$, $\hat{y}$ and $\hat{z}$ are unit vectors along the coordinate axes; $\tilde{E}_i$ and δ_i represent the amplitude and phase, respectively, of the linear vibration along the ith coordinate axis and ω represents the angular frequency.

We will prove that if a vector is drawn from the fixed observation point as an origin to represent the instantaneous electric vector E, the end-point of

---

[2] Examples of different types of light waves of general spatial structure include travelling waves guided by thin films or fibers, standing waves in resonant cavities and the fields in the focal volume of a lens illuminated by a plane wave. In ellipsometry, a polarized plane wave is often reflected from a film-coated substrate. The total field in the film consists of a superposition of a travelling wave and a standing wave.

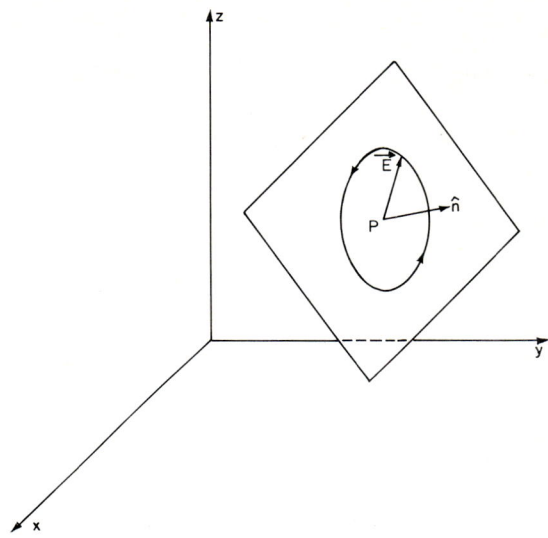

Fig. 1.1. For a time-harmonic optical wave, the end-point of a vector drawn from the fixed observation point P to represent the instantaneous electric field E periodically traces an ellipse in space. $\hat{n}$ is the unit normal to the plane of the ellipse.

that vector will trace an ellipse in space, fig. 1.1.[3] Such an ellipse is periodically described at a repetition rate equal to the optical frequency $f = \omega/2\pi$.

We resolve each of the three linear vibrations in eq. (1.2) into two colinear vibrations, in *time-quadrature*[4] with one another, along the same axis

$$E_i = (\tilde{E}_i \cos \delta_i) \cos \omega t - (\tilde{E}_i \sin \delta_i) \sin \omega t, \quad i = x, y, z. \tag{1.3}$$

The significance of this step is explained with the help of fig. 1.2. The group of three oscillations $E_i \cos(\omega t + \delta_i)$ along the three coordinate axes $i = x, y$ and z is split into two subgroups. One subgroup, $(\tilde{E}_i \cos \delta_i) \cos \omega t$ [$i = x, y, z$], represents three in-phase linear vibrations along the coordinate axes that vary with time according to $\cos \omega t$. The second subgroup, $-(\tilde{E}_i \sin \delta_i) \sin \omega t$

---

[3] The proof given here is not the only one. An alternative approach is to construct a 3×1 column vector E whose elements are given by eq. (1.2) and to search for a coordinate transformation $x, y, z \to x', y', z'$, represented by a 3×3 rotation matrix R, that would produce a new vector $E' = RE$ whose elements take the form $E_{x'} = a \cos \omega t$, $E_{y'} = \pm b \sin \omega t$ and $E_{z'} = 0$. The elements of the first and the second rows of R give the direction cosines of the major and minor axes of the ellipse whose lengths are $2a$ and $2b$, respectively, while the elements of the third row of R give the direction cosines of the unit normal $\hat{n}$ to the plane of the ellipse.

[4] This term is frequently used in electrical engineering to signify a phase displacement of $\frac{1}{2}\pi$ between two sinusoidal alternating quantities.

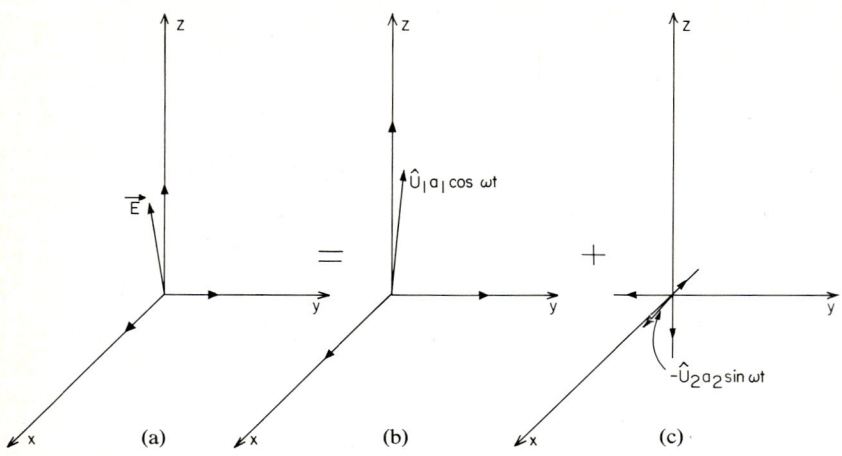

Fig. 1.2. A group of three linear oscillations $\tilde{E}_i \cos(\omega t + \delta_i)$ along the three coordinate axes x, y, z in (a) can be resolved into two subgroups: $(\tilde{E}_i \cos \delta_i) \cos \omega t$ in (b) and $-(\tilde{E}_i \sin \delta_i) \sin \omega t$ in (c) where $i = x, y$ and z. $\hat{u}_1 a_1 \cos \omega t$ and $-\hat{u}_2 a_2 \sin \omega t$ are the linear resultants of subgroups (b) and (c) respectively.

[$i = x, y, z$], represents three in-phase linear vibrations that vary with time according to $-\sin \omega t$, i.e., in time-quadrature with the first subgroup. It is a simple matter [fig. 1.3] to see that the resultant of two (hence any number of) in-phase linear vibrations along different directions is also a linear vibration that is in phase with the component vibrations. Therefore, we can

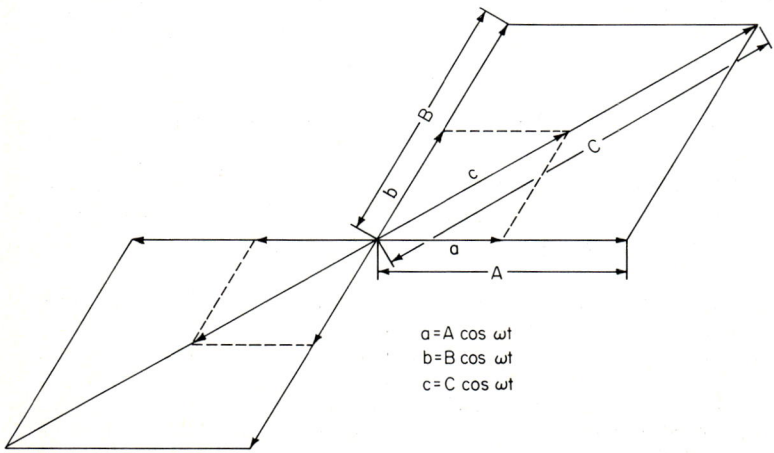

Fig. 1.3. The resultant of two in-phase linear vibrations $A \cos \omega t$ and $B \cos \omega t$ along different directions is also a linear vibration, $C \cos \omega t$, in phase with the component vibrations.

replace each of the cos ωt and the sin ωt subgroups by a resultant linear vibration. If $\hat{u}_1$ and $\hat{u}_2$ represent unit vectors along the directions of the resultants and a_1 and a_2 represent their respective amplitudes, the total electric vector E can be written as

$$E = (a_1 \cos \omega t)\hat{u}_1 - (a_2 \sin \omega t)\hat{u}_2. \tag{1.4}$$

The amplitudes a_1 and a_2 and the unit vectors $\hat{u}_1$ and $\hat{u}_2$ in eq. (1.4) are easily determined from the three linear vibrations of each subgroup, evaluated at the instant of time when their resultant reaches maximum length. Therefore a_1, a_2, $\hat{u}_1$ and $\hat{u}_2$ are given by

$$\begin{aligned} a_1 &= [(\tilde{E}_x \cos \delta_x)^2 + (\tilde{E}_y \cos \delta_y)^2 + (\tilde{E}_z \cos \delta_z)^2]^{\frac{1}{2}}, \\ a_2 &= [(\tilde{E}_x \sin \delta_x)^2 + (\tilde{E}_y \sin \delta_y)^2 + (\tilde{E}_z \sin \delta_z)^2]^{\frac{1}{2}}, \end{aligned} \tag{1.5}$$

$$\begin{aligned} \hat{u}_1 &= \frac{1}{a_1}(\tilde{E}_x \cos \delta_x, \tilde{E}_y \cos \delta_y, \tilde{E}_z \cos \delta_z), \\ \hat{u}_2 &= \frac{1}{a_2}(\tilde{E}_x \sin \delta_x, \tilde{E}_y \sin \delta_y, \tilde{E}_z \sin \delta_z). \end{aligned} \tag{1.6}$$

From eq. (1.4) it is evident that the locus of the terminus of the electric vector E is confined to the plane formed by the unit vectors $\hat{u}_1$ and $\hat{u}_2$. The unit normal $\hat{n}$ to the plane of the locus of E is given by

$$\hat{n} = (\hat{u}_1 \times \hat{u}_2)/|\hat{u}_1 \times \hat{u}_2|. \tag{1.7}$$

It remains for us to prove that the locus of E is an ellipse. In fig. 1.4 we consider the plane of the unit vectors $\hat{u}_1$ and $\hat{u}_2$ to coincide with the plane of the paper and we use γ to denote the angle between these unit vectors where

$$\cos \gamma = (\hat{u}_1 \cdot \hat{u}_2), \quad \sin \gamma = |\hat{u}_1 \times \hat{u}_2|. \tag{1.8}$$

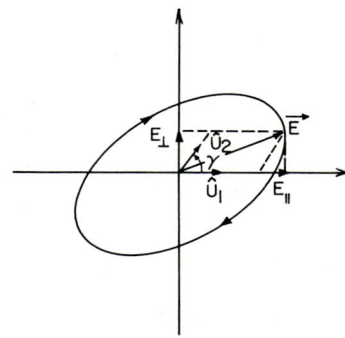

Fig. 1.4. The superposition of $\hat{u}_1 a_1 \cos \omega t$ and $-\hat{u}_2 a_2 \sin \omega t$ of fig. 1.2(b), (c) yields the ellipse of polarization.

Taking the projections of E of eq. (1.4) parallel ($\parallel$) and perpendicular ($\perp$) to $\hat{u}_1$ we get

$$E_\parallel = a_1 \cos \omega t - (a_2 \cos \gamma) \sin \omega t,$$
$$E_\perp = -(a_2 \sin \gamma) \sin \omega t. \tag{1.9}$$

By elimination of t from eq. (1.9), we find that the coordinates $(E_\parallel, E_\perp)$ of the end-point of the electric vector satisfy the following equation

$$\frac{E_\parallel^2}{a_1^2} + \frac{E_\perp^2}{(a_2 \sin \gamma)^2} - \frac{2 E_\parallel E_\perp \cot \gamma}{a_1^2} = 1, \tag{1.10}$$

which is the equation of an ellipse in the plane of the unit vectors $\hat{u}_1$ and $\hat{u}_2$. By a coordinate rotation, the azimuth (orientation) of the major axis of the ellipse and the lengths of the semi-major and the semi-minor axes can all be determined in terms of a_1, a_2 and γ.[5]

1.3.1. Elliptical polarization

A light wave whose electric vector at a fixed point in space traces the same ellipse in a regular repetitive fashion is described as elliptically polarized *at that point*. The discussion above shows that the elliptical (or elliptic) polarization is the most general state of polarization of any optical field that is *strictly* monochromatic.

For complete specification of the elliptical polarization we need to know

(1) The orientation in space of the plane of the ellipse of polarization;
(2) The orientation of the ellipse in its plane, its shape and the sense in which it is described;
(3) The amplitude (size) of the ellipse;
(4) The absolute temporal phase.

The orientation in space of the plane of the ellipse of polarization can be specified by the unit vector $\hat{n}$, directed normal to that plane. Referenced to a Cartesian coordinate system, $\hat{n}$ can be given by its three direction cosines. Because the sum of the squared direction cosines is unity, only *two* real quantities are required to determine the orientation of the plane of the ellipse. For a monochromatic wave of arbitrary spatial structure, there are

---

[5]An alternative method for determining the parameters of the polarization ellipse from eq. (1.9) is to write this equation in the form $E_\parallel = \tilde{E}_\parallel \cos(\omega t + \delta_\parallel)$, $E_\perp = \tilde{E}_\perp \cos(\omega t + \delta_\perp)$ where $\tilde{E}_\parallel = [a_1^2 + (a_2 \cos \gamma)^2]^{1/2}$, $\delta_\parallel = \tan^{-1}(a_2 \cos \gamma / a_1)$ and $\tilde{E}_\perp = a_2 \sin \gamma$, $\delta_\perp = \frac{1}{2}\pi$. If we construct the complex number $\chi = (\tilde{E}_\perp / \tilde{E}_\parallel) \exp[j(\delta_\perp - \delta_\parallel)]$, the azimuth θ and ellipticity $\tan \epsilon$ [eqs. (1.11)–1.15)] are directly obtained from χ by eqs. (1.86) and (1.87). The amplitude A of the ellipse of polarization [fig. 1.5 and eq. (1.16)] is given by $(\tilde{E}_\parallel^2 + \tilde{E}_\perp^2)^{1/2}$.

two distinct possible choices of the unit normal, either $\hat{n}$ or $-\hat{n}$. In other words, the *sense* of the unit normal has to be defined. For definiteness we may take the direction of $\hat{n}$ to be that of the *average* energy flow along the normal (the component of the time-averaged Poynting vector along the normal).[6] Because we will be interested mostly in travelling plane waves, the sense of $\hat{n}$ is uniquely and unambiguously determined parallel to the direction of propagation.

If we consider the plane of the ellipse of polarization to coincide with the plane of the paper, and the direction of $\hat{n}$ to be that of the outward normal pointing to the reader, the orientation of the ellipse in its plane, its shape and sense can be easily defined with reference to fig. 1.5. In that figure X and Y

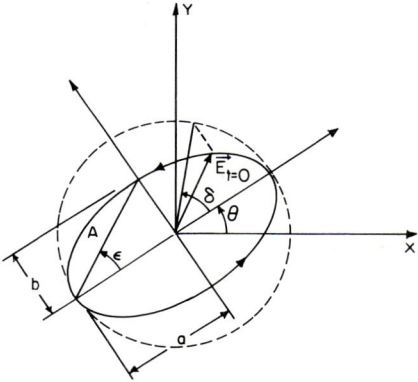

Fig. 1.5. The four parameters that define the ellipse of polarization in its plane are (1) the azimuth θ of the major axis from a fixed reference direction X, (2) the ellipticity $e = \pm b/a = \pm \tan \epsilon$ where the $+$ and $-$ signs correspond to right- and left-handed polarizations, respectively, (3) the total amplitude $A = (a^2 + b^2)^{\frac{1}{2}}$ and (4) the absolute phase δ. The indicated polarization is left-handed. Note the way the absolute phase δ is defined by the initial electric field $\boldsymbol{E}_{t=0}$ and the (dashed) auxiliary circle.

are two fixed reference directions which together with $\hat{n}$ form an orthogonal right-handed coordinate system. The parameters that describe the ellipse of polarization in its plane are defined as follows.

(1) The *azimuth* θ is the angle between the *major axis* of the ellipse and the positive direction of the X axis and defines the orientation of the ellipse in its plane. All physically distinguishable azimuths can be obtained by limiting θ to the range

$$-\tfrac{1}{2}\pi \leq \theta < \tfrac{1}{2}\pi. \tag{1.11}$$

---

[6] For the case of a standing wave the time-averaged Poynting vector is zero and the choice of the arrow of $\hat{n}$ remains ambiguous.

(2) The *ellipticity* e is the ratio of the length of the semi-minor axis of the ellipse b to the length of its semi-major axis a,

$$e = \frac{b}{a}. \tag{1.12}$$

(3) The *handedness* of the ellipse of polarization determines the *sense* in which the ellipse is described. It is a parameter that can assume only one of two discrete "values". The polarization is *right handed* if the ellipse is traversed in a *clockwise* sense when looking *against* the direction of $\hat{n}$. (This means looking "into the beam" for a travelling wave and, in general, looking into the page in fig. 1.5.) The polarization is *left handed* if the ellipse is traversed in a *counter-clockwise* sense when looking against the direction of $\hat{n}$.

It is mathematically convenient to incorporate the handedness in the definition of the ellipticity e by allowing the ellipticity to assume positive and negative values, to correspond to *right-handed* and *left-handed* polarizations, respectively. Using this notation, it is easily seen that all physically distinguishable ellipticity values can be obtained by limiting e to the range

$$-1 \leq e \leq 1. \tag{1.13}$$

We will find it convenient to introduce an *ellipticity angle* ϵ such that,

$$e = \tan \epsilon. \tag{1.14}$$

From eq. (1.13), it follows that ϵ is limited to the range

$$-\tfrac{1}{4}\pi \leq \epsilon \leq \tfrac{1}{4}\pi. \tag{1.15}$$

(4) The *amplitude* (size) of an elliptical vibration can be conveniently defined in terms of the lengths a and b of the semi-major and the semi-minor axes as[7]

$$A = (a^2 + b^2)^{\frac{1}{2}}. \tag{1.16}$$

The amplitude A, as defined by eq. (1.16), is a measure of the strength of the elliptical vibration and its square is proportional to the energy density of the wave at the point of observation of the field.

(5) The *absolute phase* δ determines the angle between the initial position of the electric vector at $t = 0$ and the major axis of the ellipse according to fig. 1.5. All possible values of the absolute phase δ will be limited to the range

$$-\pi \leq \delta < \pi. \tag{1.17}$$

---

[7] In terms of the amplitude A and the ellipticity angle ϵ, the semi-major and semi-minor axes are given by $a = A \cos \epsilon$, $b = A \sin \epsilon$ [see fig. 1.5].

1.3.2. Linear and circular polarizations

The circular and linear states of polarization are special cases of the more general state of elliptical polarization and are generated when the ellipticity e assumes the special values of ± 1 and zero, respectively. The value $e = +1$ corresponds to the *right-handed circularly polarized state* whereas $e = -1$ corresponds to the *left-handed circularly polarized state*. When $e = 0$ the light wave is *linearly* polarized. These special cases of

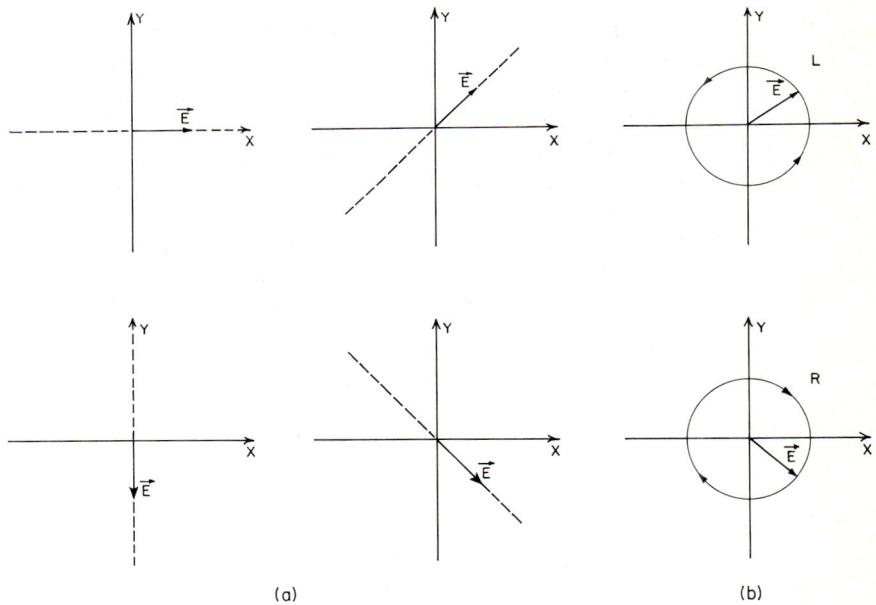

Fig. 1.6. The special cases of linear (a) and circular (b) polarization. In (a), the dashed line indicates the locus of the terminus of the electric vector E. In (b), L and R represent the left- and right-circular polarizations, respectively.

circular and linear polarization are shown in fig. 1.6. Note that the two circular polarizations have indeterminate azimuth and that all the linear polarizations [there is an infinity of such states that correspond to all of the different possible values of the azimuth θ] have no handedness.

Before concluding this section, it is important to note that we have dealt with two different, but equivalent, methods of specifying the elliptical state of polarization of a monochromatic optical wave. In one specification, the elliptical vibration of the electric vector is represented by the superposition of three linear simple-harmonic vibrations of independent amplitudes and phases along three mutually orthogonal directions in space [eq. (1.2)]. In the second, the elliptical polarization is represented by (1) the unit normal $\hat{n}$ to the plane of the ellipse, (2) the azimuth θ and ellipticity e, which define the orientation of the ellipse in its plane and its shape and sense, respectively, (3) the amplitude A and (4) the absolute phase δ. Both specifications require *six* real parameters for a complete description.

1.4. The phasor (complex) representation of time-harmonic fields

In a time-harmonic (monochromatic) optical-wave field each Cartesian component of the electric vector varies sinusoidally with time, at all points in space. Under this condition, the amplitude and phase of oscillation of each component at the different points in space represent the only information to be sought, because the time dependence is understood. The amplitude and phase information about a sinusoidal *scalar* quantity can be lumped in a *single complex number* which is referred to as the *phasor* representation of that quantity. For example, the x component of the electric vector at a fixed point in space varies with time as $\tilde{E}_x \cos(\omega t + \delta_x)$ [see eq. (1.2)]. The representative phasor of this alternating quantity is given by

$$E_{xc} = \tilde{E}_x\, e^{j\delta_x}.$$

In general, the representative phasor, Q_c, of the alternating scalar quantity, $Q = \tilde{Q} \cos(\omega t + \delta_Q)$, is defined by

$$Q_c = \tilde{Q}\, e^{j\delta_Q}. \tag{1.18}$$

To restore the time-dependence of an alternating quantity Q from its phasor representation Q_c we perform the following operation

$$Q = \mathrm{Re}\,[Q_c\, e^{j\omega t}] \tag{1.19}$$

where Re signifies "the real part of". Geometrically, eq. (1.19) represents the projection on the real axis of the complex plane of a vector of length $\tilde{Q}$ that rotates counter-clockwise around the origin at an angular speed of ω, starting from an initial position at $t = 0$ that makes an angle δ_Q with the real axis, fig. 1.7.

When each of the three alternating components of the electric vector given by eq. (1.2) is represented by its appropriate phasor, the vector that results from substituting these phasors into eq. (1.1) is given by

$$\boldsymbol{E}_c = E_{xc}\hat{\boldsymbol{x}} + E_{yc}\hat{\boldsymbol{y}} + E_{zc}\hat{\boldsymbol{z}}. \tag{1.20}$$

Equation (1.20) defines a *complex vector* which specifies the state of

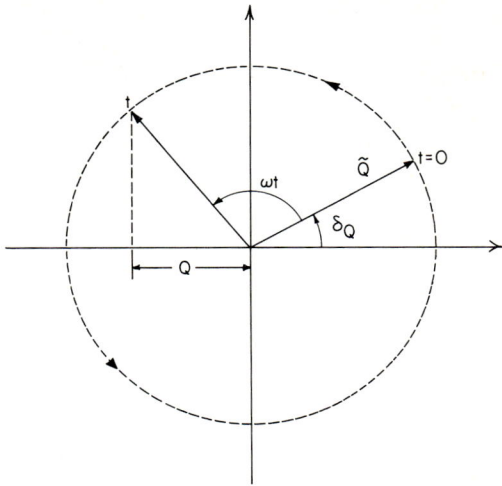

Fig. 1.7. The phasor (complex-number) representation $\tilde{Q}e^{j\delta_Q}$ of the sinusoidal time-varying real quantity $Q = \tilde{Q}\cos(\omega t + \delta_Q)$.

polarization of the field completely. To restore the time dependence from a complex vector, eq. (1.19) is still applicable with E_c replacing Q_c.

The phasor representation of time-harmonic fields will be extensively used throughout this book. *Note that we have chosen the $e^{j\omega t}$ time-dependence instead of $e^{-j\omega t}$.* This is the convention adopted by the participants of the 1968 International Conference on Ellipsometry held at the University of Nebraska [1, 2]. We will adhere to all the conventions agreed upon in that meeting. For simplicity, the subscript c, which is used in this section to distinguish a phasor from the corresponding alternating quantity, will be dropped. The distinction between the two will always be made clear.

1.5. Uniform TE plane waves of light

For the sake of generality, we have assumed in the previous sections that the time-harmonic light wave has an arbitrary spatial structure. A special case of great interest is that of a uniform TE (transverse-electric) travelling plane wave. The electric vector of a *linearly* polarized wave of this type varies with position *r* and time *t* according to

$$E(r, t) = [\tilde{E}\cos(\omega t - k \cdot r)]\hat{u}, \quad (1.21)$$

$$\hat{u} \cdot \hat{u} = 1, \; \hat{u} \cdot k = 0.$$

In eq. (1.21) $\hat{u}$ represents a constant unit vector in the direction of the linear polarization, transverse (orthogonal) to the direction of wave propagation given by the constant *wave-vector* **k**. $\tilde{E}$ is the amplitude of oscillation which is independent of **r** and t. The loci of points in space at which the electric field oscillates in phase (*i.e.* for which $\mathbf{k} \cdot \mathbf{r}$ = constant) constitute a family of parallel-plane *wave-fronts* perpendicular to the wave-vector **k**, fig. 1.8. The

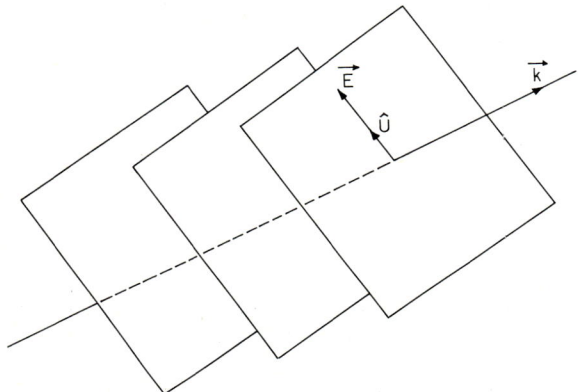

Fig. 1.8. The wave-fronts of a plane wave of wave-vector **k** as a family of parallel planes. The wave is assumed linearly polarized parallel to $\hat{u}$.

distance that separates two such constant-phase planes of phases that differ by 2π defines the *wavelength* λ which is related to **k** by

$$|\mathbf{k}| = \frac{2\pi}{\lambda}. \tag{1.22}$$

The *phase velocity* of the wave is defined as the velocity by which a point in space has to move, along the direction of propagation, so that the instantaneous magnitude of the field at that point remains constant. If s denotes distance measured along the direction of **k** (from a fixed origin), the argument of the cosine function in eq. (1.21) (the total phase) can be written as $[\omega t - (2\pi/\lambda)s]$. The condition of constant instantaneous magnitude of the field requires that s varies with t such that $[\omega t - (2\pi/\lambda)s]$ is constant or

$$\frac{\mathrm{d}}{\mathrm{d}t}\left(\omega t - \frac{2\pi}{\lambda}s\right) = 0,$$

which gives the phase velocity v as

$$v \equiv \frac{\mathrm{d}s}{\mathrm{d}t} = \frac{\omega \lambda}{2\pi} = f\lambda, \tag{1.23}$$

where f is the frequency of the monochromatic light in Hertz.

The wave given by eq. (1.21) is described as uniform because the fields at all points over a wave-front are identical in phase, amplitude and polarization. Obviously the state of polarization of that wave can be generalized from linear to elliptic by the superposition of two linearly polarized waves of different polarizations and phases. Thus eq. (1.21) can be generalized to become

$$\boldsymbol{E}(\boldsymbol{r}, t) = [\tilde{E} \cos(\omega t - \boldsymbol{k} \cdot \boldsymbol{r} + \delta)]\hat{\boldsymbol{u}} + [\tilde{E}' \cos(\omega t - \boldsymbol{k} \cdot \boldsymbol{r} + \delta')]\hat{\boldsymbol{u}}', \qquad (1.24)$$

$$\hat{\boldsymbol{u}} \cdot \hat{\boldsymbol{u}} = \hat{\boldsymbol{u}}' \cdot \hat{\boldsymbol{u}}' = 1, \qquad \hat{\boldsymbol{u}} \cdot \hat{\boldsymbol{u}}' = \hat{\boldsymbol{u}} \cdot \boldsymbol{k} = \hat{\boldsymbol{u}}' \cdot \boldsymbol{k} = 0.$$

Equation (1.24) describes a uniform TE elliptically polarized plane wave. $\hat{\boldsymbol{u}}$ and $\hat{\boldsymbol{u}}'$ represent two orthogonal unit vectors in the wave-front along which the electric vector is resolved into components of amplitudes $\tilde{E}$ and $\tilde{E}'$ and phases δ and δ', respectively.

Because the ellipse of polarization lies in the plane wave-front, the unit normal $\hat{\boldsymbol{n}}$ is parallel to the wave-vector $\boldsymbol{k}$. Complete specification of the elliptical polarization of a uniform TE plane wave thus requires four quantities only, instead of the six quantities that are needed to specify the polarization of a wave with arbitrary spatial structure. The four quantities are the azimuth θ, ellipticity e, amplitude A and phase δ [see §1.3].

In an ellipsometer a beam of polarized light (which can be approximated by a uniform TE plane wave) is passed through a sequence of elements each of which modifies the state of polarization of the beam in a specific manner. In order to describe the polarization-modifying interaction between the light beam and the optical elements (to be discussed in ch. 2), more convenient and concise representations of polarization than those described heretofore are required. The subsequent sections of this chapter are devoted to this.[8]

1.6. The Jones vector of a uniform TE plane wave

The electric vector of a single monochromatic, uniform, TE travelling plane wave of *arbitrary* polarization is given by eq. (1.24). If such a wave is assumed to propagate along the positive direction of the z axis of an xyz orthogonal, right-handed, Cartesian coordinate system and if, in addition, the unit vectors $\hat{\boldsymbol{u}}$ and $\hat{\boldsymbol{u}}'$ are chosen parallel to the positive directions of the x and y axes, eq. (1.24) becomes

---

[8]In addition to the original work by Jones [3], the reader will find it useful, with regard to the material discussed in this chapter, to consult the books by Shurcliff [4], O'Neill [5], Shurcliff and Ballard [6], Simmons and Guttmann [7], Clarke and Grainger [8] as well as the review articles by Ramachandran and Ramaseshan [9] and by Bennett and Bennett [10].

$$E(z, t) = \left[\tilde{E}_x \cos\left(\omega t - \frac{2\pi}{\lambda} z + \delta_x\right)\right]\hat{x}$$
$$+ \left[\tilde{E}_y \cos\left(\omega t - \frac{2\pi}{\lambda} z + \delta_y\right)\right]\hat{y}. \tag{1.25}$$

In eq. (1.25) $\tilde{E}_x$ and $\tilde{E}_y$ represent the amplitudes of the linear, simple-harmonic oscillations of the electric-field components along the x and y axes, and δ_x and δ_y represent the respective phases of these oscillations. $\hat{x}$ and $\hat{y}$ are unit vectors in the positive directions of the x and y axes. Because the wave is uniform, the electric field is the same at all points in a $z =$ constant plane wave-front, and because it is transverse-electric (TE), there is no field-component along the direction of propagation, $k = (2\pi/\lambda)\hat{z}$.

In considering the wave polarization and its modification by an optical device, we do not need the *full* expression of the wave given by eq. (1.25). It is therefore important to seek a more *concise* mathematical description for the wave. To achieve this result we progress through a sequence of steps.

(1) First, we note that once the *fixed* unit vectors $\hat{x}$ and $\hat{y}$ of the linearly polarized components of the wave have been chosen, there is no need to retain these unit vectors in the mathematical expression of the wave. This can be achieved by grouping the scalar (non-vector) components in the form of a 2×1 column vector (matrix) as follows:

$$E(z, t) = \begin{bmatrix} \tilde{E}_x \cos\left(\omega t - \frac{2\pi}{\lambda} z + \delta_x\right) \\ \tilde{E}_y \cos\left(\omega t - \frac{2\pi}{\lambda} z + \delta_y\right) \end{bmatrix}. \tag{1.26}$$

The reverse step from eq. (1.26) to the full expression of eq. (1.25) is to multiply the element in the 1, 1 matrix position of eq. (1.26) by $\hat{x}$, the element in the 2, 1 matrix position by $\hat{y}$ and to add the results.

(2) Because the field components at all points in space for a monochromatic field are known to oscillate sinusoidally with time at the same frequency, such *temporal* information can also be suppressed. Using the phasor notation explained in §1.4, eq. (1.26) can be replaced by

$$E(z) = e^{-j2\pi z/\lambda} \begin{bmatrix} \tilde{E}_x\, e^{j\delta_x} \\ \tilde{E}_y\, e^{j\delta_y} \end{bmatrix}. \tag{1.27}$$

The reverse step from eq. (1.27) to eq. (1.26) (that restores the time-dependence) is to multiply the former by $e^{j\omega t}$ and take the real part of the product

$$E(z, t) = \text{Re}\,[E(z)\, e^{j\omega t}]. \tag{1.28}$$

(3) The final step towards the desired mathematical description of a uniform TE plane wave is to drop the *spatial* information about the wave by considering the field over one fixed transverse plane for example, the plane $z = 0$ of the xyz coordinate system. If $z = 0$ is substituted in eq. (1.27), we obtain

$$\boldsymbol{E}(0) = \begin{bmatrix} \tilde{E}_x\, e^{j\delta_x} \\ \tilde{E}_y\, e^{j\delta_y} \end{bmatrix}. \tag{1.29}$$

The reverse step from eq. (1.29) to eq. (1.27) (that restores the space-dependence) is to multiply the former by $e^{-j2\pi z/\lambda}$

$$\boldsymbol{E}(z) = e^{-j2\pi z/\lambda}\, \boldsymbol{E}(0). \tag{1.30}$$

The vector $\boldsymbol{E}(0)$ of eq. (1.29) is the desired concise representation of a single plane wave *which is known to be monochromatic, uniform and transverse-electric.*[9] This vector is called the *Jones vector* of the wave [3]. The Jones vector contains complete information about the amplitudes and phases of the field components, hence about the polarization of the wave. From the Jones vector $\boldsymbol{E}(0)$ we can reconstruct the time and space-dependence of the entire wave by combining eqs. (1.30) and (1.28)

$$\boldsymbol{E}(z,t) = \text{Re}\,[\boldsymbol{E}(0)\, e^{j(\omega t - 2\pi z/\lambda)}]. \tag{1.31}$$

Subsequently, the full explicit expression of the wave [eq. (1.25)] can be obtained by reintroducing the unit vectors $\hat{x}$ and $\hat{y}$, as explained in step (1) above.

It should be pointed out that the Jones vector is a *complex* vector, because its elements are complex numbers or phasors. These phasors represent two sinusoidal linear oscillations along two mutually perpendicular directions in the wavefront. The Jones vector is not a vector in the real physical space, rather it is a vector in an abstract mathematical space formed by all the vectors that are obtained by considering all possible pairs of complex numbers for E_x and E_y. The reader should exercise caution not to confuse between the concept of *a representation* of the wave and the wave *itself.* Thus, the Jones vector of eq. (1.29) provides *one* representation of the wave whose full expression is given by eq. (1.25). A simple example of a physical

---

[9] A *non-uniform*, monochromatic and transverse-electric *plane* wave can be characterized by a Jones vector distribution $\boldsymbol{E}(x, y, 0)$ over the plane $z = 0$, where the z axis is along the direction of propagation. The phasor elements of the 2×1 Jones vector $\boldsymbol{E}(x, y, z)$ represent the oscillating components of the transverse electric field at the point (x, y, z), parallel to the directions of the x and y axes. The equation of propagation of the transverse Jones vector distribution is $\boldsymbol{E}(x, y, z) = e^{-j2\pi z/\lambda}\, \boldsymbol{E}(x, y, 0)$, of which eq. (1.30) is a special case. In general, the wave can have a non-uniform distribution of both intensity and polarization.

quantity and a particular representation of it that we have already encountered is that of using a complex number (phasor) to represent a real sinusoidal oscillation [§1.4].

Although we have defined $E(0)$ of eq. (1.29) as *the* Jones vector of the wave, it is obvious that the choice of the plane $z = 0$ is arbitrary and that *any* vector $E(z)$ obtained from $E(0)$ by eq. (1.30) is an equally valid representation of the wave. The important point to remember is that one Jones vector is adequate for the reconstruction of the entire plane wave. The significance of eq. (1.30) is that it relates the oscillations of the x and y components of the field in one plane wave-front to these oscillations in another wave-front which is separated from the first by a distance z along the direction of propagation.

In subsequent discussions we will use the following simplified notation for the Jones vector

$$E = \begin{bmatrix} E_x \\ E_y \end{bmatrix}, \tag{1.32}$$

where

$$E_x = |E_x| e^{j\delta_x}, \qquad E_y = |E_y| e^{j\delta_y}, \tag{1.33}$$

and the dependence on z will be explicitly indicated only when the need arises.

1.6.1. Wave intensity

At no place in this book will we be interested in the *absolute* intensity of an optical wave. This justifies the use of a *definition* for the intensity of the wave that overlooks a constant multiplicative factor. Thus we may express the intensity I simply as the sum of the squared amplitudes of the component oscillations along two mutually orthogonal directions

$$I = |E_x|^2 + |E_y|^2 = E_x^* E_x + E_y^* E_y. \tag{1.34}$$

In a more compact form, I can be obtained by pre-multiplying the Jones vector E of eq. (1.32) by its Hermitian adjoint $E^\dagger$

$$I = E^\dagger E. \tag{1.35}$$

The Hermitian adjoint of a matrix is defined as the complex conjugate of the transpose of the matrix, thus $E^\dagger$ is the row vector

$$E^\dagger = [E_x^* \ E_y^*]. \tag{1.36}$$

A wave of *unit intensity* is said to be *normalized* and its Jones vector is said to be *normal*. Such a vector satisfies the condition that

$$E^\dagger E = 1. \tag{1.37}$$

1.6.2. Jones vectors of some states of polarization

If we allow the phasor components E_x and E_y of the Jones vector to assume all possible values as two independent complex numbers, all possible states of polarization of all possible values of intensity and phase are generated. Considering a normalized wave ($I = 1$), we will inspect the Jones vectors of some typical states of polarization. For example, the Jones vector

$$\mathscr{E}_x = \begin{bmatrix} 1 \\ 0 \end{bmatrix}, \tag{1.38a}$$

represents a linearly polarized wave whose electric vector executes a simple-harmonic oscillation along the x axis, with unit amplitude ($A = 1$) and zero phase ($\delta = 0$). Similarly, the Jones vector

$$\mathscr{E}_y = \begin{bmatrix} 0 \\ 1 \end{bmatrix}, \tag{1.38b}$$

represents a linearly polarized wave whose electric vector executes a simple-harmonic oscillation along the y axis of unit amplitude and zero phase. Equations (1.38a) and (1.38b) represent a pair of orthogonal linearly polarized waves each of which is of unit intensity. (The symbols $\mathscr{E}_i$ and $\mathscr{E}_j$ are reserved to denote a pair of basis Jones vectors considered in a subsequent subsection. The subscripts i and j will be indicative of the type of polarization involved.) For an arbitrary linearly polarized light wave, the electric vector oscillates along a general direction x' in the wave-front, inclined to the fixed direction of the x axis by an azimuth angle α. For such a wave, the Jones vector is given by

$$\mathscr{E}_{x'} = \begin{bmatrix} \cos \alpha \\ \sin \alpha \end{bmatrix}, \tag{1.39a}$$

where the linear oscillation is again of unit amplitude and zero phase. The state of linear polarization that is orthogonal to the state represented by eq. (1.39a) can be obtained by the substitution $\alpha \to \alpha - \tfrac{1}{2}\pi$

$$\mathscr{E}_{y'} = \begin{bmatrix} \sin \alpha \\ -\cos \alpha \end{bmatrix}, \tag{1.39b}$$

the subscript y' indicates that the linear oscillation is along the y' axis which is orthogonal to the x' axis. (Clearly, the orthogonal pair $\mathscr{E}_x$, $\mathscr{E}_y$ of eq. (1.38) is a special case of the orthogonal pair $\mathscr{E}_{x'}$, $\mathscr{E}_{y'}$ of eq. (1.39) obtained by setting $\alpha = 0$.)

Another pair of orthogonal waves of interest are the left- and right-circularly polarized waves whose Jones vectors are given by

$$\mathscr{E}_\ell = \frac{1}{\sqrt{2}} \begin{bmatrix} 1 \\ -j \end{bmatrix}, \tag{1.40a}$$

$$\mathscr{E}_r = \frac{1}{\sqrt{2}} \begin{bmatrix} 1 \\ j \end{bmatrix}. \tag{1.40b}$$

The linear oscillations along the x and y coordinate axes from which $\mathscr{E}_\ell$ and $\mathscr{E}_r$ are constructed are of equal amplitude ($1/\sqrt{2}$) and are in time-quadrature with one another. In the case of the left-circular state the y component *lags* the x component by a phase of $\frac{1}{2}\pi$, whereas for the right-circular state the y component *leads* the x component by $\frac{1}{2}\pi$.

The states of linear and circular polarization are only limiting cases of the more general state of elliptical polarization. The straightforward procedure to determine the ellipse of polarization that corresponds to an arbitrary Jones vector of the form of eq. (1.32) is to restore the time-dependence. This is done by multiplying the complex Jones vector by $e^{j\omega t}$ and taking the real part of the product. The time-dependent field components will take the form $|E_x| \cos(\omega t + \delta_x)$, $|E_y| \cos(\omega t + \delta_y)$ parallel to the x and y coordinate axes in the wave-front $z = 0$. The two time-dependent Cartesian components give the parametric equations (where the time t is the parameter) of the locus of the terminus of the instantaneous electric vector. By elimination of the parameter t, the quadratic equation for the ellipse of polarization is obtained, from which the azimuth θ and the axial ratio $|e|$ of the ellipse can be determined. The procedure is similar to that used in §1.3.[10]

Figure 1.9 shows the four linear polarization states whose Jones vectors appear in eqs. (1.38) and (1.39), the two circular states whose Jones vectors are given by eq. (1.40), and two representative orthogonal (see below) elliptic states. Under each state of polarization shown in fig. 1.9 the appropriate Jones vector is indicated.

---

[10] The quadratic equation that relates the Cartesian coordinates of a point on the ellipse of polarization does not contain information on the sense in which the terminus of the electric vector traces the ellipse (the sign of e), neither does this equation have information on the initial position of the electric vector at $t = 0$. These two pieces of information are contained in the time-dependent expressions $|E_x| \cos(\omega t + \delta_x)$, $|E_y| \cos(\omega t + \delta_y)$ of the field components and are suppressed when the time t is eliminated.

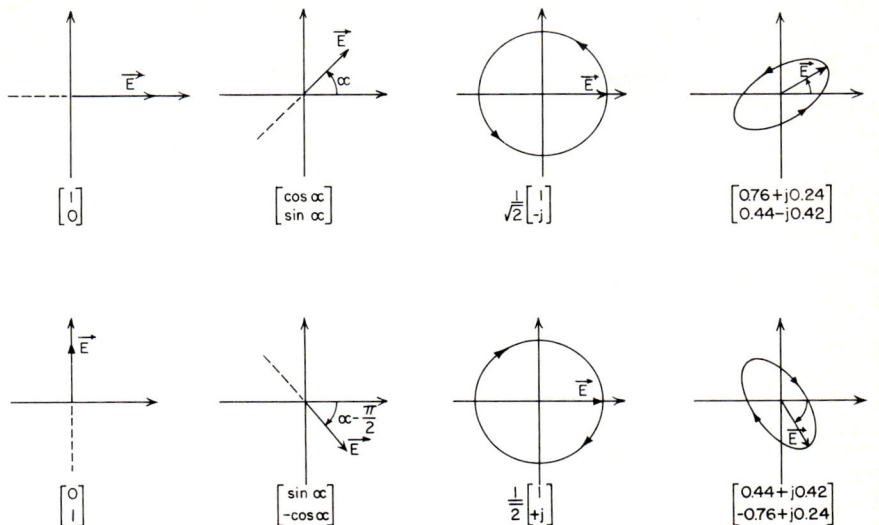

Fig. 1.9. The Jones vectors of some states of polarization.

1.6.3. Orthogonal and orthonormal pairs of Jones vectors

Two Jones vectors E_1 and E_2 are said to be *orthogonal* if they satisfy the condition that

$$E_1^\dagger E_2 = E_2^\dagger E_1 = 0, \tag{1.41}$$

where the dagger ($\dagger$) superscript indicates the Hermitian adjoint, as before.

The ellipses of polarization that correspond to two orthogonal Jones vectors have equal and opposite ellipticities (handedness) and their major axes are mutually orthogonal [see the development that follows eq. (1.82) in the next section].

If, in addition to being mutually orthogonal, each of the two vectors E_1 and E_2 is normal [*i.e.* satisfy eq. (1.37)], the pair of Jones vectors E_1 and E_2 is called an *orthonormal* pair of vectors. It is a simple matter to check that $\mathscr{E}_x$, $\mathscr{E}_y$ of eq. (1.38), $\mathscr{E}_{x'}$, $\mathscr{E}_{y'}$ of eq. (1.39) and $\mathscr{E}_\ell$, $\mathscr{E}_r$ of eq. (1.40) all represent orthonormal pairs of vectors. The same applies to the Jones vectors that correspond to the two elliptic states that are shown in fig. 1.9.

1.6.4. Transformation of the Jones vector under the effect of a coordinate rotation

We have already seen how the Jones vector is transformed under the effect of a coordinate translation parallel to the direction of propagation, eq. (1.30). The 2×1 complex Jones vector which represents a given uniform TE plane wave [eq. (1.32)] is also dependent upon the choice of the x and y coordinate axes in the plane of the wave-front. Referring to fig. 1.10, let $x'y'z'$ be a *new*

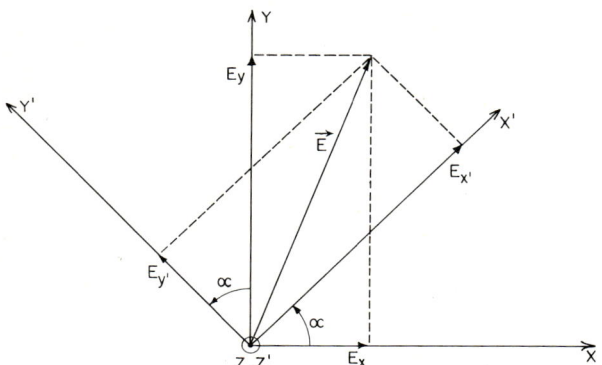

Fig. 1.10. Resolution of the electric vector $\boldsymbol{E}$ along two non-coincident coordinate systems (x, y) and (x', y').

coordinate system whose z' axis is coincident with the z axis of the *old xyz* coordinate system and whose x' and y' axes are obtained from x and y by a counter-clockwise rotation of an angle α about the z axis. By taking the projections of the electric-field vector along the old x and y coordinate axes and the new x' and y' coordinate axes it becomes obvious that these projections are interrelated by

$$E_{x'} = (\cos \alpha) E_x + (\sin \alpha) E_y,$$
$$E_{y'} = (-\sin \alpha) E_x + (\cos \alpha) E_y. \tag{1.42}$$

Equation (1.42) holds when E_x, E_y, $E_{x'}$ and $E_{y'}$ are considered as the projections (components) of the *instantaneous* electric field and also when these symbols represent the *phasors* of the simple-harmonic linear component oscillations along the corresponding coordinate axes. Casting eqs. (1.42) in matrix form we get

$$\begin{bmatrix} E_{x'} \\ E_{y'} \end{bmatrix} = \begin{bmatrix} \cos \alpha & \sin \alpha \\ -\sin \alpha & \cos \alpha \end{bmatrix} \begin{bmatrix} E_x \\ E_y \end{bmatrix}, \tag{1.43}$$

or, more concisely,

$$\boldsymbol{E}_{x',y'} = \boldsymbol{R}(\alpha) \boldsymbol{E}_{x,y}, \tag{1.44}$$

where

$$R(\alpha) = \begin{bmatrix} \cos\alpha & \sin\alpha \\ -\sin\alpha & \cos\alpha \end{bmatrix}. \tag{1.45a}$$

Equations (1.43) and (1.44) represent the law of transformation of the Jones vector under the effect of a coordinate rotation. The effect is described by the *rotation matrix* $R(\alpha)$ of eq. (1.45a) which has the following properties

$$\begin{aligned}R^{-1}(\alpha) &= R^{\dagger}(\alpha) = R(-\alpha),\\ R(\alpha_1)R(\alpha_2) &= R(\alpha_1 + \alpha_2),\end{aligned} \tag{1.45b}$$

as can be proved using eq. (1.45a) and the rules of matrix algebra. The inverse transformation, that gives the Jones vector in the (x, y) coordinate system in terms of its value in the (x', y') coordinate system, can be obtained by premultiplying eq. (1.44) by the inverse of the rotation matrix $R^{-1}(\alpha)$. This gives

$$E_{x,y} = R(-\alpha)E_{x',y'}, \tag{1.46}$$

where $R(-\alpha)$ is referred to as the *counter-rotation matrix*.

The intensity of the wave as calculated from the Jones vector $E_{x',y'}$ in the new (x', y') coordinate system is given by

$$I_{x',y'} = E^{\dagger}_{x',y'}E_{x',y'},$$

which upon substitution from eq. (1.44) becomes

$$\begin{aligned}I_{x',y'} &= [R(\alpha)E_{x,y}]^{\dagger}[R(\alpha)E_{x,y}],\\ &= E^{\dagger}_{x,y}R^{\dagger}(\alpha)R(\alpha)E_{x,y}\\ &= E^{\dagger}_{x,y}E_{x,y}\\ &= I_{x,y}.\end{aligned} \tag{1.47}$$

Equation (1.47) shows that the intensity is *invariant* under a coordinate transformation, which is expected. In reaching eq. (1.47) use has been made of the matrix identity $(AB)^{\dagger} = B^{\dagger}A^{\dagger}$, eq. (1.45b), and the fact that $R(0)$ is the 2×2 identity matrix. If the x' and y' axes are chosen to coincide with the major and minor axes of the ellipse of polarization, respectively, the intensity of the wave is given by

$$I = I_{x,y} = I_{x',y'} = a^2 + b^2$$

or

$$I = A^2, \tag{1.48}$$

where A is the amplitude of the elliptic vibration as defined by eq. (1.16) and fig. 1.5.

1.6.5. Basis Jones vectors and basis states of polarization

1.6.5.1. Cartesian basis vectors
Each and every Jones vector of the form of eq. (1.32) can be constructed as a linear superposition of the orthonormal Jones vectors $\mathscr{E}_x$ and $\mathscr{E}_y$ of eq. (1.38) in the following way

$$\boldsymbol{E}_{x,y} = E_x \mathscr{E}_x + E_y \mathscr{E}_y. \tag{1.49}$$

In eq. (1.49) $\mathscr{E}_x$ and $\mathscr{E}_y$ play the role of *basis vectors* [11] in much the same way as the unit vectors $\hat{x}$ and $\hat{y}$ play the role of basis vectors in the relation $\boldsymbol{r} = x\hat{x} + y\hat{y}$. The entire Jones-vector space is scanned by allowing the pair of complex numbers E_x and E_y to assume all possible values, similar to the way in which the entire space of a real plane can be scanned by allowing the components x and y of the position vector $\boldsymbol{r}$ to take all possible real values. The polarization states that correspond to the basis Jones vectors $\mathscr{E}_x$ and $\mathscr{E}_y$ are called the *basis polarization states* of the Jones-vector representation. In eq. (1.49) these are the two orthogonal linear polarizations of unit amplitude and zero phase that are directed along the x and y coordinate axes in the wave-front. The multiplication of $\mathscr{E}_x$ by E_x changes the amplitude of the linear x basis vibration by $|E_x|$ and shifts its phase by δ_x. Similarly, the multiplication of $\mathscr{E}_y$ by E_y changes the amplitude of the linear y basis vibration by $|E_y|$ and shifts its phase by δ_y. The superposition of these amplitude- and phase-modified basis polarizations generates a wave of any desired polarization, intensity and phase.

A more general choice of *linear* basis Jones vectors is obtained by rewriting eq. (1.49) in the form

$$\boldsymbol{E}_{x,y} = E_{x'} \mathscr{E}_{x'} + E_{y'} \mathscr{E}_{y'}, \tag{1.50}$$

where $\mathscr{E}_{x'}$ and $\mathscr{E}_{y'}$ are the orthonormal Jones vectors that appear in eq. (1.39). In this case, the basis polarizations of the Jones-vector representation are the two orthogonal linear polarizations of unit amplitude and zero phase that are directed along the x' and y' coordinate axes in the wave-front, fig. 1.10. Again, $E_{x'}$ and $E_{y'}$ adjust the amplitude and phase of the respective basis linear polarizations so that their superposition will result in any desired state of polarization of any intensity and phase.

1.6.5.2. Circular basis vectors
There are certain problems for which it is more suitable to use, as basis vectors, the orthonormal Jones vectors $\mathscr{E}_\ell$ and $\mathscr{E}_r$ of eq. (1.40) which correspond to the left- and right-circular polarizations [11, 12]. In this case, the equation that replaces eqs. (1.49) and (1.50) is

$$\boldsymbol{E}_{x,y} = E_\ell \mathscr{E}_\ell + E_r \mathscr{E}_r. \tag{1.51}$$

§1.6] JONES VECTOR OF A UNIFORM TE PLANE WAVE

The basis Jones vector $\mathscr{E}_\ell$ represents an electric vector of unit length that rotates in a counter-clockwise sense around a unit circle with its position parallel to the x axis at $t = 0$. Multiplication of $\mathscr{E}_\ell$ by the *complex amplitude* $E_\ell(= |E_\ell| e^{j\delta_\ell})$ means that the circle locus of the electric vector has a radius equal to $|E_\ell|$ instead of unity and that the position of the electric vector at $t = 0$ makes an angle δ_ℓ with the x axis. Similarly, the basis Jones vector $\mathscr{E}_r$ represents a clockwise-rotating electric vector of unit length whose terminus traces the unit circle and whose initial position at $t = 0$ is parallel to the x axis. Multiplication of $\mathscr{E}_r$ by the complex amplitude $E_r(= |E_r| e^{j\delta_r})$ signifies that the radius of the circle locus of the electric vector becomes equal to $|E_r|$ instead of unity and that the initial position of the electric vector (at $t = 0$) is inclined at an angle δ_r with respect to the x axis. The superposition of these amplitude- and phase-modified basis states generates all other possible states of polarization, fig. 1.11.

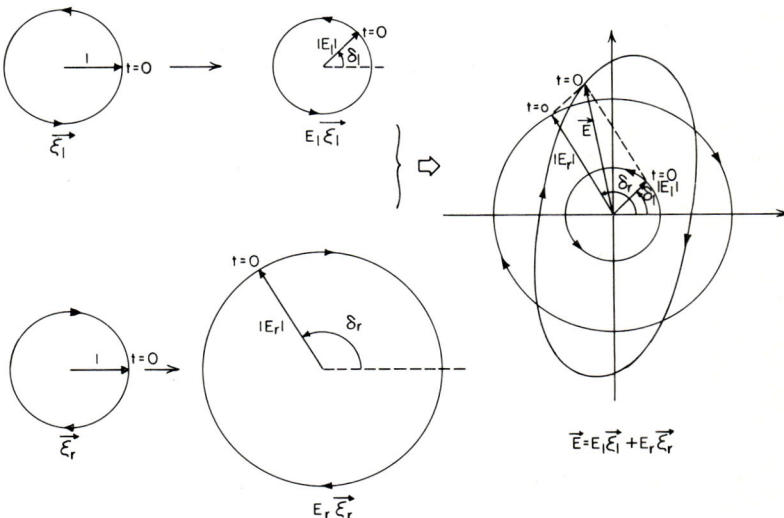

Fig. 1.11. Synthesis of an arbitrary polarization from the left- and right-circular basis states.

By the use of the explicit expressions for the basis vectors $\mathscr{E}_\ell$ and $\mathscr{E}_r$ from eq. (1.40), we can write eq. (1.51) in the form

$$\begin{bmatrix} E_x \\ E_y \end{bmatrix} = \frac{1}{\sqrt{2}} \begin{bmatrix} 1 & 1 \\ -j & j \end{bmatrix} \begin{bmatrix} E_\ell \\ E_r \end{bmatrix}, \quad (1.52)$$

or

$$\mathbf{E}_{x,y} = \mathbf{F} \mathbf{E}_{\ell, r}, \quad (1.53)$$

where

$$\boldsymbol{E}_{\ell,r} = \begin{bmatrix} E_\ell \\ E_r \end{bmatrix}, \tag{1.54}$$

and

$$\boldsymbol{F} = \frac{1}{\sqrt{2}} \begin{bmatrix} 1 & 1 \\ -j & j \end{bmatrix}. \tag{1.55}$$

$\boldsymbol{E}_{\ell,r}$ of eq. (1.54) is the Jones vector of the wave when the left- and right-circular polarizations are used as basis states. The procedure by which the ellipse of polarization is obtained from the complex elements E_ℓ and E_r of the Jones vector $\boldsymbol{E}_{\ell,r}$ is explained above with reference to fig. 1.11. Equations (1.52) and (1.53) represent the transformation of the Jones vector due to *a change of basis* from the Cartesian x and y linear polarizations to the left- and right-circular polarizations. Where a distinction has to be made, $\boldsymbol{E}_{x,y}$ will be called the *Cartesian* Jones vector and $\boldsymbol{E}_{\ell,r}$ will be called the *circular* Jones vector. Equation (1.53) can be inverted to read

$$\boldsymbol{E}_{\ell,r} = \boldsymbol{F}^{-1} \boldsymbol{E}_{x,y}, \tag{1.56}$$

where $\boldsymbol{F}^{-1}$ is the inverse of the transformation matrix $\boldsymbol{F}$ of eq. (1.55) and is given by

$$\boldsymbol{F}^{-1} = \frac{1}{\sqrt{2}} \begin{bmatrix} 1 & j \\ 1 & -j \end{bmatrix}. \tag{1.57}$$

From the expressions of $\boldsymbol{F}$ and $\boldsymbol{F}^{-1}$ in eqs. (1.55) and (1.57) respectively, it becomes clear that $\boldsymbol{F}$ is *unitary* because it satisfies the condition

$$\boldsymbol{F}^{-1} = \boldsymbol{F}^\dagger. \tag{1.58}$$

To calculate the intensity of the wave I in terms of the circular Jones vector $\boldsymbol{E}_{\ell,r}$, we substitute the Cartesian Jones vector $\boldsymbol{E}_{x,y}$ from eq. (1.53) into eq. (1.35)

$$I = (\boldsymbol{F}\boldsymbol{E}_{\ell,r})^\dagger (\boldsymbol{F}\boldsymbol{E}_{\ell,r})$$
$$= \boldsymbol{E}_{\ell,r}^\dagger \boldsymbol{F}^\dagger \boldsymbol{F} \boldsymbol{E}_{\ell,r},$$

and using the unitary property of $\boldsymbol{F}$ [eq. (1.58)] we get

$$I = \boldsymbol{E}_{\ell,r}^\dagger \boldsymbol{E}_{\ell,r} = E_\ell^* E_\ell + E_r^* E_r. \tag{1.59}$$

Equation (1.59) shows that the same expression for the intensity [namely that given by eq. (1.35)] is applicable whether $\boldsymbol{E}$ is a Cartesian or a circular Jones vector. The alternate expression for the intensity, eq. (1.48), is correct for any choice of basis vectors since the amplitude A is a property of the elliptic vibration that is obviously invariant under any transformation.

1.6.5.3. Arbitrary basis vectors

Now we will generalize the above results to the case when two *arbitrary, but different*, elliptical polarizations u and v are used as basis states. Let $\mathscr{E}_u$ and $\mathscr{E}_v$ be the basis Jones vectors that correspond to these two states. $\mathscr{E}_u$ and $\mathscr{E}_v$ need *not* be normal nor orthogonal. The Cartesian Jones vector $\boldsymbol{E}_{x,y}$ can be expressed as a linear superposition of $\mathscr{E}_u$ and $\mathscr{E}_v$

$$\boldsymbol{E}_{x,y} = E_u \mathscr{E}_u + E_v \mathscr{E}_v \tag{1.60}$$

where E_u and E_v are complex numbers that adjust the amplitudes and phases of the elliptically polarized basis states that are described by $\mathscr{E}_u$ and $\mathscr{E}_v$ so that any other state can be generated (of any ellipse of polarization, of any intensity or phase). $\mathscr{E}_u$ and $\mathscr{E}_v$ can be written explicitly as

$$\mathscr{E}_u = \begin{bmatrix} f_{11} \\ f_{21} \end{bmatrix}, \quad \mathscr{E}_v = \begin{bmatrix} f_{12} \\ f_{22} \end{bmatrix}, \tag{1.61}$$

where f_{ij} are arbitrary complex numbers. If we substitute eq. (1.61) into eq. (1.60) the Cartesian Jones vector $\boldsymbol{E}_{x,y}$ can be written as

$$\begin{bmatrix} E_x \\ E_y \end{bmatrix} = \begin{bmatrix} f_{11} & f_{12} \\ f_{21} & f_{22} \end{bmatrix} \begin{bmatrix} E_u \\ E_v \end{bmatrix}, \tag{1.62}$$

or, more concisely,

$$\boldsymbol{E}_{x,y} = \boldsymbol{F} \boldsymbol{E}_{u,v}, \tag{1.63}$$

where

$$\boldsymbol{E}_{u,v} = \begin{bmatrix} E_u \\ E_v \end{bmatrix}, \tag{1.64}$$

and

$$\boldsymbol{F} = \begin{bmatrix} f_{11} & f_{12} \\ f_{21} & f_{22} \end{bmatrix}. \tag{1.65}$$

According to eq. (1.64), the *generalized* Jones vector $\boldsymbol{E}_{u,v}$ is constructed from the complex superposition coefficients E_u and E_v in the fundamental relation of eq. (1.60) and represents the Jones vector of the wave in the representation whose basis vectors are $\mathscr{E}_u$ and $\mathscr{E}_v$. $\boldsymbol{F}$ is the transformation matrix that links the generalized Jones vector $\boldsymbol{E}_{u,v}$ to the Cartesian Jones vector $E_{x,y}$ according to eq. (1.63). Obviously, eq. (1.63) can be inverted to read

$$E_{u,v} = F^{-1} E_{x,y}, \tag{1.66}$$

where F^{-1} is the inverse of F

$$F^{-1} = (f_{11}f_{22} - f_{12}f_{21})^{-1} \begin{bmatrix} f_{22} & -f_{12} \\ -f_{21} & f_{11} \end{bmatrix}. \tag{1.67}$$

When $\mathscr{E}_u$ and $\mathscr{E}_v$ are orthonormal they can be expressed in the form

$$\mathscr{E}_u = \begin{bmatrix} m \\ n \end{bmatrix}, \quad \mathscr{E}_v = \begin{bmatrix} -n^* \\ m^* \end{bmatrix}, \quad m^*m + n^*n = 1. \tag{1.68}$$

It can be easily checked that $\mathscr{E}_u$ and $\mathscr{E}_v$ of eq. (1.68) satisfy the orthogonality condition of eq. (1.41). The transformation matrix F and its inverse F^{-1} are constructed from the basis vectors $\mathscr{E}_u$ and $\mathscr{E}_v$ as implied by eqs. (1.61), (1.65) and (1.67)

$$F = \begin{bmatrix} m & -n^* \\ n & m^* \end{bmatrix}, \tag{1.69}$$

$$F^{-1} = \begin{bmatrix} m^* & n^* \\ -n & m \end{bmatrix}, \tag{1.70}$$

$$F^{-1} = F^{\dagger}. \tag{1.71}$$

Therefore, the transformation matrix F is unitary and the expression for the intensity

$$I = E_{u,v}^{\dagger} E_{u,v} = E_u^* E_u + E_v^* E_v \tag{1.72}$$

becomes valid for any arbitrary choice of an orthonormal pair of basis vectors. The proof parallels that which led to eqs. (1.47) and (1.59) because the transformation matrices $R(\alpha)$ [eq. (1.45a)] and F [eq. (1.55)] are only special cases of eq. (1.69).

A construction similar to that of fig. 1.11 can be used in the general case of two elliptic basis states to display how an arbitrary polarization state can be synthesized from the electric-field vibrations of the basis states.

1.6.6. *The Cartesian and circular Jones vectors of a given elliptical polarization state*

When referenced to a coordinate system (x', y') that coincides with the major and minor axes, the Cartesian Jones vector of an elliptic vibration of unit amplitude ($A = 1$), zero phase ($\delta = 0$), zero azimuth ($\theta = 0$) and ellipticity angle ϵ is given by

$$\mathbf{E}_{x',y'} = \begin{bmatrix} \cos \epsilon \\ j \sin \epsilon \end{bmatrix}, \tag{1.73}$$

as can be easily checked after restoring the time-dependence. Multiplication of eq. (1.73) by $A\,e^{j\delta}$ changes the amplitude of the elliptic vibration from unity to A and shifts the initial position of the electric vector at $t = 0$ from a position parallel to the major axis to a position inclined by an angle $\tan^{-1}(\tan \epsilon \tan \delta)$ from the major axis.[11] Subsequent premultiplication of eq. (1.73) by the counter-rotation matrix $\mathbf{R}(-\theta)$ produces the Jones vector

$$\mathbf{E}_{x,y} = A\,e^{j\delta}\mathbf{R}(-\theta)\begin{bmatrix} \cos \epsilon \\ j \sin \epsilon \end{bmatrix}, \tag{1.74}$$

which describes an elliptic vibration of amplitude A, phase δ, azimuth θ and ellipticity angle ϵ. The above two-step synthesis procedure is illustrated in fig. 1.12. The definitions of A, δ, θ and ϵ have already been given in §1.3.

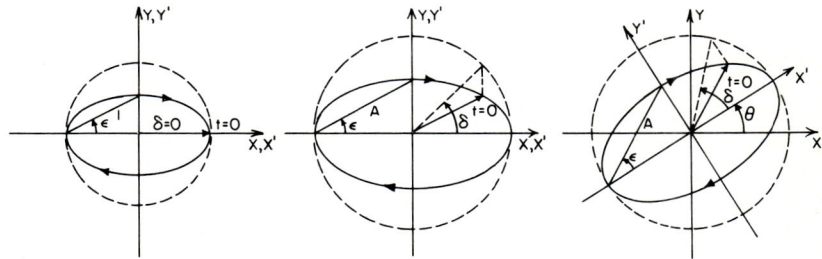

Fig. 1.12. Two-step synthesis procedure of the Jones vector of an arbitrary state of polarization.

Using the rotation matrix of eq. (1.45a) (with α replaced by $-\theta$), the expanded form of eq. (1.74) becomes

$$\begin{bmatrix} E_x \\ E_y \end{bmatrix} = A\,e^{j\delta}\begin{bmatrix} (\cos \theta \cos \epsilon - j \sin \theta \sin \epsilon) \\ (\sin \theta \cos \epsilon + j \cos \theta \sin \epsilon) \end{bmatrix}. \tag{1.75}$$

Equation (1.75) is important in that it shows how a Jones vector can be constructed that describes an elliptical vibration whose amplitude A, phase δ, azimuth θ and ellipticity angle ϵ are given. A straightforward procedure has been described earlier in this section for the opposite problem whereby the ellipse of polarization is determined from the components E_x and E_y of

---

[11] See fig. 1.5.

the Jones vector. [Equation (1.75) suggests an elegant alternate method that will be described in detail in the next section (§1.7).]

To determine the circular Jones vector of a given elliptic vibration in terms of its amplitude A, phase δ, azimuth θ and ellipticity angle ϵ, we make use of the transformation, eq. (1.56), that links the circular Jones vector of a given vibration to the Cartesian Jones vector of the same vibration. Premultiplication of eq. (1.75) by the inverse-transformation matrix $\boldsymbol{F}^{-1}$ of eq. (1.57) gives

$$\begin{bmatrix} E_\ell \\ E_r \end{bmatrix} = \frac{A\,e^{j\delta}}{\sqrt{2}} \begin{bmatrix} (\cos\epsilon - \sin\epsilon)e^{j\theta} \\ (\cos\epsilon + \sin\epsilon)e^{-j\theta} \end{bmatrix}. \tag{1.76}$$

The above results can be extended to determine the generalized Jones vector in terms of A, δ, θ and ϵ when arbitrary basis vectors are used. This can be achieved [see eq. (1.66)] by the premultiplication of eq. (1.75) by the generalized inverse-transformation matrix $\boldsymbol{F}^{-1}$ of eq. (1.67).

1.7. Representation of the polarization states of light by complex numbers[12]

1.7.1. *The Cartesian complex-plane representation of polarized light*

The two-component complex Jones vector discussed in the previous section (§1.6) carries information about the amplitude A, absolute phase δ, azimuth θ and ellipticity angle ϵ of the elliptic vibration of the electric-field vector of a uniform monochromatic TE plane wave of light. This fact is explicitly expressed by eq. (1.75).

Although the Jones vector provides a concise representation of the electric vibration of a TE plane wave, there are occasions when a *simpler* representation is adequate. This is the case when the amplitude and absolute-phase information about the elliptic vibration of the electric vector [represented by A and δ, respectively] are of secondary interest, which

---

[12] In 1892, H. Poincaré [13] introduced two important and elegant representations to describe the states of polarization of a light wave: the complex-plane (this section) and the Poincaré-sphere (§1.8) representations. Poincaré used the complex plane primarily as a stepping stone to the spherical representation, which he subsequently used without further reference to the complex plane. As if to follow the footsteps of Poincaré, subsequent investigators (see previous work as referenced in refs. 3–11) continued to use the Poincaré sphere neglecting the complex-plane representation from which it was originally derived. Examining the optics literature (for example, refs. 3–11 and 23) reveals that the complex-plane representation has remained largely unrecognized since the time of Poincaré. (Exceptions are the papers by Jerrard [14] who re-exposed some of Poincaré's work in English, Holmes and Feucht [15] who found the complex plane a convenient tool to study ellipsometry.) In the electrical engineering literature the complex-plane representation owes its rebirth to papers by Rumsey [16] and Deschamps [17].

§1.7] REPRESENTATION OF POLARIZATION BY COMPLEX NUMBERS

justifies the suppression of such information. Information about the ellipse of polarization proper [represented by the azimuth θ and ellipticity angle ϵ] can be extracted from the two-component Cartesian Jones vector of eq. (1.32) if we take the ratio

$$\chi = E_y/E_x, \tag{1.77}$$

of its phasor components. When E_x and E_y are expressed in terms of their magnitudes and angles, χ is given by

$$\chi = \frac{|E_y|}{|E_x|} e^{j(\delta_y - \delta_x)}, \tag{1.78a}$$

$$|\chi| = |E_y|/|E_x|, \quad \arg(\chi) = \delta_y - \delta_x. \tag{1.78b}$$

Equations (1.77) and (1.78) express the simple fact that when the electric vector is resolved into two components along two reference orthogonal directions in the wave-front, the ellipse of polarization is completely determined by the *relative* amplitude and the *relative* phase of one oscillating component with respect to the other orthogonal component. Using eq. (1.75), we find that the polarization variable χ, defined by eq. (1.77), is a complex function of two real arguments, the azimuth θ and the ellipticity angle ϵ,

$$\chi = \frac{\tan\theta + j\tan\epsilon}{1 - j\tan\theta\tan\epsilon}. \tag{1.79}$$

Equation (1.79) shows explicitly how each elliptic state of polarization of azimuth θ and ellipticity angle ϵ is represented by a single complex number χ. With each complex number χ, we can associate a *representative point* in the complex plane which thus provides a *space* in which to represent the states of polarization of light. This representation is called the *Cartesian complex-plane representation*, because the definition of χ in eq. (1.77) is based on the use of the phasor components of the *Cartesian* Jones vector [18].

When θ and ϵ are limited to the ranges $-\frac{1}{2}\pi \leq \theta < \frac{1}{2}\pi$ and $-\frac{1}{4}\pi \leq \epsilon \leq \frac{1}{4}\pi$ [eqs. (1.11) and (1.15)], eq. (1.79) provides *one-to-one correspondence* between points in the complex plane and the various polarization states of light. Both of eqs. (1.78) and (1.79) can be used to examine the assignment of the different ellipses of polarization to the different points in the complex plane. The following correspondence properties are readily derived.

(1) The origin ($\chi = 0$) and the point-at-infinity ($\chi = \infty$) of the complex plane represent the x-directed (horizontal) and the y-directed (vertical) linear polarizations, respectively. (Recall that the x and y linear polarizations are the basis states of the Cartesian complex-plane representation.) This follows

because, in eq. (1.77) [and eq. (1.78a)], $\chi = 0$ when $E_y = 0$, and $\chi = \infty$ when $E_x = 0$.

(2) All linear vibrations with azimuths θ from $-\frac{1}{2}\pi$ to $+\frac{1}{2}\pi$ (measured from the reference x direction) correspond to points on the real axis of the complex χ plane between $-\infty$ and $+\infty$. This is easily proved by setting the phase difference $(\delta_y - \delta_x) = 0, \pi$ in eq. (1.78a) or by substituting zero ellipticity, $\tan \epsilon = 0$, in eq. (1.79).

(3) The two points $R(\chi = j)$ and $L(\chi = -j)$ on the imaginary axis correspond to the right- and left-handed circular polarization states respectively. This is obtained by setting $|E_x| = |E_y|$, and $(\delta_y - \delta_x) = \pm\frac{1}{2}\pi$ in eq. (1.78a) or by substituting $\epsilon = \pm\frac{1}{4}\pi$ in eq. (1.79), where the plus sign corresponds to R and the minus sign corresponds to L.

(4) Excluding the real axis and the points R and L, all other points in the complex χ plane represent elliptical polarization states. Points in the upper-half plane represent right-handed polarizations and points in the lower-half plane represent left-handed polarizations. This is because, in eq. (1.78a), right-handed states are produced when $(\delta_y - \delta_x) > 0$ and left-handed states are produced when $(\delta_y - \delta_x) < 0$.

(5) The locus of points in the complex plane that correspond to elliptical vibrations whose y components lead (or lag) their x components by a constant phase difference $[(\delta_y - \delta_x) = \text{constant}]$ is a straight line through the origin. Similarly, the locus of points in the complex plane that correspond to elliptical vibrations with a constant ratio of the amplitudes of the y and x components $[|E_y|/|E_x| = \text{constant}]$ is a circle whose center is the origin. This follows directly from eq. (1.78b).

Figure 1.13 shows the complex χ plane with various states of polarization assigned to a selected sample of points and fig. 1.14 shows the loci of polarization states of constant phase difference $(\delta_y - \delta_x)$ and of constant amplitude ratio $|E_y|/|E_x|$, as two orthogonal families of straight lines through and concentric circles around the origin, respectively. At the origin $(\chi = 0)$ and the point at infinity $(\chi = \infty)$ the phase difference $(\delta_y - \delta_x)$ is indeterminate, in agreement with the fact that these two points represent vibrations for which one component (y or x, respectively) is of zero amplitude. Note also that the phase difference $(\delta_y - \delta_x)$ suffers a jump of π as the origin or the point at infinity is crossed along a line through either one of these two points.

1.7.1.1. Orthogonal states

With each state of polarization described by θ and ϵ we associate an *orthogonal* state described by $\theta \pm \frac{1}{2}\pi$ and $-\epsilon$. The ellipses of polarization that correspond to two orthogonal states have equal axial ratios, are traced

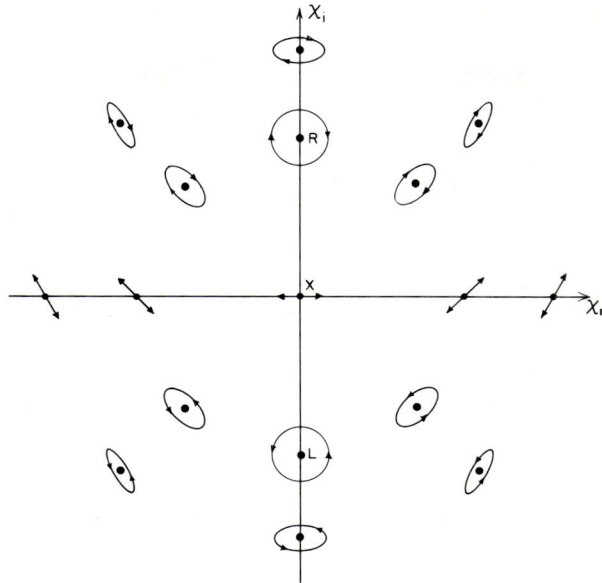

Fig. 1.13. Assignment of polarization states to different points in the Cartesian complex plane of polarization.

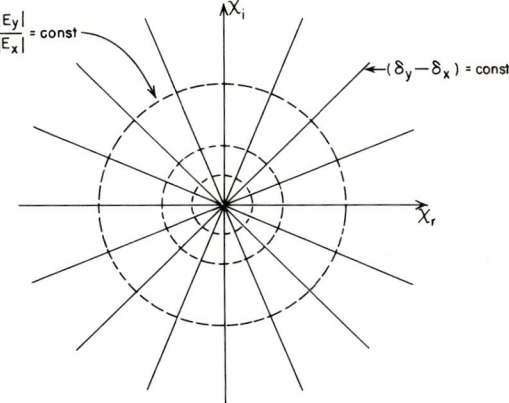

Fig. 1.14. The loci of polarization states of constant phase difference $(\delta_y - \delta_x)$ and those of constant amplitude ratio $|E_y|/|E_x|$ as two orthogonal families of straight lines (——) and concentric circles (---) through and around the origin of the χ plane, respectively.

in opposite senses and their major axes are mutually orthogonal. The complex variable χ_{orth} that represents a state orthogonal to the state χ described by eq. (1.79) can be obtained by the substitutions $\theta \to \theta \pm \frac{1}{2}\pi$ and $\epsilon \to -\epsilon$ which give

$$\chi_{\text{orth}} = -\frac{1 + j \tan \theta \tan \epsilon}{\tan \theta - j \tan \epsilon}. \tag{1.80}$$

From eqs. (1.79) and (1.80) it is evident that

$$\chi \chi^*_{\text{orth}} = \chi^* \chi_{\text{orth}} = -1, \tag{1.81}$$

so that

$$\chi_{\text{orth}} = \frac{-1}{\chi^*} = \frac{-\chi}{|\chi|^2}. \tag{1.82}$$

Substituting $\chi = E_y/E_x$, and $\chi_{\text{orth}} = E_{y\,\text{orth}}/E_{x\,\text{orth}}$ [eq. (1.77)] into eq. (1.81) we obtain

$$(E_x E^*_{x\,\text{orth}} + E_y E^*_{y\,\text{orth}}) = (E^*_x E_{x\,\text{orth}} + E^*_y E_{y\,\text{orth}}) = 0$$

or

$$\boldsymbol{E}^\dagger_{\text{orth}} \boldsymbol{E} = \boldsymbol{E}^\dagger \boldsymbol{E}_{\text{orth}} = 0,$$

which is identical to the condition of orthogonality of two Jones vectors as stated earlier in eq. (1.41).

If P and P_{orth} are the two points in the complex plane that represent the orthogonal states χ and χ_{orth}, respectively, then, according to eqs. (1.81) and (1.82), the straight line from P to P_{orth} must pass through the origin O. In addition, if a circle is drawn through R and L (representing the orthogonal circular polarizations) to pass through P, it must also pass through P_{orth} [see fig. 1.16], because

$$|OP| \times |OP_{\text{orth}}| = |OR| \times |OL| = 1. \tag{1.83}$$

Thus, in the complex plane, the opposite end-points of a chord through the origin of a circle that passes through R and L represent a pair of orthogonal polarizations. The above suggests a geometrical operation to determine the orthogonal state P_{orth} for any given state P. The transformation $P \to P_{\text{orth}}$ can be seen as the superposition of two basic geometrical transformations in succession. As illustrated in fig. 1.15, P is first inverted with respect to the unit circle to give $P_i[|OP_i| \times |OP| = 1]$ then P_i is subsequently inverted with respect to the origin to give $P_{\text{orth}}[|OP_{\text{orth}}| = |OP_i|]$. The inversion with respect to the unit circle, $P \to P_i$, changes $\chi \to 1/\chi^*$ and the inversion with respect to the origin, $P_i \to P_{\text{orth}}$, changes $1/\chi^* \to -1/\chi^*$, to produce the orthogonal state [eq. (1.82)].

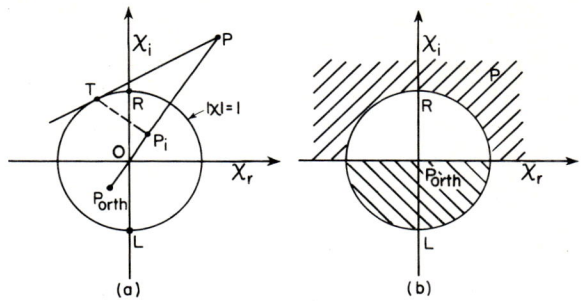

Fig. 1.15. In the χ plane each state of polarization P has an associated orthogonal state P_{orth}. In (a), P_{orth} is obtained from P by the inversion of P with respect to the unit circle to give P_i followed by the inversion of P_i with respect to the origin to give P_{orth}. In (b), polarization states represented by points P above the real axis and outside the unit circle have their associated orthogonal states P_{orth} below the real axis inside the unit circle.

1.7.1.2. Equi-azimuth and equi-ellipticity contours

In order to determine the azimuth θ and the ellipticity angle ϵ of the ellipse of polarization which is represented by a given point χ in the complex plane, eq. (1.79) has to be inverted. This can be done by taking the real part, the imaginary part and the absolute value of both sides of eq. (1.79), respectively,

$$\text{Re}(\chi) = \frac{\tan\theta(1-\tan^2\epsilon)}{1+\tan^2\theta\tan^2\epsilon}, \tag{1.84a}$$

$$\text{Im}(\chi) = \frac{\tan\epsilon(1+\tan^2\theta)}{1+\tan^2\theta\tan^2\epsilon}, \tag{1.84b}$$

$$|\chi|^2 = \frac{\tan^2\theta+\tan^2\epsilon}{1+\tan^2\theta\tan^2\epsilon}. \tag{1.84c}$$

From eq. (1.84c) it immediately follows that

$$(1-|\chi|^2) = \frac{(1-\tan^2\theta)(1-\tan^2\epsilon)}{1+\tan^2\theta\tan^2\epsilon}, \tag{1.85a}$$

$$(1+|\chi|^2) = \frac{(1+\tan^2\theta)(1+\tan^2\epsilon)}{1+\tan^2\theta\tan^2\epsilon}. \tag{1.85b}$$

Division of eq. (1.84a) by eq. (1.85a) and eq. (1.84b) by eq. (1.85b) yields, respectively,

$$\tan 2\theta = \frac{2\text{Re}(\chi)}{1-|\chi|^2}, \tag{1.86}$$

$$\sin 2\epsilon = \frac{2\mathrm{Im}(\chi)}{1+|\chi|^2}. \tag{1.87}$$

Equations (1.86) and (1.87) provide the necessary relations that are required to determine the azimuth θ and the ellipticity angle ϵ from the Cartesian complex polarization variable χ. If we substitute

$$\chi = \chi_r + j\chi_i, \tag{1.88}$$

into eqs. (1.86) we obtain

$$(2\chi_r) + \tan 2\theta \, (\chi_r^2 + \chi_i^2 - 1) = 0, \tag{1.89a}$$

which after rearrangement becomes

$$(\chi_r + \cot 2\theta)^2 + \chi_i^2 = \operatorname{cosec}^2 2\theta. \tag{1.89b}$$

Similarly, if we substitute eq. (1.88) into eq. (1.87) we get

$$2\chi_i - \sin 2\epsilon \, (\chi_r^2 + \chi_i^2 + 1) = 0, \tag{1.90a}$$

or

$$\chi_r^2 + (\chi_i - \operatorname{cosec} 2\epsilon)^2 = \cot^2 2\epsilon. \tag{1.90b}$$

For a constant value of θ, eq. (1.89b) describes a circle[13] whose center is located on the real axis at $(-\cot 2\theta, 0)$ and whose radius is equal to $|\operatorname{cosec} 2\theta|$, fig. 1.16. This circle passes through the points R(0, 1) and L(0, −1), as the substitution of $\chi_r = 0$ and $\chi_i = \pm 1$ into eq. (1.89b) reduces it to a trigonometric identity. In addition, if θ is replaced by $\theta \pm \frac{1}{2}\pi$, eq. (1.89b) remains unchanged, because $\cot 2\theta = \cot 2(\theta \pm \frac{1}{2}\pi)$ and $\operatorname{cosec} 2\theta = \operatorname{cosec} 2(\theta \pm \frac{1}{2}\pi)$. From the above, we see that the locus of points $\chi(\chi_r, \chi_i)$ in the complex plane (our polarization space) that represent elliptic vibrations whose major axes are at the same azimuth θ or at the orthogonal azimuth $\theta \pm \frac{1}{2}\pi$, is a circle through the points R and L that correspond to the right- and left-circular polarizations, respectively. It is important to note that one complete circle through R and L is assigned *two* constant azimuth values that differ by $\frac{1}{2}\pi$. If θ is limited to the interval $-\frac{1}{2}\pi \leq \theta < \frac{1}{2}\pi$, as before, it can be

---

[13] An alternative proof that the equi-azimuth and equi-ellipticity contours are represented by circles can be immediately seen from eq. (1.79). If we put $j \tan \epsilon = \zeta$ in eq. (1.79) we obtain a bilinear relation between χ and ζ, $\chi = (\zeta + a)/(1 + b\zeta)$, where $a = -b = \tan \theta$. For a constant value of θ, as ϵ varies from $-\frac{1}{4}\pi$ to $+\frac{1}{4}\pi$, ζ moves on the imaginary axis of the ζ plane from $-j$ to $+j$. Because of the circle-preserving property of the bilinear transformation [20], χ will move along an arc of a circle in the χ plane between L($\chi = -j$) and R($\chi = +j$). Similarly, if we set $\tan \theta = \zeta$ in eq. (1.79) we obtain a bilinear transformation between χ and ζ of the same form as above with $a = -b = j \tan \epsilon$. For a constant value of ϵ, as θ varies from $-\frac{1}{2}\pi$ to $+\frac{1}{2}\pi$, ζ moves on the real axis of the ζ plane from $-\infty$ to $+\infty$ and its image χ will trace a complete equi-ellipticity circle.

§1.7] REPRESENTATION OF POLARIZATION BY COMPLEX NUMBERS   35

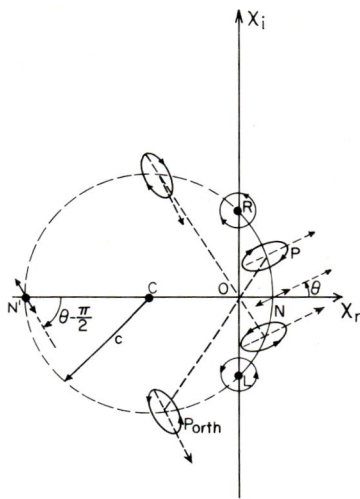

Fig. 1.16. An equi-azimuth contour in the complex χ plane. Points along the arc LNR represent elliptic vibrations with major axes oriented at the same azimuth $\theta = \tan^{-1}|ON|$ whose ellipticities vary from -1 to 0 to $+1$ from L to N to R, respectively. This arc is part of a circle with center $C(-\cot 2\theta, 0)$ and radius $c = \csc 2\theta$. The points $L(\chi = -j)$ and $R(\chi = j)$ represent the left- and right-handed circular polarizations, respectively. Ellipses corresponding to the arc LN′R have the orthogonal azimuth $\theta - \tfrac{1}{2}\pi$.

easily seen that a circle through the points R and L is divided by these two points into two branches. The branch (arc) in the right-half plane represents an *equi-azimuth contour* of ellipses with major axes aligned at the same *positive* azimuth $0 < \theta < \tfrac{1}{2}\pi$, whereas the branch (arc) in the left-half plane represents *another* equi-azimuth contour of vibrations oriented at the *negative* orthogonal azimuth $-\tfrac{1}{2}\pi < (\theta - \tfrac{1}{2}\pi) < 0$. That the sign of θ follows the sign of the real part of χ is clear from eq. (1.84a). In fig. 1.16, N(tan θ, 0) and N′(tan $(\theta - \tfrac{1}{2}\pi)$, 0) represent the points of intersection of the θ and $\theta - \tfrac{1}{2}\pi$ equi-azimuth contours with the real axis, respectively. As we move along the arc LNR, the ellipticity varies continuously and *most steeply* (as will be shown shortly) from -1 (at L), to zero (when the real axis is crossed at N), to $+1$ (at R), while the azimuth remains constant at θ. Along the complementary arc LN′R (shown as a dashed line), the azimuth stays constant at $(\theta - \tfrac{1}{2}\pi)$ while the ellipticity varies most rapidly from -1 (at L), to zero (when the real axis is crossed at N′), to $+1$ (at R). The ellipses of polarization at a number of points along the equi-azimuth contours LNR and LN′R are schematically shown in fig. 1.16. Figure 1.16 also illustrates the orthogonality of the polarization states represented by the end-points of a chord in the circle LNRN′L as has already been discussed in connection with eq. (1.83).

Now that we have discussed the equi-azimuth contours we turn our attention to the *equi-ellipticity contours* which are the loci of points in the complex plane that represent elliptic vibrations having the same ellipticity (axial ratio and handedness). According to eq. (1.90b), it is seen that an equi-ellipticity contour that corresponds to a constant ellipticity angle ϵ is a circle in the complex plane[14] with its center on the imaginary axis at (0, cosec 2ϵ) whose radius is equal to $|\cot 2\epsilon|$, fig. 1.17. Because eq. (1.90b) is obtained

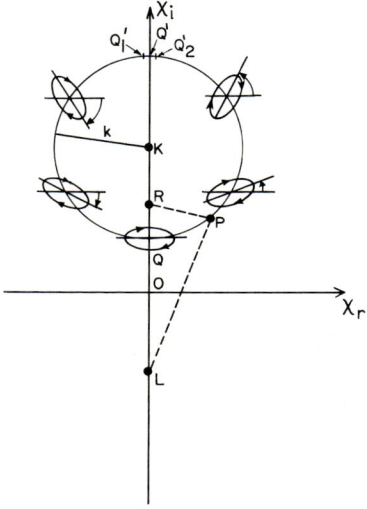

Fig. 1.17. An equi-ellipticity contour in the complex χ plane is a circle that encloses R or L depending on whether the ellipticity is positive or negative, respectively. The indicated circle with center K (0, csc 2ϵ) and radius $k = \cot 2\epsilon$ corresponds to constant ellipticity $\tan \epsilon = |OQ|$. Q'_1 and Q'_2 are points on the circle immediately to the left and right of Q' the upper intersection of the circle with the imaginary axis. Along the circle from Q'_1 to Q to Q'_2 the ellipses have the same form but change azimuth from $-\frac{1}{2}\pi$ to 0 to $+\frac{1}{2}\pi$, respectively. The equi-ellipticity contour is the locus of a point P that moves in the χ plane such that the ratio of its distances to L and R, $|PL|/|PR|$, remains constant.

from eq. (1.87), we can use eq. (1.87) to examine further properties of the equi-ellipticity circles. An equi-ellipticity contour has the significant property that it represents the locus of a point $P(\chi)$ that moves in the complex plane such that the ratio of its distances to L (the left-circular polarization) and R (the right-circular polarization) remains constant. This can be proved by manipulating eq. (1.87) as follows

[14]See footnote 13 on page 34.

$$\frac{2\chi_i}{\chi_r^2 + \chi_i^2 + 1} = \sin 2\epsilon,$$

$$\frac{\chi_r^2 + \chi_i^2 + 2\chi_i + 1}{\chi_r^2 + \chi_i^2 - 2\chi_i + 1} = \frac{1 + \sin 2\epsilon}{1 - \sin 2\epsilon},$$

$$\frac{(\chi_r - 0)^2 + (\chi_i + 1)^2}{(\chi_r - 0)^2 + (\chi_i - 1)^2} = \frac{(\cos \epsilon + \sin \epsilon)^2}{(\cos \epsilon - \sin \epsilon)^2},$$

$$\frac{|\mathrm{PL}|^2}{|\mathrm{PR}|^2} = \left(\frac{1 + \tan \epsilon}{1 - \tan \epsilon}\right)^2,$$

$$\frac{|\mathrm{PL}|}{|\mathrm{PR}|} = \left|\tan\left(\frac{\pi}{4} + \epsilon\right)\right|. \tag{1.91}$$

For a given value of ϵ in the range $-\frac{1}{4}\pi \leq \epsilon \leq \frac{1}{4}\pi$, eq. (1.91) defines a circle that encloses either L or R depending on whether ϵ is negative or positive, respectively. For the limiting values of ϵ of $-\frac{1}{4}\pi$ and $+\frac{1}{4}\pi$, eq. (1.91) shows that $|\mathrm{PL}| = 0$ in the first case, and $|\mathrm{PR}| = 0$ in the second, corresponding to coincidence of the point P with L and R, respectively. These *null* circles are also predicted by eq. (1.90b) which indicates circles of zero radius when $\epsilon = \pm\frac{1}{4}\pi$. When $\epsilon = 0$, P moves on the real axis maintaining equal distances ($|\mathrm{PL}| = |\mathrm{PR}|$) to the points L and R. This is consistent with the fact that linear polarizations, which are represented by points on the real axis, must be "equally near" to the left- and right-circular polarizations represented by L and R. In fig. 1.17, Q $(0, \tan \epsilon)$ and Q'$(0, \cot \epsilon)$ represent the lower (below R) and upper (above R) points of intersection with the imaginary axis of an equi-ellipticity contour that corresponds to a positive value of ϵ, $0 < \epsilon < \frac{1}{4}\pi$. The points Q'_1 and Q'_2 are taken on the equi-ellipticity circle just to the left and to the right of Q'. As we move along the equi-ellipticity circle from Q'_1 to Q to Q'_2, the ellipse of polarization remains of the same shape (axial ratio) and handedness (right); while its azimuth varies continuously and *most steeply* (as will be shown shortly) from $-\frac{1}{2}\pi$ (at Q'_1), to zero (when the imaginary axis is crossed at Q), to $+\frac{1}{2}\pi$ (at Q'_2). This is schematically illustrated in fig. 1.17 by displaying the ellipse of polarization at a sample of points along the equi-ellipticity circle.

If the azimuth θ is allowed to scan the interval $-\frac{1}{2}\pi \leq \theta < \frac{1}{2}\pi$, the entire family of equi-azimuth contours is generated. As we have already seen, each contour is simply an arc of a circle from L to R. That the points L and R belong to all equi-azimuth contours is not unexpected because a circular polarization (or simply a circle) has an indeterminate azimuth. The family of equi-azimuth contours is represented by solid lines in fig. 1.18. The arrow marked on each contour indicates the direction in which the ellipticity angle ϵ increases. On all contours the arrows point from L to R. For $\theta = 0$, the

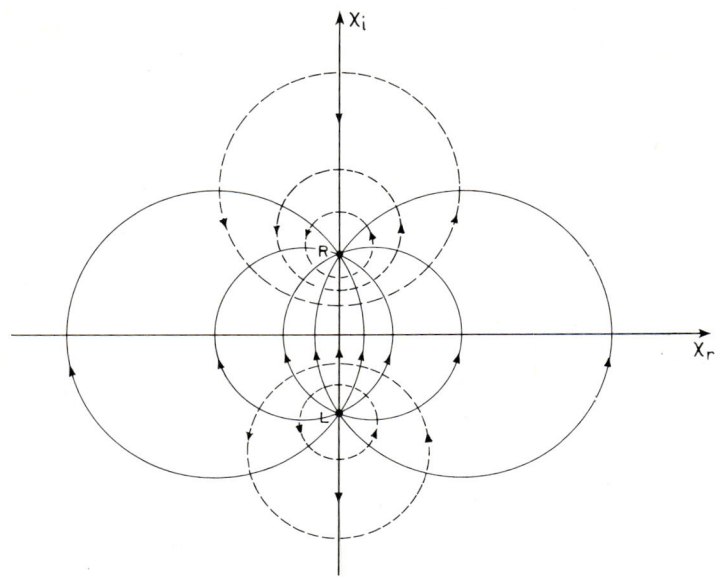

Fig. 1.18. The equi-azimuth (——) and equi-ellipticity (- - -) contours as two orthogonal families of circle arcs and circles. The arrows indicate the direction of increasing ellipticity or azimuth.

equi-azimuth contour coincides with the segment of the imaginary axis that goes from L to the origin to R. The equi-azimuth contour that corresponds to $\theta = -\frac{1}{2}\pi$ coincides with the complementary segment of the imaginary axis from L to the point-at-infinity to R. In fig. 1.18 we also show the family of equi-ellipticity contours (the dashed circles) generated by letting ϵ assume different values in the interval $-\frac{1}{4}\pi \leq \epsilon \leq \frac{1}{4}\pi$. The equi-azimuth and equi-ellipticity contours constitute *two orthogonal sets of curves* in the complex plane. To prove that the equi-azimuth and equi-ellipticity contours that pass through a given point $\chi(\chi_r, \chi_i)$ are orthogonal, we connect χ both to the center $(-\cot 2\theta, 0)$ of the equi-azimuth circle arc and to the center $(0, \operatorname{cosec} 2\epsilon)$ of the equi-ellipticity circle. The product of the slopes of these two circles' radii is

$$\frac{\chi_i}{\chi_r + \cot 2\theta} \times \frac{\chi_i - \operatorname{cosec} 2\epsilon}{\chi_r}.$$

Upon substituting $\cot 2\theta = (1 - \chi_r^2 - \chi_i^2)/2\chi_r$ and $\operatorname{cosec} 2\epsilon = (1 + \chi_r^2 + \chi_i^2)/2\chi_i$, [which follow directly from eqs. (1.86) and (1.87), respectively], the above product of slopes reduces to -1 which proves the orthogonality of the equi-azimuth and the equi-ellipticity contours. The arrows that are indicated on the equi-ellipticity contours that appear in fig. 1.18 point in the direction of increasing values of azimuth.

It is interesting to consider the azimuth θ and the ellipticity angle ϵ as two real scalar "potential" functions of position χ in the polarization space of the complex χ plane. The "equi-potentials" of the azimuth function $\theta(\chi)$ are the equi-azimuth contours while its "gradient-field lines" are the orthogonal equi-ellipticity trajectories. Likewise, the equi-potentials of the ellipticity function $\epsilon(\chi)$ are the equi-ellipticity contours while its gradient-field lines are the orthogonal equi-azimuth trajectories. This clarifies why we referred earlier to the equi-azimuth contours as paths of steepest change in ellipticity and to the equi-ellipticity contours as paths of steepest change in azimuth. In the above description, each set of contours plays the dual role of equi-potentials and gradient-field lines when both the azimuth and ellipticity functions are considered at the same time.

Table 1.1 summarizes the correspondence between different domains in the complex χ plane and their associated subsets of polarization states.

Table 1.1. The significance of the various domains of the χ plane

Domain	Azimuth θ	Ellipticity, $\tan \epsilon$	Remarks
$\operatorname{Re} \chi > 0$	$0 < \theta < \frac{1}{2}\pi$	$-1 < \tan \epsilon < 1$	
$\operatorname{Re} \chi = 0, |\chi| < 1$	$\theta = 0$	$-1 < \tan \epsilon < 1$	
$\operatorname{Re} \chi = 0, |\chi| > 1$	$\theta = -\frac{1}{2}\pi$	$-1 < \tan \epsilon < 1$	
$\operatorname{Re} \chi < 0$	$-\frac{1}{2}\pi < \theta < 0$	$-1 \leq \tan \epsilon \leq 1$	
$\operatorname{Im} \chi > 0$	$-\frac{1}{2}\pi \leq \theta < \frac{1}{2}\pi$	$0 < \tan \epsilon \leq 1$	right-handed polarizations
$\operatorname{Im} \chi = 0$	$-\frac{1}{2}\pi \leq \theta < \frac{1}{2}\pi$	$\tan \epsilon = 0$	linear polarizations
$\operatorname{Im} \chi < 0$	$-\frac{1}{2}\pi \leq \theta < \frac{1}{2}\pi$	$-1 \leq \tan \epsilon < 0$	left-handed polarizations
$|\chi| \leq 1$	$\theta \leq \frac{1}{4}\pi$	$-1 \leq \tan \epsilon \leq 1$	
$|\chi| > 1$	$|\theta| > \frac{1}{4}\pi$	$-1 < \tan \epsilon < 1$	

Table 1.2 and fig. 1.19 summarize the relationships between the ellipses of polarization associated with a set of points generated from one initial point $\chi(A)$ by the application of some basic transformations (symmetry operations).

1.7.2. *The circular complex-plane representation of polarized light*

If instead of the x and y linear polarizations we choose the left-handed (L) and right-handed (R) circular polarizations as basis states, another important

Table 1.2. Effect of the various symmetry operations on a point in the χ plane[a]

Symmetry Operation	Point in fig. 1.19	Representative Complex Number	Azimuth θ	Ellipticity angle ϵ
Identity, **E**	A	χ	θ	ϵ
Mirror reflection in the real axis, $\mathbf{M}_r$	A'_1	χ^*	θ	$-\epsilon$
Mirror reflection in the imaginary axis, $\mathbf{M}_i$	A'_2	$-\chi^*$	$-\theta$	ϵ
$\pm \pi$-rotation around the origin, $\mathbf{R}\,(\pm \pi)$	A'_3	$-\chi$	$-\theta$	$-\epsilon$
Inversion with respect to the unit circle, **I**	A'_4	$1/\chi^*$	$\frac{1}{2}\pi - \theta$	ϵ

[a]This table is intended to be used in conjunction with fig. 1.19. In fig. 1.19, there are three more points A'_5, A'_6 and A'_7 representing polarization states that are also simply related to A. They can be obtained by the successive application of some of the symmetry operations given in the above table. Of particular interest is the point A'_5 which represents the orthogonal polarization and is obtained by the successive application of **I** and $\mathbf{R}\,(\pm \pi)$.

complex-plane representation of polarized light can be introduced.[15] In this case, the information about the ellipse of polarization is extracted from the two-component *circular* Jones vector of eq. (1.54) by taking the ratio

$$\chi = E_r/E_\ell, \tag{1.92}$$

of its phasor components. With E_ℓ and E_r expressed in terms of their magnitudes and angles, eq. (1.92) becomes

$$\chi = \frac{|E_r|}{|E_\ell|}\, e^{j(\delta_r - \delta_\ell)}, \tag{1.93a}$$

$$|\chi| = |E_r|/|E_\ell|, \qquad \arg(\chi) = \delta_r - \delta_\ell. \tag{1.93b}$$

The relation between the circular complex polarization variable χ and the azimuth θ and ellipticity angle ϵ of the associated ellipse of polarization can be obtained from the definition of eq. (1.92) and the expression of the circular Jones vector in eq. (1.76) as follows

$$\chi = \tan(\epsilon + \tfrac{1}{4}\pi)\, e^{-j2\theta}. \tag{1.94}$$

The inversion of eq. (1.94), to determine the azimuth θ and the ellipticity

---

[15]A brief discussion of this representation can be found in refs. 16 and 19.

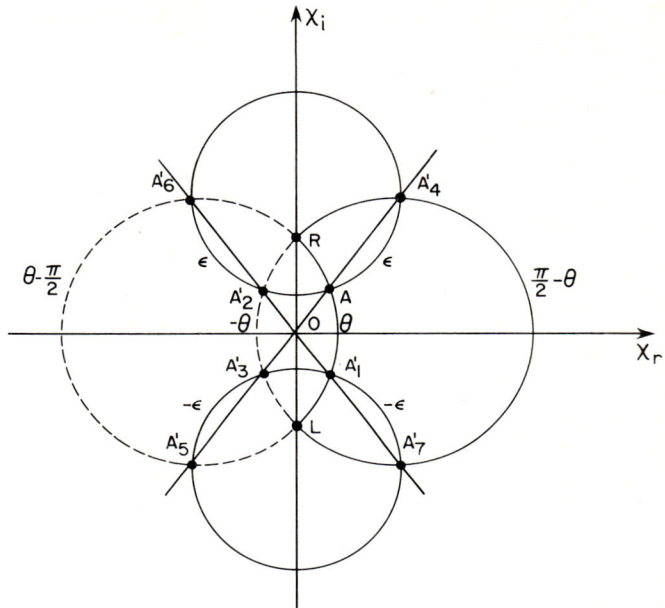

Fig. 1.19. Points of simply-related polarization states. Starting with A the points $A'_1 - A'_7$ are obtained by the application of simple symmetry operations as explained in table 1.2. For example, A'_4 is related to A by inversion with respect to the unit circle (not shown) and A'_5, the orthogonal polarization, is obtained from A'_4 by a $\pm \pi$-rotation around the origin.

angle ϵ for a polarization state represented by a given point χ in the circular complex plane, is direct and produces the following relations

$$\theta = -\tfrac{1}{2} \arg(\chi), \tag{1.95}$$

$$\tan \epsilon = \frac{|\chi| - 1}{|\chi| + 1}. \tag{1.96}$$

The assignment of the different states of polarization to the different points of the circular complex plane is described in a straightforward manner by eqs. (1.94)–(1.96). Additional insight is also gained if we consult the graphical construction of fig. 1.11 which shows how an arbitrary state of polarization can be synthesized from the left-circular and right-circular basis states using the phasor components $E_\ell = |E_\ell| e^{j\delta_\ell}$ and $E_r = |E_r| e^{j\delta_r}$ of the circular Jones vector. The following correspondence properties between points in the circular complex χ-plane and the states of polarization of light are readily obtained from eqs. (1.94)–(1.96).

(1) The origin ($\epsilon = -\tfrac{1}{4}\pi$, $\chi = 0$) and the point-at-infinity ($\epsilon = +\tfrac{1}{4}\pi$, $\chi = \infty$)

represent the left-circular (L) and right-circular (R) basis polarization states, respectively.

(2) Each point on the unit circle of the complex plane ($\epsilon = 0$, $-\frac{1}{2}\pi \leq \theta < \frac{1}{2}\pi$, $\chi = e^{-j2\theta}$) corresponds to a different linear state of polarization.

(3) Excluding the origin, the unit circle, and the point-at-infinity, all other points in the complex plane represent elliptical states of polarization. Inside the unit circle [$-\frac{1}{4}\pi \leq \epsilon < 0$, $0 \leq |\chi| < 1$] all polarizations are left-handed. Outside the unit circle [$0 < \epsilon \leq \frac{1}{4}\pi$, $1 < |\chi| \leq \infty$] all polarizations are right-handed.

(4) The equi-azimuth contours (θ = constant) consist of radial lines (rays) emanating from the origin L and ending at the point-at-infinity R. The positive real axis is an equi-azimuth contour for which $\theta = 0$, and the negative real axis is a *different* equi-azimuth contour corresponding to the orthogonal azimuth $\theta = -\frac{1}{2}\pi$. Generally, a straight line drawn to pass through the origin L (and the point-at-infinity R) consists of two branches separated by L (and R). These branches represent two equi-azimuth contours of orthogonal azimuths θ and $\theta - \frac{1}{2}\pi$. As the point L (or R) is crossed, a discontinuous jump in azimuth equal to $\frac{1}{2}\pi$ takes place.

(5) The equi-ellipticity contours (ϵ, $|\chi|$ = constant) are represented by a

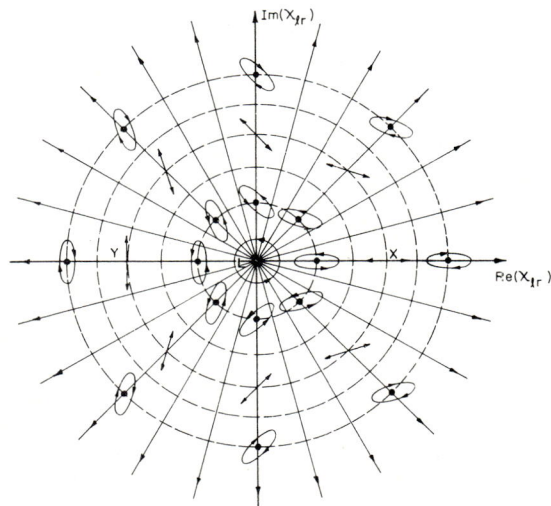

Fig. 1.20. The circular complex-plane representation of polarization whose basis states are the left (L)- and right (R)-circular polarizations. Shown are the equi-azimuth (——) and equi-ellipticity (– – –) contours as two orthogonal families of straight lines and concentric circles through and around the origin, respectively. The assignment of the different states of polarization to the points of the χ_{lr} plane is also indicated.

family of concentric circles centered on the origin L. This family is orthogonal to the family of equi-azimuth contours discussed above.

Figure 1.20 shows a map of equi-ellipticity and equi-azimuth contours and indicates the ellipses of polarization that correspond to several points in the complex plane. The arrows marked on the equi-azimuth and the equi-ellipticity contours indicate the directions of increasing ellipticity and azimuth, respectively. Along an equi-azimuth line, starting from the origin (L) and moving outwards, the ellipticity increases from -1 (at L), to zero (when the unit circle is crossed), to $+1$ (as the point-at-infinity R is approached). Starting from the point of intersection with the negative real axis and moving clockwise along an equi-ellipticity circle, the azimuth increases from $-\frac{1}{2}\pi$ (on the negative real axis), to zero (when the positive real axis is crossed), to $+\frac{1}{2}\pi$ (immediately below the starting point on the negative real axis).

From the above discussion it is evident that the use of the x and y linear polarizations and the L and R circular polarizations as basis states yield two complex-plane representations of polarized light that are quite similar. In both representations the equi-azimuth and equi-ellipticity contours consist of two orthogonal sets of curves (circle arcs and circles). We will investigate the relationship between the two representations after generalizing the definition of the complex polarization variable χ.

1.7.3. *The generalized complex-plane representation of polarized light*

The Cartesian and the circular complex-plane representations of polarized light are only two members of an infinite group of complex-plane representations that can be generated by changing the pair of reference basis states [21]. In general, we can choose *any two* fixed elliptic states u and v to synthesize *all other* states of polarizations. The *generalized* basis states u and v represent two *different* ellipses of polarization (the case when $\theta_u = \theta_v$ and $\epsilon_u = \epsilon_v$ is obviously excluded) whose amplitudes A_u and A_v ($A_u \neq 0$, $A_v \neq 0$) and absolute phases δ_u and δ_v are arbitrary. The process of generating any state of polarization from the basis of polarizations u and v involves (1) the adjustment of the amplitudes of u by v by two factors $|E_u|$ and $|E_v|$, (2) the adjustment of their absolute phases by δ_u and δ_v, and (3) the superposition of these amplitude- and phase-modified basis polarizations. This procedure has been explained in §1.6 with special reference to the linear and circular basis states. The complex quantities $E_u = |E_u| e^{j\delta_u}$ and $E_v = |E_v| e^{j\delta_v}$ represent the phasor components of the generalized Jones vector $\boldsymbol{E}_{u,v}$ [eq. (1.64)]. The ratio

$$\chi = E_v/E_u, \tag{1.97}$$

$$\chi = \frac{|E_v|}{|E_u|} e^{j(\delta_v - \delta_u)}, \tag{1.98a}$$

$$|\chi| = |E_v|/|E_u|,$$
$$\arg(\chi) = \delta_v - \delta_u, \tag{1.98b}$$

defines a single complex number that completely and uniquely describes the ellipse of polarization. Equation (1.97) is the generalization of eqs. (1.77) and (1.92) that correspond to the special choices of basis states $(u, v) = (x, y)$ and (ℓ, r), respectively. If we associate the complex number in eq. (1.97) with a point in the complex plane, the resulting representation is called the *generalized* complex-plane representation of polarized light. The origin $(\chi = 0)$ and the point-at-infinity $(\chi = \infty)$ in the χ plane represent the basis states u and v, respectively.

The various complex-plane representations that arise from a change of the basis polarization states are interrelated. Because the complex polarization variable is obtained from the Jones vector by taking the ratio of its phasor components, the transformation of this complex variable must be determined from the transformation of the corresponding Jones vector. Let $\chi_{u,v}(= E_v/E_u)$ and $\chi_{x,y}(= E_y/E_x)$ represent the generalized and the Cartesian complex polarization variables, respectively. The transformation $\chi_{x,y} \to \chi_{u,v}$ follows from the transformation of the corresponding Jones vectors $\boldsymbol{E}_{x,y} \to \boldsymbol{E}_{u,v}$ of eq. (1.66), namely,

$$\chi_{u,v} = \frac{f_{11}\chi_{x,y} - f_{21}}{-f_{12}\chi_{x,y} + f_{22}}, \tag{1.99}$$

and the inverse transformation $\chi_{u,v} \to \chi_{x,y}$ is given by

$$\chi_{x,y} = \frac{f_{22}\chi_{u,v} + f_{21}}{f_{12}\chi_{u,v} + f_{11}}. \tag{1.100}$$

Equations (1.99) and (1.100) express the important conclusion that the change of the complex polarization variable χ as a result of switching the basis polarizations between the x, y linear polarizations and the generalized u, v elliptical polarizations is governed by a *bilinear transformation* [20]. The coefficients f_{ij} of the bilinear transformation, $\chi_{x,y} \leftrightarrow \chi_{u,v}$, are the same, up to a constant multiplier, as the elements of the transformation matrix that transforms the corresponding Jones vectors $\boldsymbol{E}_{x,y} \leftrightarrow \boldsymbol{E}_{u,v}$.

The effect of two successive bilinear transformations is equivalent to a single bilinear transformation. Therefore, the end result of carrying out the two transformations $\chi_{u',v'} \to \chi_{x,y} \to \chi_{u,v}$ is the bilinear transformation

$$\chi_{u',v'} = \frac{g_{22}\chi_{u,v} + g_{21}}{g_{12}\chi_{u,v} + g_{11}}, \qquad (1.101)$$

whose coefficients g_{ij} can be obtained from the coefficients of the component bilinear transformations $(\chi_{u',v'} \to \chi_{x,y})$ and $(\chi_{x,y} \to \chi_{u,v})$ by simple matrix multiplication. Equation (1.101) represents the law of transformation of the complex polarization variable χ under the most general change of basis $(u, v) \to (u', v')$, where u, v, u' and v' correspond to four arbitrary polarization states.

Perhaps the most important property of a bilinear transformation that connects two complex variables is that circles in the complex plane of one variable are mapped into circles in the complex plane of the other variable [20]. (In this mapping, a straight line is considered as a degenerate circle of infinite radius which is transformed into a circle.) In addition, the bilinear transformation is conformal and preserves the angle between two intersecting curves in magnitude and sense. Because of its circle-preserving property, the bilinear transformation $\chi_{x,y} \to \chi_{u,v}$ of eq. (1.99) maps the equi-azimuth circle arcs and the equi-ellipticity circles through and around the points L and R in the $\chi_{x,y}$ plane into equi-azimuth circle arcs and equi-ellipticity

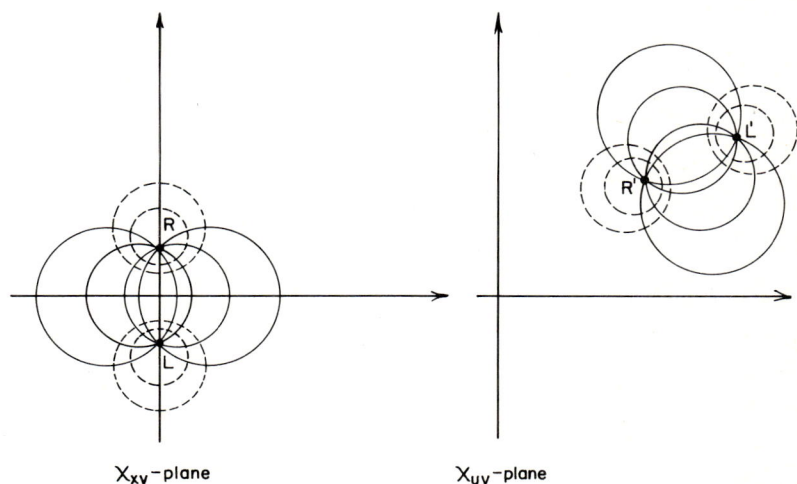

Fig. 1.21. The bilinear transformation $\chi_{x,y} \to \chi_{u,v}$ that corresponds to a change of basis [eq. (1.99)] maps the equi-azimuth circle arcs and the equi-ellipticity circles through and around the points L and R in the Cartesian complex $\chi_{x,y}$ plane into equi-azimuth circle arcs and equi-ellipticity circles through and around the image points L' and R' in the generalized complex $\chi_{u,v}$ plane.

circles through and around the image points L' and R' in the $\chi_{u,v}$ plane, fig. 1.21. Because of its conformality, the orthogonality of the equi-azimuth and equi-ellipticity contours is preserved in the generalized $\chi_{u,v}$ complex-plane representation. The points L' and R' represent the left- and right-circular polarizations in the $\chi_{u,v}$ plane and are obtained from eq. (1.99) by setting $\chi_{x,y} = \mp j$.

$$\chi_{u,v}(L') = \frac{-jf_{11} - f_{21}}{jf_{12} + f_{22}},$$

$$\chi_{u,v}(R') = \frac{jf_{11} - f_{21}}{-jf_{12} + f_{22}}.$$

(1.102)

Note that an equi-ellipticity circle in the generalized $\chi_{u,v}$ complex-plane representation still represents the locus of a point P that moves so that the distance ratio $|P\,L'|/|P\,R'|$ remains constant. The value of this constant in terms of the ellipticity angle ϵ is dependent on the choice of u and v and is given by eq. (1.91) when u and v represent the x and y linear polarizations. In general, not all properties remain invariant when we go from the Cartesian $\chi_{x,y}$ to the generalized $\chi_{u,v}$ complex-plane representation. For example, the orthogonality condition of eq. (1.81) for two polarizations in the $\chi_{x,y}$ plane does not preserve form in the $\chi_{u,v}$ plane except when u and v are chosen orthogonal. This, and other properties of the $\chi_{u,v}$ plane, may be verified by the interested reader.

We have seen that eq. (1.99) follows from the transformation $\boldsymbol{E}_{u,v} = \boldsymbol{F}^{-1}\boldsymbol{E}_{x,y}$ of the Jones vector where $\boldsymbol{F}^{-1}$ is given by eq. (1.67). When the basis states u and v are orthogonal $\boldsymbol{F}^{-1}$ takes the form of eq. (1.70) and eq. (1.99) becomes

$$\chi_{u,v} = \frac{m\chi_{x,y} - n}{n^*\chi_{x,y} + m^*}.$$

(1.103)

If the left- and right-circular polarizations are chosen as basis states $\boldsymbol{F}^{-1}$ is given by eq. (1.57), so that

$$\chi_{\ell,r} = \frac{\chi_{x,y} + j}{-\chi_{x,y} + j}.$$

(1.104)

That eq. (1.104) is a special case of eq. (1.103) can be seen by multiplying the numerator and denominator of eq. (1.104) by $e^{-j\pi/4}$. The expressions of $\chi_{x,y}$ and $\chi_{\ell,r}$ in terms of θ and ϵ that appear in eqs. (1.79) and (1.94), respectively, satisfy eq. (1.104) as may be proved by direct substitution. Equation (1.104) provides all the correspondence properties between the Cartesian and the circular complex-plane representations of polarized light. It is easily seen that the points L($\chi_{x,y} = -j$) and R($\chi_{x,y} = +j$) in the $\chi_{x,y}$ plane are mapped to the origin ($\chi_{\ell,r} = 0$) and the point-at-infinity ($\chi_{\ell,r} = \infty$) in the $\chi_{\ell,r}$ plane and that points on the real axis of the $\chi_{x,y}$ plane ($\chi_{x,y} = \tan\theta$) are transformed onto points on the unit circle ($\chi_{\ell,r} = e^{-j2\theta}$), and so forth. Another transformation of interest is that which corresponds to a coordinate rotation. For the transformation of the complex polarization variable due to a coordinate rotation $(x, y) \rightarrow (x', y')$, $\boldsymbol{F}^{-1} = \boldsymbol{R}(-\alpha)$ and

$$\chi_{x',y'} = \frac{(\cos\alpha)\chi_{x,y} - (\sin\alpha)}{(\sin\alpha)\chi_{x,y} + (\cos\alpha)}. \tag{1.105}$$

The transformation of eq. (1.105) leaves the equi-azimuth and equi-ellipticity contour map of fig. 1.18 unchanged. The only effect of such a transformation is a reassignment of azimuth values to the circle arcs joining L and R. In particular each azimuth value θ is replaced by $\theta - \alpha$, where α is the angle of the counter-clockwise rotation from x to x'.

1.8. The Poincaré-sphere representation of polarized light

In the previous section (§1.7) we have introduced the important concept of the use of a single complex variable χ to describe the state of polarization of light. We have seen how the information about the ellipse of polarization, including the azimuth θ and the ellipticity angle ϵ, can be neatly lumped in χ. The complex χ plane provides a space whose points are in one-to-one correspondence with the different possible states of polarization. By changing the pair of reference basis states, an infinite group of complex-plane representations is generated. Two such representations have been selected for detailed discussion in view of their special importance, namely, the Cartesian and the circular complex-plane representations whose basis states are the x and y linear polarizations and the left (L)- and right (R)-circular polarizations, respectively. We have proved that *all* complex-plane representations can be produced by carrying out bilinear transformations of *one* original representation.

In the present section we discuss another important polarization space in the form of a spherical surface whose points are also in one-to-one correspondence with the different states of polarization of light. This spherical polarization space is referred to as the Poincaré sphere, after H. Poincaré who introduced it in 1892[13]. Although there is more than one approach by which the Poincaré sphere can be introduced, the procedure that we will use is similar to that used by Poincaré.[16]

---

[16]Instead of using the circular $\chi_{\ell,r}$ complex-plane representation, as we do here, Poincaré [13] used the Cartesian $\chi_{x,y}$ complex-plane representation to introduce his spherical representation. Our choice of the circular, rather than the Cartesian, complex plane allows all the properties of the Poincaré sphere to be obtained more clearly and more readily. The introduction of the Poincaré sphere through the Stokes parameters, as is commonly done, is explained in the next section (§1.9).

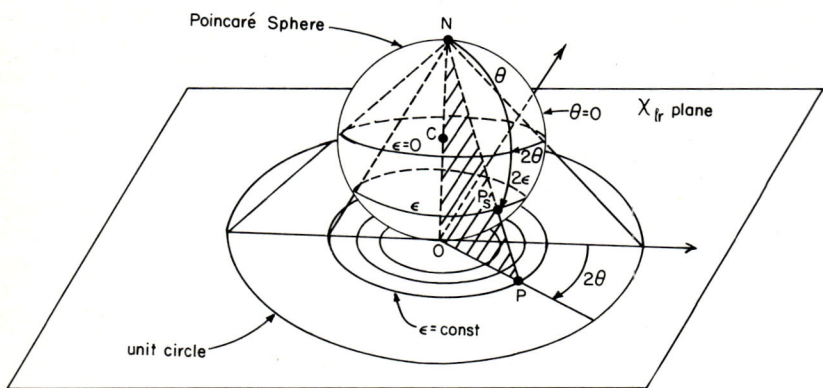

Fig. 1.22. Derivation of the Poincaré-sphere representation of polarization from the circular complex-plane representation [the $\chi_{\ell,r}$ plane] through a stereographic projection.

By reference to fig. 1.22, consider a sphere of *unit* diameter that rests tangent to the *circular* $\chi_{\ell,r}$ complex plane at its origin O. Each point P in the $\chi_{\ell,r}$ plane can be connected to the pole of the sphere diametrically opposite to the origin, the "north" pole N, by a straight line that intersects the sphere at *one* point P_s between P and N. This projective transformation, that maps $P \to P_s$, is known as a stereographic projection. If we associate with the point P_s on the sphere the *same* ellipse of polarization that corresponds to the point P in the $\chi_{\ell,r}$ complex plane, the entirety of the sphere will represent a polarization space, the Poincaré-sphere space, whose points are in one-to-one correspondence with the different states of polarization of light. The assignment of the different states of polarization to the different points of the Poincaré sphere can be readily examined from fig. 1.22 and the previously derived properties of the circular complex-plane representation (§1.7). The following correspondence properties are obtained.

(1) The "south" and "north" poles of the Poincaré sphere represent the left (L)- and right (R)-circular polarization states, respectively. (This follows because the origin O of the complex plane, which corresponds to the left-circular polarization, is projected onto itself and the point-at-infinity in the complex plane, which represents the right-circular polarization, is projected onto the north pole N (the center of projection).

(2) Each point on the equator of the Poincaré sphere represents a distinct linear state of polarization. (Because the diameter of the sphere has been chosen unity, the unit circle in the complex $\chi_{\ell,r}$ plane, representing all the linear polarizations, is projected onto the equator.)

(3) Excluding the south pole, the equator and the north pole all other points on the Poincaré sphere represent elliptical states of polarization.

Below the equator (the southern hemisphere) all polarizations are left handed. Above the equator (the northern hemisphere) all polarizations are right handed. (This follows because the inside and the outside of the unit circle in the complex plane, representing left- and right-handed polarizations, are projected onto the lower and upper hemispheres, respectively.)

(4) The equi-azimuth contours on the surface of the Poincaré sphere constitute a family of semi-great-circles (lines of longitude) drawn through the south and north poles. The (angle of) longitude is double the azimuth angle of the ellipse of polarization. The azimuth is measured positive $(0 < \theta < \tfrac{1}{2}\pi)$ in a clockwise sense and is measured negative $(-\tfrac{1}{2}\pi < \theta < 0)$ in a counter-clockwise sense (starting from the point that represents the zero-azimuth linear polarization) when looking at the sphere from above N along the extension of the diameter ON. (These results are obvious from the nature of the equi-azimuth contours as radial lines from the origin in the complex $\chi_{e,r}$ plane.)

(5) The equi-ellipticity contours on the surface of the Poincaré sphere are represented by a family of coaxial circles (latitudes) whose common axis is the polar axis from the south to the north pole. These equi-ellipticity latitudes are obviously orthogonal to the equi-azimuth longitudes. To determine the relationship between the (angle of) latitude and the ellipticity angle ϵ, we refer to fig. 1.23 which represents a cross section of the Poincaré

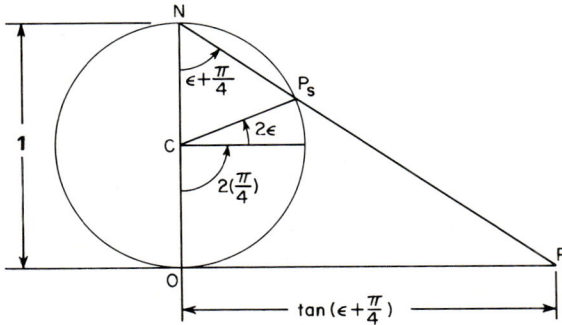

Fig. 1.23. This figure proves that the latitude of a point P_s on the Poincaré sphere is equal to double the ellipticity angle (2ϵ).

sphere by the plane formed by the origin O (the south pole), a general point P in the complex plane and the center of the projection N (the north pole). It is clear that $\tan(\angle ONP) = |OP|/|ON|$. Because $|OP| = |\chi_{e,r}| = \tan(\epsilon + \tfrac{1}{4}\pi)$ [eq. (1.94)] and $|ON| = 1$ we find that $\angle ONP = (\epsilon + \tfrac{1}{4}\pi)$. The central angle $\angle OCP_s$ and the circumferential angle $\angle ONP_s$ are both constructed on the same arc $\widehat{OP_s}$ from which $\angle OCP_s = 2\epsilon + \tfrac{1}{2}\pi$. By definition, the latitude of the point P_s

is given by the angle between the radius vector CP_s (connecting the center of the Poincaré sphere C to the point P_s) and the equatorial plane. Hence, the latitude is equal to $(2\epsilon + \frac{1}{2}\pi) - \frac{1}{2}\pi = 2\epsilon$, or simply, double the ellipticity angle ϵ. The latitude is measured positive above the equatorial plane $(0 < \epsilon \leq \frac{1}{4}\pi)$ and is measured negative below the equatorial plane $(-\frac{1}{4}\pi \leq \epsilon < 0)$ and is zero on the equator $(\epsilon = 0)$.

The above properties of the Poincaré sphere can be summarized in a single statement: a state of polarization of azimuth θ and ellipticity angle ϵ is represented on the surface of the Poincaré sphere by a point whose longitude is double the azimuth, 2θ, and whose latitude is double the ellipticity angle, 2ϵ. It can be easily seen that diametrically opposite points on the sphere represent pairs of orthogonal polarizations. Figure 1.24 shows a map of the equi-azimuth and the equi-ellipticity contours and the polarization states that correspond to a sample of points on the Poincaré sphere.

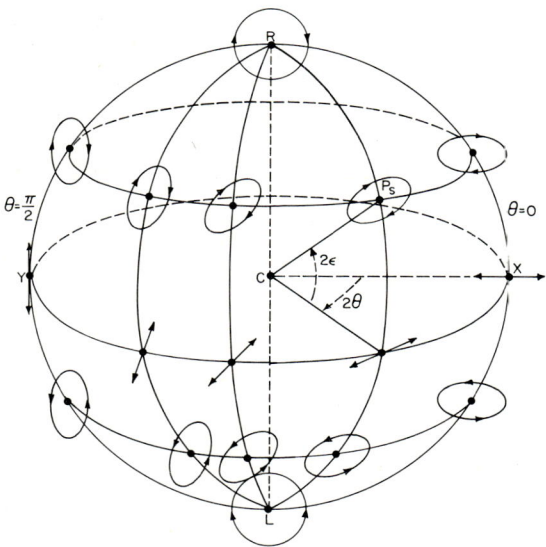

Fig. 1.24. The Poincaré-sphere representation of polarization. The longitude 2θ and latitude 2ϵ determine a point P_s that represents an ellipse of polarization with azimuth θ and ellipticity angle ϵ. The lines of longitude and latitude represent the equi-azimuth and equi-ellipticity contours, respectively. The correspondence between polarization states and points on the sphere is indicated.

If a tangent plane to the Poincaré sphere is drawn at the point X (representing the x linear polarization) and the diametrically opposite point Y (representing the y linear polarization) is considered as the center of a new stereographic projection, each point P_s on the sphere can be projected onto a point on the new tangent plane. By inspection, the resulting planar representation can be easily recognized as the Cartesian $\chi_{x,y}$ complex-plane representation whose equi-azimuth and equi-ellipticity contours appear in fig. 1.18.[17] This process can be extended to apply to the complex-plane representation whose basis states consist of any pair of orthogonal polarizations u and v. The complex $\chi_{u,v}$ plane can be identified as the stereographic projection of the Poincaré sphere onto a tangent plane at the point u (representing one basis state) with the diametrically opposite point v (representing the other orthogonal basis state) used as the center of the projection. The stereographic projection has two important properties (1) circles in the tangent plane are carried over onto circles on the sphere (a straight line is a degenerate circle) and vice versa, (2) angular relationships between intersecting curves are preserved [20, 21]. Because of these two properties, the longitudes and the latitudes of the Poincaré sphere always project as two orthogonal families of circle arcs and circles in the $\chi_{u,v}$ complex plane. This is another proof of the statement given earlier in §1.7 concerning the nature of the equi-azimuth and equi-ellipticity contours, applicable to the special case when the basis polarizations are chosen orthogonal. The transformation $\chi_{u,v} \leftrightarrow \chi_{u',v'}$ between two complex-plane representations [given by eq. (1.101)] acquires a geometrical meaning on the Poincaré sphere when (u, v) and (u', v') represent two pairs of orthogonal basis states. In this case two tangent planes to the Poincaré sphere at the points u and u' are constructed and the points v and v' are used as projection centers. Each point on the sphere has two images in the $\chi_{u,v}$ and $\chi_{u',v'}$ planes and the two planes are thus linked *through* the sphere.

Consider a Cartesian coordinate system $S_1 S_2 S_3$ at the center C of the sphere with the S_1 and S_2 axes parallel to the real and imaginary axes of the $\chi_{\ell,r}$ complex plane and with the S_3 axis directed from the south pole (the left-circular polarization) to the north pole (the right-circular polarization). Referenced to the above $S_1 S_2 S_3$ coordinate system, the polar coordinates (r, θ, ϕ) of the point P_s on the sphere are given by $(\frac{1}{2}, \frac{1}{2}\pi - 2\epsilon, 2\theta)$. Denoting the radius of the Poincaré sphere by S_0, the set of parameters (S_0, S_1, S_2, S_3) is given by

$$S_0 = \tfrac{1}{2},$$
$$S_1 = \tfrac{1}{2} \cos 2\epsilon \cos 2\theta,$$
$$S_2 = \tfrac{1}{2} \cos 2\epsilon \sin 2\theta,$$
$$S_3 = \tfrac{1}{2} \sin 2\epsilon.$$

(1.106)

According to the discussion in this section, the diameter of the Poincaré

---

[17] See footnote 16 on p. 47.

sphere is chosen equal to unity so that the Poincaré sphere and the complex-plane representations are properly inter-related. Other choices of the sphere diameter can be considered, as long as the sphere is used alone, without reference to the complex plane. When the radius is multiplied by 2, eq. (1.106) gives the Stokes parameters of light of unit intensity, as will be discussed in the following section.

1.9. The polarization of quasi-monochromatic waves

The major part of this chapter (§1.3–§1.8) has already been devoted to the polarization of monochromatic waves. We have found that the polarization of a monochromatic wave is in general elliptic, because the electric vector E, observed at a fixed point in space, precesses in a regular repetitive fashion around an ellipse which may become a circle or a straight line as limiting special cases. This type of polarization is described as *total*.

In the case of a quasi-monochromatic wave, the nature of the vibration of the electric vector E becomes more complicated. Because the spectrum of the quasi-monochromatic wave consists of a narrow band of frequencies, instead of a single discrete frequency [see §1.2], the time-variation of a Cartesian component of the electric vector is no longer represented by an exact sine wave of infinite length. Let $\Delta\omega$ denote the bandwidth of the quasi-monochromatic wave and ω_o its center frequency. This spectrum may be considered equivalent to that of a sine-wave carrier of frequency ω_o whose amplitude and phase are modulated. For a quasi-monochromatic wave of arbitrary spatial structure the electric vector E at a fixed point in space can be resolved into three Cartesian components

$$E_i = \tilde{E}_i(t) \cos[\omega_o t + \delta_i(t)], \qquad i = x, y, z. \tag{1.107}$$

In eq. (1.107) the amplitudes $\tilde{E}_i(t)$ and phases $\delta_i(t)$ are functions of time which, in general, are independent for the three Cartesian components. The six functions $[\tilde{E}_i(t), \delta_i(t)]$ $i = x, y, z$, have noise-like properties and are determined by the *source* of the quasi-monochromatic radiation. Their frequency bandwidth is of the order of $\Delta\omega/2$. Obviously, eq. (1.2) is a special case of eq. (1.107) when $\tilde{E}_i(t)$ and $\delta_i(t)$ are constant independent of time, appropriate to the case of a monochromatic wave.

If a vector is drawn from the fixed observation point to represent the resultant of the three components in eq. (1.107), the end point of that vector will describe a complicated trajectory. To gain an understanding of this trajectory, consider the variation of E over an interval of time $\tau = 4\pi/\Delta\omega$. Let this interval of time τ be subdivided into a large number, N, of equal subintervals $\Delta\tau = \tau/N$. N is chosen such that $\tilde{E}_i(t)$ and $\delta_i(t)$ remain essentially constant during $\Delta\tau$. Because $\Delta\omega \ll \omega_0$, each sub-interval $\Delta\tau$ accommodates a large number of cycles of a regular sine wave of frequency ω_0.[18] Within a sub-interval $\Delta\tau$ the discussion of §1.3 on the polarization of monochromatic waves becomes applicable and the electric vector regularly describes an ellipse. In general, the ellipses of polarization that correspond to the N different sub-intervals have different properties. Observed over the entire interval $\tau = N\Delta\tau$, the unit normal $\hat{n}$ to the plane of the ellipse [see §1.4] may precess in space; the orientation of the ellipse in its plane, its shape and size may also vary with time continuously. Therefore, although the *short-term* polarization is elliptic, the *long-term* polarization is represented by an irregular trajectory.

The case of a TE plane wave is simpler because the unit normal $\hat{n}$ remains constant (oriented along the direction of propagation) and the electric field E is entirely transverse to $\hat{n}$. Hence, we need only consider the addition of two field components of the form of eq. (1.107), *e.g.*, E_x and E_y, with $E_z = 0$. Such addition produces an ellipse whose azimuth, ellipticity (including the sense of description) and size vary continuously with time during a long observation period ($\geq \tau$). If no specific, short-term, elliptic polarization is preferred the light is described as *natural* or *unpolarized*. In the presence of preference to a specific, short-term, elliptic polarization the light is in general *partially polarized*. Depending upon the nature of the privileged polarization, the quasi-monochromatic wave may be *elliptically*, *circularly* or *linearly* partially polarized.

The existence of preference to any particular polarization in the transverse plane wave-front depends upon the *degree of correlation* between the time-variation of the Cartesian components E_x and E_y of the transverse-electric field given by eq. (1.107). The case of natural (unpolarized) light corresponds to total lack of correlation (*coherence*) between E_x and E_y. Partial correlation yields partial polarization and total correlation gives total polarization.

1.9.1. *The Jones vector of a quasi-monochromatic wave*

Over an interval of time $\Delta\tau$, short enough to make the variation of $\tilde{E}_i(t)$ and $\delta_i(t)$ negligible and long enough to accommodate a large number of optical-frequency cycles, a quasi-monochromatic wave behaves like a

---

[18] If a value of $\Delta\omega/\omega_0 = 10^{-6}$ is assumed (which is readily attainable from a laser source) and ω_0 is taken in the visible ($\sim 5 \times 10^{15}$ rad/sec), the reader can verify for himself that the choice of $N = 10^3$ leads to each sub-interval $\Delta\tau$ accommodating about a thousand optical frequency cycles.

monochromatic wave and therefore can be described by a Jones vector. From the x and y components of the electric vector of a uniform TE quasi-monochromatic plane wave the 2×1 time-dependent Cartesian complex Jones vector is constructed as follows

$$\boldsymbol{E}_{x,y}(t) = \begin{bmatrix} \tilde{E}_x(t) \, e^{j\delta_x(t)} \\ \tilde{E}_y(t) \, e^{j\delta_y(t)} \end{bmatrix} \tag{1.108}$$

where the $\omega_o t$ time-dependence [eq. (1.107)] has been suppressed and may be restored by remultiplication by $e^{j\omega_o t}$ and taking the real part of the product.[19] The orientation, shape and handedness of the short-term ellipse of polarization are determined by the single complex variable

$$\chi_{x,y}(t) = \frac{\tilde{E}_y(t)}{\tilde{E}_x(t)} \exp\{j[\delta_y(t) - \delta_x(t)]\} \tag{1.109}$$

while the amplitude (size) is given by

$$A(t) = [\tilde{E}_x^2(t) + \tilde{E}_y^2(t)]^{\frac{1}{2}}. \tag{1.110}$$

The azimuth $\theta(t)$ (specifying the orientation) and the ellipticity angle $\epsilon(t)$ (specifying the shape and handedness) of the short-term ellipse of polarization are obtained by substituting $\chi_{x,y}(t)$ of eq. (1.109) into eqs. (1.86) and (1.87) of §1.7, respectively.

Let the short-term ellipse of polarization be represented by a point P in the complex plane. Because the complex plane and the Poincaré-sphere representations are linked by a process of stereographic projection [§1.8], the short-term ellipse of polarization can also be represented by a point P_s on the surface of the Poincaré sphere. The motion of the representative points P in the complex plane (the trajectory of $\chi(t)$) and P_s on the Poincaré sphere provides an elegant method of visualizing the long-term state of polarization of quasi-monochromatic waves. We have the following three possibilities

(1) The case of unpolarized light corresponds to a "random walk" of the representative point P_s on the surface of the Poincaré sphere such that it is equally likely to find that point at any locality in this polarization space. The trajectory described by P_s is completely erratic.

(2) For partially polarized light the representative point P_s spends more time in the neighborhood of a particular point P_{s0} that corresponds to a specific privileged polarization. The probability of finding P_s is no longer uniform over the surface of the Poincaré sphere but rather peaks at P_{s0}. The

---

[19]The time-dependent phasor $\tilde{E}_x(t) e^{j\delta_x(t)}$ is called the complex envelope of the signal $\tilde{E}_x(t) \cos[\omega_o t + \delta_x(t)]$, (Helstrom, [22]).

§1.9]   POLARIZATION OF QUASI-MONOCHROMATIC WAVES   55

state of partial polarization may be linear, circular or elliptic depending on whether the position of P_{s0} is on the equator, the south or north pole, or elsewhere, respectively.

(3) For the case of a totally polarized quasi-monochromatic wave the point P_s does not move at all and the probability distribution of finding P_s assumes the form of a unit delta-function centered on one point P_{s0} on the sphere. According to eq. (1.109), this takes place if

$$\frac{\tilde{E}_y(t)}{\tilde{E}_x(t)} = \text{constant}, \quad [\delta_y(t) - \delta_x(t)] = \text{constant}, \tag{1.111}$$

which indicates complete correlation (coherence) between the Cartesian components E_x and E_y of the transverse electric field. Note that although eq. (1.111) predicts an ellipse of polarization whose orientation, shape and handedness remain constant, the amplitude (size) of that ellipse [given by eq. (1.110)] may fluctuate with time during the period of observation. *In all experiments that involve the interaction of polarized light with optical devices the two cases of monochromatic and totally polarized quasi-monochromatic light waves are indistinguishable.* (This justifies the assumption of monochromaticity for a wave which is actually quasi-monochromatic but totally polarized.)

Figure 1.25 shows schematically the difference between the three cases of

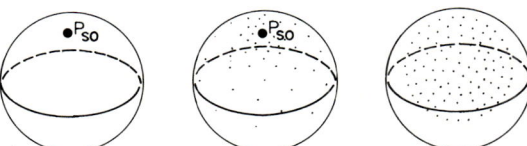

Fig. 1.25. Schematic representation of totally polarized, partially polarized and unpolarized light (in order, from left to right) on the Poincaré sphere, see text.

unpolarized, partially polarized and totally polarized quasi-monochromatic waves by means of dot patterns in the polarization space. Each dot represents a short-term ellipse of polarization that corresponds to a subinterval of time $\Delta\tau$ of an equally-divided long observation interval $\tau = N \Delta\tau$, where N is a very large number.

1.9.2. The Stokes parameters and the Stokes vector

The different possible states of polarization of a quasi-monochromatic (or monochromatic) TE plane wave of light can be represented by a set of four real quantities, called the Stokes parameters, each of which has the dimensions of intensity. In terms of the Cartesian components of the transverse electric field, the four Stokes parameters, denoted by S_0, S_1, S_2 and S_3, are defined as follows

$$\begin{aligned}S_0 &= \langle \tilde{E}_x^2(t) \rangle + \langle \tilde{E}_y^2(t) \rangle, \\ S_1 &= \langle \tilde{E}_x^2(t) \rangle - \langle \tilde{E}_y^2(t) \rangle, \\ S_2 &= 2 \langle \tilde{E}_x(t) \tilde{E}_y(t) \cos[\delta_y(t) - \delta_x(t)] \rangle, \\ S_3 &= 2 \langle \tilde{E}_x(t) \tilde{E}_y(t) \sin[\delta_y(t) - \delta_x(t)] \rangle.\end{aligned} \qquad (1.112)$$

In eq. (1.112) $\langle v \rangle$ signifies the time average of v,

$$\langle v \rangle = \frac{1}{T} \int_0^T v \, dt, \qquad (1.113)$$

where T is an interval of time long enough to make the time-average integral independent of T itself.

From eq. (1.112) it is evident that S_0 gives the total intensity of the light wave and therefore is always positive. S_1 gives the difference between the intensities of the x and y components and can be either positive, negative, or zero depending on whether the wave has stronger preference to the x linear polarization, to the y linear polarization, or to neither one of these two states, respectively. If we premultiply the Jones vector of the quasi-monochromatic wave of eq. (1.108) by the rotation matrix $\boldsymbol{R}(-\frac{1}{4}\pi)$ [eq. (1.45a)] we get

$$\begin{bmatrix} E_{-\frac{1}{4}\pi} \\ E_{+\frac{1}{4}\pi} \end{bmatrix} = \frac{1}{\sqrt{2}} \begin{bmatrix} E_x - E_y \\ E_x + E_y \end{bmatrix}, \qquad (1.114)$$

which gives the Jones vector referenced to the $-\frac{1}{4}\pi$ and $+\frac{1}{4}\pi$ bisectors of the x and y coordinate axis in the wave-front. In eq. (1.114), E_x, E_y, $E_{-\frac{1}{4}\pi}$ and $E_{+\frac{1}{4}\pi}$ should be taken to represent the *time-dependent* phasor components of the respective Jones vectors [e.g. $E_x = \tilde{E}_x(t) e^{j\delta_x(t)}$]. From eq. (1.114), the difference between the intensities of the linearly polarized components of the wave along the $+\frac{1}{4}\pi$ and $-\frac{1}{4}\pi$ bisectors is given by

$$\frac{1}{2} \langle (E_x + E_y)(E_x + E_y)^* \rangle - \frac{1}{2} \langle (E_x - E_y)(E_x - E_y)^* \rangle$$
$$= 2 \langle \mathrm{Re}\,(E_x^* E_y) \rangle = 2 \langle \tilde{E}_x(t) \tilde{E}_y(t) \cos[\delta_y(t) - \delta_x(t)] \rangle = S_2. \qquad (1.115)$$

Equation (1.115) shows that the Stokes parameter S_2, as defined in eq. (1.112), represents the preference of the wave to either the $+\frac{1}{4}\pi$ or the $-\frac{1}{4}\pi$ linearly polarized component. If S_2 is positive (negative) the wave has stronger preference to the $+\frac{1}{4}\pi$ $(-\frac{1}{4}\pi)$ linear polarization. If S_2 is zero the wave has no preference to either one of these two polarizations. In a similar manner, if we premultiply the Jones vector of the quasi-monochromatic wave of eq. (1.108) by the (inverse) transformation matrix $\boldsymbol{F}^{-1}$ of eq. (1.57) we get

$$\begin{bmatrix} E_\ell \\ E_r \end{bmatrix} = \frac{1}{\sqrt{2}} \begin{bmatrix} E_x + jE_y \\ E_x - jE_y \end{bmatrix} \qquad (1.116)$$

which gives the circular Jones vector of the wave in the representation whose basis states are the left- and right-circular polarizations [§1.6]. From eq. (1.116), the difference between the intensities of the right-handed and the left-handed circularly-polarized components of the quasi-monochromatic wave is given by

$$\tfrac{1}{2}\langle (E_x - jE_y)(E_x - jE_y)^*\rangle - \tfrac{1}{2}\langle (E_x + jE_y)(E_x + jE_y)^*\rangle$$
$$= 2\langle \mathrm{Re}\,(-jE_x^*E_y)\rangle = 2\langle \tilde{E}_x(t)\tilde{E}_y(t)\sin[\delta_y(t) - \delta_x(t)]\rangle = S_3. \qquad (1.117)$$

Equation (1.117) shows that the Stokes parameter S_3, as defined in eq. (1.112), represents the preference of the wave to either the right-handed or to the left-handed circularly polarized component (into which the wave can be resolved). S_3 is positive, negative or zero dependent on the wave possessing stronger preference to the right-circular state, the left-circular state or to neither one of these two states, respectively.

The above discussion of the significance of the Stokes parameters suggests a simple experiment by which they may be measured for a given wave. In particular, let I_0 denote the total intensity of the wave and let I_x, I_y, $I_{+\frac{1}{4}\pi}$, $I_{-\frac{1}{4}\pi}$, I_ℓ and I_r represent the intensities transmitted by an ideal variable polarizer placed in the path of the wave and adjusted to transmit the x, y, $+\tfrac{1}{4}\pi$, $-\tfrac{1}{4}\pi$ linear polarizations and the left (ℓ)- and right (r)-circular polarizations, respectively. In terms of these intensities, the Stokes parameters are given by

$$\begin{aligned} S_0 &= I_0 = (I_x + I_y) = (I_{+\frac{1}{4}\pi} + I_{-\frac{1}{4}\pi}) = (I_\ell + I_r), \\ S_1 &= I_x - I_y, \\ S_2 &= I_{+\frac{1}{4}\pi} - I_{-\frac{1}{4}\pi}, \\ S_3 &= I_r - I_\ell. \end{aligned} \qquad (1.118)$$

The Stokes vector

The Stokes parameters of a quasi-monochromatic light wave can be grouped in a 4×1 column vector

$$\mathbf{S} = \begin{bmatrix} S_0 \\ S_1 \\ S_2 \\ S_3 \end{bmatrix}, \qquad (1.119\mathrm{a})$$

called the Stokes vector of the wave. The primary advantage of this step is that it allows the use of a compact matrix formalism to handle the interaction between the light wave and the optical devices that constitute an optical system, as will be discussed in the next chapter. To save space, the 4×1 column Stokes vector is often written horizontally between curly brackets { } and the elements are separated by commas as follows

$$S = \{S_0, S_1, S_2, S_3\}. \tag{1.119b}$$

For unpolarized light, there is no preference to any particular polarization so that $S_1 = S_2 = S_3 = 0$ and the Stokes vector assumes the simple form

$$S_{un.} = \{S_0, 0, 0, 0\}, \tag{1.120}$$

where S_0 represents the total intensity of the wave. On the other extreme, a totally polarized wave possesses complete preference to one specific polarization which may be linear, circular or elliptic. In this case, the time-dependent phasor representation of the x and y linearly polarized components of the wave can be written as

$$\tilde{E}_x(t)\, e^{j\delta_x(t)} = A(t)\, e^{j\delta(t)}[\cos\theta \cos\epsilon - j \sin\theta \sin\epsilon],$$
$$\tilde{E}_y(t)\, e^{j\delta_y(t)} = A(t)\, e^{j\delta(t)}[\sin\theta \cos\epsilon + j \cos\theta \sin\epsilon], \tag{1.121}$$

which follows from eq. (1.75), keeping the azimuth θ and the ellipticity angle ϵ constant and allowing the total amplitude A and absolute phase δ to fluctuate with time. Equation (1.121) can be easily seen to satisfy the conditions of total polarization that appear in eq. (1.111). Direct substitution of eq. (1.121) into eq. (1.112) yields the following Stokes vector

$$S_{t.p.} = \{S_0, S_0 \cos 2\theta \cos 2\epsilon, S_0 \sin 2\theta \cos 2\epsilon, S_0 \sin 2\epsilon\}, \tag{1.122}$$

where $S_0 = \langle A^2(t) \rangle$ represents the total intensity of the wave. From eq. (1.122) it can be easily verified that the Stokes parameters of a totally polarized wave satisfy the following condition

$$S_0^2 = S_1^2 + S_2^2 + S_3^2. \tag{1.123}$$

Using the basic definitions of the Stokes parameters in eq. (1.112), we can prove that in place of the equality of eq. (1.123) the Stokes parameters of partially polarized or unpolarized light satisfy the inequality

$$S_0^2 > S_1^2 + S_2^2 + S_3^2. \tag{1.124}$$

Equations (1.120), (1.122)–(1.124) suggest that the general case of partially polarized light can be treated by splitting that wave into two components, a totally polarized component and an unpolarized component [23]

$$S = S_{un.} + S_{t.p.}, \tag{1.125}$$

where

$$S = \{S_0, S_1, S_2, S_3\},$$
$$S_{\text{un.}} = \{[S_0 - (S_1^2 + S_2^2 + S_3^2)^{\frac{1}{2}}], 0, 0, 0\},$$
$$S_{\text{t.p.}} = \{(S_1^2 + S_2^2 + S_3^2)^{\frac{1}{2}}, S_1, S_2, S_3\}. \quad (1.126)$$

Note that each Stokes parameter of the original beam is obtained by *adding* the corresponding parameters of the component beams.

An important quantity in the description of partially polarized light is the *degree of polarization* which is defined as the ratio of the intensity of the totally polarized component to the total intensity of the wave

$$\mathcal{P} = (S_1^2 + S_2^2 + S_3^2)^{\frac{1}{2}}/S_0. \quad (1.127)$$

The degree of polarization $\mathcal{P}$ varies from zero for unpolarized light to unity in the case of totally polarized light and assumes intermediate (fractional) values for partially polarized light. The azimuth θ and ellipticity angle ϵ of the ellipse of polarization of the totally polarized component can be obtained from a comparison of the Stokes parameters of that component in eq. (1.126) with eq. (1.122)

$$\theta = \tfrac{1}{2} \arctan(S_2/S_1), \quad \epsilon = \tfrac{1}{2} \arcsin\left[\frac{S_3}{(S_1^2 + S_2^2 + S_3^2)^{\frac{1}{2}}}\right]. \quad (1.128)$$

In terms of the total intensity I, the degree of polarization $\mathcal{P}$, the azimuth θ and ellipticity angle ϵ of the totally polarized component, the Stokes vector of any quasi-monochromatic wave can be cast in the form

$$S = I\{1, \mathcal{P} \cos 2\epsilon \cos 2\theta, \mathcal{P} \cos 2\epsilon \sin 2\theta, \mathcal{P} \sin 2\epsilon\}. \quad (1.129)$$

For a given value of the total intensity I (for example unity), the state of polarization is determined by $\mathcal{P}$, θ and ϵ. It is interesting and important to note that in the Stokes subspace $(S_1 S_2 S_3)$ the state of polarization of a quasi-monochromatic wave is represented by a point whose *polar* coordinates are $(\mathcal{P}, \tfrac{1}{2}\pi - 2\epsilon, 2\theta)$. This is shown in fig. 1.26. The following properties of this Stokes polarization subspace can be readily arrived at.

(1) The origin, $\mathcal{P} = 0$, represents the unpolarized state.

(2) Each point on the surface of the unit sphere, $\mathcal{P} = 1$, represents a distinct totally-polarized state. By comparison of the Stokes vector of eq. (1.129), when $\mathcal{P} = 1$, and the Stokes vector formed by the parameters of eq. (1.106) it becomes clear that the unit sphere in the Stokes subspace is the Poincaré sphere.

(3) Excluding the origin, any point *inside* the unit sphere, $0 < \mathcal{P} < 1$, represents a partially polarized wave. Points outside the unit sphere, $\mathcal{P} > 1$, do not represent any physical state of polarization.

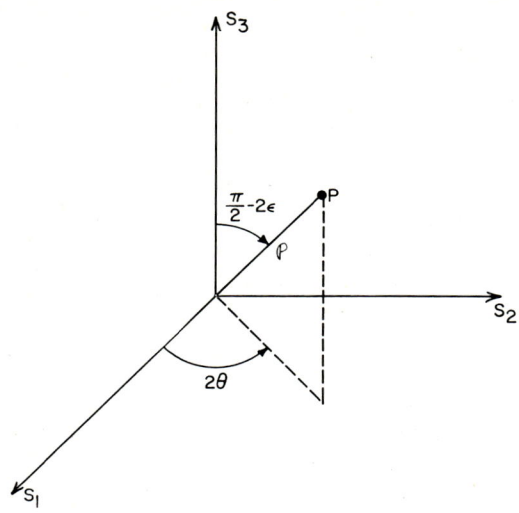

Fig. 1.26. Representation of partially polarized light by a point P in Stokes subspace S_1, S_2, S_3.

1.9.3. *The coherency-matrix representation*

One of the best known alternative representations of the polarization of quasi-monochromatic waves is that based on the coherency matrix [5, 23]. In terms of the 2×1 time-dependent Cartesian Jones vector of eq. (1.108) the 2×2 coherency matrix is defined by

$$\begin{aligned} \boldsymbol{J} &= \langle \boldsymbol{E}_{x,y}(t) \times \boldsymbol{E}_{x,y}^{\dagger}(t) \rangle \\ &= \begin{bmatrix} \langle E_x E_x^* \rangle & \langle E_x E_y^* \rangle \\ \langle E_y E_x^* \rangle & \langle E_y E_y^* \rangle \end{bmatrix} \\ &= \begin{bmatrix} J_{xx} & J_{xy} \\ J_{yx} & J_{yy} \end{bmatrix}. \end{aligned} \qquad (1.130)$$

where $\times$ denotes the direct product of the Jones vector times its Hermitian adjoint and E_x and E_y represent $\tilde{E}_x(t) \, \mathrm{e}^{\mathrm{j}\delta_x(t)}$ and $\tilde{E}_y(t) \, \mathrm{e}^{\mathrm{j}\delta_y(t)}$, respectively.

The diagonal elements J_{xx} and J_{yy} of the coherency matrix $\boldsymbol{J}$, as defined by eq. (1.130), are real and represent the intensities of the x and y linearly polarized components of the wave. The total intensity is given by the trace of $\boldsymbol{J}$

$$\begin{aligned} I &= \langle E_x E_x^* \rangle + \langle E_y E_y^* \rangle = J_{xx} + J_{yy} \\ &= \mathrm{Tr} \, \boldsymbol{J}. \end{aligned} \qquad (1.131)$$

The off-diagonal elements J_{xy} and J_{yx} define the cross correlation between the x and y components of the electric vector. These elements are complex conjugates of one another so that the coherency matrix is *Hermitian* (i.e., $\boldsymbol{J}^\dagger = \boldsymbol{J}$). A normalized cross-correlation function μ_{xy} may be defined as

$$\mu_{xy} = J_{xy}/(J_{xx}J_{yy})^{\frac{1}{2}}. \tag{1.132}$$

Schwarz's inequality[20] sets the limit on the modulus of μ_{xy} and leads to a positive real determinant

$$|\mu_{xy}| \leq 1, \tag{1.133}$$
$$\det \boldsymbol{J} = J_{xx}J_{yy} - J_{xy}J_{yx} = J_{xx}J_{yy}(1 - |\mu_{xy}|^2) \geq 0.$$

The limiting cases of unpolarized and totally polarized light correspond to complete decorrelation and correlation between E_x and E_y and lead to $|\mu_{xy}| = 0$ and 1, respectively. Values of $|\mu_{xy}|$ between 0 and 1 correspond to the intermediate more general case of partial polarization.[21]

The law by which the coherency matrix transforms $\boldsymbol{J} \to \boldsymbol{J}'$ under the effect of a coordinate rotation $x, y \to x', y'$ can be easily derived from the corresponding transformation of the Jones vector that has been discussed earlier in §1.6. Substituting from eq. (1.44) into eq. (1.130) we get

$$\boldsymbol{J}' = \langle \boldsymbol{E}_{x',y'} \times \boldsymbol{E}^\dagger_{x',y'} \rangle$$
$$= \langle \boldsymbol{R}(\alpha)\boldsymbol{E}_{x,y} \times \boldsymbol{E}^\dagger_{x,y}\boldsymbol{R}^\dagger(\alpha) \rangle,$$

or

$$\boldsymbol{J}' = \boldsymbol{R}(\alpha)\boldsymbol{J}\boldsymbol{R}^\dagger(\alpha), \tag{1.134}$$

where $\boldsymbol{R}(\alpha)$ is the 2×2 rotation matrix given by eq. (1.45a). Because $\boldsymbol{R}^\dagger(\alpha) = \boldsymbol{R}^{-1}(\alpha) = \boldsymbol{R}(-\alpha)$ [eq. (1.45b)], the coherency matrix is subjected to a *unitary transformation* [eq. (1.134)] as the x, y coordinate axes are rotated around the beam axis (the z axis) by an angle α. Such a unitary transformation leaves the determinant ($\det \boldsymbol{J}$) and the trace ($\operatorname{Tr} \boldsymbol{J}$) of the coherency matrix $\boldsymbol{J}$ invariant.

It is possible to generalize the definition of the coherency matrix by resolving the electric-field vibration into two *orthogonal* component vibrations that are circular or elliptic, instead of linear. The *generalized* coherency matrix (in the representation whose basis states are the orthogonal elliptic polarizations u and v) can be expressed in terms of the generalized

---

[20] For two complex functions $f(t)$ and $g(t)$ of the same real argument t, Schwarz's inequality takes the form $|\int_a^b f(t)g(t)\,dt|^2 \leq \int_a^b |f(t)|^2\,dt \cdot \int_a^b |g(t)|^2\,dt$. The equality occurs when $f(t)$ and $g(t)$ are proportional.

[21] When the x and y linearly-polarized components of a quasi-monochromatic light wave are totally incoherent, $|\mu_{xy}| = 0$. If the intensities of these components are not equal (i.e., $J_{xx} \neq J_{yy}$), the wave is partially polarized. Otherwise (i.e., when $J_{xx} = J_{yy}$), the wave is unpolarized.

Jones vector $E_{u,v}$ [eq. (1.64)] as

$$J = \langle E_{u,v} \times E_{u,v}^\dagger \rangle$$

$$= \begin{bmatrix} \langle E_u E_u^* \rangle & \langle E_u E_v^* \rangle \\ \langle E_v E_u^* \rangle & \langle E_v E_v^* \rangle \end{bmatrix}$$

$$= \begin{bmatrix} J_{uu} & J_{uv} \\ J_{vu} & J_{vv} \end{bmatrix}, \tag{1.135}$$

in direct analogy with the definition of the *Cartesian* coherency matrix in eq. (1.130). The transformation from the Cartesian coherency matrix of eq. (1.130) (denoted by J_{car}) to the generalized coherency matrix of eq. (1.135) (denoted by J_{gen}) is given by

$$J_{gen} = F^{-1} J_{car} F, \tag{1.136}$$

where F is the transformation matrix that links the corresponding Jones vectors, $E_{u,v} = F^{-1} E_{x,y}$ [eq. (1.66)]. The proof of eq. (1.136) is similar to that leading to eq. (1.134). Note that the basis Jones vectors that are used to define $E_{u,v}$ have been assumed orthonormal [eq. (1.68)], so that F is unitary [eq. (1.71)]. The unitary transformation of eq. (1.136) is more general than that of eq. (1.134) and also preserves the trace (the total intensity) and the determinant of the coherency matrix.

As may be expected from eqs. (1.112) and (1.130), the Stokes parameters and the elements of the coherency matrix are interrelated. In particular, the Stokes parameters are simple linear combinations of the elements of the coherency matrix given by

$$\begin{aligned} S_0 &= J_{xx} + J_{yy}, \\ S_1 &= J_{xx} - J_{yy}, \\ S_2 &= J_{xy} + J_{yx}, \\ S_3 &= +j(J_{xy} - J_{yx}). \end{aligned} \tag{1.137}$$

Written in matrix form, eq. (1.137) reads

$$\begin{bmatrix} S_0 \\ S_1 \\ S_2 \\ S_3 \end{bmatrix} = \begin{bmatrix} 1 & 0 & 0 & 1 \\ 1 & 0 & 0 & -1 \\ 0 & 1 & 1 & 0 \\ 0 & +j & -j & 0 \end{bmatrix} \begin{bmatrix} J_{xx} \\ J_{xy} \\ J_{yx} \\ J_{yy} \end{bmatrix}. \tag{1.138a}$$

or

$$S = AJ \tag{1.138b}$$

where J is a 4×1 *coherency vector* whose elements are the same as the elements of the coherency matrix J of eq. (1.130).

The degree of polarization $\mathscr{P}$ can be obtained in terms of the elements of the coherency matrix J by direct substitution from eq. (1.137) into eq. (1.127). The result can be put in the form

$$\mathscr{P} = \left[1 - \frac{4 \det J}{(\text{Tr } J)^2}\right]^{\frac{1}{2}}, \tag{1.139}$$

where the positive value of the square root is to be taken. $\det J$ and $\text{Tr } J$ are both invariant under a unitary transformation and so is the degree of polarization $\mathscr{P}$, as may be expected.

In a similar manner, the azimuth θ and ellipticity angle ϵ of the ellipse of polarization of the totally polarized component of a partially polarized beam can be obtained in terms of the elements of the coherency matrix J by direct substitution from eq. (1.137) into eq. (1.128)

$$\theta = \tfrac{1}{2} \arctan \left(\frac{J_{xy} + J_{yx}}{J_{xx} - J_{yy}}\right),$$
$$\epsilon = \tfrac{1}{2} \arcsin \left[\frac{+ j(J_{xy} - J_{yx})}{(J_{xx} + J_{yy})\mathscr{P}}\right], \tag{1.140}$$

where $\mathscr{P}$ is given by eq. (1.139).

The total intensity I, the degree of polarization $\mathscr{P}$ and the polarization form θ and ϵ (of the totally polarized component) represent the physically significant information that characterizes a beam of partially polarized light. Equations (1.131), (1.139) and (1.140) show how such information can be extracted from the coherency-matrix representation of the wave.

We conclude by presenting the coherency matrices of light polarized in various states. For totally linearly polarized light along the x axis, E_y is identically zero and eq. (1.130) leads to the simple coherency matrix

$$J = I \begin{bmatrix} 1 & 0 \\ 0 & 0 \end{bmatrix}. \tag{1.141a}$$

Application of the transformation law of the coherency matrix under a coordinate rotation [given in eq. (1.134)] generates from eq. (1.141a) the following coherency matrix

$$J = I \begin{bmatrix} \cos^2 \theta & \cos \theta \sin \theta \\ \sin \theta \cos \theta & \sin^2 \theta \end{bmatrix}, \tag{1.141b}$$

for light that is linearly polarized at an arbitrary azimuth θ. For right- and left-circularly polarized light

$$J = \tfrac{1}{2}I \begin{bmatrix} 1 & \mp j \\ \pm j & 1 \end{bmatrix}, \tag{1.141c}$$

as can be obtained from eq. (1.130). In general, the Cartesian coherency matrix of totally polarized monochromatic or quasi-monochromatic light of intensity $S_0 = I$, azimuth θ and ellipticity angle ϵ can be obtained by substituting the Stokes parameters of eq. (1.122) into eq. (1.137). This gives the following coherency matrix

$$J_{\text{car}} = \tfrac{1}{2}I \begin{bmatrix} (1 + \cos 2\theta \cos 2\epsilon) & (\sin 2\theta \cos 2\epsilon - j \sin 2\epsilon) \\ (\sin 2\theta \cos 2\epsilon + j \sin 2\epsilon) & (1 - \cos 2\theta \cos 2\epsilon) \end{bmatrix}, \tag{1.142}$$

of which eqs. (1.141a), (1.141b) and (1.141c) are special cases. The corresponding circular coherency matrix can be obtained by applying the transformation indicated by eq. (1.136) where F is given by (1.55). This yields the simpler coherency matrix

$$J_{\text{cir}} = \tfrac{1}{2}I \begin{bmatrix} (1 - \sin 2\epsilon) & \cos 2\epsilon \, e^{j2\theta} \\ \cos 2\epsilon \, e^{-j2\theta} & (1 + \sin 2\epsilon) \end{bmatrix}, \tag{1.143}$$

For unpolarized light of intensity I, $J_{xx} = J_{yy} = J_{\ell\ell} = J_{rr} = \tfrac{1}{2}I$, and $J_{xy} = J_{yx} = J_{\ell r} = J_{r\ell} = 0$, so that both the Cartesian and the circular coherency matrices are given by

$$J_{\text{cir}}^{\text{car}} = \tfrac{1}{2}I \begin{bmatrix} 1 & 0 \\ 0 & 1 \end{bmatrix}. \tag{1.144}$$

We have discussed earlier how partially polarized light can be decomposed into two independent components: a totally polarized component and an unpolarized component. The Stokes vectors of the original beam and of the component beams are given by eqs. (1.125) and (1.126). This decomposition can also be expressed in terms of the coherency matrices

$$J = J_{\text{un.}} + J_{\text{t.p.}}, \tag{1.145}$$

where J is the coherency matrix of the total beam, $J_{\text{un.}}$ and $J_{\text{t.p.}}$ are the coherency matrices of its unpolarized and totally polarized components, respectively. Equation (1.145) can be used to construct the coherency matrix of a partially polarized light beam from the coherency matrices of the unpolarized and totally polarized components of the beam. Multiplication of eq. (1.142) by $\mathscr{P}$ and eq. (1.144) by $(1 - \mathscr{P})$, and subsequent addition of the results give

$$J_{\text{car}} = \tfrac{1}{2}I \begin{bmatrix} (1 + \mathscr{P} \cos 2\theta \cos 2\epsilon) & \mathscr{P}(\sin 2\theta \cos 2\epsilon - j \sin 2\epsilon) \\ \mathscr{P}(\sin 2\theta \cos 2\epsilon + j \sin 2\epsilon) & (1 - \mathscr{P} \cos 2\theta \cos 2\epsilon) \end{bmatrix} \tag{1.146}$$

which is the Cartesian coherency matrix of partially polarized light of total intensity I, degree of polarization $\mathcal{P}$ and polarization form θ and ϵ. The corresponding circular coherency matrix can similarly be obtained from eqs. (1.143) and (1.144). The result is given by

$$\mathbf{J}_{\text{cir}} = \tfrac{1}{2}I \begin{bmatrix} (1 - \mathcal{P} \sin 2\epsilon) & \mathcal{P} \cos 2\epsilon \; e^{j2\theta} \\ \mathcal{P} \cos 2\epsilon \; e^{-j2\theta} & (1 + \mathcal{P} \sin 2\epsilon) \end{bmatrix}. \tag{1.147}$$

CHAPTER 2

Propagation of Polarized Light through Polarizing Optical Systems

2.1. Polarizing optical systems

In an ellipsometer a beam of polarized light is propagated through a succession of optical devices each of which produces a specific change in the state of polarization. In this respect, ellipsometers belong to a class of optical systems in which polarization represents the fundamental property of the wave that is processed by the optical components of the system. Such systems are described as polarizing, to distinguish them from other classes of optical systems where a property of the wave other than its polarization is affected. For example, in an imaging optical system the amplitude (intensity) distribution over an optical wave-front is transformed by optical components placed in the path of the light wave. The devices encountered in the different classes of optical systems are different. Whereas a polarizing optical system may consist of polarizers, retarders and rotators, an imaging optical system is composed primarily of lenses and spatial filters. Although the classification of optical systems in accordance with the fundamental property of the wave that they process is attractive, difficulties arise in the case of systems that significantly affect more than one property at the same time. This may be illustrated by the Šolc and Lyot filters which are examples of linear birefringent networks [24]. These filters appropriately belong to the class of polarizing optical systems, but they also serve as spectral-filtering systems. The state of polarization and the spectral distribution of intensity are the two properties of the wave that are simultaneously operated upon by such systems. Other examples include monochromators, spectrometers and interferometers which do not fit within the class of polarizing optical systems, although polarization plays an important role that has to be considered in the design of these instruments [25–27]. Also, interest has been shown recently in the processing of a non-uniform distribution of polarization over an optical wave-front by systems that have to be described as both polarizing and imaging at the same time [28–30].

In this chapter we will concern ourselves with the polarization-modifying interaction between a beam of polarized light and the optical components that constitute a polarizing optical system. The following simplifying assumptions are made.

(1) The light beam is approximated by a uniform TE infinite plane wave that may be either monochromatic or quasi-monochromatic.

(2) The interaction between an incident beam and an optical device is linear and frequency-conserving. This excludes non-linear optical effects, inelastic light scattering of Raman or Brillouin type and allied effects. In addition, the interaction may scatter the incident plane wave into one or more plane waves and yet preserve the transversality of the field.

(3) Only the *external* (terminal) properties of an optical device or an optical system are emphasized, with less attention paid to the details of the *internal* polarization-modifying processes that are responsible for such terminal behavior.

The description of a device as well as its polarization-modifying effect are determined principally by the mathematical representation that is used to describe the state of polarization of the wave. Several such representations have been developed in ch. 1, namely, the Jones-vector, complex-plane, Poincaré-sphere, Stokes-vector and coherency-matrix representations. In this chapter we will examine the propagation of light through polarizing optical systems employing these different representations.

2.2. The Jones-matrix formulation

Consider a uniform monochromatic TE plane wave incident on a non-depolarizing[1] optical system that consists of either a single optical device or successive series of such devices. As a result of interaction between the incident wave and the optical system one or more modified plane waves emerge from the system. Figure 2.1 shows a schematic diagram of the optical system, the incident wave and one of the outgoing (emergent) modified plane waves. Two space-fixed right-handed Cartesian coordinate

---

[1] See the discussion in the early part of §2.11 in connection with eq. (2.222).

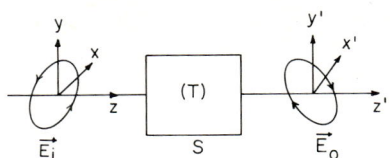

Fig. 2.1. Incident on and emergent from the optical system S are plane waves of Jones vectors E_i and E_o referenced to Cartesian coordinate systems (x, y, z) and (x', y', z'), at the input and output of S, respectively. $E_o = (T)E_i$ where (T) is the Jones matrix of S.

systems (x, y, z) and (x', y', z') are associated with the incident and the outgoing plane waves, with the directions z and z' taken parallel to their wave-vectors k and k', respectively. (k and k' need not be parallel.) The location of the input and output reference coordinate planes $z = 0$ and $z' = 0$ is arbitrary, although these planes are ordinarily chosen immediately before and immediately after the entrance and exit surfaces of the system (device), respectively.

The transverse (x, y) and (x', y') coordinate axes at the input and output of the optical system may have any suitable azimuthal orientation around the incident and the outgoing wave-vectors k and k', respectively.[2] Let the incident and the outgoing plane waves be described by their appropriate Jones vectors E_i and E_o, referenced to the input and output coordinate systems, respectively. As explained in §1.6, the input Jones vector E_i is a complex 2×1 column vector whose elements (phasor components) E_{ix} and E_{iy} represent the sinusoidal oscillations of the Cartesian projections of the electric vector of the incident light wave along the transverse x and y coordinate axes, respectively. Similarly the elements $E_{ox'}$ and $E_{oy'}$ of the output Jones vector E_o represent the sinusoidal oscillations of the Cartesian projections of the electric vector of the outgoing wave along the transverse x' and y' coordinate axes, respectively. In the absence of non-linearity and other frequency-changing processes, the pair of oscillations $E_{ox'}$ and $E_{oy'}$ at the output of the optical system are related to the pair of oscillations E_{ix} and E_{iy} at the input of the optical system by the following linear equations

$$E_{ox'} = T_{11}E_{ix} + T_{12}E_{iy}, \tag{2.1a}$$

$$E_{oy'} = T_{21}E_{ix} + T_{22}E_{iy}. \tag{2.1b}$$

---

[2] A unique orientation is usually adopted in which the directions of x and x' are chosen in the plane formed by the incident and the scattered wave-vectors k and k' and y and y' are oriented in parallel perpendicular to that plane. Figure 2.5 shows this geometry in the case of reflection from a surface with x and x' denoted by p and y and y' denoted by s. If k and k' are parallel, the coordinate systems (x, y) and (x', y') are chosen parallel unless other considerations (such as a simpler Jones matrix) favor otherwise.

Equations (2.1a) and (2.1b) can be combined in matrix form

$$\begin{bmatrix} E_{ox'} \\ E_{oy'} \end{bmatrix} = \begin{bmatrix} T_{11} & T_{12} \\ T_{21} & T_{22} \end{bmatrix} \begin{bmatrix} E_{ix} \\ E_{iy} \end{bmatrix}, \tag{2.2}$$

or, more concisely,

$$\boldsymbol{E}_o = \boldsymbol{T}\boldsymbol{E}_i \tag{2.3}$$

where

$$\boldsymbol{T} = \begin{bmatrix} T_{11} & T_{12} \\ T_{21} & T_{22} \end{bmatrix}. \tag{2.4}$$

Equation (2.3) expresses the law of interaction between the incident wave and the optical system as a simple linear matrix transformation of the Jones-vector representation of the wave. *This is the fundamental step upon which the entire Jones-matrix formulation is based* [3]. The 2×2 transformation matrix $\boldsymbol{T}$ is called the *Jones matrix* of the optical system (or the optical device) and its elements T_{ij} are, in general, complex. The Jones matrix $\boldsymbol{T}$ is a function of

(1) The optical system (device) under consideration;

(2) The frequency of the incident wave (the polarization-modifying processes inside the system are, in general, dispersive);

(3) The orientation of the system with respect to the incident wave-vector;

(4) The location of the input and output reference coordinate planes $z = 0$ and $z' = 0$, respectively;

(5) The azimuthal orientation of the input and output transverse coordinate axes (x, y) and (x', y'), around the incident and the outgoing wave-vectors, respectively; and

(6) The particular outgoing plane wave, if more than one such wave is generated as a result of the interaction between the incident wave and the optical system.

2.2.1. *The significance of the elements T_{ij} of the Jones matrix*

The Jones matrix $\boldsymbol{T}$ describes the overall effect of the optical system on the incident wave. It is interesting to examine the meaning of each individual element T_{ij} of this Jones matrix. According to eqs. (2.1a) and (2.1b), if the incident plane wave is linearly polarized with its electric vector vibrating along the x axis (*i.e.* $E_{iy} = 0$), the elements T_{11} and T_{21} are given by

$$T_{11} = \left[\frac{E_{ox'}}{E_{ix}}\right]_{E_{iy}=0}, \tag{2.5a}$$

$$T_{21} = \left[\frac{E_{oy'}}{E_{ix}}\right]_{E_{iy}=0}, \tag{2.5b}$$

respectively. The linear oscillation E_{ix} of the incident electric vector along the x coordinate axis at the input of the system has produced *two* orthogonal linear oscillations $E_{ox'}$ and $E_{oy'}$ along the x' and y' coordinate axes at the output of the system, respectively. Equation (2.5a) shows that T_{11} is determined by the relative amplitude and phase of the output vibration $E_{ox'}$ with respect to those of the input vibration E_{ix}; while eq. (2.5b) shows that T_{21} is determined by the relative amplitude and phase of the output vibration $E_{oy'}$ with respect to those of the input vibration E_{ix}. If, instead of being linearly polarized along the x axis, the incident wave is linearly polarized along the y axis, the elements T_{12} and T_{22} are determined from eqs. (2.1a) and (2.1b) by

$$T_{12} = \left[\frac{E_{ox'}}{E_{iy}}\right]_{E_{ix}=0}, \tag{2.6a}$$

$$T_{22} = \left[\frac{E_{oy'}}{E_{iy}}\right]_{E_{ix}=0}. \tag{2.6b}$$

In this case, the incident linear vibration E_{iy} (along the y axis) has generated two orthogonal linear vibrations $E_{ox'}$ and $E_{oy'}$ (along the x' and y' axes, respectively), at the output of the system. Equations (2.6a) and (2.6b) tell us that T_{12} and T_{22} are determined by the relative amplitude and phase of the output vibrations $E_{ox'}$ and $E_{oy'}$ with respect to those of the input linear vibration E_{iy} respectively. Stated differently, the diagonal elements T_{11} and T_{22} of the Jones matrix of the optical system are determined by the $x \rightarrow x'$ and $y \rightarrow y'$ input-to-output mappings of "similar" linear polarizations, whereas the off-diagonal elements T_{21} and T_{12} are determined by the $x \rightarrow y'$ and $y \rightarrow x'$ input-to-output mappings of "crossed" linear polarizations, respectively.[3] The dependence of the Jones matrix T on the location and orientation of the (x, y) and (x', y') reference coordinate axes should now be evident from the expressions of the elements T_{ij} in eqs. (2.5) and (2.6).

2.2.2. Cascade of optical systems

Referring to fig. 2.2a, consider the *combined* effect of two separate optical systems I and II placed in the path of a plane wave of light. For simplicity, consider *one* intermediate and *one* final emergent plane wave. The interaction between the incident wave and the first optical system, I, produces the

---

[3] x and x' are assumed to lie in the plane of k and k' while y and y' are perpendicular to this plane (see footnote 2 on p. 68).

§2.2] THE JONES-MATRIX FORMULATION

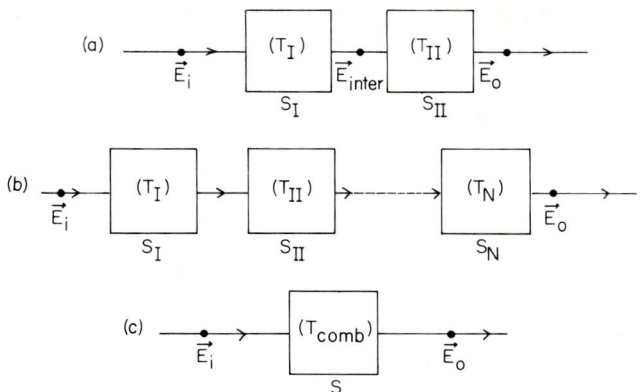

Fig. 2.2. The combined effect of N optical systems $S_I, S_{II}, \ldots, S_N$, of Jones matrices (T_I), $(T_{II}), \ldots, (T_N)$ is equivalent to one system with Jones matrix $(T_{comb}) = (T_N)(T_{N-1}) \ldots (T_{II})(T_I)$.

intermediate plane wave whose associated Jones vector is given by

$$\bm{E}_{inter} = \bm{T}_I \bm{E}_i, \tag{2.7a}$$

while the interaction between the intermediate wave and the second optical system II produces the final emergent wave whose Jones vector is given by

$$\bm{E}_o = \bm{T}_{II} \bm{E}_{inter}. \tag{2.7b}$$

Because we are interested in the combined effect of the two systems I and II, we can eliminate $\bm{E}_{inter}$ by substituting from eq. (2.7a) into eq. (2.7b). This gives

$$\bm{E}_o = \bm{T}_{II} \bm{T}_I \bm{E}_i, \tag{2.8a}$$

or

$$\bm{E}_o = \bm{T}_{comb} \bm{E}_i, \tag{2.8b}$$

where

$$\bm{T}_{comb} = \bm{T}_{II} \bm{T}_I, \tag{2.9}$$

is the Jones matrix of the combined system.

The above result can be readily extended to cover the general case when a plane wave of light is sequentially processed by a cascade of N optical systems (devices), fig. 2.2b. If we focus our attention on *one* incident and *one* emergent plane wave, the overall effect of the entire system is described by

$$\boldsymbol{E}_\text{o} = \boldsymbol{T}_N \boldsymbol{T}_{N-1} \ldots \boldsymbol{T}_\text{II} \boldsymbol{T}_\text{I} \boldsymbol{E}_\text{i}, \tag{2.10a}$$

or, more simply, by

$$\boldsymbol{E}_\text{o} = \boldsymbol{T}_\text{comb} \boldsymbol{E}_\text{i}, \tag{2.10b}$$

where

$$\boldsymbol{T}_\text{comb} = \boldsymbol{T}_N \boldsymbol{T}_{N-1} \ldots \boldsymbol{T}_\text{II} \boldsymbol{T}_\text{I}. \tag{2.11}$$

Equations (2.10) and (2.11) (which are the generalized forms of eq. (2.8) and (2.9), respectively) show that a cascade of N optical systems (devices) can be replaced by one *composite* optical system whose overall (external or terminal) Jones matrix, $\boldsymbol{T}_\text{comb}$, is obtained by multiplying the Jones matrices of the separate systems, fig. 2.2c. Because the incident plane wave encounters the optical system I first, the Jones matrix $\boldsymbol{T}_\text{I}$ must be the first to operate on the incident Jones vector $\boldsymbol{E}_\text{i}$, which explains why $\boldsymbol{T}_\text{I}$ appears to the right of the chain of matrices in eq. (2.10a) and (2.11), an order which should always be observed.

2.2.3. *Jones matrices of basic optical devices*

2.2.3.1. *Transmission-type devices*

(1) Perhaps the simplest Jones matrix is that which describes the free propagation of a plane wave of light over a distance d without encountering any interaction.[4] In this case, the matrix transformation of the Jones vector is given by

$$\begin{bmatrix} E_\text{ox} \\ E_\text{oy} \end{bmatrix} = \begin{bmatrix} e^{-j2\pi nd/\lambda} & 0 \\ 0 & e^{-j2\pi nd/\lambda} \end{bmatrix} \begin{bmatrix} E_\text{ix} \\ E_\text{iy} \end{bmatrix}, \tag{2.12}$$

which expresses the simple fact that the wave, whose vacuum wavelength is λ, has been retarded by $2\pi nd/\lambda$. Our optical device in this case is a plane-parallel section of an isotropic medium of refractive index n through which the plane wave is freely propagating. The thickness of this section is d; the input and output coordinate systems (x, y) and (x', y') coincide with the bounding planes and are aligned in parallel, fig. 2.3. Equation (2.12) is easily written by inspection, but may also be obtained, formally, if eq. (1.30) is premultiplied by the identity 2×2 matrix. The above device can be appropriately called an *isotropic retarder* (or *phase-plate*).

---

[4] We overlook the fact that the wave is slowed down to a speed of c/n (instead of c in vacuum) in an isotropic transparent medium of refractive index n.

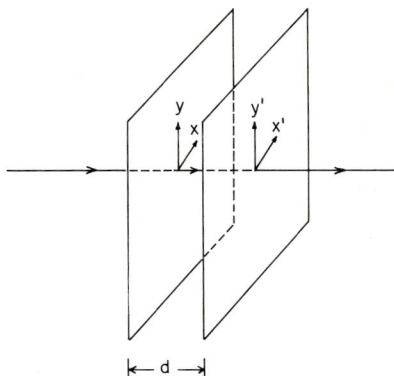

Fig. 2.3. A plane-parallel section of an isotropic medium acts as an *isotropic* phase-plate.

(2) Suppose that the medium through which the wave is propagating is not isotropic but rather *uniaxially linearly birefringent*, and that the wave is travelling in a direction perpendicular to the *optic axis* of this medium.[5] If the wave is linearly polarized parallel to the optic axis, the wave will experience a specific refractive index n_e and will travel at a speed c/n_e, where c is the speed of light in vacuum. If, on the other hand, the wave is linearly polarized orthogonal to the optic axis, the wave will experience a *different* refractive index n_o and will travel at a speed c/n_o. n_o and n_e are commonly known as the *ordinary* and the *extraordinary* refractive indices of the medium, respectively. When the transverse electric vector of the wave has components both parallel (x) and perpenicular (y) to the optic axis, the effect of wave propagation through a distance d is described by

$$\begin{bmatrix} E_{ox} \\ E_{oy} \end{bmatrix} = \begin{bmatrix} e^{-j2\pi n_e d/\lambda} & 0 \\ 0 & e^{-j2\pi n_o d/\lambda} \end{bmatrix} \begin{bmatrix} E_{ix} \\ E_{iy} \end{bmatrix}, \qquad (2.13)$$

which is a straightforward generalization of eq. (2.12). A section of the medium of thickness d, of the above description, acts as a *linear retarder*, because the components of the electric vector along the two privileged directions parallel and perpendicular to the optic axis are retarded by different amounts $2\pi n_e d/\lambda$ and $2\pi n_o d/\lambda$, respectively. The Jones matrix of this linear retarder may be written as

$$T = \begin{bmatrix} e^{-j\delta_1} & 0 \\ 0 & e^{-j\delta_2} \end{bmatrix}, \qquad (2.14)$$

---

[5]For discussions of the propagation of polarized light in crystals see Ramachandran and Ramaseshan[9] and Born and Wolf ([23], ch. 14).

where

$$\delta_1 = 2\pi n_e d/\lambda, \qquad \delta_2 = 2\pi n_o d/\lambda. \tag{2.15}$$

An alternative equivalent form of eq. (2.14) is

$$T = e^{-j\delta_1} \begin{bmatrix} 1 & 0 \\ 0 & e^{-j\delta} \end{bmatrix}, \tag{2.16}$$

where

$$\delta = \frac{2\pi d}{\lambda}(n_o - n_e), \tag{2.17}$$

is the relative (y with respect to x) retardation. The quantity $(n_e - n_o)$ is commonly called the *birefringence* of the medium. If $n_e < n_o$ (negative birefringence), the x axis (the optic axis) is called the fast axis of the linear retarder and the y axis is called its slow axis. On the other hand, if $n_e > n_o$ (positive birefringence) the x axis (the optic axis) and the y axis become the slow and fast axes, respectively. This nomenclature follows from the fact that the wave travels faster or slower if it is linearly polarized along the direction of lower or higher refractive index, respectively.

It is important to note that we have considered the effect on a travelling plane wave of a section of an anisotropic medium without actually (physically) isolating that section. In practice, a plane-parallel slab is prepared from the medium with the optic axis parallel to the faces, to make the linear retarder. If a plane wave is incident from vacuum (or air) onto such a device, the overall effect of the device is described only approximately by eqs. (2.14)–(2.17). The reason for this is the presence of multiple reflections[6] between the parallel faces of the retarder (which acts like an anisotropic Fabry–Perot etalon), that make the simple discussion based on a single forward-travelling wave inapplicable [31].

(3) In a medium which is both isotropically-refracting and isotropically-absorbing at the same time, a plane wave is retarded and attenuated, after travelling a distance d, by an amount that is independent of the direction of propagation or the state of polarization of the wave. A section of this medium, of thickness d, acts as an isotropic retarder and absorber and is described by a Jones matrix of the form

$$T = \begin{bmatrix} e^{-j2\pi d(n-jk)/\lambda} & 0 \\ 0 & e^{-j2\pi d(n-jk)/\lambda} \end{bmatrix}, \tag{2.18}$$

---

[6] The effect of these multiple reflections is particularly important when the thickness d of the retardation plate is small as in mica sheets [10].

§2.2] THE JONES-MATRIX FORMULATION 75

which is simply obtained from the Jones matrix of the isotropic phase-plate [eq. (2.12)] by substituting the *complex* index of refraction $(n - jk)$ in place of the real index of refraction n. k is called the *extinction coefficient* of the medium. The amplitude of oscillation of the electric field (linear, circular or elliptic) decays with distance according to[7]

$$A(d) = A(0) e^{-\frac{1}{2}\alpha d}, \quad (2.19a)$$

while the intensity of the wave $I = A^2$ falls off as

$$I(d) = A^2(0) e^{-\alpha d}, \quad (2.19b)$$

where $A(0)$ is the initial amplitude and

$$\alpha = \frac{4\pi k}{\lambda}, \quad (2.20)$$

is called the *absorption coefficient*. Because the absorption and refraction properties of the medium are both isotropic, the ellipse of polarization remains unchanged.

(4) If the medium is *uniaxially linearly dichroic*[8], a linearly-polarized wave travelling in a direction perpendicular to the optic axis (of absorption) will be attenuated by different amounts depending upon the direction of the vibration of the transverse electric field with respect to the optic axis. This absorption is maximum or minimum if the direction of vibration is parallel or perpendicular to the optic axis. Let k_e and k_o denote the "extraordinary" and the "ordinary" extinction coefficients of the medium for light linearly polarized parallel and perpendicular to the optic axis, respectively. For a section of the medium of thickness d, the Jones matrix can be obtained by generalizing eq. (2.18) as follows

$$T = e^{-j2\pi d n/\lambda} \begin{bmatrix} e^{-2\pi d k_e/\lambda} & 0 \\ 0 & e^{-2\pi d k_o/\lambda} \end{bmatrix}, \quad (2.21)$$

where the outside scalar factor represents an isotropic phase delay. In a slightly different form, eq. (2.21) can be written as

$$T = e^{-j2\pi n d/\lambda} \begin{bmatrix} e^{-\frac{1}{2}\alpha_1 d} & 0 \\ 0 & e^{-\frac{1}{2}\alpha_2 d} \end{bmatrix} = e^{-j2\pi n d/\lambda} e^{-\frac{1}{2}\alpha_1 d} \begin{bmatrix} 1 & 0 \\ 0 & e^{-\frac{1}{2}\alpha d} \end{bmatrix}, \quad (2.22)$$

---

[7] The Jones matrix of eq. (2.18) is equal to a complex constant, $e^{-j2\pi d(n-jk)/\lambda}$, times the 2×2 identity matrix. When operating on an incident Jones vector of the form of eq. (1.75), the Jones matrix of eq. (2.18) changes the amplitude of the wave from A to $A e^{-2\pi dk/\lambda}$, changes its phase by $-2\pi d n/\lambda$ and leaves the ellipse of polarization (θ and ϵ) unchanged.

[8] See, for example, ref. 4, ch. 4.

where

$$\alpha_1 = \frac{4\pi k_e}{\lambda}, \quad \alpha_2 = \frac{4\pi k_o}{\lambda}, \quad \alpha = \frac{4\pi}{\lambda}(k_o - k_e), \tag{2.23}$$

represent the extraordinary, the ordinary and the relative absorption coefficients, respectively. The quantity $(k_e - k_o)$ is called the *dichroism* of the medium, and may be positive or negative depending upon k_e being greater or less than k_o. The section of the linearly-dichroic medium described above acts as a *linear partial polarizer*. The term partial polarizer is used because incident unpolarized light will be transmitted as partially polarized light by this device. (In general, the device increases the degree of polarization (§1.9) of incident partially polarized light.) Note that the above discussion is based on assuming a section of an infinite medium without actually slicing that section out to form the device. Therefore, for a device made in the form of a plane-parallel slab of the linearly dichroic medium (with the optic axis parallel to the faces), the Jones matrix of eq. (2.22) should be considered only as an approximation.

(5) A special case of the linearly-dichroic device is that of an *ideal linear polarizer*. In this case $k_e = 0$, $k_o = \infty$ and the Jones matrix assumes the simple form

$$\mathbf{T} = e^{-j2\pi nd/\lambda} \begin{bmatrix} 1 & 0 \\ 0 & 0 \end{bmatrix}. \tag{2.24}$$

where the first multiplicative term expresses the isotropic refracting properties of the medium. A practical linear polarizer will have $k_e \approx 0$, $k_o \gg 1$, and the matrix of eq. (2.24) becomes applicable only as an approximation. The ideal linear polarizer transmits freely or extinguishes completely incident linearly polarized light, depending upon whether the direction of its linear polarization is parallel to or crossed with the *transmission axis* of the polarizer. The transmission axis of a linear polarizer is the axis along which the absorption coefficient is zero (small), while its *extinction axis* is associated with the infinite (large) absorption coefficient. In the above example, the transmission axis coincides with the optic axis of dichroism. The best known type of a dichroic polarizer is the Polaroid sheet [4].

(6) Within certain frequency bands, a given medium may exhibit both linear birefringence and linear dichroism concurrently. A plane-parallel slab of a uniaxial linearly birefringent and dichroic material, in which the optic axes of birefringence and dichroism are coincident and parallel with the bounding planar faces, acts as a *linearly-dichroic retarder*. This device has a Jones matrix of the form

$$T = \begin{bmatrix} e^{-j2\pi dn_e/\lambda} \, e^{-2\pi dk_e/\lambda} & 0 \\ 0 & e^{-j2\pi dn_o/\lambda} \, e^{-2\pi dk_o/\lambda} \end{bmatrix} \qquad (2.25a)$$

or, simply,

$$T = \begin{bmatrix} \rho_1 e^{-j\delta_1} & 0 \\ 0 & \rho_2 e^{-j\delta_2} \end{bmatrix}, \qquad (2.25b)$$

where

$$\rho_1 = e^{-2\pi dk_e/\lambda}, \qquad \rho_2 = e^{-2\pi dk_o/\lambda};$$
$$\delta_1 = 2\pi dn_e/\lambda, \qquad \delta_2 = 2\pi dn_o/\lambda. \qquad (2.26)$$

(7) Another important type of optical behavior is encountered when polarized light propagates through a medium exhibiting *optical activity* (natural or induced) [32]. In this case, linearly polarized light remains linearly polarized as the light progresses through the medium but the direction of the vibration of the electric vector rotates uniformly and continuously with distance. (This is why optically active materials are said to possess *optical rotatory power*.) In general, if the light is elliptically polarized, the major axis is rotated but the axial ratio and the sense in which the ellipse of polarization is traced remain unaffected. If α denotes the magnitude of rotation of the major axis of the ellipse of polarization per unit path length in the medium, a section of thickness d will be described by a Jones matrix of the form

$$T = e^{-j2\pi nd/\lambda} \begin{bmatrix} \cos(\alpha d) & -\sin(\alpha d) \\ \sin(\alpha d) & \cos(\alpha d) \end{bmatrix} \qquad (2.27)$$
$$= e^{-j2\pi nd/\lambda} \, \boldsymbol{R}(-\alpha d)$$

where $\boldsymbol{R}$ is the rotation matrix [eq. (1.45a)] and the multiplicative scalar outside this matrix corresponds to an *overall* phase delay, with n representing an "average" index of refraction. The Jones matrix of eq. (2.27) can be obtained in a straightforward manner by resolving the linearly vibrating input ($\boldsymbol{E}_i$) and output ($\boldsymbol{E}_o$) electric vectors along the *same* coordinate axes x and y, fig. 2.4, where $\boldsymbol{E}_o$ is obtained from $\boldsymbol{E}_i$ by a rotation through an angle (αd) (and a phase retardation of $2\pi nd/\lambda$). A device whose Jones matrix is of the form of eq. (2.27) is called an *optical rotator*.

2.2.3.2. Reflection-type devices

All the optical devices described above are *transmission-type* devices because the wave is modified as it propagates through the bulk of a medium

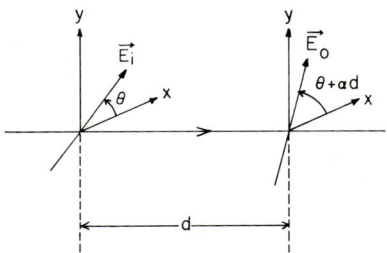

Fig. 2.4. Rotation of the electric-field vibration of a linearly polarized plane wave travelling through an optically active medium. α is the rotation per unit path length.

(isotropic or anisotropic). There is still another class of devices, *reflection-type* devices, in which the wave is modified by reflection from the surface of a medium (isotropic or anisotropic). The basic distinction between the two lies in the fact that the wave is modified *continuously* as it progresses through the medium of a transmission-type device, whereas it is changed *abruptly* as the wave bounces off the surface (once or more) of a reflection-type device.[9] The simplest reflection-type device is one in which the wave bounces once from a single interface between two dissimilar media. Although the details of the polarization-modifying interaction by reflection will be described later (ch. 4), we will consider the case of oblique reflection from a two-media interface, across which the change in the index of refraction from one medium to the other is assumed to be sharp (a step function). The media are both isotropic. If the incident plane wave has an electric vibration parallel (p) to the plane of incidence, the reflected wave will also have an electric vibration parallel to the plane of incidence. Likewise, an electric vibration perpendicular (s) to the plane of incidence is reflected into a similar vibration perpendicular to the plane of incidence. By resolving the incident and the reflected electric vectors $\boldsymbol{E}_i$ and $\boldsymbol{E}_r$, respectively, along the p and s directions, fig. 2.5, the effect of reflection can be expressed as follows

$$\boldsymbol{E}_r = \boldsymbol{R}\,\boldsymbol{E}_i, \tag{2.28}$$

where the reflection Jones matrix $\boldsymbol{R}$ is given by

$$\boldsymbol{R} = \begin{bmatrix} R_{pp} & 0 \\ 0 & R_{ss} \end{bmatrix} = \begin{bmatrix} |R_{pp}|e^{j\delta_{pp}} & 0 \\ 0 & |R_{ss}|e^{j\delta_{ss}} \end{bmatrix}. \tag{2.29}$$

---

[9] The distinction might not be clear in certain cases. For example, a Glan-Thompson prism polarizer operates in a transmission mode but polarizes the incident wave by abrupt double refraction followed by abrupt reflection to eliminate the unwanted beam. The state of polarization of light does not change continuously inside a prism polarizer.

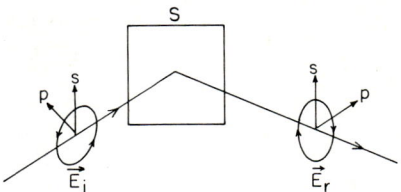

Fig. 2.5. Oblique reflection of light by a surface S. p and s are transverse to the direction of propagation and are parallel and perpendicular to the plane of incidence, respectively.

The p→p and s→s diagonal reflection coefficients R_{pp} and R_{ss} are given by Fresnel's formulae and both are, in general, complex.[10]

All transmission-type devices discussed earlier have their reflection-type counter-parts. The case of *total* internal reflection, where the wave is incident on a dielectric-dielectric interface from the side of the more optically dense medium, provides the example of a reflection-type retarder. Under the condition of total reflection, $|R_{pp}| = |R_{ss}| = 1$ and the phase jumps δ_{pp} and δ_{ss} are controllable by the dielectric constants (refractive indices) of the two media and the angle of incidence. This is the basis of the Fresnel-rhomb retarders. If the reflection from the dielectric-dielectric interface is *external*, instead of internal, the phase jumps δ_{pp} and δ_{ss} assume the "trivial" values of either 0 or π, while $|R_{pp}|$ and $|R_{ss}|$ are controlled by the indices of refraction and the angle of incidence. For any angle of incidence other than the Brewster angle, we have a reflection-type linear partial polarizer. At the Brewster angle, $|R_{pp}| = 0$, and we have an ideal linear polarizer. This is the basis of the Brewster-angle reflection polarizers [10]. If either one of the two media is absorbing, we have a reflection-type linear partially polarizing retarder. Reflection-type optical rotators are obtained when light is reflected from the surfaces of optically active media. However, the effect is too small to make a practically useful device.

For a general classification of optical devices, we will wait until we introduce the concept of eigenpolarizations and eigenvalues that provides a suitable basis for such classification, §2.6.

2.2.4. Transformation of the Cartesian Jones matrix under the effect of a coordinate rotation

In the discussion of the Jones matrix of an optical system, the electric vectors of the incident and outgoing plane waves are referenced to Cartesian coordinate systems (x, y) and (x', y') at the input and output of the optical

---

[10] See §4.2.

system. Such a Jones matrix can be appropriately called the *Cartesian Jones matrix* of the optical system, in conformity with the fact that this matrix links together the Cartesian Jones vectors E_i and E_o of the incident and the outgoing plane waves according to eqs. (2.2) and (2.3). Let the (x, y) input coordinate system be rotated around the incident wave-vector k by an angle α in a counter-clockwise sense when looking against k, and let, at the same time, the (x', y') output coordinate system be rotated by the *same* angle in the *same* sense around and looking against k'. From eq. (1.44), the effect of this rotation is to transform $E_i \to R(\alpha)E_i$ and $E_o \to R(\alpha)E_o$. Pre-multiplication of eq. (2.3) by $R(\alpha)$ gives

$$R(\alpha)E_o = R(\alpha)TE_i.$$

and using the fact that $R(-\alpha) R(\alpha) =$ identity matrix, we may write

$$[R(\alpha)E_o] = [R(\alpha)TR(-\alpha)][R(\alpha)E_i]. \tag{2.30}$$

From eq. (2.30), we conclude that the effect of rotating *both* the input *and* output coordinate systems by the same angle α in the same sense (looking against the direction of propagation) is to transform the Cartesian Jones matrix of the optical system according to

$$T_{\text{new}} = R(\alpha) T_{\text{old}} R(-\alpha), \tag{2.31}$$

where T_{old} and T_{new} are the Jones matrices of the optical system before and after the rotation, respectively.

As an example, consider the case of an ideal linear polarizer with the x and x' axes of the input and output coordinate axes chosen parallel to its transmission axis, while the y and y' axes are directed along its extinction axis. In this orientation, the Jones matrix of the ideal linear polarizer is given by eq. (2.24). Assume that both the input and output coordinate systems are rotated together by an angle $-\alpha$ (*i.e.* a clockwise rotation looking into the beam). Applying the transformation of eq. (2.31), the new Jones matrix of the polarizer is obtained from the old Jones matrix of eq. (2.24) by

$$\begin{aligned}T_{\text{new}} &= e^{-j2\pi nd/\lambda} \begin{bmatrix} \cos\alpha & -\sin\alpha \\ \sin\alpha & \cos\alpha \end{bmatrix} \begin{bmatrix} 1 & 0 \\ 0 & 0 \end{bmatrix} \begin{bmatrix} \cos\alpha & \sin\alpha \\ -\sin\alpha & \cos\alpha \end{bmatrix} \\ &= e^{-j2\pi nd/\lambda} \begin{bmatrix} \cos^2\alpha & \sin\alpha\cos\alpha \\ \sin\alpha\cos\alpha & \sin^2\alpha \end{bmatrix}.\end{aligned} \tag{2.32}$$

(Note that the transmission axis of the polarizer subtends a *positive* azimuth α measured from the reference x and x' coordinate axes.)

2.2.5. Circular and generalized Jones matrices

The effect of the coordinate rotation described above is equivalent to the effect of a change of the basis states of polarization at both the input and output of the optical system from one pair of orthogonal linear states to another pair of orthogonal linear states. Suppose we perform a more general change of the input basis states from the (x, y) orthogonal linear polarizations to an arbitrary pair (u, v) of elliptic polarizations, and that we make a similar change of basis from (x', y') to (u', v') at the output of the optical system. According to eq. (1.66) of §1.6, the effect of the change of basis $(x, y) \to (u, v)$ and $(x', y') \to (u', v')$ is to transform the input and output Cartesian Jones vectors $\boldsymbol{E}^i_{x,y}$, $\boldsymbol{E}^o_{x',y'}$, into the generalized input and output Jones vectors[11] $\boldsymbol{E}^i_{u,v}$, $\boldsymbol{E}^o_{u',v'}$, where

$$\boldsymbol{E}^i_{u,v} = \boldsymbol{F}^{-1} \boldsymbol{E}^i_{x,y}, \qquad \boldsymbol{E}^o_{u',v'} = \boldsymbol{F}^{-1} \boldsymbol{E}^o_{x',y'}. \tag{2.33}$$

In eq. (2.33), $\boldsymbol{F}$ is the appropriate transformation matrix which is assumed the *same* for the change of basis at both the input and output of the optical system. The mapping of the Cartesian Jones vector between the input and output of the optical system is described by

$$\boldsymbol{E}^o_{x',y'} = \boldsymbol{T}_{\text{car}} \boldsymbol{E}^i_{x,y},$$

which can be pre-multiplied by $\boldsymbol{F}^{-1}$ to read

$$\boldsymbol{F}^{-1} \boldsymbol{E}^o_{x',y'} = \boldsymbol{F}^{-1} \boldsymbol{T}_{\text{car}} \boldsymbol{E}^i_{x,y}.$$

Subsequent use of the fact that $\boldsymbol{F}\boldsymbol{F}^{-1}$ is equal to the identity matrix gives

$$[\boldsymbol{F}^{-1} \boldsymbol{E}^o_{x',y'}] = [\boldsymbol{F}^{-1} \boldsymbol{T}_{\text{car}} \boldsymbol{F}] [\boldsymbol{F}^{-1} \boldsymbol{E}^i_{x,y}],$$

$$\boldsymbol{E}^o_{u',v'} = \boldsymbol{T}_{\text{gen}} \boldsymbol{E}^i_{u,v}, \tag{2.34}$$

where

$$\boldsymbol{T}_{\text{gen}} = \boldsymbol{F}^{-1} \boldsymbol{T}_{\text{car}} \boldsymbol{F}. \tag{2.35}$$

$\boldsymbol{T}_{\text{car}}$ and $\boldsymbol{T}_{\text{gen}}$ refer to the Cartesian and the *generalized* Jones matrices of the optical system, respectively. Equation (2.34) is the basis of the *generalized Jones matrix formulation* in the same way as eq. (2.3) is the basis of the Cartesian Jones matrix formulation. Equation (2.35) shows how to obtain the generalized Jones matrix from the Cartesian Jones matrix. (The development that led to eq. (2.35) is a generalization of that which led to eq. (2.31).)

A special case of considerable interest is that of the *circular* Jones matrix

---

[11] $\boldsymbol{E}^i_{u,v}$ reads "the input Jones vector in the representation whose basis states are the (elliptic) polarizations u and v." For an output Jones vector, the superscript o is used.

formulation. In this case, we use the left- and right-circular polarizations as basis states at both the input and output of the optical system. The transformation matrix F becomes that given by eq. (1.55). The input and output circular Jones vectors $E^i_{\ell,r}$ and $E^o_{\ell',r'}$ are interrelated by

$$E^o_{\ell',r'} = T_{\text{cir}} E^i_{\ell,r}, \tag{2.36}$$

where the circular Jones matrix T_{cir} is obtained from the Cartesian Jones matrix T_{car} by

$$T_{\text{cir}} = F^{-1} T_{\text{car}} F. \tag{2.37}$$

(Obviously eqs. (2.36) and (2.37) are special cases of eqs. (2.34) and (2.35), respectively.) Substituting F and F^{-1} from eqs. (1.55) and (1.57) of §1.6 into eq. (2.36) gives

$$T_{\text{cir}} = \frac{1}{2}\begin{bmatrix} 1 & j \\ 1 & -j \end{bmatrix}\begin{bmatrix} T^{\text{car}}_{11} & T^{\text{car}}_{12} \\ T^{\text{car}}_{21} & T^{\text{car}}_{22} \end{bmatrix}\begin{bmatrix} 1 & 1 \\ -j & j \end{bmatrix}. \tag{2.38}$$

Carrying out the matrix multiplication in eq. (2.38), we obtain the elements of the circular Jones matrix in terms of the elements of the Cartesian Jones matrix

$$\begin{aligned}
T^{\text{cir}}_{11} &= \tfrac{1}{2}[T^{\text{car}}_{11} - jT^{\text{car}}_{12} + jT^{\text{car}}_{21} + T^{\text{car}}_{22}], \\
T^{\text{cir}}_{12} &= \tfrac{1}{2}[T^{\text{car}}_{11} + jT^{\text{car}}_{12} + jT^{\text{car}}_{21} - T^{\text{car}}_{22}], \\
T^{\text{cir}}_{21} &= \tfrac{1}{2}[T^{\text{car}}_{11} - jT^{\text{car}}_{12} - jT^{\text{car}}_{21} - T^{\text{car}}_{22}], \\
T^{\text{cir}}_{22} &= \tfrac{1}{2}[T^{\text{car}}_{11} + jT^{\text{car}}_{12} - jT^{\text{car}}_{21} + T^{\text{car}}_{22}].
\end{aligned} \tag{2.39}$$

The inversion of eq. (2.39) can be easily done by using, $T_{\text{car}} = F T_{\text{cir}} F^{-1}$, which is derived by pre- and post-multiplying both sides of eq. (2.37) by F and F^{-1} respectively. This gives

$$\begin{aligned}
T^{\text{car}}_{11} &= \tfrac{1}{2}[T^{\text{cir}}_{11} + T^{\text{cir}}_{12} + T^{\text{cir}}_{21} + T^{\text{cir}}_{22}], \\
T^{\text{car}}_{12} &= \tfrac{1}{2}j[T^{\text{cir}}_{11} - T^{\text{cir}}_{12} + T^{\text{cir}}_{21} - T^{\text{cir}}_{22}], \\
T^{\text{car}}_{21} &= -\tfrac{1}{2}j[T^{\text{cir}}_{11} + T^{\text{cir}}_{12} - T^{\text{cir}}_{21} - T^{\text{cir}}_{22}], \\
T^{\text{car}}_{22} &= \tfrac{1}{2}[T^{\text{cir}}_{11} - T^{\text{cir}}_{12} - T^{\text{cir}}_{21} + T^{\text{cir}}_{22}].
\end{aligned} \tag{2.40}$$

As an example, let us find the circular Jones matrix of an optical rotator whose Cartesian Jones matrix is given by eq. (2.27). Substituting

$$\begin{aligned}
T^{\text{car}}_{11} &= T^{\text{car}}_{22} = \cos(\alpha d)\, e^{-j2\pi nd/\lambda}, \\
T^{\text{car}}_{12} &= -T^{\text{car}}_{21} = \sin(\alpha d)\, e^{-j2\pi nd/\lambda},
\end{aligned} \tag{2.41}$$

into eq. (2.39), we immediately get

$$T^{\text{cir}}_{11} = e^{-j(\alpha d)}\, e^{-j2\pi nd/\lambda}, \quad T^{\text{cir}}_{22} = e^{+j(\alpha d)}\, e^{-j2\pi nd/\lambda}, \quad T^{\text{cir}}_{12} = T^{\text{cir}}_{21} = 0. \tag{2.42}$$

Therefore, the circular Jones matrix of an optical rotator

$$T_{\text{cir}} = e^{-j2\pi n d/\lambda} \begin{bmatrix} e^{-j(\alpha d)} & 0 \\ 0 & e^{+j(\alpha d)} \end{bmatrix}, \qquad (2.43)$$

is diagonal. The transformation of the circular Jones vector between the input and output of the optical rotator becomes

$$\begin{bmatrix} E_{\ell'} \\ E_{r'} \end{bmatrix} = e^{-j2\pi n d/\lambda} \begin{bmatrix} e^{-j(\alpha d)} & 0 \\ 0 & e^{+j(\alpha d)} \end{bmatrix} \begin{bmatrix} E_\ell \\ E_r \end{bmatrix} \qquad (2.44)$$

The diagonal nature of the circular Jones matrix tells us that left-circularly-polarized incident light is transmitted as left-circularly-polarized light by the optical rotator and that right-circularly-polarized incident light is likewise transmitted unchanged.[12] Expanding eq. (2.44) gives

$$\begin{aligned} E_{\ell'} &= [e^{-j2\pi n_\ell d/\lambda}] E_\ell, \\ E_{r'} &= [e^{-j2\pi n_r d/\lambda}] E_r, \end{aligned} \qquad (2.45)$$

where

$$\begin{aligned} n_\ell &= n + \frac{\alpha \lambda}{2\pi}, \\ n_r &= n - \frac{\alpha \lambda}{2\pi}. \end{aligned} \qquad (2.46)$$

From eqs. (2.45) and (2.46) we see that when a left-circularly-polarized wave propagates through an optically active medium it experiences an index of refraction, $n_\ell = n + \alpha\lambda/2\pi$, and hence travels at a speed c/n_ℓ, whereas a right-circularly-polarized wave experiences a different index of refraction, $n_r = n - \alpha\lambda/2\pi$ and travels at a different speed c/n_r. By analogy with the case of linear birefringence, the optically active medium is said to exhibit *circular birefringence* of magnitude

$$(n_\ell - n_r) = \frac{\alpha \lambda}{\pi}. \qquad (2.47)$$

($\alpha\lambda$ is equal to the optical rotation when the wave travels through the medium a distance equal to the vacuum wavelength of light.) We have referred to n in eq. (2.27) as an "average" index of refraction of the medium. The meaning of this index of refraction becomes evident from eq. (2.46) because

$$n = \tfrac{1}{2}(n_\ell + n_r), \qquad (2.48)$$

---

[12] The left- and right-circular polarizations are called the *eigenpolarizations* of the optical rotator (see §2.6).

which does represent the average of the indices of refraction of the optically-active medium for left- and right-circularly-polarized light. The above discussion gives an account of Fresnel's phenomenological theory of optical activity [33].

2.3. Separability of information on the ellipse of polarization in the Jones-matrix formulation – The polarization transfer function

The two-component complex Jones vector of a uniform TE plane wave of light carries information about the amplitude A, absolute phase δ, azimuth θ and ellipticity angle ϵ of the elliptic vibration of its electric vector [fig. 1.5]. If such a plane wave is allowed to interact with an optical system, all of the above four properties of the wave $(A, \delta, \theta, \epsilon)$ are changed. In many instances (e.g., in ellipsometry) we are interested in the transformation of the ellipse of polarization (represented by θ and ϵ) by the optical system disregarding the effect of the system on the amplitude A and phase δ of the elliptic vibration of the incident wave.[13] In §1.7 we have seen that the information on the ellipse of polarization proper is contained in a single complex variable χ which is obtained by taking the ratio of the phasor components of the Jones vector. The question now arises as to whether the ellipses of polarization of the incident and the outgoing plane waves are directly related to each other, independent of their associated amplitudes or absolute phases.

Consider the optical system shown in fig. 2.1, with one incident and one outgoing plane wave. When the input and output electric vibrations are resolved into orthogonal linear components (E_{ix}, E_{iy}) and $(E_{ox'}, E_{oy'})$, along coordinate systems (x, y) and (x', y'), respectively, the relationship between $(E_{ox'}, E_{oy'})$ and (E_{ix}, E_{iy}) is given by eq. (2.1). According to eq. (1.77) of §1.8, the input and output ellipses of polarization are described by the complex numbers χ_i, χ_o where

$$\chi_i = E_{iy}/E_{ix}, \qquad \chi_o = E_{oy'}/E_{ox'} \tag{2.49}$$

Dividing eq. (2.1b) by eq. (2.1a), we obtain

$$\frac{E_{oy'}}{E_{ox'}} = \frac{T_{21}E_{ix} + T_{22}E_{iy}}{T_{11}E_{ix} + T_{12}E_{iy}}$$

$$= \frac{T_{22}(E_{iy}/E_{ix}) + T_{21}}{T_{12}(E_{iy}/E_{ix}) + T_{11}}$$

---

[13] Even when the amplitude and phase information is important, the procedure whereby the effect of the optical system on the ellipse of polarization (θ, ϵ) is studied separately from its effect on the amplitude A and phase δ is significant.

or, using eq. (2.49),

$$\chi_o = \frac{T_{22}\chi_i + T_{21}}{T_{12}\chi_i + T_{11}}. \qquad (2.50)$$

Equation (2.50) is an important useful result because it shows that the information on the ellipse of polarization is *separable* in the Jones-matrix formulation, so that the ellipse of polarization at the output of the system χ_o is determined *only by* the ellipse of polarization at the input of the system χ_i, irrespective of the amplitude A_i and phase δ_i of the incident wave [18, 34, 35]. The functional relationship $\chi_o = f(\chi_i)$ in eq. (2.50) can be appropriately called the *Polarization Transfer Function*[14] (PTF) of the polarizing optical system. According to eq. (2.50), the polarization transfer function is a *bilinear transformation* whose coefficients T_{ij} are the same as the elements of the Jones matrix T of the optical system.[15]

The above result can be generalized to apply to the case when arbitrary basis polarization states (u, v) and (u', v') are used at the input and output of the optical system, respectively. According to eq. (1.97) of §1.7, the ellipses of polarization at the input and output of the system can be described by the generalized complex polarization variables

$$\chi_i = E_{iv}/E_{iu}, \qquad \chi_o = E_{ov'}/E_{ou'}, \qquad (2.51)$$

where (E_{iu}, E_{iv}) and $(E_{ou'}, E_{ov'})$ are the phasor components of the generalized input and output Jones vectors, respectively. The transformation of the generalized Jones vector between the input and output of the optical system, described by eq. (2.34), together with eq. (2.51), immediately lead to the conclusion that the polarization transfer function is given by eq. (2.50) where T_{ij} now represent the elements of the generalized Jones matrix of the optical system [21].

2.4. Transformation of the ellipse of polarization by an optical system in the complex-plane representation

The polarization transfer function [eq. (2.50)] gives an overall view of how

---

[14] This is in direct analogy with the modulation or optical transfer functions (MTF or OTF) of imaging optical systems.

[15] Note that the complex polarization variable χ that describes the ellipse of polarization of a light wave is also bilinearly transformed due to *merely changing the basis states of polarization*, eq. (1.101). In eq. (1.101), $\chi_{u,v}$ and $\chi_{u',v'}$ describe the *same* ellipse of polarization of the *same* wave. In eq. (2.50), χ_i and χ_o describe the ellipses of polarization of two *different* waves, the incident wave and the outgoing wave, respectively. The bases with respect to which χ_i and χ_o are referenced may but need not be the same. Equations (1.101) and (2.50) should not be confused.

an optical system responds to an incident light beam, as the polarization of that beam is allowed to assume all possible states. The inverse of eq. (2.50),

$$\chi_i = \frac{T_{11}\chi_o - T_{21}}{-T_{12}\chi_o + T_{22}}, \tag{2.52}$$

gives the incident ellipse of polarization χ_i necessary to produce a prescribed polarization ellipse χ_o at the output of the system. The inverse transformation in eq. (2.52) is also bilinear.

Important conclusions on the polarization-mapping properties of optical systems can be reached by use of the PTF of eq. (2.50) [18]. The system is quite general and is subject only to the limitation that it *can* be represented by a Jones matrix.[16] There are two limiting cases of eq. (2.50) that have to be considered separately.

(a) The determinant $(T_{22}T_{11} - T_{21}T_{12})$ of the transformation vanishes. Under this condition, eq. (2.50) becomes

$$\chi_o = \frac{T_{22}}{T_{12}} = \frac{T_{21}}{T_{11}}, \tag{2.53}$$

i.e., all incident polarizations χ_i are mapped into one emergent polarization χ_o given by eq. (2.53). In this case, the system acts as an ideal polarizer that is in general elliptic.

(b) $T_{12} = 0$. In this case, the bilinear transformation in eq. (2.50) degenerates into a linear form

$$\chi_o = \frac{T_{22}}{T_{11}}\chi_i + \frac{T_{21}}{T_{11}}. \tag{2.54}$$

Ordinarily T_{12} and T_{21} vanish together, so that a realization of the transfer function of eq. (2.54) can be found in a linear retardation plate or an isotropic reflecting surface.

The following conclusions are obtained from known properties of the bilinear transformation [20].

I–If the response of an optical system to *three* different incident polarizations is known, its response to *all* other incident polarizations is completely determined. This is because a unique bilinear transformation can be found that carries any three different points $(\chi_{i1}, \chi_{i2}, \chi_{i3})$ in the χ_i plane over into three points $(\chi_{o1}, \chi_{o2}, \chi_{o3})$ in the χ_o plane. This transformation provides the basis of *generalized ellipsometry* (§3.2) and is given by

$$\frac{(\chi_o - \chi_{o1})(\chi_{o3} - \chi_{o2})}{(\chi_o - \chi_{o2})(\chi_{o3} - \chi_{o1})} = \frac{(\chi_i - \chi_{i1})(\chi_{i3} - \chi_{i2})}{(\chi_i - \chi_{i2})(\chi_{i3} - \chi_{i1})}. \tag{2.55}$$

---

[16]This is equivalent to the statement that the optical system is linear and non-depolarizing (see §2.11).

II – The polarization transfer function, eq. (2.50), transforms any circle on the χ_i plane into a circle on the χ_o plane. This circle-to-circle mapping property is the most important feature of bilinear mappings. Applying this to the equi-azimuth and equi-ellipticity contours, fig. 1.18, we find that

(1) If the input polarization χ_i to an optical system corresponds to a vibration of fixed azimuth and variable ellipticity, the output polarization χ_o moves along a circle in the χ_o plane. This circle always passes through the two points L' and R'

$$\chi_o(L') = \frac{-jT_{22} + T_{21}}{-jT_{12} + T_{11}}, \qquad \chi_o(R') = \frac{jT_{22} + T_{21}}{jT_{12} + T_{11}}, \qquad (2.56)$$

which correspond to the system's response to incident left (L)- and right (R)-handed circular polarizations, respectively.

(2) Similarly, if the incident vibration χ_i has a fixed ellipticity, the emergent vibration χ_o will track a circle in the χ_o plane, as the azimuth of χ_i is continuously varied. This circle will enclose either one of the two points in eq. (2.56).

Because the bilinear transformation is conformal, the two families of circles in the χ_o plane that correspond to the system's response to constant azimuth and constant ellipticity incident vibrations χ_i are orthogonal. The statements in (1) and (2) above could be reversed. For example, if the output ellipse of polarization χ_o remains of fixed azimuth, the incident polarization χ_i follows a circle in the χ_i plane passing through the points L" and R"

$$\chi_i(L'') = \frac{-jT_{11} - T_{21}}{jT_{12} + T_{22}}, \qquad \chi_i(R'') = \frac{jT_{11} - T_{21}}{-jT_{12} + T_{22}} \qquad (2.57)$$

which represent the incident polarization states necessary to produce an exiting beam which is left (L)- and right (R)-handed circularly polarized, respectively.

When a circle in the χ_i plane is mapped into a circle in the χ_o plane, the domain inside the first circle is mapped onto the interior or the exterior domains of the second. A polarization transfer function can always be found that transforms incident polarization states in a circular domain of the χ_i plane onto a prescribed circular domain of the χ_o plane. Equation (2.55) provides the necessary transformation in terms of three points on each of the two circles bounding these domains. This provides a simple example of system synthesis.

III – The Eigenpolarizations of an Optical System: Each optical system has two eigenpolarizations that correspond to the two invariant points of the bilinear transformation. The eigenpolarizations are the two polarizations that pass through the system unchanged and are obtained from eq. (2.50) by setting $\chi_o = \chi_i = \chi$

$$\chi = \frac{T_{22}\chi + T_{21}}{T_{12}\chi + T_{11}},$$

$$T_{12}\chi^2 + (T_{11} - T_{22})\chi - T_{21} = 0. \qquad (2.58)$$

The roots of eq. (2.58) are the eigenpolarizations χ_{e1} and χ_{e2}

$$\chi_{e1,2} = \frac{1}{2T_{12}}\{(T_{22} - T_{11}) \pm [(T_{22} - T_{11})^2 + 4T_{12}T_{21}]^{\frac{1}{2}}\}. \qquad (2.59)$$

Note that χ_{e1} and χ_{e2} are different unless

$$(T_{22} - T_{11})^2 + 4T_{12}T_{21} = 0, \qquad (2.60)$$

in which case they become equal.[17] If we superimpose the χ_o and χ_i planes, any circle through the eigenpolarizations (the invariant points) χ_{e1} and χ_{e2} is transformed into another circle through these same points. The orthogonal family of loci (which are also circles that enclose χ_{e1} and χ_{e2}) is similarly transformed into itself under the bilinear transformation.

IV – The Loci of Invariant-Azimuth and Invariant-Ellipticity Polarization States of an Optical System [37]: The eigenpolarizations refer to incident elliptic vibrations that propagate through and emerge from the system with both ellipticity and azimuth unaffected.[18] It is interesting to search for the polarization states for which either ellipticity alone or azimuth alone remains unchanged after propagation through the system. For any one of the invariant-ellipticity states (IES) the polarization ellipses at input and output of the optical system have the same axial ratio and the same sense of description, although the major axis at the output is, in general, rotated with respect to that at input. On the other hand, for any one of the invariant-azimuth states (IAS) the ellipses of polarization at input and output have their major axes oriented "parallel" to one another and, in general, they have different ellipticities. The eigenpolarizations of the optical system correspond to two points of intersection of the locus of IES and that of IAS.

The circle-to-circle transformation property of an optical system can be used to introduce the loci of invariant-ellipticity states (IES) and invariant-azimuth states (IAS). Consider, for example, the locus of IES. If the emergent polarization χ_o is to describe an equi-ellipticity circle Γ in the complex plane, the incident polarization χ_i should describe a circle γ.

---

[17] An example of the significance of an optical system which possesses this type of *degenerate anisotropy* is described by de Lang [36].

[18] Since the azimuth of the ellipse of polarization is measured from the x and x' axes at the input and output of the optical system, it is implicitly understood that x and x' are chosen in the plane formed by the incident and the scattered wave-vectors k and k' (see footnote 2, p. 68). This choice is essential in defining the eigenpolarizations according to eq. (2.59) and the loci of IAS and IES.

When the two circles Γ and γ intersect one another the points of intersection will represent two incident polarization states whose ellipticity is preserved after propagation through the system. This is schematically shown in fig. 2.6. The points L'' and R'' represent the polarization states at the

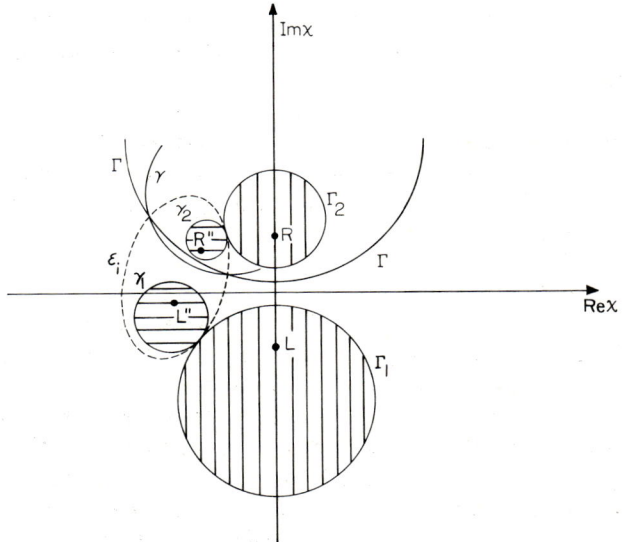

Fig. 2.6. Introduces the locus $\mathscr{E}_i$ of the incident invariant-ellipticity states (IES) of an optical system in the complex plane of polarization.

input of the optical system which produce left (L)- and right (R)-circularly polarized light at its output, respectively, eq. (2.57). The circle γ represents one member of a family of circles that enclose L'' and R'' which is mapped by the optical system onto the family of equi-ellipticity circles Γ. There is a one-to-one correspondence between the circles of the two families γ and Γ. Of particular interest are the two limiting circles γ_1 and γ_2 which touch their images Γ_1 and Γ_2, respectively. The interior of γ_1 and γ_2 (horizontally hatched) is mapped into the interior of Γ_1 and Γ_2 (vertically hatched), respectively. Thus circles that enclose L'' or R'' and lie inside γ_1 or γ_2 do not intersect with their images. However, each circle γ outside γ_1 and γ_2 (as the one shown) intersects its image Γ (outside Γ_1 and Γ_2) in two points belonging to the locus $\mathscr{E}_i$ of the incident IES. Therefore, for each ellipticity angle ϵ, in the range $\epsilon_1 \leq \epsilon \leq \epsilon_2$ (where ϵ_1 and ϵ_2 are determined by the equi-ellipticity circles Γ_1 and Γ_2, respectively), there are two incident vibrations which emerge from the system with their ellipticity unchanged. The locus $\mathscr{A}_i$ of the incident invariant-azimuth states (IAS) can be introduced in a similar manner. Reference [37] shows how to determine the Cartesian equations of the loci $\mathscr{E}_i$ and $\mathscr{A}_i$ of the incident IES and IAS for any optical system in terms of the elements of its Jones matrix.

V–Response of an Optical System to Polarization-Modulated Incident Light [38]: In a number of important applications the polarization of a light beam is modulated in accordance with some function of time (*e.g.*, a signal to be transmitted) [39]. The propagation of such a beam through an optical system is of interest.

Let $\chi_i(t)$ represent the time-varying polarization state of the incident light beam. The highest Fourier frequency associated with this time variation is expected to be many orders of magnitude less than any frequency in the optical spectrum. Therefore, the polarization states of the emergent light $\chi_o(t)$ can be obtained from

$$\chi_o(t) = f(\chi_i(t)), \qquad (2.61)$$

where f is the system's PTF, so far associated with time-independent incident polarization states only. According to eq. (2.61) it can be seen that if $\chi_i(t)$ describes some trajectory in the complex χ_i plane, $\chi_o(t)$ will describe the image trajectory in the complex χ_o plane. Because f is bilinear, $\chi_o(t)$ will describe a circle in the χ_o plane if $\chi_i(t)$ moves on a circle in the χ_i plane, for any circle $\chi_i(t)$. The preservation of form for *large* changes in the state of polarization is restricted to the case when the locus of $\chi_i(t)$ is a circle. On the other hand, all *small* changes in the state of polarization of an incident beam around a quiescent state preserve form upon passing through the optical system because the transformation f is conformal. For this case let

$$\chi_i(t) = \chi_{io} + \Delta\chi_i(t), \qquad (2.62a)$$

where

$$|\Delta\chi_i(t)| \le \epsilon \ll 1, \forall t. \qquad (2.62b)$$

Thus $\chi_i(t)$ is limited to a small circular domain of center χ_{io} (representing the quiescent state of polarization) and radius ϵ. Correspondingly $\chi_o(t)$ will vary as

$$\chi_o(t) = \chi_{oo} + \Delta\chi_o(t), \qquad (2.63a)$$

where

$$|\Delta\chi_o(t)| < \epsilon' \le 1, \forall t, \qquad (2.63b)$$

with similar meanings for χ_{oo} and ϵ'. χ_{oo} and χ_{io} are related by

$$\chi_{oo} = f(\chi_{io}), \qquad (2.64)$$

and $\Delta\chi_o(t)$ and $\Delta\chi_i(t)$ by

$$\Delta\chi_o(t) = f'(\chi_{io})\Delta\chi_i(t), \qquad (2.65)$$

where

$$f'(\chi_{io}) = \left.\frac{df}{d\chi_i}\right|_{\chi_i = \chi_{io}}. \qquad (2.66)$$

The preservation of form is evident from eq. (2.65) which gives the law of propagation of small changes in the state of polarization through an optical system. To get $\Delta\chi_o(t)$ we need only multiply $\Delta\chi_i(t)$ by $f'(\chi_{io})$. This simple multiplication involves a magnification of $|f'(\chi_{io})|$ and a solid rotation through an angle of $\arg f'(\chi_{io})$. By simple differentiation of eq. (2.50) we get

$$f'(\chi_i) = \frac{(T_{22}T_{11} - T_{21}T_{12})}{(T_{12}\chi_i + T_{11})^2}. \qquad (2.67)$$

Note that the determinant of the transformation appears in the numerator of eq. (2.67). Except for the case of a vanishing determinant (*i.e.* excluding a system that acts as an ideal polarizer), $f'(\chi_i) \neq 0$ and the mapping is conformal for all quiescent polarization states χ_{io}.

VI – Analogy Between Linear Optical Systems and Linear Two-Port Electrical Networks [40]: There is much to be gained from the analogy between the description of the terminal behavior of linear optical systems and linear two-port electrical networks. In both cases we deal with the transformation of a pair of oscillating quantities between the input and output of the system. These are the two oscillating components of the electric vector of the light wave in the case of an optical system and the terminal (port) voltage and current oscillations in the case of a two-port electrical network. The complex polarization variable χ corresponds to the port impedance Z (or admittance Y). The mapping between input and output of the ellipse of polarization by an optical system and of impedance by a two-port network is described by a bilinear transformation. For each terminal characteristic of a system of one type there is a similar characteristic for the system of the other type. For example, the two eigenpolarizations of the optical system are the analogs of the two interative impedances of the two-port network. The loci of polarization states that preserve ellipticity or azimuth correspond to the loci of impedances (or admittance) that preserve magnitude or angle respectively, and both are described by the same equations in the complex plane. Two-port electrical networks can be synthesized whose impedance- (or admittance-) mapping properties are the same as the polarization-mapping properties of a given optical system and vice versa. This analogy, which is examined in more detail in ref. [40], should be useful because (1) it unifies the methods of treating both kinds of systems, (2) it leads to the reciprocal simulation of systems of one type by systems of the other.

2.5. Transformation of polarization by an optical system in the Poincaré-sphere representation

The idea of associating a (complex) sphere with the complex plane is known in the theory of complex variables and is used to visualize the behavior of a function over regions in the complex plane which include the point at infinity. Such a sphere can have an arbitrary diameter[19] and rests tangent to the complex plane at its origin. If the pole of the sphere which is diametrically above the origin is considered as a center of a stereographic projection, all details in the complex plane can be projected onto the sphere and vice versa. The Poincaré sphere and complex-plane representations for polarized light are related exactly in this manner, as has already been discussed in §1.8. The stereographic projection has two important properties: (1) circles in the plane are carried over onto circles on the sphere (a straight line is a degenerate circle), (2) angular relationships between intersecting curves are preserved, *i.e.*, the process is isogonal. The conclusions reached below derive directly from those two properties.

Let the polarization states of light at the input and output of an optical system be represented by points $P_i(\chi_i)$ and $P_o(\chi_o)$ in the same complex plane and let P_{is} and P_{os} be the corresponding points on the Poincaré sphere, fig. 2.7. According to property (1) above, if P_i is allowed to describe a circle in the plane, so will P_{is} on the sphere. Because the transformation of polarization $\chi_i \rightarrow \chi_o$ by the optical system is bilinear [eq. (2.50)], P_o will track a circular path in the complex plane following the motion of P_i and with it P_{os} on the sphere. This leads to the following important theorem. On the Poincaré sphere, the effect of any linear non-depolarizing optical system is to transform incident polarization states on a circle onto outgoing states on another circle.[20]

---

[19]When the basis states of polarization u and v of the complex-plane representation have equal amplitudes, the diameter of the sphere has to be chosen unity (See §1.8 and §1.9).

[20]As a simple example, consider linearly polarized light incident on a linear retarder. If the azimuth of the incident linearly polarized light is varied, the representative point moves on the equator of the Poincaré sphere. The linear retarder must transform the equator into another circle that represents the locus of the outgoing states. Such a circle must pass through the two diametrically opposite points F and S on the equator that represent incident linear vibrations parallel to the fast and the slow axes that are transmitted unchanged. In other words, the effect of the retarder is to rotate the equator around the diameter FS. To prove that this rotation is equal to the relative retardation δ, consider F and S as basis states of a complex-plane representation. Such a complex plane will be tangent to the Poincaré sphere at F. In this complex plane the relationship between incident and outgoing polarizations on and from the retarder is simply given by $\chi_o = e^{-j\delta}\chi_i$. Thus, a point that represents a general incident polarization χ_i is rotated by an angle δ around the origin F in the complex plane. The stereographic projection of this point (with S as center) is likewise rotated by δ on the Poincaré sphere around F.

Because of this circle-into-circle mapping property of optical systems the transformation of the totality of the polarization space can be seen by considering the effect of the system on a family of circles that completely covers the sphere. For example, the family of constant azimuth circles (the longitudes) through the points R and L are transformed onto another family of circles on the sphere through R' and L' where R' and L' represent the system's response to incident right (R)- and left (L)-circularly polarized states, respectively. Because of the bilinear form of eq. (2.50) and the isogonal nature of the stereographic projection, the family of constant ellipticity contours (the latitudes) are transformed onto the orthogonal family of circles enclosing R' and L'.

Besides the circle-to-circle polarization-mapping property discussed in §2.4, every other property once established in the complex plane could be subsequently restated with reference to the Poincaré sphere. This includes the loci of the invariant-ellipticity states $\mathscr{E}_i$ (IES) and the invariant-azimuth states $\mathscr{A}_i$, introduced in §2.4. For example, fig. 2.6 can be projected onto the Poincaré sphere. R and L now correspond to the north and south poles with Γ_1 and Γ_2 representing the two limiting equi-ellipticity latitudes. γ_1 and γ_2 become two circles which enclose the points L'' and R'' on the sphere and touch Γ_1 and Γ_2, respectively. Every circle γ (between γ_1 and γ_2) around L''

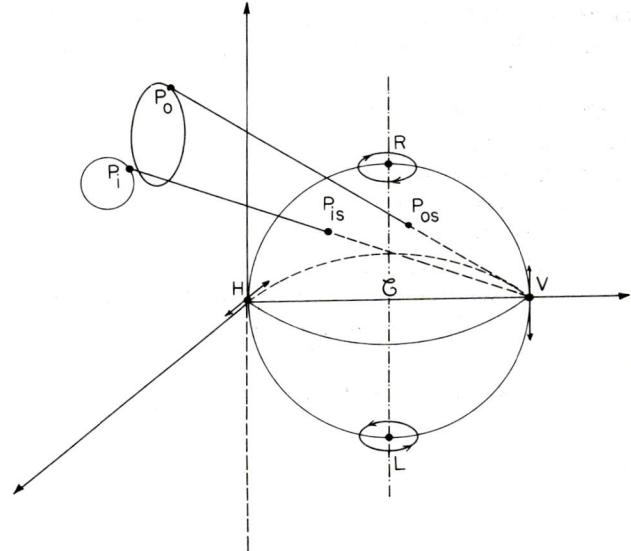

Fig. 2.7. The complex-plane and the Poincaré-sphere representations are linked by a stereographic projection. The points P_i and P_o represent the polarization states at the input and output of an optical system in the complex plane and their projections P_{is} and P_{os} represent the same states on the sphere.

or R″ intersects its image Γ (between Γ_1 and Γ_2) in two points belonging to $\mathscr{E}_i$. Obviously, the analytic equation of $\mathscr{E}_i$ on the sphere for a general optical system must be determined from the corresponding Cartesian equation in the plane [37].

The response of an optical system to polarization-modulated incident light has been considered in §2.4, using the idea of the PTF. The conclusions reached therein can now be restated with reference to the Poincaré sphere. Referring to fig. 2.7, note that if the incident light is modulated such that $P_{is}(t)$ (which represents the incident light) describes any circle large or small, then $P_{os}(t)$ (representing the outgoing light) will likewise describe a circle on the Poincaré sphere. Such preservation of form for large changes in the polarization state is restricted to the case when the path of $P_{is}(t)$ is a circle. Similarly, the theorem on the preservation of form of any small changes of the polarization state around some quiescent point holds on the Poincaré sphere. This follows from the isogonal property of the process of stereographic projection. Thus, if $P_{is}(t)$ describes a small contour in the vicinity of P_{iso}, $P_{os}(t)$ will describe a contour of the same shape in the vicinity P_{oso} where P_{iso} and P_{oso} represent the quiescent states at the input and output of the system respectively.

2.6. Separability of information on amplitude and phase in the Jones-matrix formulation – The Amplitude and Phase Transfer Functions

Because χ, which describes the polarization ellipse alone, is extracted from the two-component complex Jones vector E, it becomes apparent that the information that remains in E must involve the total amplitude and absolute phase of the wave. Here we concentrate on this amplitude and phase information and in so doing introduce the useful concepts of polarization-dependent Amplitude and Phase Transfer Functions (ATF and PHTF) [41]. The ATF and PHTF are both real functions of the complex polarization variable χ and are obtained from a single Complex-Amplitude Transfer Function (CATF).

The two-component Jones vector E_i of the incident wave is ordinarily written as

$$E_i = \begin{bmatrix} E_{iu} \\ E_{iv} \end{bmatrix}, \qquad (2.68)$$

where E_{iu} and E_{iv} denote the complex amplitudes of the two projections (components) of E_i along a chosen pair of basis polarization states u and v (typically two orthogonal linear or circular polarizations). When the basis

Jones vectors are *orthonormal* the intensity of the wave is given by[21]

$$E_i^\dagger E_i = E_{iu}^* E_{iu} + E_{iv}^* E_{iv}. \tag{2.69}$$

The Jones vector in eq. (2.68) can be recast in the form

$$E_i = \frac{A_{ci}}{(1+\chi_i\chi_i^*)^{\frac{1}{2}}} \begin{bmatrix} 1 \\ \chi_i \end{bmatrix} \tag{2.70}$$

where

$$\chi_i = E_{iv}/E_{iu}, \tag{2.71}$$

$$A_{ci} = E_{iu}[1+(E_{iv}^* E_{iv}/E_{iu}^* E_{iu})]^{\frac{1}{2}}. \tag{2.72}$$

The complex variable χ_i [eq. (2.71)] has been shown in §1.7 to carry information on the polarization ellipse alone. In fact, χ is a complex function $\chi(\theta, \epsilon)$ of two real arguments, the azimuth θ and the ellipticity angle ϵ. The form of $\chi(\theta, \epsilon)$, the way in which χ is structured from θ and ϵ, depends on the choice of the two basis polarization states [eqs. (1.79) and (1.94) of §1.7].

By substitution of eq. (2.70) into eq. (2.69) the expression for the intensity of the wave becomes

$$E_i^\dagger E_i = A_{ci}^* A_{ci}. \tag{2.73}$$

Equation (2.73) shows that A_{ci} represents the total complex amplitude of the incident wave. The significance of writing the Jones vector in the form of eq. (2.70) now becomes evident. *It is a form in which the information on the polarization ellipse alone (represented by χ_i) has been separated from the information on the total amplitude and absolute phase of the wave (represented by A_{ci}).* The complex amplitude A_{ci} can be expressed in terms of the real amplitude A_i and phase δ_i

$$A_{ci} = A_i e^{j\delta_i}. \tag{2.74}$$

The effect of a linear nondepolarizing optical system on the light wave is described by the following transformation of the Jones vector

$$E_o = TE_i, \tag{2.75}$$

where $T = (T_{ij})$ is the 2×2 Jones matrix of the system. To see the individual effects of the system on the polarization χ_i and complex amplitude A_{ci}, eq. (2.70) is substituted into eq. (2.75). This gives

$$E_o = \frac{A_{co}}{(1+\chi_o\chi_o^*)^{\frac{1}{2}}} \begin{bmatrix} 1 \\ \chi_o \end{bmatrix}, \tag{2.76}$$

[21] See eq. (1.72).

where

$$\chi_o = (T_{22}\chi_i + T_{21})/(T_{12}\chi_i + T_{11}), \qquad (2.50)$$

$$A_{co} = \left[\left(\frac{1+\chi_o\chi_o^*}{1+\chi_i\chi_i^*}\right)^{\frac{1}{2}}(T_{12}\chi_i + T_{11})\right]A_{ci}. \qquad (2.77)$$

Equation (2.50) gives the polarization transfer function (PTF) of the optical system as a bilinear transformation that relates the input and output ellipses of polarization described by χ_i and χ_o respectively. This result has already been discussed in §2.3.

The result in eq. (2.77) gives the law of transformation of the complex amplitude of the wave upon passing through the optical system. It can be written in the form

$$A_{co} = F(\chi_i)A_{ci}, \qquad (2.78)$$

where

$$F(\chi_i) = \left(\frac{1+\chi_o\chi_o^*}{1+\chi_i\chi_i^*}\right)^{\frac{1}{2}}(T_{12}\chi_i + T_{11}), \qquad (2.79)$$

is the *Complex-Amplitude Transfer Function (CATF)*. Because χ_o is related to χ_i by eq. (2.50), the CATF is a complex function of χ_i only. In other words, the effect of the optical system on the amplitude and phase of an incident light wave depends on the elliptic polarization state of that wave. The explicit form of $F(\chi_i)$ (obtained from the above-mentioned substitution) is

$$F(\chi_i) = \left[\frac{(|T_{12}\chi_i + T_{11}|^2 + |T_{22}\chi_i + T_{21}|^2)}{(1+|\chi_i|^2)}\right]^{\frac{1}{2}} \frac{(T_{12}\chi_i + T_{11})}{|T_{12}\chi_i + T_{11}|}. \qquad (2.80)$$

The function $F(\chi_i)$ in eq. (2.80) is not analytic in χ_i and hence the transformation of the complex amplitude of the wave is not conformal. This is in contrast with the transformation of polarization $\chi_o(\chi_i)$ [eq. (2.50)] which is analytic, hence conformal.

The complex-amplitude transfer function CATF can be separated into real *Amplitude and Phase Transfer Functions (ATF and PHTF)* by writing $F(\chi_i)$ in the form

$$F(\chi_i) = f_A(\chi_i)\exp[jf_\delta(\chi_i)]. \qquad (2.81)$$

Using eqs. (2.74) and (2.81) in eq. (2.78) gives

$$A_o = f_A(\chi_i)A_i, \qquad (2.82)$$

$$\delta_o = \delta_i + f_\delta(\chi_i). \qquad (2.83)$$

Equations (2.82) and (2.83) show that $f_A(\chi_i)$ and $f_\delta(\chi_i)$ represent the ATF and PHTF, respectively. The explicit forms of these two real functions of the complex polarization variable χ_i are obtained from eq. (2.80)

$$f_A(\chi_i) = \left[\frac{(|T_{12}\chi_i + T_{11}|^2 + |T_{22}\chi_i + T_{21}|^2)}{(1+|\chi_i|^2)}\right]^{\frac{1}{2}}, \tag{2.84}$$

$$f_\delta(\chi_i) = \arg(T_{12}\chi_i + T_{11}), \tag{2.85}$$

where arg signifies the angle or argument of the complex number. The effect on the intensity of the wave is described by the squared ATF or

$$f_A^2(\chi_i) = \frac{(|T_{12}\chi_i + T_{11}|^2 + |T_{22}\chi_i + T_{21}|^2)}{(1+|\chi_i|^2)}. \tag{2.86}$$

The meaning of the three transfer functions (PTF, ATF and PHTF) for polarization, amplitude, and phase may be explained as follows. The complex Jones vector E describes the sinusoidal vector oscillation of the electric field of the light wave, fig. 1.5 of §1.3. The tip of the electric vector describes an ellipse at a repetition rate equal to the optical frequency. The dissociation of the Jones vector into a complex polarization variable χ and a complex amplitude variable A_c, according to eq. (2.70), corresponds to the splitting of the information on the elliptic vibration of the electric field into two parts. The first part (χ) describes the form (axial ratio and sense of rotation, $\tan\epsilon$) and orientation (azimuth of the major axis θ) of the ellipse of polarization. The second part (A_c) specifies the amplitude A (size) of the elliptic vibration as well as its phase δ (the position of the electric vector at certain moment t). The PTF determines how the ellipse of polarization changes in form, orientation and sense. The ATF determines the attenuation or amplification of the elliptic vibration. The PHTF relates the positions of the electric-field vectors at input and output on their elliptic loci.

The PTF, ATF and PHTF provide a complete set of tools that allow physically significant information to be derived on the behavior of polarizing optical systems. The usefulness of the concept of the PTF has been demonstrated [§2.3 and §2.4].

2.6.1. *The eigenvalue problem – General classification of optical devices*

Each optical system has two eigenpolarizations χ_{e1} and χ_{e2} obtained by setting $\chi_o = \chi_i$ in eq. (2.50)

$$\chi_{e1,2} = \frac{1}{2T_{12}}\{(T_{22} - T_{11}) \pm [(T_{22} - T_{11})^2 + 4T_{12}T_{21}]^{\frac{1}{2}}\}. \tag{2.59}$$

The associated eigenvalues V_{e1} and V_{e2} can be found by substituting

$\chi_i = \chi_o = \chi_{e1,2}$ into the expression for the CATF, $F(\chi_i)$, in eq. (2.79)

$$V_{e1,2} = F(\chi_{e1,2}) = T_{12}\chi_{e1,2} + T_{11}. \tag{2.87}$$

Using $\chi_{e1,2}$ from eq. (2.59) into eq. (2.87) we get

$$V_{e1,2} = \tfrac{1}{2}\{(T_{22} + T_{11}) \pm [(T_{22} - T_{11})^2 + 4T_{12}T_{21}]^{\frac{1}{2}}\}, \tag{2.88}$$

which completes the solution of the eigenvalue problem.

The reverse problem of determining the Jones matrix of a system of known eigenpolarizations χ_{e1} and χ_{e2} and known eigenvalues V_{e1} and V_{e2} is simple. Equation (2.87) provides two linear equations that can be solved for T_{11} and T_{12}. Equation (2.88) gives $V_{e1} + V_{e2} = T_{22} + T_{11}$, hence T_{22} is determined. Finally, $T_{21} = - T_{12}(\chi_{e1}\chi_{e2})$ as can be seen from eq. (2.59). This gives

$$\begin{aligned} T_{11} &= K(V_{e2}\chi_{e1} - V_{e1}\chi_{e2}), \\ T_{12} &= K(V_{e1} - V_{e2}), \\ T_{21} &= - K\chi_{e1}\chi_{e2}(V_{e1} - V_{e2}), \\ T_{22} &= K(V_{e1}\chi_{e1} - V_{e2}\chi_{e2}), \\ K &= (\chi_{e1} - \chi_{e2})^{-1}. \end{aligned} \tag{2.89}$$

It is important to observe that we have solved the eigenvalue problem using only the definitions of the polarization and complex-amplitude transfer functions (PTF and CATF). As an exercise, the reader can verify that the solution of the Jones matrix eigenvalue equation [11]

$$TE_e = V_e E_e, \tag{2.90}$$

yields the *Jones eigenvectors*

$$E_{e1} = \begin{bmatrix} 1 \\ \chi_{e1} \end{bmatrix}, \quad E_{e2} = \begin{bmatrix} 1 \\ \chi_{e2} \end{bmatrix}, \tag{2.91}$$

and their associated eigenvalues V_{e1} and V_{e2}, where χ_{e1}, χ_{e2}; V_{e1} and V_{e2} are as given by eqs. (2.59) and (2.88), respectively. If either E_{e1} or E_{e2} of eq. (2.91) is multiplied by a complex constant, the resulting Jones vector continues to be an eigenvector of the optical system corresponding to the eigenvalue V_{e1} or V_{e2}, respectively. This is in contrast with χ_{e1} and χ_{e2} which represent the unique eigenpolarizations of the optical system.

The eigenpolarizations and eigenvalues serve as a suitable basis for the following classification of optical devices.[22]

2.6.1.1. Elliptic retarders
An optical device is described as an elliptic retarder if its eigenpolarizations χ_{e1} and χ_{e2} are orthogonal and if their associated eigenvalues are given by the

---

[22] A similar classification has been proposed by Shurcliff [4] and de Lang [36].

pure phase factors $V_{e1} = e^{j\delta/2}$ and $V_{e2} = e^{-j\delta/2}$. The eigenpolarization χ_{e1} is *phase-advanced* by $\frac{1}{2}\delta$, while χ_{e2} is *phase-retarded* by $\frac{1}{2}\delta$. Therefore, χ_{e1} represents the *fast* eigenpolarization, χ_{ef}, while χ_{e2} represents the *slow* eigenpolarization, χ_{es}. The phase angle δ represents the slow-to-fast *relative retardation* or *retardance* of the elliptic retarder. The fast and slow eigenpolarizations $\chi_{ef}(=\chi_{e1})$ and $\chi_{es}(=\chi_{e2})$ are related by the orthogonality condition [eq. (1.81)]

$$\chi_{ef}\chi_{es}^* = -1, \tag{2.92a}$$

while their associated eigenvalues $V_{ef}(=V_{e1})$ and $V_{es}(=V_{e2})$ are related by

$$V_{ef} = V_{es}^*, \qquad V_{ef}V_{es} = 1. \tag{2.92b}$$

The Jones matrix of an elliptic retarder can be obtained in terms of the complex polarization variable χ_{ef} that describes the fast eigenpolarization and the relative retardation δ by substituting from eqs. (2.92a) and (2.92b) into eq. (2.89)

$$T = (1 + \chi_{ef}\chi_{ef}^*)^{-1} \begin{bmatrix} (e^{j\delta/2} + \chi_{ef}\chi_{ef}^* e^{-j\delta/2}) & 2j\chi_{ef}^* \sin\tfrac{1}{2}\delta \\ 2j\chi_{ef} \sin\tfrac{1}{2}\delta & (\chi_{ef}\chi_{ef}^* e^{j\delta/2} + e^{-j\delta/2}) \end{bmatrix}. \tag{2.93}$$

The elements of the Jones matrix in eq. (2.93) satisfy the following conditions

$$T_{11} = T_{22}^*, \qquad T_{12} = -T_{21}^*, \qquad \det T = 1, \tag{2.94}$$

therefore, the Jones matrix of an elliptic retarder is *unitary*. Equation (2.93) gives the Jones matrix of *any* elliptic retarder (linear and circular retarders (optical rotators) are special cases) for *any* choice of the basis states of polarization. The Cartesian Jones matrix of the general elliptic retarder can be obtained by substituting, $\chi_{ef} = (\tan\theta + j\tan\epsilon)/(1 - j\tan\theta\tan\epsilon)$ [eq. (1.79)], into eq. (2.93), where θ and ϵ are the azimuth and ellipticity angle of the fast eigenpolarization, respectively. The circular Jones matrix is similarly obtained by substituting $\chi_{ef} = \tan(\epsilon + \tfrac{1}{4}\pi)e^{-j2\theta}$ [eq. (1.94)] into eq. (2.93). As an example, consider a linear retarder whose fast axis (the direction of the linear fast eigenpolarization) is inclined at an azimuth θ from the x and x' axes of parallel input and output Cartesian coordinate systems. In this case, $\chi_{ef} = \tan\theta$, which when substituted into eq. (2.93) gives the Cartesian Jones matrix

$$T_{car} = \begin{bmatrix} (\cos^2\theta\, e^{j\delta/2} + \sin^2\theta\, e^{-j\delta/2}) & 2j\sin\theta\cos\theta\sin\tfrac{1}{2}\delta \\ 2j\sin\theta\cos\theta\sin\tfrac{1}{2}\delta & (\sin^2\theta\, e^{j\delta/2} + \cos^2\theta\, e^{-j\delta/2}) \end{bmatrix} \tag{2.95}$$

of any linear retarder of orientation θ and retardance δ. The circular Jones matrix of the same retarder is obtained by substituting, $\chi_{\text{ef}} = e^{-j2\theta}$, into eq. (2.93)

$$T_{\text{cir}} = \begin{bmatrix} \cos\tfrac{1}{2}\delta & j\,e^{j2\theta}\sin\tfrac{1}{2}\delta \\ j\,e^{-j2\theta}\sin\tfrac{1}{2}\delta & \cos\tfrac{1}{2}\delta \end{bmatrix}. \tag{2.96}$$

2.6.1.2. Elliptic partial polarizers

An elliptic partial polarizer has orthogonal eigenpolarizations $\chi_{e\ell}$ and χ_{eh} with associated eigenvalues $V_{e\ell} = e^{\frac{1}{2}\alpha}$, $V_{eh} = e^{-\frac{1}{2}\alpha}$ that are real and positive. The subscripts ℓ and h refer to the *low-* and *high*-absorption eigenpolarizations respectively.[23] The real parameter α defines the *relative attentuation* of the partial polarizer. The low- and high-absorption eigenpolarizations $\chi_{e\ell}$ and χ_{eh} are related by the orthogonality condition [eq. (1.81)]

$$\chi_{e\ell}\chi_{eh}^* = -1, \tag{2.97a}$$

while their associated eigenvalues are related by

$$V_{e\ell}V_{eh} = 1. \tag{2.97b}$$

Substituting from eq. (2.97a) and (2.97b) into eq. (2.89) gives

$$T = (1 + \chi_{e\ell}\chi_{e\ell}^*)^{-1} \begin{bmatrix} (e^{\frac{1}{2}\alpha} + \chi_{e\ell}\chi_{e\ell}^* e^{-\frac{1}{2}\alpha}) & 2\chi_{e\ell}^* \sinh\tfrac{1}{2}\alpha \\ 2\chi_{e\ell}\sinh\tfrac{1}{2}\alpha & (\chi_{e\ell}\chi_{e\ell}^* e^{\frac{1}{2}\alpha} + e^{-\frac{1}{2}\alpha}) \end{bmatrix}, \tag{2.98}$$

which is the Jones matrix of an elliptic partial polarizer in terms of (1) the complex polarization variable $\chi_{e\ell}$ that describes its low-absorption eigenpolarization, (2) its relative attentuation α. Note that T_{11} and T_{22} are real and that $T_{21} = T_{12}^*$ so that the Jones matrix of an elliptic partial polarizer, as given by eq. (2.98), is *Hermitian*. Equation (2.98) could have also been obtained from eq. (2.93) by the simple substitution $\delta = -j\alpha$, noting that $j\sin(-j\alpha/2) = \sinh\tfrac{1}{2}\alpha$. The Cartesian and circular Jones matrices of the general elliptic partial polarizer are obtained by substituting, $\chi_{e\ell} = (\tan\theta + j\tan\epsilon)/(1 - j\tan\theta\tan\epsilon)$ [eq. (1.79)], and $\chi_{e\ell} = \tan(\epsilon + \tfrac{1}{4}\pi)\,e^{-j2\theta}$ [eq. (1.94)], into eq. (2.98), respectively, where θ and ϵ are the azimuth and ellipticity angle of the low-absorption eigenpolarization. As an example, consider the special case of a partial circular polarizer. The eigenpolarizations of such a polarizer are the left- and right-circular polarizations. Assume that the left-circular polarization experiences lower absorption than the right-circular polarization and that the right-circular-to-left-circular relative attenuation is α. The Cartesian Jones matrix of such a polarizer is

---

[23] The choice of the eigenvalues indicated here shows that the low-absorption eigenpolarization is *amplified*. This description is valid in as much as the differential absorption, rather than the total absorption, experienced by the eigenwaves is of interest.

obtained by substituting into eq. (2.98), $\chi_{e\ell} = -j$, which is the Cartesian complex number that describes the left-circular, low-absorption, eigenpolarization [eq. (1.79)]. This gives

$$T_{\text{car}} = \begin{bmatrix} \cosh \tfrac{1}{2}\alpha & j \sinh \tfrac{1}{2}\alpha \\ -j \sinh \tfrac{1}{2}\alpha & \cosh \tfrac{1}{2}\alpha \end{bmatrix}. \tag{2.99}$$

The circular Jones matrix of the same partial polarizer is similarly obtained by substituting into eq. (2.98), $\chi_{e\ell} = 0$, which is the circular complex number that describes the left-circular, low-absorption, eigenpolarization [eq. (1.94)]. This gives the diagonal matrix

$$T_{\text{cir}} = \begin{bmatrix} e^{\tfrac{1}{2}\alpha} & 0 \\ 0 & e^{-\tfrac{1}{2}\alpha} \end{bmatrix}, \tag{2.100}$$

as expected.

2.6.1.3. Elliptic ideal polarizers

An elliptic ideal polarizer is characterized by orthogonal eigenpolarizations χ_{et} and χ_{ee} whose associated eigenvalues V_{et} and V_{ee} are unity and zero, respectively. The eigenpolarization χ_{et} is *transmitted* unchanged ($V_{et} = 1$) while χ_{ee} is completely *extinguished* ($V_{ee} = 0$). Substituting these eigenpolarizations and eigenvalues in eq. (2.89), and making use of the orthogonality condition $\chi_{et}\chi_{ee}^* = -1$, we get

$$T = (1 + \chi_{et}\chi_{et}^*)^{-1} \begin{bmatrix} 1 & \chi_{et}^* \\ \chi_{et} & \chi_{et}\chi_{et}^* \end{bmatrix}, \tag{2.101}$$

which is the Jones matrix of an elliptic ideal polarizer in terms of its transmitted eigenpolarization χ_{et}. Such a Jones matrix is *singular* because det $T = 0$. Since the ideal polarizer is a limiting case of the partial polarizer, the Jones matrix of eq. (2.101) should be obtainable from eq. (2.98). This can be seen if α is assumed large enough so that terms containing $e^{-\alpha}$ in eq. (2.98) can be neglected. This gives the Jones matrix in eq. (2.101) directly except for a multiplicative factor of $e^{\tfrac{1}{2}\alpha}$ (see below). The Cartesian Jones matrix of the general elliptic ideal polarizer can be readily obtained by substituting $\chi_{et} = (\tan \theta + j \tan \epsilon)/(1 - j \tan \theta \tan \epsilon)$ [eq. (1.79)] into eq. (2.101). where θ and ϵ are the azimuth and ellipticity angle of the transmitted eigenpolarization, respectively. The elements of the resulting Jones matrix are

$$\begin{aligned} T_{11}^{\text{car}} &= (\cos^2 \theta \cos^2 \epsilon + \sin^2 \theta \sin^2 \epsilon), \\ T_{22}^{\text{car}} &= (\sin^2 \theta \cos^2 \epsilon + \cos^2 \theta \sin^2 \epsilon), \\ T_{12}^{\text{car}} &= T_{21}^{\text{car}*} = (\cos \theta \cos \epsilon + j \sin \theta \sin \epsilon)(\sin \theta \cos \epsilon + j \cos \theta \sin \epsilon). \end{aligned} \tag{2.102}$$

For a *linear* ideal polarizer, we can set $\epsilon = 0$ in eq. (2.102). Apart from a constant multiplicative factor, this leads to the Jones matrix that we derived earlier in eq. (2.32). The circular Jones matrix of the elliptic ideal polarizer can be obtained by setting, $\chi_{et} = \tan(\epsilon + \frac{1}{4}\pi) e^{-j2\theta}$ [eq. (1.94)], into eq. (2.101). This gives

$$\boldsymbol{T}_{\text{cir}} = \begin{bmatrix} \cos^2(\epsilon + \frac{1}{4}\pi) & \frac{1}{2}\cos(2\epsilon) e^{j2\theta} \\ \frac{1}{2}\cos(2\epsilon) e^{-j2\theta} & \sin^2(\epsilon + \frac{1}{4}\pi) \end{bmatrix}, \tag{2.103}$$

where θ and ϵ are the azimuth and ellipticity angle of the transmitted eigenpolarization, as before. For a left-circular ideal polarizer, $\epsilon = -\frac{1}{4}\pi$ and eq. (2.103) leads to the diagonal circular Jones matrix

$$\boldsymbol{T}_{\text{cir}} = \begin{bmatrix} 1 & 0 \\ 0 & 0 \end{bmatrix}, \tag{2.104}$$

as expected.

2.6.1.4. Other devices

Most of the useful optical devices encountered in practice belong to one or the other of the above-mentioned types. Other devices may have orthogonal eigenpolarizations that experience different amounts of retardation and attenuation *at the same time*. The eigenpolarizations could be non-orthogonal and they may become identical (the Jones matrix for this case of *degenerate* anisotropy satisfies eq. (2.60)). The Jones matrix of the general optical device of known eigenpolarizations χ_{e1} and χ_{e2} and known eigenvalues V_{e1} and V_{e2} is obtained by direct use of eq. (2.89). Dependent upon the choice of the basis states of polarization that are used to define χ_{e1} and χ_{e2}, we get the Jones matrix in any desired representation. In particular, the Cartesian and circular Jones matrices are obtained by the use of the expressions, $\chi = (\tan\theta + j\tan\epsilon)/(1 - j\tan\theta\tan\epsilon)$ [eq. (1.79)], and $\chi = \tan(\epsilon + \frac{1}{4}\pi) e^{-j2\theta}$ [eq. (1.94)], respectively, where θ and ϵ are the azimuth and ellipticity angle of the ellipse of polarization described by χ.

It is important to point out that practical devices (elliptic retarders, partial or ideal polarizers) have Jones matrices that consist of a constant multiplier times the Jones matrices that are given above for these elements. A constant multiplying a given Jones matrix leaves its eigenpolarizations unchanged [eq. (2.59)] and multiplies its eigenvalues directly [eq. (2.87)]. Such a multiplier takes the form $e^{-(\alpha_0 + j\delta_0)}$ where α_0 and δ_0 represent the isotropic-absorption and isotropic-retardation properties of the device.

A number of devices placed in succession in the path of a polarized light wave is equivalent to a single composite device whose properties can be deduced from those of the component devices. In particular, the Jones

§2.7] POLARIZATION-DEPENDENT INTENSITY TRANSMITTANCE

matrix of the composite device is obtained by multiplying the Jones matrices of the individual devices, as indicated by eq. (2.11). The opposite approach in which a complex device may be replaced by an equivalent succession of simpler devices is also useful. A number of equivalence theorems are available by which a group of devices can be replaced by another group of fewer elements and vice versa [42–46].

For ease of reference, appendix A contains a summary of the Jones matrices that we have discussed.

2.7. Polarization-dependent intensity transmittance of optical systems

Here we investigate the intensity transmittance τ of a linear non-depolarizing optical system as a function of the state of polarization of the incident light [47]. We will show that the loci of incident polarizations that experience equal attenuation or amplification $[\tau(\chi) = \text{constant}]$ are a family of non-intersecting coaxial circles in the χ plane [48]. The zero-radius point circles of the coaxial family represent two orthogonal polarizations $\chi_{\max}$ and $\chi_{\min}$ that pass through the system with maximum ($\tau_{\max}$) and minimum ($\tau_{\min}$) transmittances, respectively. We will also obtain expressions for τ in terms of the properties $\chi_{\max}$, $\chi_{\min}$, $\tau_{\max}$ and $\tau_{\min}$ of the system. We will prove that when χ is expressed in terms of the azimuth and the ellipticity of the polarization ellipse, a generalized version of Malus' law is obtained which is applicable to any optical system. The results make it possible to resolve an arbitrary polarization into two orthogonal states and to define nearness functions which characterize the closeness of the polarization states represented by two points in the complex-plane representation [§2.8].

2.7.1. *Loci of polarization states of equal attenuation or amplification*

The polarization-dependent intensity transmittance of a linear non-depolarizing optical system is given by eq. (2.86) in terms of the complex polarization variable χ_i that describes the state of polarization of the incident light as well as the elements T_{ij} of the system's Jones matrix. For simplicity, we denote the intensity transmittance by τ, and drop the subscript i from χ so that eq. (2.86) can be rewritten as

$$\tau = [|T_{12}\chi + T_{11}|^2 + |T_{22}\chi + T_{21}|^2]/[1 + |\chi|^2]. \tag{2.105}$$

The locus, in the complex χ plane, of incident polarization states that experience the same amount of attenuation or amplification upon passing through an optical system with Jones matrix (T_{ij}) can be obtained by making the following substitution

$$\chi = x + jy, \tag{2.106}$$

in eq. (2.105). The denominator and numerator of eq. (2.105) are given by

$$1 + |\chi|^2 = 1 + x^2 + y^2,$$
$$|T_{12}\chi + T_{11}|^2 + |T_{22}\chi + T_{21}|^2 = k_1(x^2 + y^2) + 2k_2 x + 2k_3 y + k_4, \tag{2.107}$$

where

$$\begin{aligned} k_1 &= |T_{12}|^2 + |T_{22}|^2, \\ k_2 &= |T_{11}||T_{12}|\cos(\theta_{11} - \theta_{12}) + |T_{21}||T_{22}|\cos(\theta_{21} - \theta_{22}), \\ k_3 &= -|T_{11}||T_{12}|\sin(\theta_{11} - \theta_{12}) - |T_{21}||T_{22}|\sin(\theta_{21} - \theta_{22}), \\ k_4 &= |T_{11}|^2 + |T_{21}|^2, \end{aligned} \tag{2.108}$$

and θ_{ij} is the angle of T_{ij}. Using eq. (2.107), eq. (2.105) can be rewritten as

$$[k_1(x^2 + y^2) + 2k_2 x + 2k_3 y + k_4] - \tau[x^2 + y^2 + 1] = 0. \tag{2.109}$$

Equation (2.109) shows that the locus of polarization states $\chi (= x + jy)$ that experience the same intensity transmittance τ after passing through an optical system is a circle in the complex plane with center at

$$x_c = \frac{k_2}{\tau - k_1}, \quad y_c = \frac{k_3}{\tau - k_1}, \tag{2.110}$$

and radius of

$$r = [k_2^2 + k_3^2 - (\tau - k_1)(\tau - k_4)]^{\frac{1}{2}}/(k_1 - \tau). \tag{2.111}$$

When τ is varied in eq. (2.109), a family of non-intersecting coaxial circles is generated, fig. 2.8. From eq. (2.110), we have

$$y_c = (k_3/k_2)x_c, \tag{2.112}$$

which shows that the line of centers passes through the origin. The equation of the common axis of the coaxial circles is obtained by setting $\tau = k_1$ in eq. (2.109),

$$2k_2 x + 2k_3 y + (k_4 - k_1) = 0, \tag{2.113}$$

which is orthogonal to the line of centers as expected. There is a limited range of values for τ that allows eq. (2.109) to represent a real circle. The range is determined by the condition that τ produces a circle of zero radius, i.e., a null circle. If we set $r = 0$ in eq. (2.111), the minimum and maximum values of τ are found as

$$\tau_{\min \atop \max} = \tfrac{1}{2}\{(k_1 + k_4) \mp [(k_1 - k_4)^2 + 4(k_2^2 + k_3^2)]^{\frac{1}{2}}\}, \tag{2.114}$$

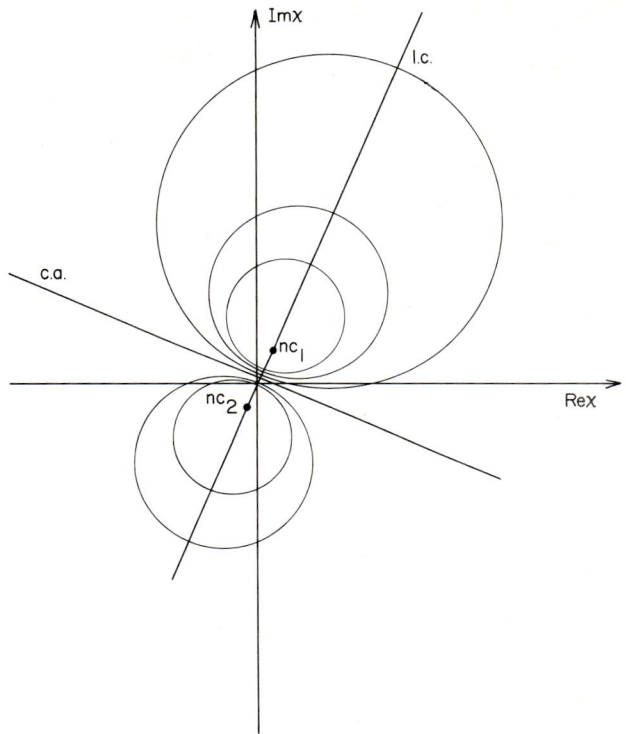

Fig. 2.8. The loci in the complex χ plane of incident polarization states that experience equal intensity transmittance $[\tau(\chi) = \text{const}]$ upon passing through an optical system constitute a family of non-intersecting coaxial circles. c.a. and l.c. represent the common axis and the line of centers, respectively; nc_1 and nc_2 represent the two null (zero-radius) circles of the family. The polarization states χ_{max} and χ_{min} that correspond to the points nc_1 and nc_2 are orthogonal and propagate through the system with maximum and minimum transmittances τ_{max} and τ_{min}, respectively.

where the $-$ and $+$ signs give the minimum and maximum values of τ, respectively. With the k's of eq. (2.114) expressed in terms of the elements T_{ij} of the Jones matrix [using eq. (2.108)], we obtain

$$\tau_{\substack{\min \\ \max}} = \frac{1}{2}\left\{\left(\sum_{ij}|T_{ij}|^2\right) \mp \left[\left(\sum_{ij}|T_{ij}|^2\right)^2 - 4|\det T|^2\right]^{\frac{1}{2}}\right\}. \tag{2.115}$$

In eq. (2.115) Σ_{ij} indicates summation over all four positions in the Jones matrix ($i, j = 1, 2$) and $\det T$ represents the determinant of that matrix. The values of τ_{min} and τ_{max} [from eq. (2.115)] are both real and positive. For each value of the system transmittance τ in the range

$$\tau_{min} \leq \tau \leq \tau_{max}, \tag{2.116}$$

eqs. (2.109) and (2.111) predict a circle locus for $\chi(x, y)$. Its radius increases from zero ($\tau = \tau_{min}$) to infinity ($\tau = k_1$) and then decreases to zero ($\tau = \tau_{max}$). Values of τ outside this range are not physically possible because they make the quantity under the square root in eq. (2.111) negative. This leads to an imaginary circle. The polarization states χ_{min} and χ_{max} that experience minimum and maximum transmittances τ_{min} and τ_{max} are represented by the centers of the two null circles. Therefore, χ_{min} and χ_{max} are obtained by substituting from eq. (2.114) into eq. (2.110)

$$\chi_{min} = (k_2 + jk_3)/(\tau_{min} - k_1), \quad \chi_{max} = (k_2 + jk_3)/(\tau_{max} - k_1). \tag{2.117}$$

From τ_{min} and τ_{max} of eq. (2.114), we find that

$$(\tau_{min} - k_1)(\tau_{max} - k_1) = -(k_2^2 + k_3^2),$$

hence, from eq. (2.117), we readily obtain

$$\chi_{min} \chi_{max}^* = -1. \tag{2.118}$$

Equation (2.118) indicates that the polarization states χ_{min} and χ_{max} that experience minimum and maximum transmittances are orthogonal.[24] In terms of the elements T_{ij} of the Jones matrix, χ_{min} and χ_{max} are given by

$$\chi_{\substack{min \\ max}} = 2(T_{11}^* T_{12} + T_{21}^* T_{22}) \left\{ \left[\sum_{ij} (-1)^{j+1} |T_{ij}|^2 \right] \right.$$
$$\left. \mp \left[\left(\sum_{ij} |T_{ij}|^2 \right)^2 - 4 |\det \mathbf{T}|^2 \right]^{\frac{1}{2}} \right\}^{-1}, \tag{2.119}$$

which is obtained by substituting from eqs. (2.108) and (2.114) into eq. (2.117).

In the above discussion, the system's intensity transmittance τ has been considered as a function, $\tau(\chi)$, of the complex polarization variable χ [eq. (2.104)]. The contours that represent polarization states χ which experience the same transmittance [$\tau(\chi) = $ constant] constitute a family of coaxial non-intersecting circles in the complex χ plane. The centers of the two null (zero-radius) circles of this family coincide with the two polarization states χ_{min} and χ_{max} that undergo minimum and maximum transmittances τ_{min} and τ_{max}, respectively. Associated with the $\tau(\chi) = $ const family of circles is another orthogonal family of coaxial circles that pass through χ_{min} and χ_{max}, fig. 2.9. These are contours of steepest change in the system transmittance $\tau(\chi)$ as χ is varied. Consequently, if incident light of constant intensity is polarized in an arbitrary state χ, the intensity of the output light varies most

---

[24] A different proof is given by Jones [11].

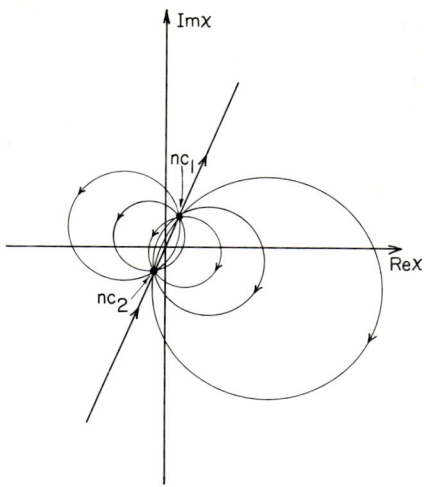

Fig. 2.9. Paths in the polarization χ space of steepest variation in the system transmittance $\tau(\chi)$. These trajectories are orthogonal to the family of circles of fig. 2.8. Along an arc of a circle from nc_1 (χ_{max}) to nc_2 (χ_{min}) the system transmittance $\tau(\chi)$ falls monotonically and most rapidly from τ_{max} to τ_{min}.

rapidly if the polarization state χ follows the arc of the circle joining χ to χ_{max} and χ_{min}. (The arrows on the arcs of circles from χ_{max} to χ_{min} indicate the direction of the negative gradient of $\tau(\chi)$). Also, the system transmittance $\tau(\chi)$ falls monotonically and most rapidly from τ_{max} to τ_{min} as χ traverses the arc from χ_{max} to χ_{min}. The arrows on all the arcs that emanate from χ_{max} point outwards because this is the point at which $\tau(\chi)$ is a maximum. Similarly, the arrows converge to χ_{min} because it is the state of minimum transmittance. The function $\tau(\chi)$ is circularly symmetric in the neighborhood of χ_{max} and χ_{min} because small circles of constant τ, around these two points, are centered on them.

2.7.2. Simplified expressions for the intensity transmittance

We can express the polarization-dependent intensity transmittance $\tau(\chi)$ in terms of the incident polarization χ and the system's χ_{max}, χ_{min}; τ_{max}, and τ_{min}. Recall that the system is most transparent for χ_{max} and most opaque for χ_{min} and offers intensity transmittance of τ_{max} and τ_{min} for these two polarizations, respectively.

The equation of a family of coaxial circles can be written in the form

$$\mathscr{C}_1 + \mu \mathscr{C}_2 = 0, \tag{2.120}$$

where $\mathscr{C}_1 = 0$ and $\mathscr{C}_2 = 0$ represent any two member circles of the family and μ is a parameter. By allowing μ to vary from $-\infty$ to $+\infty$ all circles of the family, both real and imaginary, are generated. Equation (2.120) can be written in an infinite number of ways depending upon the two chosen circles $\mathscr{C}_1 = 0$ and $\mathscr{C}_2 = 0$. Equation (2.109) is of the same form as eq. (2.120) and represents the coaxial family of circles of equal intensity transmittance in the complex χ plane. The parameter $\tau(=-\mu)$ in eq. (2.109) gives the amount of intensity transmittance and, for real circles, is restricted to the range given by eqs. (2.115) and (2.116). Another case of particular interest results when $\mathscr{C}_1 = 0$ and $\mathscr{C}_2 = 0$ represent the null circles. Then

$$[(x - x_{cmax})^2 + (y - y_{cmax})^2] - p^2[(x - x_{cmin})^2 + (y - y_{cmin})^2] = 0, \qquad (2.121)$$

or

$$\frac{|\chi - \chi_{max}|^2}{|\chi - \chi_{min}|^2} = p^2. \qquad (2.122)$$

Equations (2.121) and (2.122) show that if the intensity transmittance $\tau(\chi)$ is to remain constant, the ratio p of the distances from χ to χ_{max} and χ_{min} must remain constant. All real members of the family are generated by allowing the distance ratio p to vary from zero ($\chi = \chi_{max}$) to infinity ($\chi = \chi_{min}$).

If we normalize eqs. (2.109) and (2.121) so that the coefficient of $(x^2 + y^2)$ is unity, and equate the constant terms in the resulting two equations, we obtain

$$\frac{|\chi_{max}|^2 - p^2|\chi_{min}|^2}{1 - p^2} = \frac{\tau - k_4}{\tau - k_1}. \qquad (2.123)$$

The ratio q by which the origin internally divides the distance between χ_{max} and χ_{min} is

$$q^2 = |\chi_{max}|^2/|\chi_{min}|^2. \qquad (2.124)$$

From the orthogonality of χ_{max} and χ_{min}, eq. (2.118), it follows that

$$|\chi_{max}|^2 = q, \qquad |\chi_{min}|^2 = q^{-1}. \qquad (2.125)$$

Substituting eq. (2.125) into eq. (2.123), and solving for τ we get

$$\tau = \left[\frac{(q^2 - p^2)}{(q^2 - p^2) - q(1 - p^2)}\right] k_1 + \left[\frac{-q(1 - p^2)}{(q^2 - p^2) - q(1 - p^2)}\right] k_4. \qquad (2.126)$$

Because

$$\tau = \tau_{max}, \qquad \text{when } p = 0,$$
$$\tau = \tau_{min}, \qquad \text{when } p = \infty, \qquad (2.127)$$

we get from eq. (2.126) two linear equations that can be solved for k_1 and k_4 to give

$$k_1 = (q\tau_{max} + \tau_{min}), \qquad k_4 = (\tau_{max} + q\tau_{min}). \tag{2.128}$$

Finally, if k_1 and k_4 in eq. (2.126) are replaced by their values from eq. (2.128), we obtain

$$\tau = \frac{q}{q+p^2}\tau_{max} + \frac{p^2}{q+p^2}\tau_{min}. \tag{2.129}$$

Equation (1.129) gives a general expression for the intensity transmittance τ of an optical system as a function of the complex variable χ that describes the incident polarization. The properties of the system are given by the two incident orthogonal polarization states χ_{min} and χ_{max} that experience minimum and maximum transmittances in passing through the system and the extrema of τ, τ_{min} and τ_{max}. When χ, χ_{min} and χ_{max} are represented by three points in the polarization complex plane, fig. 2.10, the two distance ratios p and q that appear in eq. (2.129) are given by

$$p = |\chi - \chi_{max}|/|\chi - \chi_{min}|, \qquad q = |\chi_{max}|/|\chi_{min}|. \tag{2.130}$$

Equation (2.129) can be rewritten in a slightly different form as

$$\tau = \tau_{min} + \frac{q}{q+p^2}(\tau_{max} - \tau_{min}), \tag{2.131}$$

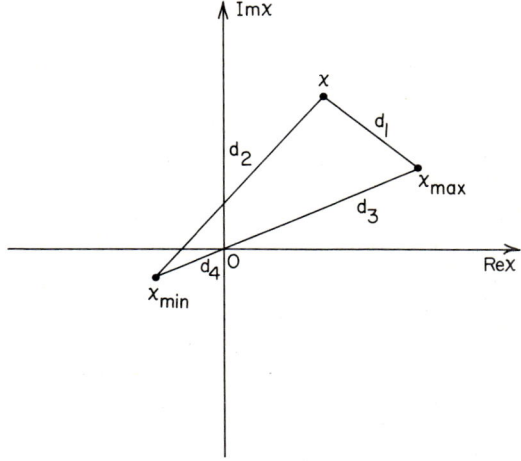

Fig. 2.10. The triangle of points χ, χ_{max}, and χ_{min}. The distance ratios p and q are given by $p = d_1/d_2$, $q = d_3/d_4$.

which has an advantage over eq. (2.129), because the dependence on the incident polarization is now contained in a single term, namely, the coefficient of $(\tau_{max} - \tau_{min})$.

2.7.3. Generalization of the Law of Malus

We have used the single complex variable χ to uniquely define the incident ellipse of polarization. We will now obtain an explicit expression for the intensity transmittance τ in terms of the azimuth θ and ellipticity $\tan \epsilon$ of the polarization ellipse at the input of the system.

While χ as a function of θ and ϵ depends on the choice of the basis polarization states, eq. (2.105) and all subsequent discussion are valid for any choice of basis states. [The coefficient $(q/q + p^2)$ of $(\tau_{max} - \tau_{min})$ in eq. (2.131) remains *invariant* under a change of the basis states.] Therefore, we can choose the basis states to make the dependence of χ on θ and ϵ simplest, namely, the right- and left-circular polarizations. Then χ is given by eq. (1.94). Substituting

$$\chi = \tan(\epsilon + \tfrac{1}{4}\pi)\, e^{-j2\theta}, \qquad \chi_{max} = \tan(\epsilon_{max} + \tfrac{1}{4}\pi)\exp[-j2\theta_{max}],$$
$$\chi_{min} = \tan(-\epsilon_{max} + \tfrac{1}{4}\pi)\exp[-j2(\theta_{max} - \tfrac{1}{2}\pi)], \tag{2.132}$$

into eq. (2.130) gives

$$p^2 = \frac{g^2 + g_{max}^2 - 2gg_{max}\cos 2(\theta - \theta_{max})}{g^2 + g_{max}^{-2} + 2gg_{max}^{-1}\cos 2(\theta - \theta_{max})},$$
$$q = g_{max}^2. \tag{2.133}$$

where

$$g = \tan(\epsilon + \tfrac{1}{4}\pi), \qquad g_{max} = \tan(\epsilon_{max} + \tfrac{1}{4}\pi). \tag{2.134}$$

Using the values of p^2 and q from eq. (2.133) in eq. (2.131), we obtain

$$\tau = \tau_{min} + \left[\frac{1 + g^2 g_{max}^2 + 2gg_{max}\cos 2(\theta - \theta_{max})}{(1 + g^2)(1 + g_{max}^2)}\right](\tau_{max} - \tau_{min}). \tag{2.135}$$

Equation (2.135) is the *generalized* law of Malus. It gives the intensity transmittance τ of an optical system as a function of: (1) the parameters θ, g of the incident ellipse of polarization, (2) the parameters θ_{max}, g_{max} of the incident ellipse of polarization that propagates through the optical system with maximum transmittance, and (3) the values of the maximum (τ_{max}) and minimum (τ_{min}) transmittances of the system. The parameter g of the ellipse of polarization is related to the ellipticity, $e = \tan \epsilon$, by

$$g = \frac{1+e}{1-e}, \tag{2.136}$$

as can be seen by expanding the tangent of the sum of two angles in the first of eq. (2.134).

To obtain the familiar cosine-squared law of Malus [49] for an *ideal linear polarizer*, let θ_{max} be the azimuth of the transmission axis of the polarizer measured from a given reference direction and τ_{max} be the maximum transmittance for light linearly polarized parallel to the transmission axis. Because the polarizer is ideal, $\tau_{min} = 0$; and because it is linear, $e_{max} = 0$, $g_{max} = 1$. If linearly polarized light ($e = 0, g = 1$) is incident on this polarizer at an azimuth θ measured from the same reference direction, the intensity transmittance is obtained by substituting $\tau_{min} = 0$ and $g = g_{max} = 1$ into eq. (2.135); this yields

$$\tau = \tau_{max} \cos^2(\theta - \theta_{max}). \tag{2.137}$$

2.7.4. Special cases of optical systems

The optical system considered so far has been described as linear and non-depolarizing but, otherwise, is quite general. Here we consider a number of special cases.

Active systems: The above analysis can be applied to linear amplifying (active) optical systems as well as linear attenuating (passive) optical systems. For the Jones matrix to represent an active system, we must have

$$\tau_{max} > 1,$$

or

$$\frac{1}{2}\left\{\left(\sum_{ij}|T_{ij}|^2\right) + \left[\left(\sum_{ij}|T_{ij}|^2\right)^2 - 4|\det \boldsymbol{T}|^2\right]^{\frac{1}{2}}\right\} > 1. \tag{2.138}$$

If τ_{min} is also greater than unity, the system will amplify all incident polarization states. If $\tau_{max} > 1$ and $\tau_{min} < 1$, the complex plane of polarization will be divided into two domains of polarization states: in one domain the incident light is attenuated and in the other it is amplified. The boundary separating the two domains is a circle that represents incident polarization states whose intensity is unaffected by the system, fig. 2.11. The laser is an example of an amplifying optical system.

Isotropic absorption or amplification: A system is isotropically absorbing or amplifying if its effect on the intensity of an incident wave is independent of the polarization of that wave. For such a system

$$\tau_{min} = \tau_{max}, \tag{2.139}$$

and the transmittance τ in eq. (2.129) is constant,

$$\tau = \tau_{max} = \tau_{min}. \tag{2.140}$$

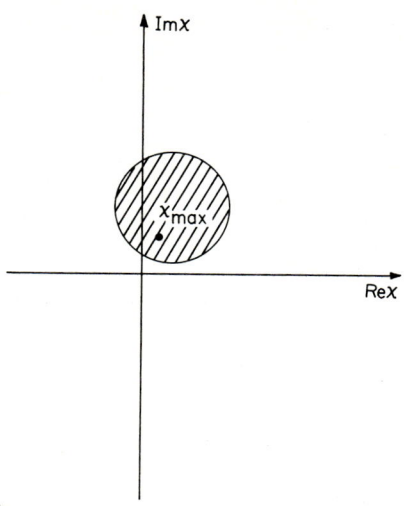

Fig. 2.11. An active optical system with $\tau_{max} > 1$ and $\tau_{min} < 1$ amplifies incident polarization states inside a circular domain in the complex plane of polarization enclosing χ_{max}. The circle boundary represents polarization states whose intensity is unaffected by the system $[\tau(\chi) = 1]$. The complementary domain outside the circle represents polarization states that are attenuated by the system $[\tau(\chi) < 1]$.

Equation (2.115) shows that for $\tau_{max} = \tau_{min}$, the Jones matrix must satisfy the condition that

$$\left(\sum_{ij} |T_{ij}|^2\right)^2 = 4|\det T|^2. \tag{2.141}$$

For an anisotropic medium the condition of eq. (2.141) is satisfied in the absence of any type of dichroism (linear, circular, or elliptic). The medium may, however, be linearly, circularly, or elliptically birefringent.

Perfect polarizers: An optical system will act as a perfect (ideal) polarizer if

$$\tau_{min} = 0. \tag{2.142}$$

From eq. (2.115), we see that $\tau_{min} = 0$ when

$$\det T = 0, \tag{2.143}$$

i.e., when the Jones matrix is singular. The transmittance of such system is obtained by setting $\tau_{min} = 0$ in eq. (2.135); this gives

$$\tau = \frac{1 + g^2 g_{max}^2 + 2gg_{max} \cos 2(\theta - \theta_{max})}{(1 + g^2)(1 + g_{max}^2)} \tau_{max}, \tag{2.144}$$

2.8. Resolution of an arbitrary polarization state into two given orthogonal states in the complex-plane representation[25]

which is the most general expression of the transmittance of an arbitrary elliptic polarizer for *any* incident polarization.

Equation (1.129) can be written as

$$\tau = \eta_{max}\tau_{max} + \eta_{min}\tau_{min},$$
$$\eta_{max} = q/(q + p^2),$$
$$\eta_{min} = p^2/(q + p^2), \quad (2.145)$$
$$\eta_{max} + \eta_{min} = 1.$$

Equations (2.145) suggest that η_{max} and η_{min} are the fractions of intensity of the incident state χ contained in the two orthogonal states χ_{max} and χ_{min}, respectively. Also, eqs. (2.145) show that after the incident state χ is resolved into two components of polarizations χ_{max} and χ_{min}, these components are propagated with transmittances τ_{max} and τ_{min}, respectively. This resolution can be correctly thought of as independent of whether there is an optical system or not and should hold for any two orthogonal states χ_1 and χ_2 [47b]. Thus, an arbitrary state χ can be resolved into two components of polarizations χ_1 and χ_2 whose intensities are fractions η_1 and η_2 of the intensity of χ given by

$$\eta_1 = q/(q + p^2), \qquad \eta_2 = p^2/(q + p^2), \quad (2.146)$$

where

$$\eta_1 + \eta_2 = 1,$$
$$p = |\chi - \chi_1|/|\chi - \chi_2|, \qquad q = |\chi_1|/|\chi_2|, \qquad \chi_1\chi_2^* = -1. \quad (2.147)$$

2.8.1. Nearness of two polarization states in the complex-plane representation

In the complex-plane representation, the length of the straight-line segment joining two points (the metric in the polarization space) is not a true measure of the nearness of the two polarization states represented by the two points. This is in contrast with the Poincaré-sphere representation, where the length of the smaller arc of the great circle joining two points is a correct expression of nearness of the corresponding polarization states. From the

---
[25] The reader may wish to skip this section on first reading.

above discussion, we propose that the fraction η_1 of the intensity of χ in the state χ_1 be used as a measure of the nearness of the two polarization states χ and χ_1. Denoting the nearness of χ and χ_1 by $\mathcal{N}(\chi, \chi_1)$, we have

$$\mathcal{N}(\chi, \chi_1) = \eta_1 = q/(q + p^2), \tag{2.148}$$

which upon substitution from eq. (2.147) gives

$$\mathcal{N}(\chi, \chi_1) = \frac{\chi\chi^*\chi_1\chi_1^* + \chi\chi_1^* + \chi^*\chi_1 + 1}{\chi\chi^*\chi_1\chi_1^* + \chi\chi^* + \chi_1\chi_1^* + 1}. \tag{2.149}$$

The nearness function $\mathcal{N}(\chi, \chi_1)$ has the following properties

$$\mathcal{N}(\chi, \chi_1) = \mathcal{N}(\chi_1, \chi); \tag{2.150a}$$

$$\mathcal{N}(\chi, \chi) = 1, \quad \mathcal{N}\left(\chi, \frac{-1}{\chi^*}\right) = 0; \tag{2.150b}$$

$$\mathcal{N}\left(\chi, \frac{-1}{\chi_1^*}\right) = 1 - \mathcal{N}(\chi, \chi_1); \tag{2.150c}$$

$$\mathcal{N}(\chi^*, \chi_1^*) = \mathcal{N}(\chi, \chi_1); \tag{2.150d}$$

$$\mathcal{N}(\chi\, e^{j\theta}, \chi_1\, e^{j\theta}) = \mathcal{N}(\chi, \chi_1); \tag{2.150e}$$

which follow directly from eq. (2.149). The significance of the properties given by eq. (2.150), in the same order, is

(a) The nearness of χ to χ_1 is the same as the nearness of χ_1 to χ.

(b) A state χ is most near to itself ($\mathcal{N} = 1$) and most far from the orthogonal state ($\mathcal{N} = 0$). The nearness function $\mathcal{N}(\chi, \chi_1)$ is limited to the range $0 \le \mathcal{N} \le 1$ for any χ and χ_1.

(c) A state χ is as near to another state χ_1 as it is far from the state orthogonal to χ_1. A "farness" function may be defined as $1 - \mathcal{N}(\chi, \chi_1)$.

(d) The nearness of χ and χ_1 is the same as the nearness of χ^* and χ_1^* which are obtained from χ and χ_1 by complex conjugation or mirror reflection in the real axis.

(e) The nearness of χ and χ_1 is the same as the nearness of $\chi\, e^{j\theta}$ and $\chi_1\, e^{j\theta}$ which are obtained from χ and χ_1 by the same rotation θ around the origin in the complex plane. When $\theta = \pi$, the rotation becomes equivalent to inversion with respect to the origin. Combining reflection in the real axis with inversion with respect to the origin, we find that the nearness of χ and χ_1 is the same as the nearness of $-\chi^*$ and $-\chi_1^*$ which are obtained from χ and χ_1 by mirror reflection in the imaginary axis. Starting from any two states χ and χ_1, fig. 2.12 shows how the application of different symmetry operations produces other pairs of states whose nearness is the same as that of χ and χ_1.

When the complex-plane representation is related to the Poincaré sphere via a stereographic projection (§2.5), the operations that lead to the

§2.8] RESOLUTION OF AN ARBITRARY POLARIZATION STATE 115

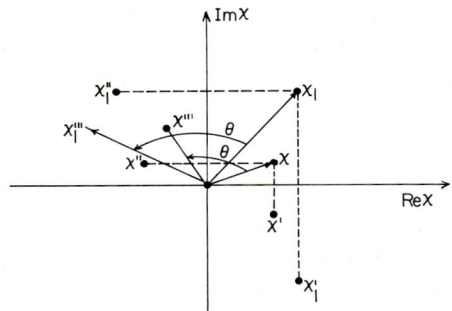

Fig. 2.12. Symmetry operations that preserve the nearness of two polarization states χ and χ_1 include reflection in the real and imaginary axes and any rotation θ.

invariance of $\mathcal{N}(\chi, \chi_1)$ can be seen to correspond to operations that preserve the distance between two points on the sphere.

The choice of the orthogonal basis polarization states for the complex-plane representation has been arbitrary so that the nearness of two polarization states χ and χ_1 given by eq. (2.149) is independent of the basis states used to define χ and χ_1. Formally, eq. (2.149) remains invariant under any bilinear transformation that corresponds to a change of basis from one pair of orthogonal states to another.

The form of the nearness function $\mathcal{N}(\chi, \chi_1)$ given by eq. (2.148) indicates that when χ_1 (hence q) is fixed, and χ moves so that it stays equally near to χ_1, p remains constant and χ describes a circle. For different values of the nearness function $\mathcal{N}(\chi, \chi_1)$, a family of coaxial non-intersecting circles is generated whose null circles coincide with χ_1 and $-1/\chi_1^*$ (the state orthogonal to χ_1). On the Poincaré sphere this family of circles becomes a coaxial family whose polar (common) axis is the diameter of the sphere joining the points that correspond to χ_1 and $-1/\chi_1^*$. This is not unexpected and, as will be shown below, could be used to obtain the law of resolution of a given polarization into two orthogonal states in the Poincaré-sphere representation.

A modified nearness function $\mathcal{N}'(\chi, \chi_1)$ can be introduced which is given by the natural logarithm of the reciprocal of $\mathcal{N}(\chi, \chi_1)$, i.e.

$$\mathcal{N}'(\chi, \chi_1) = \log \frac{1}{\mathcal{N}(\chi, \chi_1)}. \tag{2.151}$$

This modified nearness function satisfies the conditions that

$$\mathcal{N}'(\chi, \chi) = 0, \quad \mathcal{N}'\left(\chi, \frac{-1}{\chi^*}\right) = \infty, \tag{2.152}$$

i.e., its value is an indication of the separation of χ and χ_1. The separation is zero when these states are identical and infinity when they are orthogonal.

Another function that gives the same nearness of two polarization states whether they are represented in the complex plane or on the Poincaré sphere is

$$\mathcal{N}''(\chi, \chi_1) = \tfrac{1}{4} \cos^{-1} [\mathcal{N}(\chi, \chi_1)^{\frac{1}{2}}]. \tag{2.153}$$

This is the metric on the Poincaré sphere (of unit diameter), for two polarizations represented in the complex plane by χ and χ_1 [see below]. Equation (2.153) represents an important relation between the complex-plane and the Poincaré-sphere representations.

2.8.2. *Resolution of an arbitrary polarization state into two given orthogonal states in the Poincaré-sphere representation*

There is no loss of generality if we take the two given orthogonal polarizations (denoted by χ_1 and χ_2) as basis states for the complex-plane representation. The representative points for the states χ_1 and χ_2 now coincide with the origin and the point at infinity of the complex plane, respectively. The nearness of an arbitrary state χ in the complex plane to χ_1 (which is now represented by the origin) can be obtained by substituting $\chi_1 = 0$ in eq. (2.149)

$$\mathcal{N}(\chi, 0) = \frac{1}{1 + |\chi|^2}. \tag{2.154}$$

States that are equally near to χ_1 are seen from eq. (2.154) to lie on a circle centered at the origin. Remember that $\mathcal{N}(\chi, 0)$ represents the fraction of the intensity of the state χ in the state $\chi_1 = 0$.

Consider a sphere of unit diameter that rests tangent to the complex plane at its origin which is the point χ_1. Each point χ in the plane can be joined to the pole of the sphere which lies diametrically opposite to the origin by a straight line that intersects the sphere in one and only one point denoted by χ_s. (χ_s is used only to designate a point on the sphere and does not represent any complex number.) The origin (χ_1) of the complex plane is projected in this way onto itself (χ_{1s}) and the point at infinity in the plane χ_2 goes to the center of the projection (χ_{2s}). Figure 2.13 shows the great circle G that results from the intersection of the plane of χ, χ_{1s} and χ_{2s} with the Poincaré sphere. If θ is the angle subtended at the center of the sphere by the smaller arc $\chi_{1s}\chi_s$ of the great circle G, it is immediately seen from fig. 2.13 that

$$|\chi| = \tan \tfrac{1}{2}\theta. \tag{2.155}$$

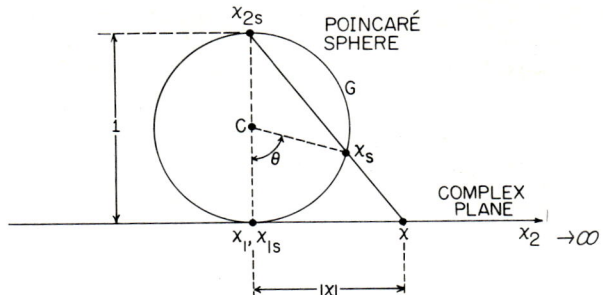

Fig. 2.13. Graphical construction for the derivation of the expression for the fraction of the intensity of one polarization state into another state on the Poincaré sphere.

Substituting from eq. (2.155) into eq. (2.154) it is seen that

$$\mathcal{N}(\chi, 0) = \mathcal{N}(\chi_s, \chi_{1s}) = \cos^2 \tfrac{1}{2}\theta. \tag{2.156}$$

Thus the fraction of the intensity of a polarization state χ_s on the sphere in another state χ_{1s} also on the sphere is given by the squared cosine of half the angle subtended at the center of the sphere by the smaller of the two arcs of the great circle through χ_s and χ_{1s} [9]. Obviously, the proportion of the intensity of χ_s in the orthogonal state χ_{2s} is given by $\cos^2(\pi - \theta/2)$ or $\sin^2 \tfrac{1}{2}\theta$.

2.9. Polarization-dependent phase shift introduced by an optical system

So far we have dealt with the transformation of the ellipse of polarization ($\chi_i \rightarrow \chi_o$) and the transformation of the intensity ($I_i \rightarrow I_o$) of a totally polarized wave upon passing through a linear non-depolarizing optical system. To complete our discussion, we need to consider the transformation of the absolute phase of the wave. It is unlikely that this absolute-phase information would be of interest if a single light beam is involved. However in the presence of several beams, which could be made to overlap or interfere, such phase information becomes essential.

According to eqs. (2.83) and (2.85) the absolute phase shift experienced by an incident light wave polarized in the state χ_i upon passing through the optical system can be written as

$$f_\delta(\chi_i) = \arg(T_{12}) + \arg(\chi_i - \chi_a), \tag{2.157}$$

where

$$\chi_a = -T_{11}/T_{12}. \tag{2.158}$$

Equation (2.157) has two terms. The first term represents an isotropic, polarization-independent, phase shift. The second term indicates a component of the phase shift that depends on the incident polarization χ_i. This dependence is rather simple. Polarization states χ_i that experience the same phase shift are seen from eq. (2.157) to lie on a straight line in the complex plane through the point χ_a. By varying the amount of phase shift a family of straight lines through χ_a is generated, fig. 2.14. Associated with this family of

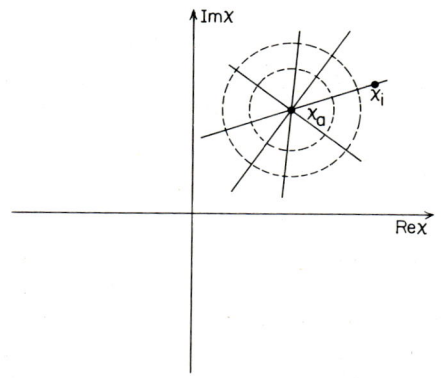

Fig. 2.14. The loci of polarization states that experience constant phase shift upon passing through an optical system are shown in the complex plane as a family of straight lines passing through a common point χ_a. The concentric circles around χ_a represent paths of steepest change of phase.

straight lines through χ_a is an orthogonal family of concentric circles centered on that point. These are shown dashed in fig. 2.14 and represent the paths in the complex plane of steepest change of phase. (In other words, they represent the gradient of the scalar phase $f_\delta(\chi)$ in the polarization space of the complex plane.)

According to eq. (2.157) the phase shift is indeterminate when $\chi_i = \chi_a$ or $\chi_i = \infty$. Besides, the phase shift varies rapidly as χ_i is moved along an arbitrarily small circle centered on χ_a, (or around a large circle centered on the point at infinity). Also, there is an abrupt change in phase of π as the point χ_a (or the point at infinity) is crossed along a curve through such a point. These seemingly non-physical conditions can be resolved by examining the meaning of the absolute phase shift. First note that χ_i and χ_o contain relative phase information. Specifically, eq. (2.71) tells us that $\arg(\chi_i)$ gives the relative phase between the two component polarizations of the incident light along the basis states. $\arg(\chi_o)$ gives the relative phase at the output of the system. The change in this relative phase introduced by the system is given by

$$\Delta \text{ (relative phase)} = \arg(\chi_o) - \arg(\chi_i)$$
$$= \arg(\chi_o/\chi_i). \tag{2.159}$$

All that remains to determine the resultant phase shift is to examine the absolute phase shift suffered by the first basis polarization state. This is indeed the phase shift given by eqs. (2.85) and (2.157), because

$$(E_{ou}/E_{iu}) = T_{12}\chi_i + T_{11}, \tag{2.160}$$

which follows from eqs. (2.75) and (2.71). In other words, the absolute phase shift is determined by comparing the phases of the components of the incident and outgoing light whose polarization is that of the first basis state. (The basis states are assumed the same for both the input and output of the optical system.) Because intensity in the second basis state is convertible to intensity in the first basis state at the output of the system (due to the non-zero values of T_{12} and T_{21}), the phase shift suffered by the first basis state is dependent on χ_i. When $\chi_i = \chi_a$, the output light is polarized in the state $\chi_o = \infty$ [eq. (2.50)]. Therefore, when

$$\chi_i = \infty, \quad \text{or} \quad \chi_o = \infty, \tag{2.161}$$

the component of the incident or the outgoing light, respectively, whose polarization is that of the first basis state is zero. Consequently, the phase jump suffered by that component upon passing through the system is indeterminate.

2.10. Propagation of polarized light in anisotropic media

We mentioned in §2.2 that the basic distinction between transmission and reflection optical devices lies in the fact that the wave is modified *continuously* as it progresses through the medium of a transmission-type device, whereas it is changed *abruptly* as the wave bounces off the surface of a reflection-type device. The over-all change in the state of polarization of a light wave upon passing through a transmission-type device is described by its "terminal" Jones matrix T. The resultant change in polarization is the cumulative effect of incremental changes that take place as light travels through successive incremental distances in the medium. In this section, we will study the continuous evolution of polarized light as it propagates through an optically anisotropic medium. We will not resort to the electromagnetic wave-equation but rather we employ an operational approach similar to that of Jones [50]. As presented by Jones, the N matrices of the operational approach are fundamentally related to the dielectric and gyrotropic properties of the medium via the electromagnetic theory [51]. A more general approach will be examined in §4.7.

Consider a plane wave of polarized light travelling in an anisotropic medium along the z axis of an xyz Cartesian right-handed orthogonal coordinate system. The properties of the medium are assumed to be uniform over any transverse plane perpendicular to the direction of propagation of the beam. Along the beam direction (the z axis) the properties of the medium may be variable. The two-component Jones vector that describes the electric field of the wave is a function, $E(z)$, of position, z, along the direction of propagation. As shown in fig. 2.15, $E(z)$ and $E(z + \Delta z)$ are the Jones vectors

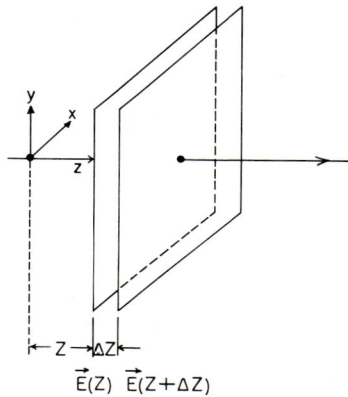

Fig. 2.15. Light propagating through an anisotropic medium along the z axis of an xyz coordinate system is described by Jones vectors $E(z)$ and $E(z + \Delta z)$ at two transverse planes separated by an incremental distance Δz.

of the wave at two transverse planes that are located at z and $z + \Delta z$ along the direction of propagation. The modification of the wave as a result of its propagation through the incremental path length Δz, between z and $z + \Delta z$, is described by

$$E(z + \Delta z) = M(z, \Delta z)E(z), \tag{2.162}$$

where $M(z + \Delta z)$ is the Jones matrix of the thin (infinitesimal) slab of the anisotropic medium of thickness Δz located at the coordinate z with both of its two plane boundaries parallel to the xy plane. Subtracting $E(z)$ from both sides of eq. (2.162) we obtain the change in the Jones vector between z and $z + \Delta z$

$$E(z + \Delta z) - E(z) = [M(z, \Delta z) - \mathscr{I}]E(z), \tag{2.163}$$

where $\mathcal{I}$ is the 2×2 identity matrix. Dividing both sides of eq. (2.163) by Δz and taking the limit as Δz approaches zero we obtain the differential equation of propagation for the Jones vector $\boldsymbol{E}$

$$d\boldsymbol{E}(z)/dz = \boldsymbol{N}(z)\boldsymbol{E}(z), \tag{2.164}$$

where

$$\boldsymbol{N}(z) = \lim_{\Delta z \to 0} \frac{\boldsymbol{M}(z, \Delta z) - \mathcal{I}}{\Delta z}. \tag{2.165}$$

The 2×2 matrix $\boldsymbol{N}(z)$, defined by eq. (2.165), is the *differential propagation Jones matrix* that characterizes the anisotropic optical properties of the medium over the transverse plane at z.

Consider an optical device in the form of a thick (non-infinitesimal) slab of an anisotropic medium. It is of interest to determine the relation between the *external*, over-all, Jones matrix $\boldsymbol{T}$ of such a device and the differential propagation Jones matrix $\boldsymbol{N}$ that describes its *internal* properties. The Jones vector $\boldsymbol{E}(z)$ of a light wave that has travelled a total distance z, starting from an *initial* Jones vector $\boldsymbol{E}(0)$ at $z = 0$, is given by

$$\boldsymbol{E}(z) = \boldsymbol{T}(z)\boldsymbol{E}(0), \tag{2.166}$$

where $\boldsymbol{T}(z)$ is the over-all Jones matrix that characterizes the section of the medium of thickness z between the two transverse planes at 0 and z. Differentiating eq. (2.166) with respect to z, we get

$$d\boldsymbol{E}(z)/dz = [d\boldsymbol{T}(z)/dz]\boldsymbol{E}(0). \tag{2.167}$$

Equations (2.164) and (2.167) provide two alternative expressions for the derivative of the Jones vector $\boldsymbol{E}(z)$ with respect to z. Because these alternative expressions must be equivalent, we have

$$[d\boldsymbol{T}(z)/dz]\boldsymbol{E}(0) = \boldsymbol{N}(z)\boldsymbol{E}(z). \tag{2.168}$$

Replacing $\boldsymbol{E}(z)$ in eq. (2.168) by its value given by eq. (2.166) and dropping $\boldsymbol{E}(0)$ we obtain

$$\frac{d\boldsymbol{T}(z)}{dz} = \boldsymbol{N}(z)\boldsymbol{T}(z). \tag{2.169}$$

Post-multiplying both sides of eq. (2.169) by $\boldsymbol{T}^{-1}(z)$, we find that

$$\boldsymbol{N}(z) = \frac{d\boldsymbol{T}(z)}{dz}\boldsymbol{T}^{-1}(z), \tag{2.170}$$

which is the desired relation between the over-all Jones matrix $\boldsymbol{T}(z)$ of an optical device in the form of a slab of anisotropic material of thickness z and the differential propagation Jones matrix $N(z)$ that describes the optical properties of the device material.

In discussing the polarization-mapping properties between the input and output of an optical system, we have seen in §2.3 and §2.6 that it is fruitful to separate the effect of the system on the ellipse of polarization proper from that on the amplitude and absolute phase of the light wave. In a similar manner, a useful approach to study the propagation of polarized light in anisotropic media would be to examine the evolution of the ellipse of polarization and that of the complex amplitude of the wave *separately*, as the light progresses through the anisotropic medium. In the following, we prove that this separability is possible and we derive the equations that govern the evolution of the ellipse of polarization and the complex amplitude of the light wave.

2.10.1. *Evolution of the ellipse of polarization*

Expanding eq. (2.164) gives two *coupled* (simultaneous) first-order ordinary differential equations

$$\frac{dE_u}{dz} = n_{11}E_u + n_{12}E_v,$$
$$\frac{dE_v}{dz} = n_{21}E_u + n_{22}E_v, \qquad (2.171)$$

where E_u and E_v are the components of the generalized Jones vector E along a pair of basis states of polarization u and v [eq. (1.64)] and n_{ij} are the elements of the Jones matrix N

$$N = \begin{bmatrix} n_{11} & n_{12} \\ n_{21} & n_{22} \end{bmatrix}. \qquad (2.172)$$

For simplicity, we have dropped the z argument of the z-dependent functions E_u, E_v and n_{ij}. The ellipse of polarization of the light beam is determined by the single complex variable χ

$$\chi = E_v/E_u. \qquad (1.97)$$

For each choice of the basis states of polarization (u, v) there is an associated complex-plane representation of polarization. These representations are interrelated as explained in §1.7 [see eq. (1.101)]. To determine the differential equation for χ, we take the first derivative of both sides of eq. (1.97)

$$\frac{d\chi}{dz} = \frac{1}{E_u}\frac{dE_v}{dz} - \frac{E_v}{E_u^2}\frac{dE_u}{dz}. \qquad (2.173)$$

Upon substitution from eqs. (2.171) and (1.97) into eq. (2.173) we get

$$\frac{d\chi}{dz} = -n_{12}\chi^2 + (n_{22} - n_{11})\chi + n_{21}. \tag{2.174}$$

Equation (2.174) represents the significant conclusion that the evolution of the ellipse of polarization *can be examined separately* [52]. It is a first-order non-linear ordinary differential equation whose solution, $\chi(z, \chi_0)$, gives the evolution of the ellipse of polarization of light as it propagates through an anisotropic medium, starting from the initial polarization state χ_0 at $z = 0$. For a given direction of propagation, the properties of the anisotropic medium are represented by the differential propagation Jones matrix (n_{ij}). This matrix can be function of distance, z, measured along the direction of propagation in which case eq. (2.174) becomes the general Riccati equation [53].

The evolution of each of the two parameters of the ellipse of polarization, the azimuth θ and ellipticity e, can be easily obtained from the function $\chi(z, \chi_0)$. It should be emphasized, however, that the relationship that links θ and e to the complex polarization variable χ depends on the basis polarization states that are used for the definition of χ in eq. (1.97). In most cases the chosen basis states (u, v) are either two orthogonal linear polarizations (x, y) or the left- and right-circular polarizations (ℓ, r). When the x and y orthogonal linear polarizations are used as basis states, θ and e are obtained from χ by the following relation (§1.7)

$$\tan 2\theta = \frac{2 \operatorname{Re}(\chi)}{1 - |\chi|^2}, \tag{1.86}$$

$$\sin 2\epsilon = \frac{2 \operatorname{Im}(\chi)}{1 + |\chi|^2}, \tag{1.87}$$

where ϵ is the ellipticity angle and is related to the ellipticity e by

$$e = \tan \epsilon.$$

When the right- and left-circularly polarized states are used instead, the expressions for θ and e become (§1.7)

$$\theta = -\tfrac{1}{2} \arg(\chi), \tag{1.95}$$

$$e = \frac{|\chi| - 1}{|\chi| + 1}. \tag{1.96}$$

2.10.1.1. The case of homogeneous anisotropic media

For a homogeneous anisotropic medium, the elements n_{ij} of the matrix N are constant independent of z. In this case eq. (2.174) is easily integrable by separation of variables. This gives

$$\chi = -\alpha - (\beta/n_{12}) \tan(\beta z + C), \tag{2.175}$$

where

$$\alpha = (n_{11} - n_{22})/2n_{12}, \tag{2.176}$$

$$\beta = [-\tfrac{1}{4}(n_{11} - n_{22})^2 - n_{12}n_{21}]^{\tfrac{1}{2}}. \tag{2.177}$$

C is a constant of integration and can be determined from the initial condition

$$\chi = \chi_0 \quad \text{when} \quad z = 0, \tag{2.178}$$

where χ_0 is the initial or incident state of polarization. Substituting eq. (2.178) into eq. (2.175) gives

$$\tan C = -(n_{12}/\beta)(\chi_0 + \alpha). \tag{2.179}$$

Expanding the tangent in eq. (2.175) and substituting for $\tan C$ its value in eq. (2.179) we obtain, after some manipulations, the desired function $\chi(z, \chi_0)$

$$\chi(z, \chi_0) = \frac{[\beta - \tfrac{1}{2}(n_{11} - n_{22}) \tan \beta z]\chi_0 + [n_{21} \tan \beta z]}{[n_{12} \tan \beta z]\chi_0 + [\beta + \tfrac{1}{2}(n_{11} - n_{22}) \tan \beta z]}. \tag{2.180}$$

Note that the dependence of $\chi(z, \chi_0)$ on χ_0 is bilinear, as expected. The coefficients of this bilinear transformation determine, up to a complex multiplier, the Jones matrix of a section of the medium between two parallel planes perpendicular to the direction of propagation whose separation is z.

For each direction of propagation in a homogeneous anisotropic medium there is an associated N matrix and with it a value for β from eq. (2.177). Direct substitution of the elements of N and β into eq. (2.180) gives the law of evolution of the ellipse of polarization for propagation starting from an initial ellipse of polarization described by χ_0. When linear or circular orthogonal basis states are used, eqs. (1.86), (1.87), (1.95) and (1.96) provide the relations that are needed to follow the evolution of the azimuth θ and ellipticity e of the polarization ellipse once $\chi(z, \chi_0)$ has been determined.

Because the trajectory of $\chi(z, \chi_0)$ in the complex plane can be stereographically projected on the surface of the Poincaré sphere, eq. (2.180) also establishes the solution in this more-familiar representation (§2.5).

To examine the nature of the trajectory described by eq. (2.180) for any (constant) N matrix, we substitute for $\tan \beta z$ in that equation the expression [54a]

$$\tan \beta z = j \frac{1 - e^{j2\beta z}}{1 + e^{j2\beta z}}, \tag{2.181}$$

which gives
$$\chi(z,\chi_0) = \frac{A + B\, e^{j2\beta z}}{C + D\, e^{j2\beta z}}, \qquad (2.182)$$
where
$$\begin{aligned}
A &= jn_{21} + [\beta - \tfrac{1}{2}j(n_{11} - n_{22})]\chi_0, \\
B &= -jn_{21} + [\beta + \tfrac{1}{2}j(n_{11} - n_{22})]\chi_0, \\
C &= [\beta + \tfrac{1}{2}j(n_{11} - n_{22})] + jn_{12}\chi_0, \\
D &= [\beta - \tfrac{1}{2}j(n_{11} - n_{22})] - jn_{12}\chi_0.
\end{aligned} \qquad (2.183)$$

Equation (2.182) is interesting because it shows that the evolution of $\chi(z,\chi_0)$ with z, starting from an initial polarization χ_0 at $z = 0$, is an image through a bilinear transformation of the evolution of $e^{j2\beta z}$ with z. The coefficients A, B, C and D of this bilinear transformation are linear functions of the initial polarization state χ_0 as evident from eq. (2.183). It can be proved that the ratios A/C and B/D are independent of χ_0 and give the two eigenpolarizations χ_{e1} and χ_{e2} of the medium[26], i.e.,

$$\chi_{e1} = A/C, \qquad \chi_{e2} = B/D. \qquad (2.184)$$

In terms of (n_{ij}) the eigenpolarizations are given by

$$\chi_{e1,2} = \frac{1}{2n_{12}}\{(n_{22} - n_{11}) \pm [(n_{22} - n_{11})^2 + 4n_{12}n_{21}]^{\frac{1}{2}}\}. \qquad (2.185)$$

Equation (2.185) has the same form as eq. (2.59) which gives the eigenpolarizations of an optical system in terms of its Jones matrix.

From the above we have seen that $\chi(z,\chi_0)$ is obtained from $e^{j2\beta z}$ by a bilinear transformation. Therefore, to determine the trajectory of $\chi(z,\chi_0)$ we consider the behavior of $e^{j2\beta z}$ as z is increased from zero to infinity. This leads to the following three cases.[27]

Case I, β real
In this case $e^{j2\beta z}$ moves uniformly around the unit circle in the complex plane starting from the point $(1, 0)$ on the real axis when $z = 0$. This motion is counter-clockwise or clockwise depending on whether β is positive or negative, respectively; fig. 2.16a. One revolution around the unit circle is completed each time z is increased a distance d given by

$$d = \pi/\beta. \qquad (2.186)$$

[26] The proof follows because the condition of the independence of A/C and B/D of χ_0 leads to the definition of β in eq. (2.177).
[27] As will be shown later in this section, the parameter β is simply determined by the difference between the complex refractive indices of the eigenwaves [eq. (2.220)].

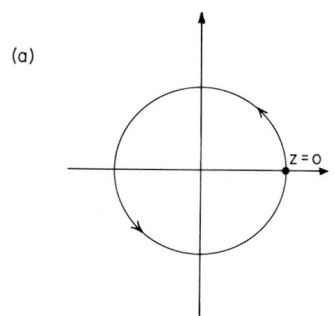

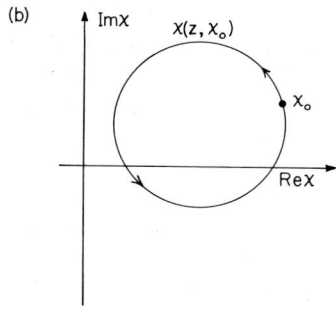

Fig. 2.16. (a) The function $\exp(j2\beta z)$ for β real and positive gives the unit circle in the complex plane. The rotational sense is reversed when β is negative. (b) The trajectory of $\chi(z, \chi_0)$ is an image through a bilinear transformation of the unit circle of (a). $\chi(z, \chi_0)$ is a circle that is traversed periodically as z is increased starting from the initial state χ_0 at $z = 0$.

Because a bilinear transformation maps a circle into a circle the locus of $\chi(z, \chi_0)$ is a circle through χ_0 that is traversed as z is increased; fig. 2.16b. The solution is obviously periodic with a period given by eq. (2.186).

Case II, β imaginary

As z is increased from zero to infinity $e^{j2\beta z}$ moves on the real axis from the point $(1, 0)$ to the origin if β is positive or to infinity if β is negative; fig. 2.17a. The real axis is a degenerate circle which is mapped by a bilinear transformation into a circle in the complex plane. Because $e^{j2\beta z}$ scans only one segment of the real axis and does that only once (as z is changed from zero to infinity), the evolution of $\chi(z, \chi_0)$ is aperiodic and proceeds along an arc of a circle, fig. 2.17b, from the initial state χ_0 to a final state χ_f given by

$$\chi_f = A/C = \chi_{e1}, \tag{2.187a}$$

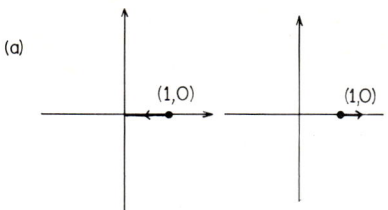

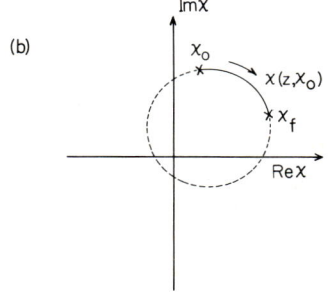

Fig. 2.17. (a) The function $\exp(j2\beta z)$ when β is pure imaginary and positive (left) or negative (right). (b) $\chi(z, \chi_0)$, which is the image of $\exp(j2\beta z)$ in (a), follows an arc of a circle that starts at the initial state $\chi_0(z=0)$ and converges asymptotically to the final state $\chi_f(z=\infty)$. χ_f is the low-absorption eigenpolarization.

or

$$\chi_f = B/D = \chi_{e2}. \qquad (2.187b)$$

Equations (2.187a) and (2.187b) are obtained by setting $e^{j2\beta z}$ equal to zero and infinity, respectively, in eq. (2.182) and directly using eq. (2.184). This shows that the ellipse of polarization converges asymptotically (in a nonoscillatory fashion) to one of the two eigenpolarizations, obviously the one of lower absorption.

Based on cases (I) and (II) above we arrive at the following conclusion. If the differential propagation Jones matrix N of a homogeneous anisotropic medium is such that β [eq. (2.177)] is either real or imaginary the evolution of the ellipse of polarization is along a circle in the complex plane, and also on the Poincaré sphere. The latter part of this conclusion results from the circle-preserving property of the stereographic projection which relates the complex-plane and Poincaré-sphere representations of polarization (§2.5). An N matrix that leads to β which is either real or imaginary satisfies the condition that

$$\text{Im}\left[\tfrac{1}{4}(\text{Tr } N)^2 - (\det N)\right] = 0, \qquad (2.188)$$

as can be seen from the expression of β in eq. (2.177). Tr N and det N stand for the trace and determinant of N, respectively, and Im signifies "the imaginary part of."

Two common types of anisotropy that satisfy eq. (2.188) are elliptic birefringence (eb) and elliptic dichroism (ed). In an elliptically birefringent medium two orthogonally polarized waves propagate with different velocities and equal attenuation with their polarization unaffected. In an elliptically dichroic medium the two eigenwaves are attenuated at different rates but propagate with the same velocity. The N matrix for these two types of anisotropy takes the form[28]

$$N_{eb} = \begin{bmatrix} A & B \\ -B^* & A^* \end{bmatrix}, \quad N_{ed} = \begin{bmatrix} p & R \\ R^* & q \end{bmatrix}, \quad (2.189)$$

where A, B and R are complex and p and q are real. Both N_{eb} and N_{ed} have real trace and real determinant and thus satisfy eq. (2.188). Furthermore, N_{eb} leads to a real value of β corresponding to case (I) and N_{ed} leads to an imaginary value for β corresponding to case (II).

Case III, β complex
Writing $\beta = \beta_r + j\beta_i$, $e^{j2\beta z}$ becomes $e^{-2\beta_i z} e^{j2\beta_r z}$. As z varies from zero to infinity $e^{j2\beta z}$ describes a logarithmic spiral (fig. 2.18a) that starts at $(1, 0)$ and converges to the origin if β_i is positive or diverges to infinity if β_i is negative. The sense of description is counter-clockwise or clockwise dependent on the sign of β_r being positive or negative. According to eq. (2.182) the image trajectory $\chi(z, \chi_0)$ starts at $\chi_0(z = 0)$ and approaches, in a spiral fashion, the eigenpolarization $\chi_{e1} = A/C$ ($z = \infty$, β_i positive) or $\chi_{e2} = B/D$ ($z = \infty$, β_i negative). The final state is necessarily the one of lower absorption. That the logarithmic spiral of $e^{j2\beta z}$ leads to a spiral for $\chi(z, \chi_0)$ (fig. 2.18b) is a consequence of the conformal nature of the bilinear transformation. From the relation between the complex plane and the Poincaré sphere it also follows that the representative point converges spirally towards the low-absorption eigenstate on the sphere. Cases (I) and (II) are special cases of Case (III) when either β_i or β_r is zero, respectively.

Figures 2.19, 2.20, and 2.21 give the computed evolution of the ellipse of polarization and its parameters (ellipticity and azimuth) for one example of cases (I), (II) and (III), respectively. In part (a) of these figures the trajectories of $\chi(z, \chi_0)$ for different initial states are drawn, and in part (b)

---

[28] These matrices do not take proper account of any isotropic absorption or phase-retardation properties of the medium. They can be obtained from eq. (2.219) by setting $\bar{n}_0 = 0$ and using the orthogonality condition $\chi_{e1}\chi_{e2}^* = -1$. When $\overline{\Delta n}$ is pure real N_{eb} is obtained. $\overline{\Delta n}$ pure imaginary leads to N_{ed}.

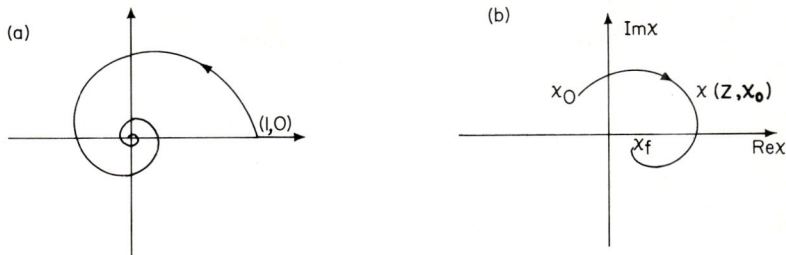

Fig. 2.18. (a) The function $\exp(j2\beta z)$ for β complex ($\beta = \beta_r + j\beta_i$) is a logarithmic spiral. Reversing the sign of β_i causes the spiral to expand out to infinity. The sense of the spiral is clockwise (instead of counter-clockwise) when β_r is negative. (b) $\chi(z, \chi_0)$, being an image of the spiral of (a), is also a spiral. The final state χ_f is the low-absorption eigenpolarization.

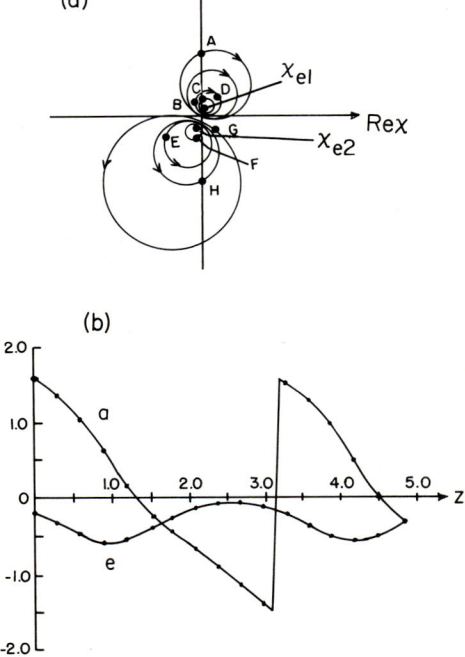

Fig. 2.19. (a) The computed evolution of the ellipse of polarization $[\chi(z, \chi_0)]$ in an elliptically birefringent medium with eigenpolarizations χ_{e1} and χ_{e2} for eight different initial polarization states $A, B, \ldots$ and H. The N matrix of the medium has the form of N_{eb} [eq. (2.189)] with $A = 0.75 + j.25$ and $B = 0.875 + j.375$. (b) The variation of the ellipticity e and azimuth a with distance z along the direction of propagation in the elliptically birefringent medium corresponding to the trajectory that starts at H in (a). Note that both functions are periodic with z.

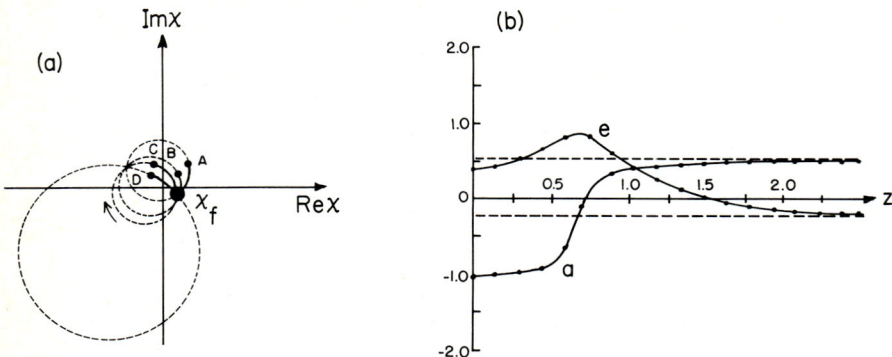

Fig. 2.20. (a) Same as in fig. 2.19a for an elliptically dichroic medium whose N matrix has the form of N_{ed} [eq. (2.189)] with $p = 2$, $q = 1$, and $R = 0.866 + j5$. A, B, C, and D indicate four different initial states and χ_f, the final state, corresponds to the low-absorption eigenpolarization. (b) Same as in fig. 2.19b for the trajectory that starts at C in (a). Note the aperiodic nature of the curves and their asymptotic approach to the ellipticity and azimuth of the low-absorption eigenpolarization.

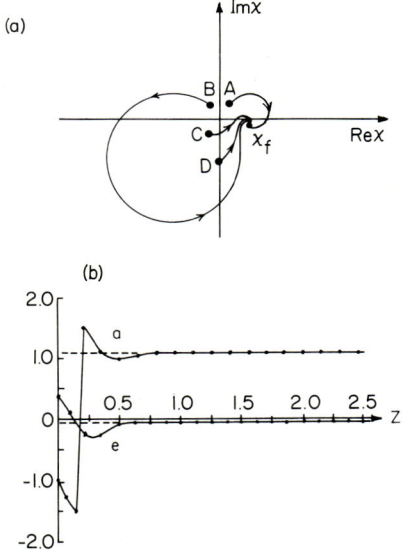

Fig. 2.21. (a) Same as in figs. 2.19a and 2.20a for the general homogeneous anisotropic medium that exhibits combined birefringence and dichroism. The N matrix was chosen arbitrarily with elements $n_{11} = 1 + j2$, $n_{12} = 2 + j3$, $n_{21} = 3 + j4$, and $n_{22} = 4 + j5$. Note the spiraling convergence to the low-absorption eigenpolarization χ_f for the four different initial polarizations A, B, C, and D. (b) Same as in figs. 2.19b and 2.20b for the trajectory that starts at B in (a). Note the damped oscillatory convergence to the final ellipticity and azimuth.

the ellipticity and azimuth for one initial state are plotted as functions for distance z along the direction of propagation.

2.10.1.2. The case of inhomogeneous anisotropic media

In the case of an inhomogeneous anisotropic medium the elements n_{11}, n_{12}, n_{21} and n_{22} of the N matrix that appear in eq. (2.174) are no longer constant, but are functions of the distance along the direction of propagation z. For arbitrary dependence of n_{ij} on z, eq. (2.174) does not have an analytical solution. Using numerical methods, however, $\chi(z, \chi_0)$ which describes the evolution of the ellipse of polarization can be determined [54b].

An example of practical importance is that of a liquid crystal in the cholesteric phase. Here the molecules of the liquid crystal orient themselves parallel to one another in planes. For the spectral regions of transparency each molecular plane may be thought of as a very thin linearly birefringent plate. The principal axes of birefringence, which are determined by the direction of molecular orientation, are gradually rotated from one molecular plane to the next as we proceed along the helical axis of the structure.

In the following we consider the propagation of totally polarized light along the helical axis of the cholesteric structure which is assumed to coincide with the z axis of a fixed xyz coordinate system. The x and y axes of this coordinate system are chosen parallel to the principal axes of birefringence of the entrance molecular plane (where the light beam enters into the liquid crystal). The angle, θ, by which the principal axes of the molecular plane at a distance z are rotated with respect to those of the entrance plane (fig. 2.22) is given by

$$\theta = \alpha z, \tag{2.190}$$

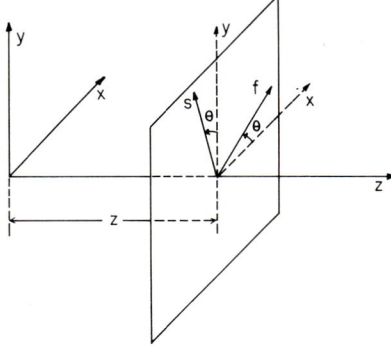

Fig. 2.22. The xyz space-fixed and fsz space-rotating coordinate systems used to discuss light propagation along the helical axis (the z axis) of a cholesteric liquid crystal. f and s designate the fast and slow principal axes of birefringence of the molecular plane located at a distance z from the origin.

where

$$\alpha = 2\pi/p, \tag{2.191}$$

and p is the pitch of the helical structure, that is, the smallest distance between two molecular planes of the same orientation.

The helicity of the cholesteric structure suggests that the right- and left-circular polarizations be used as the basis states for the complex-plane representation. It can be shown that, in this representation, the matrix N takes the form[29]

$$N = g_0 \begin{bmatrix} 0 & je^{j2\alpha z} \\ je^{-j2\alpha z} & 0 \end{bmatrix}, \tag{2.192}$$

where $g_0 = \frac{1}{2}(\eta_y - \eta_x)$, and η_y and η_x represent the principal propagation constants of a single molecular plane or more accurately, a birefringent medium formed by an infinite stack of such planes oriented in the same way. Substituting from eq. (2.192) into eq. (2.174) we get

$$d\chi/dz = -(jg_0 e^{j2\alpha z})\chi^2 + jg_0 e^{-j2\alpha z}, \tag{2.193}$$

which describes the evolution of the ellipse of polarization in the circular complex-plane representation as the light beam propagates along the helical axis.

Particular solutions

By inspection, it is seen that eq. (2.193) has a particular solution of the form

$$\chi = K e^{-j2\alpha z}, \tag{2.194}$$

where K is a complex constant to be determined. Substituting eq. (2.194) into eq. (2.193) gives a quadratic in K,

$$K^2 - 2(\alpha/g_0)K - 1 = 0, \tag{2.195}$$

whose roots

$$K_{1,2} = (\alpha/g_0) \pm [(\alpha/g_0)^2 + 1]^{\frac{1}{2}}, \tag{2.196}$$

satisfy the condition

$$K_1 K_2 = -1. \tag{2.197}$$

---

[29] This matrix can be obtained by direct application of the basic definition of N in eq. (2.165). The *circular* Jones matrix of a linearly-birefringent thin slab parallel to the molecular planes, $M(Z, \Delta z)$, is given by eq. (2.96) where δ and θ are replaced by $2g_0\Delta z$ and αz, respectively. Carrying out the limit operation immediately gives eq. (2.192). The signs of both α and g_0 are opposite to those used in ref. 52. This is because the definition of the circular complex polarization variable has been changed slightly. See eq. (1.95) of ch. 1 and eq. (10) of ref. 52.

Thus we have two particular solutions

$$\chi = K_1 e^{-j2\alpha z}, \qquad \chi = K_2 e^{-j2\alpha z}, \tag{2.198}$$

which satisfy the conditions of propagation along the helical axis of the liquid crystal. Note that K_1 and K_2 in eq. (2.198) correspond to the values of χ at $z = 0$, i.e., the initial or incident polarization states. Because α and g_0 are real quantities, K_1 and K_2, which satisfy eq. (2.197), represent two orthogonal polarizations. From the properties of the circular complex-plane representation [§1.7], it also follows that the incident orthogonal vibrations, K_1 and K_2 represent ellipses whose major and minor axes coincide with the x and y directions, that is, with the principal axes of the entrance molecular plane. The ellipticity of K_1 is obtained by substituting from eq. (2.196) into eq. (1.96)

$$e_1 = \frac{[(\alpha/g_0) - 1] + [(\alpha/g_0)^2 + 1]^{\frac{1}{2}}}{[(\alpha/g_0) + 1] + [(\alpha/g_0)^2 + 1]^{\frac{1}{2}}} \tag{2.199}$$

and, because K_1 and K_2 are orthogonal, $e_2 = -e_1$, with the understanding that $|e|$ is the ratio of the minor to the major axis of the ellipse of polarization. Note that $|e_1| = 0$ only if $\alpha = 0$; i.e., when the medium is homogeneous.

Equation (2.198) shows that for an initial polarization state represented by either K_1 or K_2 the evolution of χ, which describes the ellipse of polarization at any z, is along an equiellipticity circle in the circular complex-plane representation, fig. 2.23. Thus the two privileged polarizations K_1 and K_2 (given by eq. (2.196)) suffer pure optical rotation as they propagate along the helical axis of the cholesteric liquid crystal which is characterized by two real parameters α and g_0. The magnitude of this optical rotation is equal to α

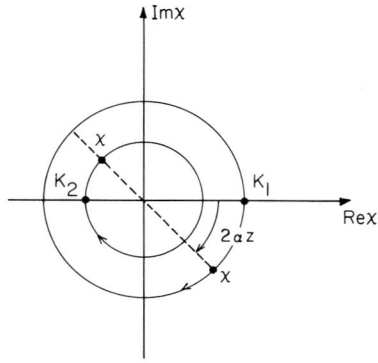

Fig. 2.23. The two particular solutions that describe the evolution of the two privileged polarizations K_1 and K_2 that remain in alignment with the principal axes of the molecular planes. This is the circular complex-plane representation whose basis states are the left- and right-circular polarizations, represented by the origin and the point at infinity, respectively.

radians per unit distance and is the same as the rotation of the principal axes of the molecular planes along the helical axis. For any value of $\alpha \neq 0$ (hence for any finite pitch p) K_1 and K_2 represent orthogonal ellipses of polarizations whose axes coincide with the principal axes of the entrance molecular plane. After propagation through a distance z along the helical axis, the ellipses of polarization that result from K_1 and K_2 as initial states continue to preserve their orthogonality and alignment with the principal axes of the molecular planes.

If L and R are the poles of the Poincaré sphere that represent left- and right-circular polarizations, respectively, the circular complex-plane representation can be obtained by a stereographic projection of the sphere onto a tangent plane at L with R as the center [§1.8]. Because K_1 and K_2 are orthogonal, the two circles in fig. 2.23 are projected on the sphere as two latitudes symmetrically located above and below the equatorial plane. Thus, in Stokes space, the vector that represents the state of polarization precesses about the axis that joins the poles L and R as light, initially polarized in either one of the two privileged states K_1 or K_2, propagates along the helical axis of the cholesteric liquid crystal. A similar conclusion has been derived using a matrix-operator description [55].

It should be mentioned that Oseen [56] has obtained solutions for the propagation of light along the helical structure of a liquid crystal by solving Maxwell's equations for the electric- and magnetic-field vectors. His solutions reveal two elliptically polarized waves (for each direction of propagation) whose ellipses of polarization behave in the same manner as we have found above. The simplicity of the present approach, which is based on the solution of Ricatti's equation for the propagation of the ellipse of polarization [eq. (2.174)], should be evident.

General solution[30]

We will now determine the solution of eq. (2.193) that describes the most general behavior for the evolution of the ellipse of polarization of light as it propagates along the helical axis. The fact that eq. (2.193) has a solution of the form $e^{-j2\alpha z}$ suggests that we put the general solution as

$$\chi = u(z) e^{-j2\alpha z}, \qquad (2.200)$$

where $u(z)$ is a complex function of the real variable z to be determined. The function $u(z)$ is interesting because it gives the complex-variable description of the polarization ellipse in the space-rotating coordinate system that coincides with the principal axes of the molecular planes of the liquid

---

[30] This part is included for completeness of the present example and may be skipped.

crystal. When $u(z)$ is a constant independent of z we obtain the particular solutions discussed above. Substituting eq. (2.200) into eq. (2.193) yields the following equation for u

$$du/dz = -jg_0 u^2 + j2\alpha u + jg_0. \tag{2.201}$$

Equation (2.201) has constant coefficients and can be solved by separation of variables. A similar equation has already been met in connection with homogeneous anisotropic media. Using the result in eq. (2.180) the solution of eq. (2.201) is

$$u(z, u_0) = \frac{(\beta + j\alpha \tan \beta z)u_0 + (jg_0 \tan \beta z)}{(jg_0 \tan \beta z)u_0 + (\beta - j\alpha \tan \beta z)}, \tag{2.202a}$$

where

$$\beta = g_0[(\alpha/g_0)^2 + 1]^{\frac{1}{2}}, \tag{2.202b}$$

and u_0 represents the initial polarization state at $z = 0$ referenced to the principal axes of the entrance molecular plane. Because the latter have been chosen to coincide with the x and y directions of the space-fixed coordinate system we have $u_0 = \chi_0$. (Note that χ is the complex-variable description of polarization in the space-fixed coordinate system.) It can be shown that the right side of eq. (2.202a) is independent of z only if u_0 (or χ_0) satisfies eq. (2.195). This is how the particular solutions of eq. (2.198) can be obtained from the general solution. Substituting eq. (2.202a) into eq. (2.200) gives

$$\chi(z, \chi_0) = \left[\frac{(\beta + j\alpha \tan \beta z)\chi_0 + (jg_0 \tan \beta z)}{(jg_0 \tan \beta z)\chi_0 + (\beta - j\alpha \tan \beta z)}\right] e^{-j2\alpha z}. \tag{2.203}$$

Equation (2.203) represents the final form of the general solution. The term $e^{-j2\alpha z}$ represents optical rotation and is periodic in z with a period equal to half the pitch p of the helical structure. The first (bracketed) term describes the variation of the polarization ellipse in the space-rotating coordinate system as light propagates along the axis of the helix and is periodic in z with a period d' shorter than half the pitch p, where

$$d' = p/2[(g_0/\alpha)^2 + 1]^{\frac{1}{2}}. \tag{2.204}$$

Equation (2.204) follows from setting $\beta d' = \pi$ where β is given by eq. (2.202b). It is interesting to note that although both factors of eq. (2.203) are periodic functions of z, the entire function $\chi(z, \chi_0)$ is not periodic because the periods of the two factors are different. Therefore for an arbitrary initial state χ_0 we have the result that this state does not repeat itself along the helical axis. The ellipticity alone, which is determined by $|\chi|$, is periodic with a period given by eq. (2.204), while the azimuth is, in general, aperiodic. Note that the maximum and minimum values of $|\chi|$ (or $|u|$) define two circles in the complex plane inside which the trajectory of the function $\chi(z, \chi_0)$ should be confined.

A trajectory that describes the evolution of the ellipse of polarization of light along the helical axis of a cholesteric liquid crystal is shown in fig. 2.24a [54a]. This is computer-plotted using eq. (2.203). Figure 2.24b shows the evolution of ellipticity and azimuth with distance z along the helical axis.

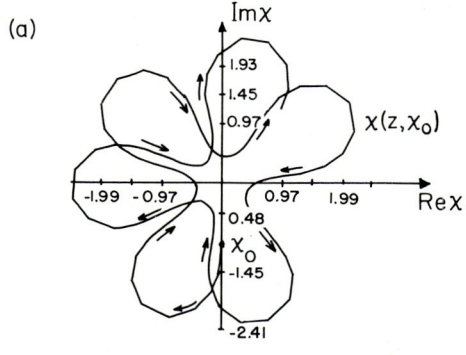

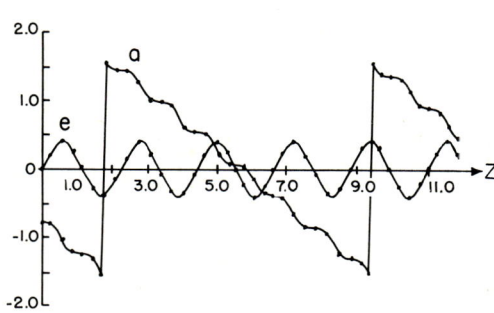

Fig. 2.24. (a) The evolution of the polarization ellipse of a light beam travelling along the helical axis of a cholesteric liquid crystal with $\alpha = g_0 = 1$ in a space-fixed coordinate system for an initial state $\chi_0 = -j$. (b) The evolution of the ellipticity e and azimuth a as a function of distance z along the helical axis for the trajectory of a.

The parameters α and g_0 of the liquid crystal are both taken as unity. Figure 2.24 is obtained by choosing an initial linear polarization state $\chi_0 = -j$ and represents the general condition of a trajectory confined between two concentric circles with center at the origin. From fig. 2.24b it is readily observable that the ellipticity of the ellipse of polarization is periodic while the azimuth is not. The azimuth appears to make several discontinuous jumps. Closer inspection shows that these jumps do not represent true discontinuities. The discontinuity occurs when the azimuth reaches a value of $\frac{1}{2}\pi$ or $-\frac{1}{2}\pi$ at which a jump of $\pm \pi$ takes place. This follows from the fact that the range of azimuth values is limited between $-\frac{1}{2}\pi$ and $+\frac{1}{2}\pi$ and that these limiting azimuths are physically indistinguishable. The interested reader may consult ref. [54] for additional discussion and more examples of trajectories that describe the evolution of the state of polarization of light travelling along the helical axis of a cholesteric liquid crystal.

2.10.2. Evolution of the complex amplitude

For complete description of the propagation of polarized light in anisotropic media we need to follow the evolution of both the ellipse of polarization and the complex amplitude of the wave. When the ellipse of polarization is described by a single complex variable χ, the evolution of χ with distance z along the direction of propagation is governed by the Riccati equation [eq. (2.174)]. It remains for us to determine the evolution of the complex amplitude A_c of the wave. Recall that the complex amplitude A_c lumps together the real amplitude A of the elliptic vibration and its absolute phase δ such that $A_c = A\,\mathrm{e}^{\mathrm{j}\delta}$ [see fig. 1.5 of §1.3 and eq. (2.74) of §2.6]. According to eqs. (2.70)–(2.73) the complex amplitude A_c is given by (the subscript i is dropped)

$$A_c = E_u(1 + \chi\chi^*)^{\frac{1}{2}}. \tag{2.205}$$

Equation (2.205), which is based on the assumption that the basis states u and v are orthogonal, indicates that the evolution of A_c can be obtained if both χ and E_u are determined as functions of z. The evolution of χ is assumed to have been found by solving eq. (2.174). The equation that governs the evolution of E_u (the complex amplitude of the projection of the electric vibration along the basis state u) can be derived by the elimination of E_v between the system of coupled eqs. (2.171). Such elimination produces the following homogeneous second-order differential equation for E_u

$$\frac{\mathrm{d}^2 E_u}{\mathrm{d}z^2} + \left[-(\mathrm{Tr}\,\mathbf{N}) - \frac{1}{n_{12}}\frac{\mathrm{d}n_{12}}{\mathrm{d}z} \right]\frac{\mathrm{d}E_u}{\mathrm{d}z} + \left[(\det \mathbf{N}) + \frac{n_{11}}{n_{12}}\frac{\mathrm{d}n_{12}}{\mathrm{d}z} - \frac{\mathrm{d}n_{11}}{\mathrm{d}z} \right] E_u = 0, \tag{2.206}$$

where $\mathrm{Tr}\,\mathbf{N} = n_{11} + n_{22}$, and $\det \mathbf{N} = n_{11}n_{22} - n_{12}n_{21}$, are the trace and the determinant of the differential propagation Jones matrix $\mathbf{N}$, respectively. In the general case of an inhomogeneous anisotropic medium, the coefficients of $\mathrm{d}E_u/\mathrm{d}z$ and E_u in eq. (2.206) become functions of z that are determined by both $\mathbf{N}$ and the first derivative of its elements n_{11} and n_{12}. Equation (2.206) does not possess a closed-form solution for arbitrarily inhomogeneous medium. In the special case of a homogeneous anisotropic medium, $dn_{11}/dz = dn_{12}/dz = 0$ and eq. (2.206) reduces to a second-order differential equation with constant coefficients

$$\frac{d^2E_u}{dz^2} - (\text{Tr } \mathbf{N})\frac{dE_u}{dz} + (\det \mathbf{N})E_u = 0. \tag{2.207}$$

The general solution of eq. (2.207) is given by

$$E_u(z) = E_{u1} e^{V_{e1}z} + E_{u2} e^{V_{e2}z}, \tag{2.208}$$

where E_{u1} and E_{u2} are the arbitrary constants of integration and V_{e1} and V_{e2} are the roots of the auxiliary equation of the differential eq. (2.207)

$$V^2 - (\text{Tr } \mathbf{N})V + (\det \mathbf{N}) = 0. \tag{2.209}$$

Solving eq. (2.209) and using the explicit form of the trace and the determinant of $\mathbf{N}$, we obtain

$$V_{e1,2} = \tfrac{1}{2}\{(n_{11} + n_{22}) \pm [(n_{11} - n_{22})^2 + 4n_{12}n_{21}]^{\frac{1}{2}}\}, \tag{2.210}$$

which can be readily recognized as the eigenvalues of $\mathbf{N}$.

If instead of E_v, E_u is eliminated between the system of coupled eqs. (2.171), the equation that results for E_v is similar to eq. (2.206) with n_{11} and n_{12} replaced by n_{22} and n_{21}, respectively. When $\mathbf{N}$ is constant (which is the case under consideration), E_v and E_u satisfy the same differential eq. (2.207) and both have a solution of the form of eq. (2.208). The general solution of eq. (2.171) has two independent arbitrary constants E_{u1} and E_{u2} and can be written as a linear superposition of two eigenwave components

$$\mathbf{E}(z) = E_{u1} \begin{bmatrix} 1 \\ \chi_{e1} \end{bmatrix} e^{V_{e1}z} + E_{u2} \begin{bmatrix} 1 \\ \chi_{e2} \end{bmatrix} e^{V_{e2}z}, \tag{2.211}$$

where χ_{e1} and χ_{e2} are the eigenpolarizations of the medium as given by eq. (2.185). Using eq. (2.211), the values of the constants of integration E_{u1} and E_{u2} can be related to the initial polarization χ_0 and initial value E_{u0} of E_u as follows

$$\begin{aligned} E_{u0} &= E_{u1} + E_{u2}, \\ \chi_0 &= \frac{\chi_{e1}E_{u1} + \chi_{e2}E_{u2}}{E_{u1} + E_{u2}} \end{aligned} \tag{2.212}$$

where the second of eqs. (2.212) follows from the definition of χ in eq. (1.97). Solving eq. (2.212) for E_{u1} and E_{u2} and substituting the result in eq. (2.208) we obtain

$$E_u(z) = E_{u0}\left[\left(\frac{\chi_{e2} - \chi_0}{\chi_{e2} - \chi_{e1}}\right)e^{V_{e1}z} + \left(\frac{\chi_0 - \chi_{e1}}{\chi_{e2} - \chi_{e1}}\right)e^{V_{e2}z}\right]. \tag{2.213}$$

When the wave is initially polarized in either one of the two eigenstates $\chi_0 = \chi_{e1}$ or χ_{e2}, eqs. (2.213) and (2.211) lead to a single propagating eigenwave whose polarization remains unchanged. Substituting eq. (2.213) into eq. (2.205), the evolution of the complex amplitude of the wave $A_c(z)$ is seen to be governed by

$$A_c(z) = A_{c0} \left(\frac{1 + \chi\chi^*}{1 + \chi_0\chi_0^*} \right)^{\frac{1}{2}} \left[\left(\frac{\chi_{e2} - \chi_0}{\chi_{e2} - \chi_{e1}} \right) e^{V_{e1}z} + \left(\frac{\chi_0 - \chi_{e1}}{\chi_{e2} - \chi_{e1}} \right) e^{V_{e2}z} \right], \quad (2.214)$$

where χ describes the ellipse of polarization after the wave has travelled a distance z and χ_0 and A_{c0} ($= A_c(0)$) are the initial polarization and initial complex amplitude of the wave at $z = 0$, respectively. χ_{e1}, χ_{e2}; V_{e1} and V_{e2} are the eigenpolarizations and eigenvalues of the homogeneous anisotropic medium and are determined by the differential propagation Jones matrix N through eqs. (2.185) and (2.210), respectively.

From eq. (2.214) it is evident that the eigenvalues V_{e1} and V_{e2} represent the propagation constants for the corresponding eigenwaves. These propagation constants can be written in terms of *generalized complex refractive indices* $\bar{n}_{e1}$, $\bar{n}_{e2}$ for the eigenwaves as follows

$$V_{e1} = -j\frac{2\pi\bar{n}_{e1}}{\lambda}, \qquad V_{e2} = -j\frac{2\pi\bar{n}_{e2}}{\lambda}, \quad (2.215)$$

where λ is the free-space wavelength of light. Each generalized complex refractive index $\bar{n}$ can, in turn, be expressed in terms of a real index of refraction n and an extinction coefficient k

$$\bar{n}_{e1} = n_{e1} - jk_{e1}, \qquad \bar{n}_{e2} = n_{e2} - jk_{e2}. \quad (2.216)$$

The N matrix of an anisotropic medium can be easily synthesized from its eigenpolarizations χ_{e1}, χ_{e2} and its eigenvalues V_{e1} and V_{e2} in exactly the same way as the Jones matrix of an optical system is constructed by eq. (2.89). Expressed in terms of the eigenpolarizations χ_{e1}, χ_{e2} and the generalized complex refractive indices of the eigenwaves $\bar{n}_{e1}$ and $\bar{n}_{e2}$, the elements of the differential propagation Jones matrix N are given by [see eq. (2.89)]

$$\begin{aligned} n_{11} &= -j\frac{2\pi}{\lambda}(\bar{n}_{e2}\chi_{e1} - \bar{n}_{e1}\chi_{e2})/(\chi_{e1} - \chi_{e2}), \\ n_{12} &= -j\frac{2\pi}{\lambda}(\bar{n}_{e1} - \bar{n}_{e2})/(\chi_{e1} - \chi_{e2}), \\ n_{21} &= j\frac{2\pi}{\lambda}\chi_{e1}\chi_{e2}(\bar{n}_{e1} - \bar{n}_{e2})/(\chi_{e1} - \chi_{e2}), \\ n_{22} &= -j\frac{2\pi}{\lambda}(\bar{n}_{e1}\chi_{e1} - \bar{n}_{e2}\chi_{e2})/(\chi_{e1} - \chi_{e2}). \end{aligned} \quad (2.217)$$

Instead of $\bar{n}_{e1}$ and $\bar{n}_{e2}$, it is convenient to introduce an *average* complex index of refraction $\bar{n}_0$ and a *differential* complex index of refraction $\overline{\Delta n}$ such that

$$\bar{n}_{e1} = \bar{n}_0 - \overline{\Delta n},$$
$$\bar{n}_{e2} = \bar{n}_0 + \overline{\Delta n};$$
(2.218a)

$$\bar{n}_0 = \tfrac{1}{2}(\bar{n}_{e1} + \bar{n}_{e2}) = n_0 - jk_0,$$
$$\overline{\Delta n} = \tfrac{1}{2}(\bar{n}_{e2} - \bar{n}_{e1}) = \Delta n - j\Delta k.$$
(2.218b)

In terms of $\bar{n}_0$, $\overline{\Delta n}$, χ_{e1} and χ_{e2}, eq. (2.217) can be rewritten as

$$n_{11} = -j\frac{2\pi}{\lambda}\left[\bar{n}_0 + \left(\frac{\chi_{e1} + \chi_{e2}}{\chi_{e1} - \chi_{e2}}\right)\overline{\Delta n}\right],$$

$$n_{12} = j\frac{4\pi}{\lambda}\frac{\overline{\Delta n}}{(\chi_{e1} - \chi_{e2})},$$
(2.219)

$$n_{21} = -j\frac{4\pi}{\lambda}\left(\frac{\chi_{e1}\chi_{e2}}{\chi_{e1} - \chi_{e2}}\right)\overline{\Delta n},$$

$$n_{22} = -j\frac{2\pi}{\lambda}\left[\bar{n}_0 - \left(\frac{\chi_{e1} + \chi_{e2}}{\chi_{e1} - \chi_{e2}}\right)\overline{\Delta n}\right].$$

The parameter β that appears in eq. (2.180) is determined by $\overline{\Delta n}$ only, as can be seen by substituting from eq. (2.219) into eq. (2.177). This gives

$$\beta = \pm\frac{2\pi}{\lambda}\overline{\Delta n}.$$
(2.220)

Equation (2.220) shows that the differential complex index of refraction $\overline{\Delta n}$ plays a most significant role in the evolution of polarized light in homogeneous anisotropic media. In fact, $\overline{\Delta n}$ represents the combined birefringent and dichroic properties of an anisotropic medium that may be linear, circular or elliptic, depending upon the eigenpolarizations χ_{e1} and χ_{e2}. The special cases when $\overline{\Delta n}$ is either pure real or pure imaginary leads to pure birefringence and pure dichroism, respectively. The first, second and third distinct possibilities for the trajectory of $\chi(z, \chi_0)$ discussed earlier in this section correspond to $\overline{\Delta n}$ being real, imaginary or complex, respectively.

It should be pointed out that eq. (2.217) gives the differential propagation Jones matrix N in *any* polarization representation, because the dependence of N on the basis states is implicit in χ_{e1} and χ_{e2}.

Jones has compiled a table of eight N matrices, called the Θ matrices, that describe the different types of optical behavior in an anisotropic medium [50]. Jones' table of the Θ matrices can be readily generated from eq. (2.219). For example, for linear birefringence at $+45°$, $\chi_{e1} = 1$, $\chi_{e2} = -1$ [see eq. (1.79)]. To express the differential phase-retarding properties of the medium, *without account for its isotropic phase-retarding properties*, we set $\bar{n}_0 = 0$ and $\overline{\Delta n} = \Delta n$ (representing pure linear birefringence) in eq. (2.219). This gives the following *Cartesian N* matrix for a medium exhibiting linear birefringence at $+45°$ to the coordinate axes

$$N = g_0 \begin{bmatrix} 0 & j \\ j & 0 \end{bmatrix}, \qquad g_0 = \frac{2\pi}{\lambda_0} \Delta n, \tag{2.221}$$

and is the same as Θ_7 of Jones [50].

2.11. Propagation of partially polarized quasi-monochromatic light through non-depolarizing optical systems—The coherency-matrix formulation

In the previous sections of this chapter (§2.2–§2.10), we have considered the propagation of *totally polarized* quasi-monochromatic or monochromatic light through linear *non-depolarizing* optical systems. The Jones-matrix formulation [§2.2] provided the basic framework within which we were able to describe the effect of the optical system on wave polarization [§2.3–§2.5], intensity [§2.6–§2.8] and absolute phase [§2.6 and §2.9]. The evolution of totally polarized light as it propagates through a non-depolarizing anisotropic medium has also been examined [§2.10].

In the present section, we deal with the propagation of *unpolarized* or *partially polarized* quasi-monochromatic light through *non-depolarizing* optical systems. A non-depolarizing optical system is one that does not randomize or decorrelate the time-dependent amplitudes and phases of the x and y electric-field components of an incident quasi-monochromatic light wave. Therefore, for a non-depolarizing optical system, the degree of polarization [§1.8] of the output light from the system $\mathcal{P}_o$ is either greater than or equal to the degree of polarization of the input light $\mathcal{P}_i$, or,

$$\mathcal{P}_o \geq \mathcal{P}_i, \tag{2.222a}$$

for *all* incident states of (total or partial) polarization. The case of depolarizing optical systems that include incoherent scattering processes and for which

$$\mathcal{P}_o \leq \mathcal{P}_i, \tag{2.222b}$$

at least for *one* incident state will be studied in §2.12.

The propagation of totally-polarized light through non-depolarizing optical systems has been adequately handled by the Jones-matrix formulation with the state of polarization of the light wave represented by a Jones vector and the effect of the system described by a Jones matrix. With partially-polarized quasi-monochromatic light, a statistical description for the wave is required, because of the stochastic nature of the time-variation of the field components. The Stokes vector and the coherency matrix offer two equivalent alternative representations of partially-polarized light that have been discussed in §1.9 of ch. 1.

The most concise description of the polarization-modifying interaction between partially-polarized light and linear non-depolarizing optical systems is based upon the representation of the state of polarization of the wave by its coherency matrix J (§1.9) and the representation of the polarization-modifying optical system by its Jones matrix T.

In §1.9, a quasi-monochromatic wave was described by a time-dependent two-component complex Jones vector $E_{x,y}(t)$ as indicated by eq. (1.108). This Jones vector is linearly transformed under the effect of a linear non-depolarizing optical system so that the output and input Jones vectors are related by

$$\begin{bmatrix} \tilde{E}_{ox'}(t) \exp[j\delta_{ox'}(t)] \\ \tilde{E}_{oy'}(t) \exp[j\delta_{oy'}(t)] \end{bmatrix} = \begin{bmatrix} T_{11} & T_{12} \\ T_{21} & T_{22} \end{bmatrix} \begin{bmatrix} \tilde{E}_{ix}(t) \exp[j\delta_{ix}(t)] \\ \tilde{E}_{iy}(t) \exp[j\delta_{iy}(t)] \end{bmatrix}, \quad (2.223)$$

where the Jones matrix T of the (dispersive) optical system is evaluated at the center frequency ω_o of the finite-bandwidth quasi-monochromatic light [23]. The choice of the input (x, y) and output (x', y') coordinate systems and other notation are the same as explained in the early part of §2.2. Written concisely eq. (2.223) becomes

$$E_o(t) = TE_i(t). \quad (2.224)$$

Taking the direct product of each side of eq. (2.224) times its Hermitian adjoint followed by time-averaging yields

$$\langle E_o(t) \times E_o(t) \rangle = \langle TE_i(t) \times E_i^\dagger(t) T^\dagger \rangle.$$

so that

$$J_o = TJ_i T^\dagger. \quad (2.225)$$

Equation (2.225) represents the basic law of propagation of partially polarized light through linear non-depolarizing optical systems [5, 46, 57]. In eq. (2.225), J_i and J_o represent the coherency matrices of the input (incident) and output (emergent) light waves, and T represents the Jones matrix of the optical system. In reaching eq. (2.225), use has been made of the matrix identity $(AB)^\dagger = B^\dagger A^\dagger$. It is interesting to observe the similarity between the transformation of the coherency matrix as a result of a mere change of basis [eq. (1.136)], and the transformation of that matrix as the wave is modified by an optical system [eq. (2.225)]. These two transformations should not be confused.

Equation (2.225) shows how the outcome of passing a quasi-monochromatic wave through a linear non-depolarizing optical system can be computed. In particular, the coherency matrix J_o of the wave that emerges from the optical system can be computed from the coherency matrix J_i of the incident wave and the Jones matrix T of the optical system by carrying out the multiplication of the 2×2 matrices that appear on the right-hand side of eq. (2.225). From the now-computed coherency matrix J_o, we may calculate such physical properties of the outgoing wave as its total intensity [eq. (1.131)], degree of polarization [eq. (1.139)] or the polarization form of its totally polarized component [eq. (1.140)]. Such a procedure is obviously applicable to any linear non-depolarizing optical system and for any state of polarization of the incident light.

Some special cases of eq. (2.225) deserve consideration as illustrative examples.

(1) When the incident light is totally polarized, $\mathscr{P}_i = 1$ and $\det J_i = 0$ [eq. (1.139)], so that the input coherency matrix J_i is singular. Because the product of any number of matrices of which one is singular results in a singular matrix, eq. (2.225) shows that J_o will be singular and the output light is totally polarized. This is expected as long as the optical system is non-depolarizing.

(2) When T is singular, J_o becomes singular for any value of J_i. The optical system in this case acts as an ideal polarizer [eq. (2.143)] that transforms any incident unpolarized or partially-polarized light into emergent totally-polarized light.

(3) If T is unitary (i.e. $T^\dagger = T^{-1}$), eq. (2.225) becomes a unitary transformation that preserves both the trace, $\operatorname{Tr} J$, and the determinant, $\det J$, of the coherency matrix. The optical system in this case acts as an elliptic retarder [eq. (2.94)] that does not affect the total intensity or the degree of polarization of the propagating wave.

(4) When the incident light is unpolarized, its coherency matrix J_i becomes equal to one-half the intensity ($\frac{1}{2}I_i$) times the 2×2 identity matrix [eq. (1.144)]. The coherency matrix of the output light becomes in this case

$$J_o = \tfrac{1}{2}I_i(TT^\dagger), \tag{2.226}$$

whose expansion gives

$$\begin{aligned} J_{oxx} &= \tfrac{1}{2}I_i(T_{11}T_{11}^* + T_{12}T_{12}^*), & J_{oxy} &= \tfrac{1}{2}I_i(T_{11}T_{21}^* + T_{12}T_{22}^*), \\ J_{oyx} &= \tfrac{1}{2}I_i(T_{21}T_{11}^* + T_{22}T_{12}^*), & J_{oyy} &= \tfrac{1}{2}I_i(T_{21}T_{21}^* + T_{22}T_{22}^*). \end{aligned} \tag{2.227}$$

The total intensity of the output light I_o is given by the trace of the output coherency matrix J_o [eq. (1.131)].

$$I_o = \text{Tr } J_o = J_{oxx} + J_{oyy} = \frac{1}{2} I_i \left(\sum_{ij} |T_{ij}|^2 \right). \qquad (2.228)$$

From eq. (2.228), the intensity transmittance to incident unpolarized light τ_{un} is given by

$$\tau_{un} = (I_o/I_i) = \frac{1}{2} \left(\sum_{ij} |T_{ij}|^2 \right). \qquad (2.229)$$

Because the determinant of the product of two matrices is equal to the product of the determinants of the two matrices, *i.e.* $\det AB = (\det A)(\det B)$, the trace of J_o of eq. (2.226) is given by

$$\det J_o = (\tfrac{1}{2} I_i)^2 |\det T|^2. \qquad (2.230)$$

Substituting Tr J_o and det J_o from eqs. (2.228) and (2.230) into eq. (1.139) we obtain the degree of polarization of the output light from an optical system of Jones matrix T when the incident light is unpolarized

$$\mathscr{P}_o = \left[\left(\sum_{ij} |T_{ij}|^2 \right)^2 - 4|\det T|^2 \right]^{\frac{1}{2}} \Big/ \left(\sum_{ij} |T_{ij}|^2 \right). \qquad (2.231)$$

Finally, the azimuth θ and ellipticity angle ϵ of the totally polarized component of the output light can be obtained by direct substitution from eqs. (2.227) and (2.231) into eq. (1.140).

2.11.1. *Intensity transmittance of a linear non-depolarizing optical system for partially polarized incident light*

When a beam of partially polarized quasi-monochromatic light is passed through an optical system its intensity is attenuated or amplified to an extent determined by the state of polarization at the input of the system. The polarization-dependent intensity transmittance of polarizing optical instruments and systems represents one important aspect of their light-transmitting properties that we will now investigate in more detail. The other two aspects are the effect of the optical system on the degree of polarization and its effect on the polarization form of the totally-polarized component of the light wave.

The coherency-matrix formulation presented in the previous subsection is capable of predicting the effect of an optical system on the intensity as well as the degree of polarization and polarization form of the incident partially polarized light. Using this formulation, the polarization-dependent intensity transmittance τ of a linear non-depolarizing optical system to partially polarized incident light is given by

$$\tau = (I_o/I_i) = \frac{\operatorname{Tr} \boldsymbol{J}_o}{\operatorname{Tr} \boldsymbol{J}_i},$$

or

$$\tau = \frac{\operatorname{Tr}(\boldsymbol{T}\boldsymbol{J}_i\boldsymbol{T}^\dagger)}{\operatorname{Tr} \boldsymbol{J}_i}, \tag{2.232}$$

which follows from eqs. (2.225) and eq. (1.131). Equation (2.232) is an elegant form for the intensity transmittance τ in terms of the coherency matrix of the incident light $\boldsymbol{J}_i$ and the Jones matrix $\boldsymbol{T}$ of the optical system. However, this equation is difficult to interpret because of the *implicit* way the coherency matrix $\boldsymbol{J}$ carries information about the intensity I, degree of polarization $\mathcal{P}$ and polarization form θ, ϵ of partially polarized light.

An expression can be derived for the intensity transmittance τ of an optical system that is explicitly dependent on the degree of polarization $\mathcal{P}_i$ and the polarization form θ_i, ϵ_i of the incident light. This is carried out by generalization of results we obtained earlier in §2.7 that dealt with the intensity transmittance of an optical system to incident totally polarized light. The major conclusion of §2.7 is that the intensity transmittance τ for incident totally-polarized light is completely determined by the two incident polarizations $\chi_{\max}$ and $\chi_{\min}$ for which the optical system is most transparent and most opaque, as well as their associated maximum and minimum transmittances $\tau_{\max}$ and $\tau_{\min}$, respectively. Equation (2.129) gives τ in terms of $\chi_{\max}, \chi_{\min}; \tau_{\max}, \tau_{\min}$. Equation (2.135) gives τ as a function of $\tau_{\max}, \tau_{\min}$; $\theta_{\max}, g_{\max}$ where the latter two quantities give the azimuth and the ellipticity of the polarization state $\chi_{\max}$.

First, we note that incident unpolarized light can be considered as a superposition of two *independent*, orthogonally polarized, component waves of equal intensities with the instantaneous amplitude and phase of one component varying with time in a completely uncorrelated fashion with those of the other component. The choice of the two orthogonal polarizations into which unpolarized light can be resolved is arbitrary and such polarizations can be chosen to coincide with the two orthogonal states $\chi_{\max}$ and $\chi_{\min}$ that pass through the system with maximum and minimum transmittance. Because the two component waves are independent their intensities at the output of the system can be added algebraically. Thus if I_i denotes the total intensity of the unpolarized incident beam and $\chi_{\max}$ and $\chi_{\min}$ are chosen as the orthogonal polarizations of the component waves, the intensity of the output light I_o will be given by

$$\begin{aligned} I_o &= (\tfrac{1}{2}I_i)\tau_{\max} + (\tfrac{1}{2}I_i)\tau_{\min}, \\ &= \tfrac{1}{2}(\tau_{\max} + \tau_{\min})I_i, \end{aligned}$$

and the intensity transmittance (I_o/I_i) for unpolarized incident light becomes

$$\tau_{un} = \tfrac{1}{2}(\tau_{max} + \tau_{min}). \tag{2.233}$$

If we substitute for τ_{max} and τ_{min} their values in terms of the elements T_{ij} of the Jones matrix that appear in eq. (2.115), we obtain

$$\tau_{un} = \frac{1}{2}\left(\sum_{ij} |T_{ij}|^2\right),$$

in agreement with eq. (2.229) that we obtained using the coherency-matrix formulation.

Now we are in a position to consider the case of partially-polarized incident light. As explained in §1.9, partially polarized light can be uniquely decomposed into two *independent* components: a totally-polarized component and an unpolarized component. Therefore, we can consider the propagation through the system of each of the totally polarized and the unpolarized components separately, adding their intensities at the output of the system.

For the purpose of this discussion, it is convenient to introduce a new set of parameters s to characterize partially-polarized light namely,

$$s = (I, \mathcal{P}, \chi). \tag{2.234}$$

The set s consists of three elements, the first two, I and $\mathcal{P}$, are real and the third, χ, is complex. I represents the total intensity of the wave and $\mathcal{P}$ represents its degree of polarization. By definition [eq. (1.139)], $\mathcal{P}$ is equal to the fraction of the total intensity that constitutes the totally-polarized component. χ is a complex number that completely describes the ellipse of polarization of the totally polarized component. Although I and $\mathcal{P}$ are uniquely defined, χ can assume different values depending upon the choice of the basis polarization states which are used to define the polarization of the totally polarized component.

The intensity I_o of the light emergent from the optical system can now be written as

$$I_o = \left[\tfrac{1}{2}(\tau_{max} + \tau_{min})\right][(1 - \mathcal{P})I_i] + \left[\tau_{max} + \left(\frac{q}{q+p^2}\right)(\tau_{max} - \tau_{min})\right][\mathcal{P}I_i]. \tag{2.235}$$

The significance of the two terms of eq. (2.235) is as follows. The first term is the product of the intensity transmittance for unpolarized light [eq. (2.233)] times the intensity of the unpolarized component of the incident light. The second term is the product of the intensity transmittance for totally-polarized light [eq. (2.131)] times the intensity of the totally polarized component of the incident light. The intensity transmittance (I_o/I_i) of an optical system to incident partially polarized light is obtained from eq. (2.235)

$$\tau_{\text{p.p.}} = \frac{1}{2}(\tau_{\max} + \tau_{\min}) + \frac{1}{2}(\tau_{\max} - \tau_{\min})\mathcal{P} + (\tau_{\max} - \tau_{\min})\mathcal{P}\frac{q}{q+p^2}. \qquad (2.236)$$

The first term in eq. (2.236) is constant, independent of the degree of polarization $\mathcal{P}$ or the ellipse of polarization which is represented by χ. The second term is proportional to the degree of polarization but is independent of the nature of the ellipse of polarization. The third term in eq. (2.236) is proportional to the degree of polarization and depends on the ellipse of polarization (χ) through the factor $q/(q+p^2)$. Remember that p and q are determined by χ, $\chi_{\max}$ and $\chi_{\min}$ according to eq. (2.130) and fig. 2.10.

Equation (2.236) reduces to eq. (2.233) when $\mathcal{P} = 0$, appropriate to the case of unpolarized incident light and reduces to eq. (2.131) when $\mathcal{P} = 1$, appropriate to the case of totally-polarized incident light.

An alternative form of eq. (2.236) can be obtained by replacing the factor $q/(q+p^2)$ by the expression in the square brackets that appears in eq. (2.135). This yields

$$\tau_{\text{p.p.}} = \frac{1}{2}(\tau_{\max} + \tau_{\min}) + \frac{1}{2}(\tau_{\max} - \tau_{\min})\mathcal{P}$$
$$+ (\tau_{\max} - \tau_{\min})\mathcal{P}\frac{1 + g^2 g_{\max}^2 + 2gg_{\max}\cos 2(\theta - \theta_{\max})}{(1+g^2)(1+g_{\max}^2)}, \qquad (2.237)$$

where θ and g give the azimuth and ellipticity [eq. (2.136)] of the ellipse of polarization of the totally polarized component of the incident light and $\theta_{\max}$ and $g_{\max}$ give the azimuth and ellipticity of the ellipse of polarization of totally-polarized incident light that passes through the system with maximum transmittance (minimum attenuation). Equation (2.237) is a useful practical form of the intensity transmittance of any linear non-depolarizing optical system for incident light which may be polarized, unpolarized or partially polarized.

An important consequence of eq. (2.237) is that the polarization-dependent intensity transmittance of a linear non-depolarizing optical system is completely determined by four real quantities $\tau_{\min}$, $\tau_{\max}$; $\theta_{\max}$ and $g_{\max}$. This is true irrespective of the nature of the incident light and is applicable when such light is totally polarized, unpolarized or partially polarized. From eq. (2.232), for example, one might reach the wrong conclusion that the entire Jones matrix, which is defined by seven real quantities (the absolute phase is neglected), is required to determine the intensity transmittance as a function of the incident polarization.

2.12. Propagation of partially polarized quasi-monochromatic light through depolarizing optical systems—The Mueller-matrix formulation

In this section, we examine the most general case when partially polarized quasi-monochromatic light is propagated through a *depolarizing* optical system that decreases the degree of polarization of the transmitted light [eq. (2.222b)]. Neither the Jones-matrix formulation [§2.2], nor the coherency-matrix formulation [§2.11] is capable of handling this general case. The reason is that a *deterministic* 2×2 complex Jones matrix can no longer be used to express the depolarizing incoherent interaction between the incident wave and the optical system. We have to resort to the more powerful Mueller-matrix formulation [58, 59].

The Mueller-matrix formulation is based upon the representation of the state of polarization of the light wave by a Stokes vector [§1.9] and the representation of the depolarizing optical system by a 4×4 Mueller matrix, all of whose elements are real. The processing of the light wave by the optical system is calculable from the pre-multiplication of the incident Stokes vector by the system's 4×4 Mueller matrix to produce the Stokes vector of the outgoing wave. To see this, consider a non-depolarizing optical system, since the Mueller-matrix formulation must be applicable to this special case. Following O'Neill [5], we use eq. (2.224) as a starting point. The time average of the direct product of each side of eq. (2.224) times its complex conjugate gives

$$\langle E_o(t) \times E_o^*(t) \rangle = \langle TE_i(t) \times T^*E_i^*(t) \rangle$$
$$= (T \times T^*) \langle E_i(t) \times E_i^*(t) \rangle, \qquad (2.238)$$

where use has been made of the matrix identity, $AB \times CD = (A \times C)(B \times D)$, and the assumption that the Jones matrix T of the optical system is time-independent. The time-averaged direct product $\langle E(t) \times E^*(t) \rangle$ produces the 4×1 *coherency vector* J

$$J = \langle E(t) \times E^*(t) \rangle = \begin{bmatrix} J_{xx} \\ J_{xy} \\ J_{yx} \\ J_{yy} \end{bmatrix}. \qquad (2.239)$$

In terms of J eq. (2.238) can be written as

$$J_o = (T \times T^*) J_i. \qquad (2.240)$$

Because the coherency vector J is linearly related to the Stokes vector S, $S = AJ$ [eq. (1.138b)], the relationship between the Stokes vectors S_o and S_i

§2.12] THE MUELLER-MATRIX FORMULATION

of the outgoing and the incident waves is given by

$$(A^{-1}S_o) = (T \times T^*)(A^{-1}S_i),$$

$$S_o = [A(T \times T^*)A^{-1}]S_i,$$

or, more concisely,

$$S_o = MS_i, \tag{2.241}$$

where

$$M = A(T \times T^*)A^{-1}, \tag{2.242}$$

and A is the 4×4 matrix that appears in eq. (1.138a).

Equation (2.241) gives the basic law of transformation of the Stokes vector of a partially polarized light wave as it propagates through an optical system. The 4×4 real matrix M is called the Mueller matrix of the optical system. Although the derivation of eq. (2.241) has been based on the assumption that the optical system is non-depolarizing, *eq. (2.241) is also applicable to the more general case of an optical system that exhibits depolarization*. For a non-depolarizing optical system, eq. (2.242) gives the Mueller matrix M in terms of the Jones matrix T. Of the sixteen elements of M, only seven are independent in this case. When the optical system exhibits depolarization, eq. (2.242) does not apply, because a Jones matrix T for such a system cannot be defined. In this case, all the sixteen real elements of the Mueller matrix can be independent.

If we carry out the matrix multiplication in eq. (2.242) (where A is given by eq. (1.138a)), we obtain [60]

$$M = \begin{bmatrix} \frac{1}{2}(E_1 + E_2 + E_3 + E_4) & \frac{1}{2}(E_1 - E_2 - E_3 + E_4) & F_{13} + F_{42} & -G_{13} - G_{42} \\ \frac{1}{2}(E_1 - E_2 + E_3 - E_4) & \frac{1}{2}(E_1 + E_2 - E_3 - E_4) & F_{13} - F_{42} & -G_{13} + G_{42} \\ F_{14} + F_{32} & F_{14} - F_{32} & F_{12} + F_{34} & -G_{12} + G_{34} \\ G_{14} + G_{32} & G_{14} - G_{32} & G_{12} + G_{34} & F_{12} - F_{34} \end{bmatrix}, \tag{2.243a}$$

where

$$E_i = T_i T_i^* = |T_i|^2, \quad i = 1, 2, 3, 4,$$
$$F_{ij} = F_{ji} = \text{Re}(T_i T_j^*) = \text{Re}(T_j T_i^*), \quad i, j = 1, 2, 3, 4, \tag{2.243b}$$
$$G_{ij} = -G_{ji} = \text{Im}(T_i^* T_j) = -\text{Im}(T_j^* T_i), \quad i, j = 1, 2, 3, 4.$$

For simplicity of notation, we have used T_1, T_2, T_3, and T_4 to represent T_{11}, T_{22}, T_{12}, and T_{21}, respectively, in eqs. (2.243b) *only*. Equations (2.243) can be used to construct the Mueller matrix of an optical device whose Jones matrix is known. As an example, the Jones matrix of an optical rotator is given by [see eq. (2.27)]

$$T = R(-\alpha) = \begin{bmatrix} \cos \alpha & -\sin \alpha \\ \sin \alpha & \cos \alpha \end{bmatrix},$$

which when substituted into eq. (2.243) gives the corresponding Mueller matrix

$$M = R(-\alpha) = \begin{bmatrix} 1 & 0 & 0 & 0 \\ 0 & \cos 2\alpha & -\sin 2\alpha & 0 \\ 0 & \sin 2\alpha & \cos 2\alpha & 0 \\ 0 & 0 & 0 & 1 \end{bmatrix}. \qquad (2.244)$$

The Mueller matrices of other optical devices are given in table A.4 of Appendix A.

$R(\alpha)$ of eq. (2.244) can also be considered as the 4×4 rotation matrix that transforms the Stokes vector from one Cartesian coordinate system (x, y) to another Cartesian coordinate system (x', y') that is rotated with respect to the first by an angle α according to

$$S' = R(\alpha)S. \qquad (2.245)$$

Following steps with which the reader is now familiar, the Mueller matrix of an optical system is affected by simultaneous equal rotations of the input and output reference Cartesian coordinate systems according to

$$M_{\text{new}} = R(\alpha) M_{\text{old}} R(-\alpha), \qquad (2.246)$$

in direct analogy with eq. (2.31) that describes the corresponding transformation of the Jones matrix.

When a light wave is sequentially processed by a cascade of N optical devices (systems), the over-all combined effect of the train is described by a Mueller matrix

$$M_{\text{comb}} = M_N M_{N-1} \ldots M_{\text{II}} M_{\text{I}}, \qquad (2.247)$$

where M_k is the Mueller matrix of the kth optical device to operate on the wave and I, II ... N is the order in which the devices are encountered by the wave. Again, eq. (2.247) is the exact counter-part of eq. (2.11) in the Jones-matrix formulation.

The basic equation of the Mueller-matrix formulation, eq. (2.241), shows how to calculate the result of passing a quasi-monochromatic light wave through a depolarizing optical system. In particular, the Stokes vector S_o of the wave outgoing from the optical system is obtained by operating on (pre-multiplying) the Stokes vector S_i of the incident wave by the Mueller matrix of the optical system. From the computed Stokes vector S_o, we may calculate such properties of the outgoing wave as its total intensity [eq. (1.119)], degree of polarization [eq. (1.127)] or the polarization form of its totally polarized component [eq. (1.128)].

The expression for the polarization-dependent intensity transmittance of an optical system is most easily obtained when the Mueller-matrix formulation is used. The reason is that the total intensity shows explicitly as the first element of the Stokes vector [eq. (1.119)]. Substituting the Stokes vector of eq. (1.129) for S_i in eq. (2.241) gives the intensity transmittance τ

$$\tau = \left(\frac{I_o}{I_i}\right) = m_{11} + \mathcal{P}[m_{12}\cos 2\epsilon \cos 2\theta + m_{13}\cos 2\epsilon \sin 2\theta + m_{14}\sin 2\epsilon],$$
(2.248)

where m_{11}, m_{12}, m_{13} and m_{14} are the elements of the first row of the Mueller matrix M of the optical system, $\mathcal{P}$ is the degree of polarization of the incident wave and θ, ϵ describe the polarization form of the totally polarized component of that wave. Equation (2.248) shows that the polarization-dependent intensity transmittance of *any* optical system (depolarizing or non-depolarizing) is completely determined by four real parameters. The same conclusion has been reached before in connection with non-depolarizing optical systems and appears in eq. (2.237). Equation (2.237) can be derived from eq. (2.248) with the help of eqs. (2.243) and (2.134).

The transformation of the Stokes vector according to eq. (2.241) can be used to derive the movement on the Poincaré sphere (or, more generally, in the Stokes subspace of §1.9) of the point that represents the state of polarization of a light wave as a result of its interaction with an optical system with a known Mueller matrix.

Although the Mueller-matrix formulation is tailored for the study of the propagation of partially-polarized light through depolarizing optical systems, its range of applicability obviously covers the special cases when the optical system is non-depolarizing and the incident light is either partially or totally polarized. In the latter special cases, the Mueller-matrix formulation is definitely more powerful (general) than we need. If the incident wave is totally polarized and the optical system is non-depolarizing, neither the Mueller-matrix nor the coherency-matrix formulation can be used to examine information on the absolute phase of the wave. This is because such absolute-phase information is suppressed in the definitions of the Stokes vector [eq. (1.112)] and the coherency matrix [eq. (1.130)]. The Jones-matrix formulation has to be used in this case. In general, the choice of a particular formulation to handle the propagation of polarized light is dependent upon the nature of the optical system and the light beam under consideration.

CHAPTER 3

Theory and Analysis of Measurements in Ellipsometer Systems

3.1. Introduction

Ellipsometry can be generally defined as the measurement of the state of polarization of a polarized vector wave. Here we are particularly interested in *optical ellipsometry* of polarized light waves.[1]

Although measurement of the state of polarization of a light wave is important in its own right, ellipsometry is generally conducted in order to obtain "information" about an "optical system" that modifies the state of polarization. In the general scheme of ellipsometry to be studied here, a polarized light-wave probe is allowed to interact with an optical system under investigation. This interaction changes the state of polarization of the wave (and quite probably, other properties too). Measurement of the initial and final states of polarization, repeated for an adequate number of different initial states, leads to the determination of the law of transformation of polarization by the system as described, say, by its Jones or Mueller matrix.[2] To extract more fundamental information about the optical system than is conveyed by its Jones or Mueller matrix, it is necessary to examine

---

[1] Polarimetry (measurement of *polarization*), rather than ellipsometry (measurement of the *ellipse* of polarization), may appear to fit the present definition better. However, because of the growing use of the term ellipsometry, we have chosen it to convey the above extended meaning.

[2] Dependent upon the optical system under measurement, the transformation of polarization may be specified by the system's Jones matrix (in the absence of depolarization) or Mueller matrix (in the presence of depolarization). See §2.2 and §2.12.

light-matter interaction within the system by the electromagnetic theory of light. In other words, we are required to study the details of the *internal* polarization-modifying processes that are responsible for the external behavior as described by the measured Jones or Mueller matrix of the system.

An operational diagram of a general ellipsometer arrangement is shown in fig. 3.1. A well-collimated monochromatic or quasi-monochromatic beam from a suitable light source (L) is passed through a variable polarizer (P) to

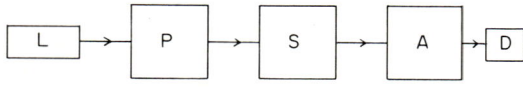

Fig. 3.1. An operational diagram of a general ellipsometer arrangement. L, P, S, A, and D represent a light source, controlled polarizer, optical system under measurement, variable polarization analyzer, and photodetector, respectively.

produce light of known controlled polarization. This light interacts with the optical system (S) under study and its polarization is modified. The modified state of polarization at the output of the system is measured (analyzed) by a variable polarization analyzer (A) followed by a photodetector (D).[3]

It is assumed that the interaction between the light wave and the optical system is linear and frequency-conserving (§2.1). The optical system can modify the state of polarization of the probe wave through one or a combination of the following basic processes.

(1) *Reflection or Refraction*: when a light wave is reflected or refracted at the interface between two optically dissimilar media, the state of polarization is changed abruptly. Such a change is caused by the difference in the Fresnel reflection or transmission coefficients for the two linear polarizations parallel (p) and perpendicular (s) to the plane of incidence.

(2) *Transmission*: here the state of polarization changes continuously as light progresses through a medium that exhibits optical anisotropy (refractive, absorptive or both).

---

[3]In astronomical ellipsometry [see, for example, refs. 61 and 62] where the state of polarization of an em signal from outer space is modified before it reaches the earth's ellipsometer or polarimeter by interacting with an intervening system (*e.g.*, the magnetic field of the radio source or the earth's ionosphere), the experimenter obviously has no control over either the source of radiation or the system that modifies its polarization. Hence, the ellipsometer, in this case, consists essentially of a polarization analyzer. In radio-echo experiments, where a signal of controlled polarization is sent out and the polarization of the echo signal is analyzed, the general ellipsometer arrangement used is the same as that discussed here.

(3) *Scattering*: this takes place when a light wave traverses a medium with spatially inhomogeneous index of refraction caused by the presence of scattering centers as in aerosols and emulsions. In contrast with reflection and transmission which do not significantly affect the collimation of a beam, scattering is usually accompanied by a redistribution of the scattered energy over a wide range of solid angles.

Dependent upon the prevalent mode of interaction that modifies the state of polarization of light, we may distinguish between

(i) Reflection or Surface Ellipsometry,

(ii) Transmission Ellipsometry[4] (Polarimetry), and

(iii) Scattering Ellipsometry.

It is interesting to observe that although many of the underlying principles are the same, the above division corresponds to three distinct disciplines of research that have remained largely separated from one another.

Reflection ellipsometry has long been recognized as an important tool for the study of surfaces and thin films since the time of Drude [63]. Among the many useful applications of reflection ellipsometry [1, 64] (to be discussed in detail in chapter 6) are (1) measurement of the optical properties of materials and their frequency dependence (wavelength dispersion); the materials may be in the liquid or solid phase, may be optically isotropic or anisotropic, and can be either in bulk or thin-film form. (2) Monitoring of phenomena on surfaces that involve either the growth of thin films starting from a submonolayer (*e.g.*, by oxidation, deposition, adsorption or diffusion of impurities), or the removal of such films (*e.g.*, by desorption, sputtering or diffusion). (3) Measurement of physical factors that affect the optical properties such as electric and magnetic fields, stress or temperature.

Transmission ellipsometry (polarimetry) is an important well-established analytical method of physical chemistry [32]. It is primarily concerned with measurements on bulk samples (gas, liquid, or solid) of (1) *natural* optical rotation (OR) and circular dichroism (CD), natural linear birefringence (LB) and linear dichroism (LD), and, more generally, natural elliptical birefringence (EB) and elliptical dichroism (ED). (2) *Induced* optical anisotropy as in streaming birefringence, photoelasticity, the Faraday, Kerr and Cotton-Mouton effects. (3) The wavelength dispersion of the above properties. The major objective in such studies is to explore and identify the molecular structure of matter.

Scattering ellipsometry [60] has a wide range of uses: (1) industrial

---

[4]In ch. 4, we will use the term transmission ellipsometry to indicate measurements on transparent ambient-film-substrate systems, to obtain information ordinarily sought in reflection ellipsometry. In that case, the media need not be optically anisotropic; the change in polarization is caused by a combination of refraction and interference in the thin transparent film.

applications as in the measurement of density and particle-size distributions of colloidal solutions and aerosols, (2) meteorological applications as in the study of haze, clouds, and rain, and (3) astronomical applications as in the study of planetary atmospheres and interplanetary dust.

The term ellipsometer (from which the technique of ellipsometry derives its name) was first coined by Rothen [65] to denote an optical instrument for the measurement of thin surface films by the reflection of polarized light. Polarimetry is the term more commonly used to denote transmission ellipsometry for the measurement of optical rotation and circular dichroism. Scattering ellipsometry refers to light-scattering measurements that do not involve a change of frequency (linear scattering). In the literature on frequency-conserving light scattering, the word ellipsometry is hardly used. The aim of a unified definition of ellipsometry, as given at the beginning of this section, is to encourage fruitful interaction between three seemingly isolated areas of research.

Throughout this and the remaining chapters of this book, our emphasis will be primarily on reflection and, to a lesser extent, on transmission ellipsometry.

3.2. Ellipsometric measurement of the Jones matrix of an optical system–The basis of generalized ellipsometry

In the following, we assume that the optical system under investigation is non-depolarizing,[5] in addition to being linear, so that a 2×2 Jones matrix T describes such a system completely. This accounts for most applications of reflection and transmission ellipsometry. In the absence of depolarization, both the incident wave on the system and the modified wave emergent from the system will be totally polarized.

To measure the Jones matrix of an optical system, we make use of the important fact that when the ellipses of polarization (with no account of total amplitude or absolute phase) of the incident and the outgoing waves are represented by the complex variables χ_i and χ_o [§1.7], the relationship between χ_o and χ_i is given by a bilinear transformation [eq. (2.50), §2.3]

$$\chi_o = \frac{T_{22}\chi_i + T_{21}}{T_{12}\chi_i + T_{11}}. \tag{3.1}$$

This bilinear transformation is uniquely determined by the mapping of three input polarizations $(\chi_{i1}, \chi_{i2}, \chi_{i3})$ into the corresponding three output

---

[5] In §3.8, we will study the effect of small amounts of depolarization (incoherent scattering) as a source of error in ellipsometric measurements.

§3.2] MEASUREMENT OF THE JONES MATRIX

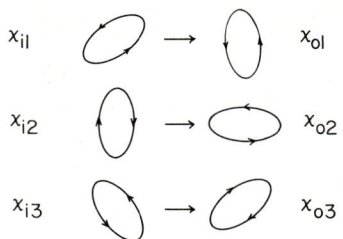

Fig. 3.2. The normalized Jones matrix of an optical system is determined by the mapping of three input polarizations $\chi_{i1}, \chi_{i2}, \chi_{i3}$ onto the corresponding three output polarizations $\chi_{o1}, \chi_{o2}, \chi_{o3}$.

polarizations $(\chi_{o1}, \chi_{o2}, \chi_{o3})$ [66]. Thus according to eq. (2.55) of §2.4, we have

$$\frac{(\chi_o - \chi_{o1})(\chi_{o3} - \chi_{o2})}{(\chi_o - \chi_{o2})(\chi_{o3} - \chi_{o1})} = \frac{(\chi_i - \chi_{i1})(\chi_{i3} - \chi_{i2})}{(\chi_i - \chi_{i2})(\chi_{i3} - \chi_{i1})}. \tag{3.2}$$

Figure 3.2 is intended to pictorially emphasize this basic fact. By rearranging eq. (3.2) to take the form of eq. (3.1), the elements of the Jones matrix can be obtained, up to a constant multiplier, as follows:

$$\begin{aligned} T_{11} &= \chi_{i2} - \chi_{i1} H, \\ T_{12} &= H - 1, \\ T_{21} &= \chi_{i2}\chi_{o1} - \chi_{i1}\chi_{o2} H, \\ T_{22} &= -\chi_{o1} + \chi_{o2} H, \\ H &= \frac{(\chi_{o3} - \chi_{o1})(\chi_{i3} - \chi_{i2})}{(\chi_{o3} - \chi_{o2})(\chi_{i3} - \chi_{i1})}. \end{aligned} \tag{3.3}$$

Let us denote the so-far-undetermined constant multiplier by g, so that the elements of the actual Jones matrix are g times those given by eq. (3.3), *i.e.*, $gT_{11}, gT_{12}, gT_{21}$, and gT_{22}. To determine $|g|$, we have to measure the effect of the optical system on the intensity of the incident wave in one known polarization state. Substitution into any of the expressions for the polarization-dependent intensity transmittance of an optical system that are given in §2.7 [*e.g.*, eq. (2.105)] gives $|g|$ directly. In order to determine the effect of the optical system on the absolute phase of the light wave, the angle of the constant multiplier g, arg (g), should also be measured. This can be done by placing the optical system in one of the two arms of an interferometer using light polarized in one of the two eigenstates of the optical system and recording the shift in the fringe pattern [67].

In most cases of interest, it will be adequate to determine the *normalized* Jones matrix obtained by dividing T_{ij} of eq. (3.3) by T_{22} ($\neq 0$), *i.e.* the matrix whose elements are T_{11}/T_{22}, T_{12}/T_{22}, T_{21}/T_{22} and 1.

3.3. Ellipsometry on optical systems with known eigenpolarizations

Let us now consider the special, but important, case when the eigenpolarizations χ_{e1} and χ_{e2} of the optical system are known *a priori*. (By definition, the eigenpolarizations are the two polarization states that pass through the optical system unchanged, eq. (2.59).) Knowledge of the eigenpolarizations may be based on symmetry and/or physical consideration. For example, when light is obliquely reflected at the interface between two isotropic and non-gyrotropic (non-optically active) media, the linear polarizations (electric-field vibrations) parallel (p) and perpendicular (s) to the plane of incidence are reflected with their polarization unchanged. This leads to a *diagonal* Cartesian reflection (Jones) matrix in the p-s frame of reference. Both symmetry with respect to the plane of incidence and absence of optical activity are necessary requirements for the eigenpolarizations of a reflector to become linear, parallel and perpendicular to the plane of incidence. Another example of a system with known eigenpolarizations is that of a medium whose molecules have helical structure, leading to natural optical rotation and circular dichroism; or the case of an optically isotropic medium placed in a magnetic field that induces circular birefringence (Faraday effect). In either of these two examples, the right- and left-circular polarizations are the eigenpolarizations of the system.

With the eigenpolarizations assumed known, all that remain to be measured are their associated eigenvalues V_{e1} and V_{e2}. As we know from §2.6, the eigenvalues V_{e1} and V_{e2} determine the effect of the optical system on the amplitude and phase of the incident light when it is polarized in the eigenstates χ_{e1} and χ_{e2}, respectively.

Optical systems of known eigenpolarizations can further be subdivided according to the nature of the eigenpolarizations into (1) systems with orthogonal *linear* eigenpolarizations, (2) systems with orthogonal *circular* eigenpolarizations, and (3) systems with arbitrary *elliptic* eigenpolarizations. Although the linear and circular polarizations are special cases of elliptic polarization, the many applications of ellipsometry on systems with orthogonal linear or circular eigenpolarizations justify separate detailed treatments for these special cases. In fact, optical systems with orthogonal linear eigenpolarizations account for most applications of reflection ellipsometry, and the applications of transmission ellipsometry (polarimetry) concerned with linear birefringence and linear dichroism. The case of orthogonal circular eigenpolarizations covers transmission ellipsometry (polarimetry) on systems that exhibit optical rotation (OR) and circular dichroism (CD), either natural or induced.

3.4. Ellipsometric measurement of the ratio of eigenvalues of an optical system with orthogonal linear eigenpolarizations

Figure 3.3 shows the ellipsometer arrangement under consideration. A well-collimated beam of monochromatic, or quasi-monochromatic, circularly polarized or unpolarized (natural) light from a suitable source is passed through the polarizing section of the instrument that consists of a linear

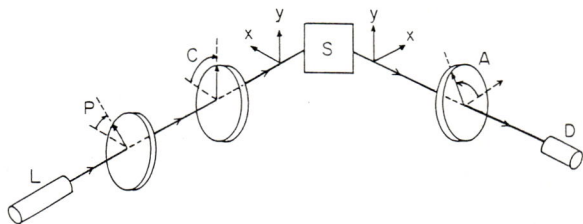

Fig. 3.3. The PCSA ellipsometer arrangement with a polarizing section that consists of a linear polarizer P and a compensator (linear retarder) C, and an analyzing section that consists of a linear analyzer A. S is the optical system under measurement with linear eigenpolarizations x and y. L and D are the light source and photodetector, respectively. P, C, and A also represent the rotational azimuth angles of polarizer, compensator and analyzer, measured from the direction of the x linear eigenpolarization.

polarizer P and a linear retarder or compensator C. Emergent from the polarizing section of the instrument, and incident on the optical system, is totally polarized light whose state of polarization is controlled by the rotational azimuth positions of the polarizer and compensator around the beam axis. The optical system S under investigation has orthogonal linear eigenpolarizations (χ_{ex} and χ_{ey}) along orthogonal coordinate axes x and y. The optical system modifies the state of polarization of the incident light beam when that state is different from either one of its eigenpolarizations χ_{ex} or χ_{ey}. The modified state of polarization of the light beam outgoing from the optical system is analyzed by the analyzing section of the instrument that consists of a linear analyzer A followed by a photodetector D. The photodetector D measures the light flux after the beam has travelled through the polarizer-compensator-system-analyzer (PCSA) sequence of elements.

The orientations of the polarizer, compensator, and analyzer around the beam axis are specified by the azimuth angles P, C, and A, respectively, fig. 3.4. For the polarizer and analyzer, the azimuths P and A define the orientation of their transmission axes (*i.e.*, the directions of the *transmitted* linear eigenpolarizations). For the compensator, the azimuth C defines the orientation of its fast axis (*i.e.*, the direction of the fast linear eigenpolarization). All azimuths P, C, and A are measured from the direction of the x

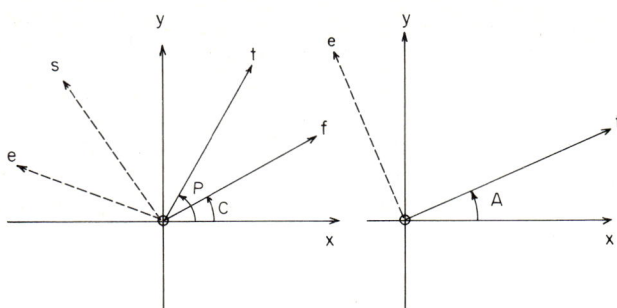

Fig. 3.4. The orientation of the polarizer and compensator (left) and that of the analyzer (right) around the beam axis are specified by the azimuth angles P, C, and A, respectively. For the polarizer and analyzer, the orientation is that of the transmission axis t; for the compensator the orientation is that of the fast axis f. (e represents the extinction axis of P and A while s represents the slow axis of C, all shown in dashed lines.) All azimuths are measured positive in a counter-clockwise sense looking into the beam (coming out of the page in this figure) from the x linear eigenpolarization of the optical system S.

linear eigenpolarization of the optical system under measurement and are taken positive in a counterclockwise sense when looking against the direction of propagation of the beam.

In order to follow the state of polarization of the light beam as it progresses through the ellipsometer, we describe the light beam by its Jones vector and the optical elements by their Jones matrices [§2.2]. Because the ellipse of polarization (as specified by its azimuth and ellipticity) is of special interest, we also characterize the wave by the complex polarization variable χ.[6]

The following notation will be used. Letter superscripts denote the coordinate system with respect to which a Jones vector or matrix is referenced. A principal frame is a coordinate system in which the (Cartesian) Jones matrix of an optical component is diagonal (it is being assumed that the *same* system is used at both the input and output of the optical component). The first letter of a subscript identifies an optical component, the second either its input or output, and, in a few cases, a third subscript will indicate a Cartesian component of the electric field. For example, $\boldsymbol{E}_{CO}^{fs}$ is the Jones vector (electric field) of the light beam at the *o*utput of the *c*ompensator in its *f*ast-*s*low principal frame of reference.

---

[6] In the subsequent analysis, we examine the transformation of the Jones vector and complex polarization variable χ as if they were independent. However, from the basic definition of eq. (1.77), we know that χ can be obtained from the Jones vector by taking the ratio of its two elements. Our objective here is to familiarize the reader with both representations.

The component of this electric vector parallel to the fast axis is denoted by $E_{\text{CO,f}}$. Switching between two coordinate systems changes the Jones vector in a manner determined by the rotation $\mathbf{R}(\alpha)$ and counter-rotation $\mathbf{R}(-\alpha)$ matrices [eqs. (1.42)–(1.46)].

First, consider the polarizing section of the ellipsometer. Unpolarized or circularly polarized light incident on the ideal linear polarizer becomes linearly polarized at its output, with a Jones vector and complex polarization variable given by

$$\mathbf{E}_{\text{PO}}^{\text{te}} = A_c \begin{bmatrix} 1 \\ 0 \end{bmatrix}, \qquad \chi_{\text{PO}}^{\text{te}} = 0, \tag{3.4}$$

[see eqs. (2.70) and (2.71), also fig. 1.9]. A_c contains information on the intensity ($I = |A_c|^2$) and absolute phase ($\delta = \arg(A_c)$) of the wave emergent from the polarizer. The superscript te refers to the *transmission-extinction* principal frame of the polarizer. The linear polarization (electric-field vibration) described by eq. (3.4) is oriented at an azimuth P measured from the direction of the x-axis of the optical system under measurement, fig. 3.4.

To examine the effect of the compensator, which follows the polarizer along the path of the light beam, we change the reference coordinate system from the transmission-extinction principal frame of the polarizer to the fast-slow principal frame of the compensator. This te→fs coordinate rotation transforms the Jones vector of eq. (3.4) according to

$$\mathbf{E}_{\text{CI}}^{\text{fs}} = \mathbf{R}(P - C)\mathbf{E}_{\text{PO}}^{\text{te}}, \tag{3.5}$$

[see eq. (1.44)] where $\mathbf{R}(P - C)$ is the rotation matrix given in expanded form by eq. (1.45a). The transformation of the associated complex polarization variable is given by

$$\chi_{\text{CI}}^{\text{fs}} = \frac{\cos(P - C)\chi_{\text{PO}}^{\text{te}} + \sin(P - C)}{-\sin(P - C)\chi_{\text{PO}}^{\text{te}} + \cos(P - C)}, \tag{3.6}$$

[see eq. (1.105)]. Upon substituting $\mathbf{E}_{\text{PO}}^{\text{te}}$ and $\chi_{\text{PO}}^{\text{te}}$ from eqs. (3.4) into eqs. (3.5) and (3.6), respectively, we obtain

$$\mathbf{E}_{\text{CI}}^{\text{fs}} = A_c \begin{bmatrix} \cos(P - C) \\ \sin(P - C) \end{bmatrix}, \qquad \chi_{\text{CI}}^{\text{fs}} = \tan(P - C). \tag{3.7}$$

The compensator is assumed to have a slow-to-fast relative complex-amplitude transmittance, *i.e.*, ratio of slow- and fast-axis eigenvalues, of $\rho_C = T_c e^{i\delta_c}$. (This means that the component of the electric vector incident on the compensator parallel to its slow axis is retarded in phase by $-\delta_C$ and is attenuated in amplitude by T_C, relative to the orthogonal component parallel to its fast axis.) The Jones matrix of the compensator referenced to coordinate systems at its input and output that are parallel to the fast and slow axes is given by (see §2.2 and the appendix)

$$\boldsymbol{T}_C^{fs} = K_C \begin{bmatrix} 1 & 0 \\ 0 & \rho_C \end{bmatrix}, \tag{3.8}$$

where K_C accounts for the equal attenuation and phase shift along the fast and the slow axes. Premultiplication of the incident Jones vector $\boldsymbol{E}_{CI}^{fs}$ by the compensator's Jones matrix $\boldsymbol{T}_C^{fs}$ gives the Jones vector of the output light. The transformation of the complex polarization variable $\chi_{CI}^{fs} \to \chi_{CO}^{fs}$, between the input and output of the compensator, is given by eq. (3.1). Thus we have

$$\boldsymbol{E}_{CO}^{fs} = \boldsymbol{T}_C^{fs} \boldsymbol{E}_{CI}^{fs}, \qquad \chi_{CO}^{fs} = (T_{C22}^{fs}/T_{C11}^{fs})\chi_{CI}^{fs}. \tag{3.9}$$

Upon substitution of $\boldsymbol{E}_{CI}^{fs}$ and χ_{CI}^{fs} from eqs. (3.7), and $\boldsymbol{T}_C^{fs}$ from eq. (3.8), into eqs. (3.9), we find that

$$\boldsymbol{E}_{CO}^{fs} = K_C A_c \begin{bmatrix} \cos(P-C) \\ \rho_C \sin(P-C) \end{bmatrix}, \qquad \chi_{CO}^{fs} = \rho_C \tan(P-C). \tag{3.10}$$

A coordinate counter-rotation by an angle C, from the f-s principal frame of the compensator to the x-y principle frame of the optical system under measurement, transforms the Jones vector of the output light from the compensator according to

$$\boldsymbol{E}_{CO}^{xy} = R(-C)\boldsymbol{E}_{CO}^{fs}. \tag{3.11}$$

The transformation of the associated complex polarization variable is given by

$$\chi_{CO}^{xy} = \frac{\cos C \chi_{CO}^{fs} + \sin C}{-\sin C \chi_{CO}^{fs} + \cos C}. \tag{3.12}$$

Substitution of $\boldsymbol{E}_{CO}^{fs}$ and χ_{CO}^{fs} from eqs. (3.10) into eqs. (3.11) and (3.12) respectively, leads to

$$\boldsymbol{E}_{CO}^{xy} = K_C A_c \begin{bmatrix} \cos C \cos(P-C) - \rho_C \sin C \sin(P-C) \\ \sin C \cos(P-C) + \rho_C \cos C \sin(P-C) \end{bmatrix}, \tag{3.13}$$

$$\chi_{CO}^{xy} = \frac{\tan C + \rho_C \tan(P-C)}{1 - \rho_C \tan C \tan(P-C)}. \tag{3.14}$$

After having analyzed the polarizing section of the instrument, we turn our attention to the modification of polarization caused by the optical system S under measurement. Because the field emergent from the compensator is incident on the optical system (neglecting inter-element propagation phase factors), $\boldsymbol{E}_{CO}^{xy} = \boldsymbol{E}_{SI}^{xy}$, and $\chi_{CO}^{xy} = \chi_{SI}^{xy}$. The Jones vector of the light wave at the output of the optical system is related to that at its input by

$$\boldsymbol{E}_{SO}^{xy} = \boldsymbol{T}_{S}^{xy}\boldsymbol{E}_{SI}^{xy}. \tag{3.15}$$

The optical system is assumed to have orthogonal linear eigenpolarizations parallel to the x and y coordinate axes so that the Jones matrix $\boldsymbol{T}_{S}^{xy}$ is diagonal

$$\boldsymbol{T}_{S}^{xy} = \begin{bmatrix} V_{ex} & 0 \\ 0 & V_{ey} \end{bmatrix}, \tag{3.16}$$

where V_{ex} and V_{ey} represent the eigenvalues associated with the x and y linear eigenpolarizations, respectively. From eqs. (3.1) and (3.16), the transformation of the complex polarization variable between the input and output of S is given by

$$\chi_{SO}^{xy} = \frac{V_{ey}}{V_{ex}} \chi_{SI}^{xy}. \tag{3.17}$$

Substituting $\boldsymbol{E}_{SI}^{xy}(=\boldsymbol{E}_{CO}^{xy})$ and $\chi_{SI}^{xy}(=\chi_{CO}^{xy})$ from eqs. (3.13) into eqs. (3.15) and (3.17), respectively (and making use of eq. (3.16)), we get

$$\boldsymbol{E}_{SO}^{xy} = K_C A_c \begin{bmatrix} V_{ex}[\cos C \cos (P-C) - \rho_C \sin C \sin (P-C)] \\ V_{ey}[\sin C \cos (P-C) + \rho_C \cos C \sin (P-C)] \end{bmatrix}, \tag{3.18}$$

$$\chi_{SO}^{xy} = \frac{V_{ey}}{V_{ex}} \frac{\tan C + \rho_C \tan (P-C)}{1 - \rho_C \tan C \tan (P-C)}.$$

The final step in the chain of operations on the light beam is performed by the analyzing section of the ellipsometer. Here we assume that the outgoing light from the optical system is analyzed by a linear analyzer that can be rotated around the beam axis, followed by a photodetector that records the light flux that impinges upon it. The ideal linear analyzer has its transmission axis oriented at an azimuth A, measured from the direction of the x linear eigenpolarization of the optical system. A coordinate rotation from the optical system's x-y principal frame to the analyzer's t-e principal frame changes the Jones vector before the analyzer according to

$$\boldsymbol{E}_{AI}^{te} = \boldsymbol{R}(A)\boldsymbol{E}_{AI}^{xy}. \tag{3.19}$$

By use of the expression for the rotation matrix $\boldsymbol{R}(A)$, eq. (1.45a), eq. (3.19) is expanded to read

$$\begin{bmatrix} E_{AI,t} \\ E_{AI,e} \end{bmatrix} = \begin{bmatrix} (\cos A\, E_{AI,x} + \sin A\, E_{AI,y}) \\ (-\sin A\, E_{AI,x} + \cos A\, E_{AI,y}) \end{bmatrix}. \tag{3.20}$$

In its principal frame, the Jones matrix of the analyzer is diagonal

$$T_A^{te} = K_A \begin{bmatrix} 1 & 0 \\ 0 & 0 \end{bmatrix}, \tag{3.21}$$

where K_A accounts for amplitude and phase changes experienced by the transmitted linear eigenpolarization. The field transmitted by the analyzer is given in terms of the incident field by

$$\boldsymbol{E}_{AO}^{te} = \boldsymbol{T}_A^{te} \boldsymbol{E}_{AI}^{te}, \qquad \chi_{AO}^{te} = (T_{A22}^{te}/T_{A11}^{te})\chi_{AI}^{te}. \tag{3.22}$$

By use of eq. (3.21), eqs. (3.22) become

$$\boldsymbol{E}_{AO}^{te} = K_A \begin{bmatrix} E_{AI,t} \\ 0 \end{bmatrix}, \qquad \chi_{AO}^{te} = 0. \tag{3.23}$$

If the photodetector response is a linear function of the total light flux that impinges upon it, its output will be proportional to the intensity of the light emergent from the analyzer. From eq. (3.23), the detected signal $\mathscr{I}_D$ is given by [see eq. (1.35)]

$$\begin{aligned} \mathscr{I}_D &= K_D(\boldsymbol{E}_{AO}^{\dagger} \boldsymbol{E}_{AO}) \\ &= K_D K_A K_A^* E_{AI,t} E_{AI,t}^* \\ &= K_D |K_A|^2 |E_{AI,t}|^2, \end{aligned} \tag{3.24}$$

where K_D is a real factor that depends on the intensity profile of the light beam and the nature of the photodetector.[7] The component $E_{AI,t}$ of the electric field incident on the analyzer parallel to its transmission axis in eq. (3.24) is obtained from eq. (3.20) as

$$E_{AI,t} = \cos A \; E_{AI,x} + \sin A \; E_{AI,y}.$$

Because $E_{AI,x}$ and $E_{AI,y}$ are, in turn, given by the first of eqs. (3.18), the above equation becomes

$$\begin{aligned} E_{AI,t} = K_C A_c \{ &V_{ex} \cos A \; [\cos C \cos (P-C) - \rho_C \sin C \sin (P-C)] \\ &+ V_{ey} \sin A \; [\sin C \cos (P-C) + \rho_C \cos C \sin (P-C)] \}. \end{aligned} \tag{3.25}$$

Substituting eq. (3.25) into eq. (3.24), and lumping together the different constant multiplicative factors into a single factor G, gives

$$\mathscr{I}_D = GLL^* = G|L|^2, \tag{3.26}$$

---

[7] Throughout our discussion, we have assumed that the properties of the optical devices are uniform over the cross section of the beam. Because the sensitivity S_D of the photocathode can be non-uniform over the area a covered by the beam, the detected signal $\mathscr{I}_D$ is related to the incident intensity distribution $I(=\boldsymbol{E}^{\dagger}\boldsymbol{E})$ by $\mathscr{I}_D = \int_a IS_D da$. With the beam uniformly polarized, eq. (3.24) is obtained.

where

$$L = V_{ex} \cos A [\cos C \cos (P - C) - \rho_C \sin C \sin (P - C)]$$
$$+ V_{ey} \sin A [\sin C \cos (P - C) + \rho_C \cos C \sin (P - C)], \quad (3.27)$$

and

$$G = |A_c|^2 |K_C|^2 |K_A|^2 K_D. \quad (3.28)$$

Equations (3.26) and (3.27) constitute the basis of subsequent discussions. These results have been derived by a procedure whereby the state of polarization of the light beam was examined at different points along its path. The same result could have been obtained, more directly, by use of the fact that the over-all effect of a train of devices placed in the path of a light beam is described by the resultant Jones matrix obtained by multiplying together the Jones matrices of the individual devices, taking proper account of their arrangement and inter-element coordinate transformations. According to this procedure, the Jones vector of the light wave after the analyzer, falling on the photodetector, is related to the Jones vector of the wave at the output of the polarizer by [see eqs. (2.10) and (2.11) of §2.2]

$$\boldsymbol{E}_{AO}^{te} = \boldsymbol{T}_A^{te} \boldsymbol{R}(A) \boldsymbol{T}_S^{xy} \boldsymbol{R}(-C) \boldsymbol{T}_C^{fs} \boldsymbol{R}(P - C) \boldsymbol{E}_{PO}^{te}, \quad (3.29)$$

from which the detected signal can be obtained using eq. (3.24). In eq. (3.29), the matrices $\boldsymbol{T}_C^{fs}$, $\boldsymbol{T}_S^{xy}$ and $\boldsymbol{T}_A^{te}$ are given by eqs. (3.8), (3.16), and (3.21), respectively, while the rotation matrix is given by eq. (1.45a). The field outgoing from the polarizer $\boldsymbol{E}_{PO}^{te}$ is given by eq. (3.4). If we carry out the matrix multiplications that appear in eq. (3.29), and use eq. (3.24), we obtain the same result as in eqs. (3.26) and (3.27). The step-by-step procedure gives detailed insight as to the state of polarization along each section of the beam between two optical components in succession and serves as an example of the use of the Jones-matrix and complex-variable formulations in the analysis of a cascade of devices.

From eqs. (3.26) and (3.27), we may write

$$\mathcal{I}_D = f(P, C, A, \rho_C, V_{ex}, V_{ey}), \quad (3.30)$$

which shows that the detected signal, as recorded by the photodetector, is a function of (1) the azimuth-angle settings P, C, and A of the polarizer, compensator, and analyzer, (2) the slow-to-fast relative complex-amplitude transmittance of the compensator ρ_C, and (3) the complex eigenvalues V_{ex} and V_{ey} of the optical system to be measured. The principle of *photometric ellipsometry*[8] is based on the fact that the detected light flux contains information about V_{ex} and V_{ey} that can be extracted by proper use of eq.

---

[8] This will be discussed in detail in §3.10.

(3.30). Note that the absolute phase angle of either V_{ex} or V_{ey} is inaccessible, however, because the final quantity to be measured is total light flux.

3.4.1. Null ellipsometry

Null ellipsometry is based on finding a set of azimuth angles for the polarizer, compensator and analyzer (P, C, A) such that the light flux falling on the photodetector is extinguished. Besides the three azimuth angles P, C, and A, the relative retardation δ_C of the compensator is a fourth parameter that can be adjusted in search for the null condition, if a variable-retardation compensator is used.[9] Ideally, the null condition corresponds to zero detected signal[10]

$$\mathscr{I}_D = 0. \tag{3.31}$$

According to eq. (3.26), this is equivalent to the requirement that

$$L = 0. \tag{3.32}$$

If we substitute into eq. (3.32) the expression for L given by eq. (3.27), we obtain

$$\rho_S = -\tan A \left[\frac{\tan C + \rho_C \tan(P-C)}{1 - \rho_C \tan C \tan(P-C)} \right], \tag{3.33}$$

where

$$\rho_S = \frac{V_{ex}}{V_{ey}}. \tag{3.34}$$

From eqs. (3.33) and (3.34), we conclude that the ratio of the eigenvalues $\rho_S = V_{ex}/V_{ey}$ of an optical system with orthogonal linear eigenpolarizations can be measured by an ellipsometer from *one set of nulling angles* (P, C, A), provided that the slow-to-fast relative complex-amplitude transmittance ρ_C of the compensator is known.

An alternative approach to eq. (3.33) is based on the relationship between the complex variables χ_{SO}^{xy} and χ_{SI}^{xy} that describe the states of polarization at the input and the output of the optical system. From eq. (3.17), it is evident

---

[9] Because the resolution to which δ_C can be measured using compensators of the common Babinet-Soleil type is about two orders of magnitude poorer than that of reading the azimuth angles of the divided circles, nulling schemes that involve adjustment of δ_C are usually avoided.

[10] In practice, the null condition corresponds to *minimum* detected light flux. The depth of the minimum is not strictly zero, and depends on the depolarizing tendencies of the optical devices encountered by the light beam, any stray-light pick-up, as well as the dark current of the photodetector.

that the ratio of the eigenvalues of the optical system is given by

$$\rho_S = \frac{V_{ex}}{V_{ey}} = \frac{\chi_{SI}^{xy}}{\chi_{SO}^{xy}}. \tag{3.35}$$

At null, the state of polarization at the output of the optical system must be linear and oriented orthogonal to the transmission axis (*i.e.*, parallel to the extinction axis) of the analyzer. The complex polarization variable χ_{SO}^{xy} that describes such linear vibration of azimuth, $\theta = A + \frac{1}{2}\pi$, is given by [see eq. (1.79) of §1.7]

$$\chi_{SO}^{xy} = \tan(A + \frac{1}{2}\pi) = -\cot A. \tag{3.36}$$

Substitution of χ_{SI}^{xy} ($= \chi_{CO}^{xy}$) from eqs. (3.13), and χ_{SO}^{xy} from eq. (3.36), into eq. (3.35) leads directly to eq. (3.33).

3.4.2. The fixed-compensator-azimuth nulling scheme with $C = \pm\frac{1}{4}\pi$, $\delta_C = -\frac{1}{2}\pi$ and $T_C = 1$

Consider the special case when the compensator acts as an ideal quarter-wave retarder so that

$$\delta_C = -\frac{1}{2}\pi, \quad T_C = 1, \quad \rho_C = T_C e^{j\delta_C} = -j. \tag{3.37a}$$

Also, assume that the compensator is set with its fast axis oriented at an azimuth of either

$$C = +\tfrac{1}{4}\pi \quad \text{or} \quad C = -\tfrac{1}{4}\pi \tag{3.37b}$$

from the direction of the x linear eigenpolarization of the optical system under measurement. Inserting $\rho_C = -j$ and $C = +\frac{1}{4}\pi$ into eq. (3.33) and making use of the identity $(1 - j\tan\theta)/(1 + j\tan\theta) = e^{-j2\theta}$, we obtain

$$\rho_S = -\tan A \, \exp[-j2(P - \tfrac{1}{4}\pi)], \quad \text{when } C = +\tfrac{1}{4}\pi. \tag{3.38}$$

Let (P', A') represent one set of nulling angles. According to eq. (3.38), this means that

$$\rho_S = -\tan A' \, \exp[-j2(P' - \tfrac{1}{4}\pi)]. \tag{3.39}$$

An associated pair

$$(P'', A'') = (P' + \tfrac{1}{2}\pi, \pi - A'), \tag{3.40}$$

represents another distinct set of nulling angles, as can be proved by direct substitution into eq. (3.39) as follows

$$-\tan A'' \exp[-j2(P'' - \tfrac{1}{4}\pi)] = -\tan(\pi - A') \exp[-j2(P' + \tfrac{1}{2}\pi - \tfrac{1}{4}\pi)]$$
$$= -\tan A' \exp[-j2(P' - \tfrac{1}{4}\pi)].$$

If instead of $C = +\frac{1}{4}\pi$, the compensator is set at an azimuth $C = -\frac{1}{4}\pi$, the equation that replaces eq. (3.38) in this case becomes

$$\rho_S = \tan A \, \exp[j2(P + \tfrac{1}{4}\pi)], \quad \text{when } C = -\tfrac{1}{4}\pi, \tag{3.41}$$

as obtained from eq. (3.33) by setting $\rho_C = -j$, $C = -\frac{1}{4}\pi$. Again, if (P', A') represents one solution of eq. (3.41), (P'', A''), which is related to (P', A') by eq. (3.40), represents another distinct solution.

A notation has been developed [68] to identify the different null pairs that are available when the polarizer and analyzer are rotated. The two distinguishable polarizer-analyzer pairs of nulling angles that are obtainable when the compensator is set at the $C = +\frac{1}{4}\pi$ azimuth are denoted by (P_2, A_2) and (P_4, A_4). These are referred to as *the nulls in zones two and four*. Similarly, the two polarizer-analyzer pairs of nulling angles that are available when the compensator is set at the $C = -\frac{1}{4}\pi$ azimuth are denoted by (P_1, A_1) and (P_3, A_3) and are referred to as *the nulls in zones one and three*. For the same value of compensator azimuth ($C = +\frac{1}{4}\pi$ or $C = -\frac{1}{4}\pi$), the two polarizer positions in the two *conjugate zones* (2, 4 or 1, 3) are orthogonal

$$P_4 = P_2 + \tfrac{1}{2}\pi, \qquad P_3 = P_1 + \tfrac{1}{2}\pi. \tag{3.42}$$

The associated analyzer-azimuth angles are related by

$$A_2 + A_4 = \pi, \qquad A_1 + A_3 = \pi. \tag{3.43}$$

It is convenient to express the ratio ρ_S of the eigenvalues of the optical system in terms of its magnitude T_S and its phase angle δ_S

$$\rho_S = T_S \, e^{j\delta_S}. \tag{3.44}$$

Use of this form of ρ_S into eq. (3.38) gives

$$\left.\begin{array}{l} T_S = |\tan A| \\ \delta_S = -2P \mp \tfrac{1}{2}\pi \end{array}\right\}, \quad \text{when } C = +\tfrac{1}{4}\pi. \tag{3.45}$$

Similarly, if eq. (3.44) is substituted into eq. (3.41), we find that

$$\left.\begin{array}{l} T_S = |\tan A| \\ \delta_S = 2P \pm \tfrac{1}{2}\pi \end{array}\right\}, \quad \text{when } C = -\tfrac{1}{4}\pi. \tag{3.46}$$

It is instructive to consider the expression for the detected signal $\mathcal{I}_D$ as a function of the polarizer and the analyzer azimuths P and A, respectively. Inserting $\rho_C = -j$, $C = +\frac{1}{4}\pi$, and $V_{ex}/V_{ey} = T_S \, e^{j\delta_S}$ into eq. (3.27), we find that

$$L = \frac{V_{ey}}{\sqrt{2}}[T_S \, e^{j\delta_S} \cos A \, \exp[j(P - \tfrac{1}{4}\pi)] + \sin A \, \exp[-j(P - \tfrac{1}{4}\pi)]],$$

$$|L|^2 = \frac{|V_{ey}|^2}{4}[(T_S^2 + 1) + (T_S^2 - 1)\cos 2A + 2T_S \sin 2A \, \sin(2P + \delta_S)]. \tag{3.47}$$

Substituting this result into eq. (3.26) gives

$$\mathcal{I}_D = G'[(T_S^2 + 1) + (T_S^2 - 1)\cos 2A + 2T_S \sin 2A \sin(2P + \delta_S)],$$
$$\text{when } C = +\tfrac{1}{4}\pi, \qquad (3.48)$$

where $G' = \tfrac{1}{4}G|V_{ey}|^2$ is a constant independent of P or A. The expression that corresponds to eq. (3.48) when $C = -\tfrac{1}{4}\pi$, instead of $+\tfrac{1}{4}\pi$, is

$$\mathcal{I}_D = G'[(T_S^2 + 1) + (T_S^2 - 1)\cos 2A - 2T_S \sin 2A \sin(2P - \delta_S)],$$
$$\text{when } C = -\tfrac{1}{4}\pi. \qquad (3.49)$$

Equations (3.48) and (3.49) show that the detected signal $\mathcal{I}_D$ is a periodic function of the polarizer azimuth P and the analyzer azimuth A, with period equal to π. Thus if a null is obtained at (P, A), an infinite number of trivially related nulls are obtained at $(P \pm m\pi, A \pm n\pi)$, where m and n are integers.[11] This multiplicity of nulls that are generated from one null by rotating the polarizer or the analyzer by an integral multiple of π will be overlooked.

3.4.3. Alternate ellipsometer arrangement with the compensator placed after the optical system

In the polarizer-compensator-system-analyzer (PCSA) ellipsometer arrangement discussed above and shown in fig. 3.3, linearly polarized light from the polarizer P becomes, in general, elliptically polarized after the compensator C, and is subsequently transformed back into linearly polarized light by the optical system S under measurement. The output linear polarization from S is detected by the crossed linear analyzer A resulting in zero detected light flux.

An alternate ellipsometer arrangement that can be used is obtained by moving the compensator from the polarizing section of the instrument to its analyzing section. This new arrangement, with the compensator and optical system transposed, is shown in fig. 3.5. The polarizing section of the instrument now contains the linear polarizer P only. The state of polarization of the probe light wave incident on the optical system S under measurement is linear and its azimuth is controlled by rotation of the polarizer around the beam axis. After interacting with the optical system under measurement, the polarization of the wave is modified and becomes, in general, elliptic. This elliptic state of polarization is analyzed by the

---

[11] By differentiation of eqs. (3.48) and (3.49), it can be readily seen that

$$\frac{\partial \mathcal{I}_D}{\partial P} = \frac{\partial \mathcal{I}_D}{\partial A} = \mathcal{I}_D = 0,$$

when P and A satisfy eqs. (3.45) and (3.46), respectively.

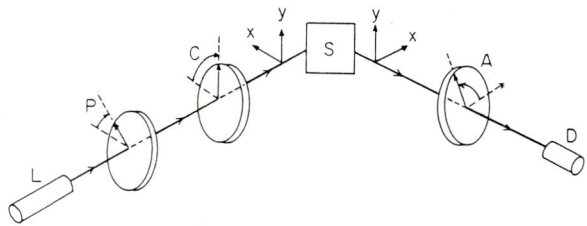

Fig. 3.5. The PSCA ellipsometer arrangement with a polarizing section consisting of a linear polarizer P only, and an analyzing section consisting of a compensator (linear retarder) C and a linear analyzer A. S is the optical system under measurement with linear eigenpolarizations x and y. L and D are the light source and photodetector, respectively. P, C, and A also represent the rotational azimuth angles of the polarizer, compensator, and analyzer measured positive in a counterclockwise sense looking into the beam from the direction of the x linear eigenpolarization.

compensator C and the linear analyzer A which, together with the photodetector D, make the analyzing section of the instrument in its present arrangement. If the instrument is operated in the nulling mode, the action of the compensator will be to compensate (cancel) the ellipticity introduced by the optical system, producing linearly polarized light at its output that is extinguishable by the linear analyzer. The photodetector acts as a null detector.

Analysis of the state of polarization of the light beam as it propagates through the polarizer-system-compensator-analyzer (PSCA) sequence of elements can be carried out in a step-by-step fashion, as has been conducted for the PCSA sequence. Instead, we use the short-cut method of combining the effect of the individual elements and the inter-element coordinate transformations in a single Jones matrix. The notation is similar to that used earlier. Starting with the Jones vector $\boldsymbol{E}_{PO}^{te}$ of the light emergent from the polarizer, the Jones vector $\boldsymbol{E}_{AO}^{te}$ of the light emergent from the analyzer and falling on the photodetector is given by

$$\boldsymbol{E}_{AO}^{te} = \boldsymbol{T}_A^{te} \boldsymbol{R}(A - C) \boldsymbol{T}_C^{fs} \boldsymbol{R}(C) \boldsymbol{T}_S^{xy} \boldsymbol{R}(-P) \boldsymbol{E}_{PO}^{te}, \tag{3.50a}$$

or

$$\boldsymbol{E}_{AO}^{te} = \boldsymbol{T}^{comb} \boldsymbol{E}_{PO}^{te}, \tag{3.50b}$$

where

$$\boldsymbol{T}^{comb} = \boldsymbol{T}_A^{te} \boldsymbol{R}(A - C) \boldsymbol{T}_C^{fs} \boldsymbol{R}(C) \boldsymbol{T}_S^{xy} \boldsymbol{R}(-P). \tag{3.51a}$$

In expanded form, the combined Jones matrix $\boldsymbol{T}^{comb}$ is obtained by inserting the explicit expressions of the matrices that appear in eq. (3.51a),

§3.4] ELLIPSOMETRIC MEASUREMENT 171

$$T^{comb} = \begin{bmatrix} T_{11}^{comb} & T_{12}^{comb} \\ T_{21}^{comb} & T_{22}^{comb} \end{bmatrix},$$

$$= K_A K_C \begin{bmatrix} 1 & 0 \\ 0 & 0 \end{bmatrix} \begin{bmatrix} \cos(A-C) & \sin(A-C) \\ -\sin(A-C) & \cos(A-C) \end{bmatrix} \begin{bmatrix} 1 & 0 \\ 0 & \rho_C \end{bmatrix}$$

$$\times \begin{bmatrix} \cos C & \sin C \\ -\sin C & \cos C \end{bmatrix} \begin{bmatrix} V_{ex} & 0 \\ 0 & V_{ey} \end{bmatrix} \begin{bmatrix} \cos P & -\sin P \\ \sin P & \cos P \end{bmatrix}. \quad (3.51b)$$

From the rules of matrix multiplication, it is simple to prove that when a 2×2 matrix with zero elements in the second row *premultiplies* *any* other 2×2 matrix, the result is a 2×2 matrix with zero elements in the second row. Starting matrix multiplication from left to right, on the right-hand side of eq. (3.51b), it is immediately concluded that $T_{21}^{comb} = T_{22}^{comb} = 0$. (This is to be expected because the field exiting an *ideal* linear analyzer must have zero component parallel to its extinction axis, *i.e.*, $E_{AO,e} \equiv 0$.) With the elements in the second row of the combined Jones matrix T^{comb} zeros, eq. (3.50b) reduces to a scalar equation

$$E_{AO,t} = T_{11}^{comb} E_{PO,t}, \quad (3.52)$$

that relates the amplitude of the linear electric vibration at the output of the analyzer, $E_{AO,t}$, to that at the output of the polarizer, $E_{PO,t}$, in the ellipsometer sequence PSCA. In other words, the effect of the ellipsometer is completely described by the (1, 1) element of the combined Jones matrix of eq. (3.51b)

$$T_{11}^{comb} = K_A K_C \{ V_{ex} \cos P [\cos C \cos(A-C) - \rho_C \sin C \sin(A-C)] + V_{ey} \sin P [\sin C \cos(A-C) + \rho_C \cos C \sin(A-C)] \}. \quad (3.53)$$

The detected signal $\mathscr{I}_D$ is given by

$$\mathscr{I}_D = K_D |E_{AO,t}|^2,$$
$$= K_D |E_{PO,t}|^2 |T_{11}^{comb}|^2,$$
$$= G|L|^2,$$

which is identical to eq. (3.26). K_D and G are exactly the same as before, and L is given by

$$L = V_{ex} \cos P [\cos C \cos(A-C) - \rho_C \sin C \sin(A-C)] + V_{ey} \sin P [\sin C \cos(A-C) + \rho_C \cos C \sin(A-C)]. \quad (3.54)$$

The null condition corresponds to zero detected signal ($\mathscr{I}_D = 0$, and $L = 0$, as in eqs. (3.31) and (3.32)); this leads to

$$\rho_S = -\tan P \frac{\tan C + \rho_C \tan (A - C)}{1 - \rho_C \tan C \tan (A - C)}, \qquad (3.55)$$

where $\rho_S = V_{ex}/V_{ey}$ is the ratio of the eigenvalues V_{ex} and V_{ey} of the optical system that are associated with the x and y linear eigenpolarizations.

Comparison between eqs. (3.55) and (3.33) shows that the expression of ρ_S in the PSCA ellipsometer arrangement is structurally the same as that in the PCSA ellipsometer arrangement with the only difference being an interchange of P and A.

Equation (3.55) can be rearranged to read as [see eqs. (3.35) and (3.17)]

$$\rho_S = \frac{\chi_{SI}^{xy}}{\chi_{SO}^{xy}},$$

where

$$\begin{aligned}\chi_{SI}^{xy} &= (E_{SI,y}/E_{SI,x}) = \tan P, \\ \chi_{SO}^{xy} &= (E_{SO,y}/E_{SO,x}) = -\frac{1 - \rho_C \tan C \tan (A - C)}{\tan C + \rho_C \tan (A - C)},\end{aligned} \qquad (3.56)$$

represent the complex variables that describe the ellipses of polarization at the input and output of the optical system, respectively. The reader may verify that χ_{SO}^{xy} of eq. (3.56) is the extinguished eigenpolarization (i.e., the eigenpolarization associated with the eigenvalue zero) of the compound elliptic analyzer that consists of the compensator C followed by the linear analyzer A.

With an ideal quarter-wave compensator, set at an azimuth C of either $+\frac{1}{4}\pi$ or $-\frac{1}{4}\pi$, the ratio ρ_S of the eigenvalues of the optical system becomes

$$\rho_S = -\tan P \exp[-j2(A - \tfrac{1}{4}\pi)], \quad \text{when } C = +\tfrac{1}{4}\pi, \qquad (3.57)$$

$$\rho_S = \tan P \exp[j2(A + \tfrac{1}{4}\pi)], \quad \text{when } C = -\tfrac{1}{4}\pi, \qquad (3.58)$$

as may be checked by setting $\rho_C = -j$ and $C = \pm\frac{1}{4}\pi$ in eq. (3.55). For $C = +\frac{1}{4}\pi$, there are two different sets of nulling angles denoted by (P_2, A_2) and (P_4, A_4). The corresponding sets when $C = -\frac{1}{4}\pi$ are (P_1, A_1) and (P_3, A_3). The analyzer positions at null in zones 2 and 4 ($C = +\frac{1}{4}\pi$) are orthogonal, and so are the analyzer positions at null in zones 1 and 3 ($C = -\frac{1}{4}\pi$),

$$A_4 = A_2 + \tfrac{1}{2}\pi, \qquad A_3 = A_1 + \tfrac{1}{2}\pi. \qquad (3.59)$$

The associated polarizer-azimuth angles are related by

$$P_2 + P_4 = \pi, \qquad P_1 + P_3 = \pi. \qquad (3.60)$$

The magnitude T_S and phase angle δ_S of the ratio of eigenvalues ρ_S ($= T_S e^{j\delta_S}$)

of the optical system are derived from eqs. (3.57) and (3.58) as

$$\left.\begin{array}{l}T_S = |\tan P| \\ \delta_S = -2A \mp \tfrac{1}{2}\pi\end{array}\right\} \quad \text{when } C = +\tfrac{1}{4}\pi, \qquad (3.61)$$

$$\left.\begin{array}{l}T_S = |\tan P| \\ \delta_S = 2A \pm \tfrac{1}{2}\pi\end{array}\right\} \quad \text{when } C = -\tfrac{1}{4}\pi. \qquad (3.62)$$

Finally, the expressions for the detected signal $\mathscr{I}_D$ as functions of P and A are given by

$$\mathscr{I}_D = G'[(T_S^2 + 1) + (T_S^2 - 1)\cos 2P + 2T_S \sin 2P \sin(2A + \delta_S)],$$
$$\text{when } C = +\tfrac{1}{4}\pi, \qquad (3.63)$$

and

$$\mathscr{I}_D = G'[(T_S^2 + 1) + (T_S^2 - 1)\cos 2P - 2T_S \sin 2P \sin(2A - \delta_S)],$$
$$\text{when } C = -\tfrac{1}{4}\pi. \qquad (3.64)$$

Notice that eqs. (3.57)–(3.64) could also have been obtained simply by interchanging P and A in the corresponding expressions derived earlier for the PCSA ellipsometer arrangement.

3.4.4. *The ratio of eigenvalues ρ_S in reflection and transmission ellipsometry*

In the above discussion, the optical system S under measurement was assumed to have orthogonal linear eigenpolarizations parallel to two orthogonal directions x and y that are fixed with respect to the optical system. For generality, no attempt has been made to indicate what such an optical system might actually be. Two important examples are optically isotropic reflectors (mirrors) and samples of linearly-birefringent, linearly-dichroic media tested in transmission.

For isotropic reflectors, the eigenpolarizations are the linear vibrations parallel (p) and perpendicular (s) to the plane of incidence. (x and y are now replaced by p and s, respectively.) The associated eigenvalues V_{ex} and V_{ey} become the complex-amplitude reflection coefficients R_{pp} and R_{ss}

$$\begin{aligned} V_{ex} &= R_{pp} \\ &= |R_{pp}|\, e^{j\delta_{pp}}, \\ V_{ey} &= R_{ss} \\ &= |R_{ss}|\, e^{j\delta_{ss}}. \end{aligned} \qquad (3.65)$$

(Refer to the discussion connected with fig. 2.5 and eq. (2.29), §2.2.) The ratio ρ_S of these reflection coefficients,

$$\rho_S = \frac{R_{pp}}{R_{ss}} = \frac{|R_{pp}|}{|R_{ss}|} e^{j(\delta_{pp} - \delta_{ss})}, \tag{3.66}$$

is measured by the ellipsometer. This ratio is often written as

$$\rho_S = \tan \psi \, e^{j\Delta}, \tag{3.67}$$

where

$$\tan \psi = \frac{|R_{pp}|}{|R_{ss}|}, \tag{3.68}$$

$$\Delta = \delta_{pp} - \delta_{ss}. \tag{3.69}$$

The angles ψ and Δ are called the *ellipsometric angles* of the reflector. From eq. (3.68), ψ is the angle whose tangent gives the ratio of the amplitude attenuation (or magnification) upon reflection for the p and s polarizations, respectively. According to eq. (3.69), Δ gives the difference between the phase shifts experienced upon reflection by the p and s polarizations, respectively. Previous relations that were expressed in terms of $\rho_S = T_S e^{j\delta_S}$ can be rewritten in terms of ψ and Δ using the substitutions

$$T_S = \tan \psi, \qquad \delta_S = \Delta. \tag{3.70}$$

For samples of linearly birefringent and linearly dichroic media, for which the optic axes of birefringence and dichroism are coincident, the Jones matrix, in diagonal form, is given by eqs. (2.25) and (2.26). Such samples are assumed to be tested in transmission with the light beam travelling in the direction perpendicular to two of the principal axes of birefringence and dichroism. The ratio of eigenvalues measured by the ellipsometer is obtained from eq. (2.26) as

$$\rho_S = T_S e^{j\delta_S},$$

where

$$T_S = \exp\left[-\frac{2\pi d}{\lambda}(k_e - k_o)\right], \qquad \delta_S = -\frac{2\pi d}{\lambda}(n_e - n_o). \tag{3.71}$$

$(n_o - jk_o)$ and $(n_e - jk_e)$ are the complex ordinary and extraordinary refractive indices respectively, d is the total distance travelled by the probe light wave in the medium, and λ is the vacuum wavelength of light. The quantities $(n_e - n_o)$ and $(k_e - k_o)$ represent the birefringence and dichroism of the medium, respectively.

3.4.5. Unified treatment of the PCSA and PSCA ellipsometer arrangements – Example of the application of the bilinear polarization transfer function

Consider a system of two optical elements, each has two orthogonal linear eigenpolarizations. Assume that the input (x_1, y_1) and output (x_2, y_2) Cartesian coordinate systems coincide with the linear eigenpolarizations of the first and second elements, respectively, fig. 3.6. Let $T_1 e^{j\delta_1}$ and $T_2 e^{j\delta_2}$ be the

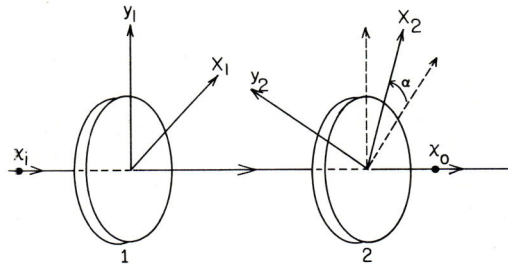

Fig. 3.6. A sequence of two optical elements 1 and 2 with orthogonal eigenpolarizations x_1, y_1 and x_2, y_2, respectively, transforms the state of polarization of a light beam from χ_i to χ_o. The angle α specifies the relative orientation of element 2 with respect to element 1 around the beam axis.

relative (y-to-x) complex-amplitude transmittances of these elements (*i.e.*, the ratio of the eigenvalues associated with the y and x eigenpolarizations, respectively). Also, let α be the angle of a counter-clockwise rotation (looking into the beam) that brings the x_1-axis of the first element into coincidence with the x_2-axis of the second element. The Jones matrix of this system is given by[12]

$$T = \begin{bmatrix} 1 & 0 \\ 0 & T_2 e^{j\delta_2} \end{bmatrix} \begin{bmatrix} \cos \alpha & \sin \alpha \\ -\sin \alpha & \cos \alpha \end{bmatrix} \begin{bmatrix} 1 & 0 \\ 0 & T_1 e^{j\delta_1} \end{bmatrix},$$

whose expansion yields

$$\begin{aligned} T_{11} &= \cos \alpha, & T_{12} &= T_1 \sin \alpha \; e^{j\delta_1}, \\ T_{21} &= -T_2 \sin \alpha \; e^{j\delta_2}, & & \\ T_{22} &= T_1 T_2 \cos \alpha \; e^{j(\delta_1+\delta_2)}. & & \end{aligned} \quad (3.72)$$

---

[12]The double-subscripted elements T_{ij} of the combined Jones matrix T should not be confused with the magnitudes T_1 and T_2 of the relative transmittances of the first and second optical elements.

Substituting from eqs. (3.72) into eq. (3.1), we get

$$\chi_o = \frac{T_1 T_2 \, e^{j(\delta_1+\delta_2)} \chi_i - T_2 \tan\alpha \, e^{j\delta_2}}{T_1 \tan\alpha \, e^{j\delta_1} \chi_i + 1}, \tag{3.73}$$

which represents the bilinear transformation relating the input and output polarizations χ_i and χ_o, respectively. Equation (3.73) can be written as

$$T_1 \, e^{j\delta_1} = \frac{1}{\chi_i} \frac{T_2 \, e^{j\delta_2} \tan\alpha + \chi_o}{T_2 \, e^{j\delta_2} - \tan\alpha \chi_o}, \tag{3.74}$$

or

$$T_2 \, e^{j\delta_2} = \chi_o \frac{1 + T_1 \, e^{j\delta_1} \tan\alpha \chi_i}{-\tan\alpha + T_1 \, e^{j\delta_1} \chi_i}. \tag{3.75}$$

Equations (3.74) and (3.75) express the fact that the relative complex-amplitude transmittance of either one of the two elements can be determined in terms of that of the other, if the output polarization χ_o that corresponds to *one* input polarization χ_i is known.[13]

For ellipsometry, two aspects of the polarization transfer by the system described above are of interest.

(1) The output polarizations χ_o that correspond to input *linear* polarizations χ_i with azimuths from $-\frac{1}{2}\pi$ to $+\frac{1}{2}\pi$, measured from the x_1-axis of the first element. According to §2.4, the locus of χ_o is a circle in the complex χ_o-plane that passes through the two points

$$\chi_o(\chi_i = 0) = -T_2 \tan\alpha \, e^{j\delta_2}, \qquad \chi_o(\chi_i = \pm\infty) = T_2 \cot\alpha \, e^{j\delta_2}. \tag{3.76}$$

These points represent the system's response to incident linear polarizations, $\chi_i = 0$ and $\chi_i = \pm\infty$, along the x_1- and y_1-axes of the first element, respectively. This circle, denoted by Γ_1' in fig. 3.7, is the image in the χ_o-plane of the real axis Γ_1 of the χ_i-plane. Because the two points of eq. (3.76) lie on opposite sides of the origin along a line passing through it, the circle Γ_1' always intersects the real axis of the χ_o-plane at two points that are denoted by χ_{o1} and χ_{o2}.

[13] Alternatively, the relative complex-amplitude transmittances of *both* elements can be determined from the mapping of *two* incident polarization χ_{i1} and χ_{i2} into the corresponding two output polarizations χ_{o1} and χ_{o2}, respectively. Writing eq. (3.74) twice, in terms of (χ_{i1}, χ_{o1}) and (χ_{i2}, χ_{o2}), and equating the results, we obtain the following quadratic equation in $T_2 \, e^{j\delta_2}$

$$(T_2 \, e^{j\delta_2})^2 + (T_2 \, e^{j\delta_2})\left[\left(\frac{\chi_{i2}\chi_{o1} - \chi_{i1}\chi_{o2}}{\chi_{i2} - \chi_{i1}}\right)\cot\alpha + \left(\frac{\chi_{i1}\chi_{o2} - \chi_{i2}\chi_{o2}}{\chi_{i2} - \chi_{i1}}\right)\tan\alpha\right] - \chi_{o1}\chi_{o2} = 0.$$

Applying the same to eq. (3.75), we obtain the following quadratic equation in $T_1 \, e^{j\delta_1}$

$$(T_1 \, e^{j\delta_1})^2 + (T_1 \, e^{j\delta_1})\left[\left(\frac{\chi_{i2}\chi_{o1} - \chi_{i1}\chi_{o2}}{\chi_{o1} - \chi_{o2}}\right)\frac{\cot\alpha}{\chi_{i1}\chi_{i2}} + \left(\frac{\chi_{i2}\chi_{o2} - \chi_{i1}\chi_{o2}}{\chi_{o1} - \chi_{o2}}\right)\frac{\tan\alpha}{\chi_{i1}\chi_{i2}}\right] - \frac{1}{\chi_{i1}\chi_{i2}} = 0.$$

Similar equations have been derived by Johnson and Bashara [83].

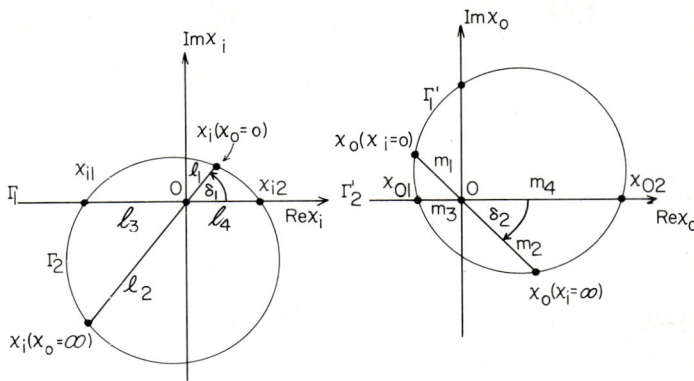

Fig. 3.7. Mapping of polarization between the input (χ_i) and output (χ_o) of the two optical elements 1, 2 shown in fig. 3.6. The real axis Γ_1 of the complex χ_i plane is mapped onto the circle Γ'_1 in the complex χ_o plane. The circle Γ_2 in the χ_i plane is mapped into the real axis Γ'_2 of the χ_o plane. The points χ_{i1}, χ_{i2} on the real axis of the χ_i plane represent the two linearly polarized input states that are mapped by the two elements 1 and 2 of fig. 3.6 into two linearly polarized output states χ_{o1} and χ_{o2} on the real axis of the χ_o plane. The distances ℓ's and m's are given by eqs. (3.76)–(3.78).

(2) The input polarizations χ_i required to produce output linear polarizations χ_o of azimuths from $-\frac{1}{2}\pi$ to $\frac{1}{2}\pi$ measured from the x_2-axis of the second element. From §2.4, the locus of χ_i is a circle Γ_2 in the χ_i-plane that passes through the two points

$$\chi_i(\chi_o = 0) = \frac{1}{T_1} \tan \alpha \, e^{-j\delta_1}, \qquad \chi_i(\chi_o = \pm \infty) = \frac{-1}{T_1} \cot \alpha \, e^{-j\delta_1}. \tag{3.77}$$

These are the input polarizations required to produce output linear polarizations, $\chi_o = 0$ and $\chi_o = \pm \infty$, along the x_2- and the y_2-axes of the second element, respectively. The circle Γ_2 always intersects the real axis of the χ_i-plane at the two points denoted by χ_{i1} and χ_{i2} in fig. 3.7. Note that Γ_2 is transformed by eq. (3.73) onto the real axis Γ'_2 of the χ_o-plane.

It is convenient to express the linear polarizations χ_{i1}, χ_{i2}; χ_{o1}, and χ_{o2} in terms of their azimuths θ_{i1}, θ_{i2}; θ_{o1}, and θ_{o2}, respectively. Thus we can write[14]

$$\chi_{i1} = \tan \theta_{i1}, \qquad \chi_{i2} = \tan \theta_{i2},$$
$$\chi_{o1} = \tan \theta_{o1}, \qquad \chi_{o2} = \tan \theta_{o2}. \tag{3.78}$$

From fig. 3.7, we have

$$\ell_1 \ell_2 = \ell_3 \ell_4, \qquad m_1 m_2 = m_3 m_4, \tag{3.79}$$

[14] The Cartesian complex polarization variable χ that describes an elliptic state of azimuth θ, and ellipticity angle ϵ, is given by eq. (1.79) of §1.7. For linear polarizations, we have $\epsilon = 0$ and $\chi = \tan \theta$.

where the ℓ's and m's are as indicated, and the equations follow from properties of the circle. Substituting the values of the ℓ's and m's from eqs. (3.76)–(3.78) into eq. (3.79), and taking proper account of signs, we obtain

$$\tan\theta_{i1}\tan\theta_{i2} = -1/T_1^2,$$
$$\tan\theta_{o1}\tan\theta_{o2} = -T_2^2. \qquad (3.80)$$

In the ellipsometer, the combination of the compensator C and the optical system S under measurement (whose eigenpolarizations are linear) can be identified with the above pair of elements. The process of obtaining zero detected light flux by adjustment of the polarizer and analyzer is one of hunting for the condition in which the axes of transmission of these two polarizing elements are oriented parallel and perpendicular to the linear polarizations $\chi_{i1}(\chi_{i2})$ and $\chi_{o1}(\chi_{o2})$, respectively. Note that we have proved the existence of the two nonequivalent nulls (zones) using geometrical considerations. These two nulls exist for all possible eigenvalues of the compensator and optical system under measurement, and for any value of their relative orientation angle.

It should be emphasized that null ellipsometry, as discussed above, involves special use of eqs. (3.74) and (3.75). Specifically, χ_i is chosen to be one of the two linear polarizations χ_{i1} and χ_{i2} that are mapped by the combination of the compensator and the optical system into two linear polarizations χ_{o1} and χ_{o2}. Substituting for χ_i and χ_o in eqs. (3.74) and (3.75) the values $\chi_{i1}(\chi_{i2})$ and $\chi_{o1}(\chi_{o2})$, respectively, from eq. (3.78), we obtain

$$T_1 e^{j\delta_1} = \frac{1}{\tan\theta_i} \frac{T_2 e^{j\delta_2}\tan\alpha + \tan\theta_o}{T_2 e^{j\delta_2} - \tan\alpha\tan\theta_o},$$
$$T_2 e^{j\delta_2} = \tan\theta_o \frac{1 + T_1 e^{j\delta_1}\tan\alpha\tan\theta_i}{-\tan\alpha + T_1 e^{j\delta_1}\tan\theta_i}, \qquad (3.81)$$

where

$$(\theta_i, \theta_o) = (\theta_{i1}, \theta_{o1}) \text{ or } (\theta_{i2}, \theta_{o2}). \qquad (3.82)$$

Obviously, the above formulation applies to both of the PCSA and PSCA ellipsometer arrangements, with the compensator before and after the optical system, respectively. For the PCSA arrangement, the quantities that appear in eqs. (3.72)–(3.82) have the following meanings[15]

$$T_1 = T_C, \quad \delta_1 = \delta_C, \quad \alpha = -C,$$
$$T_2 = 1/T_S, \quad \delta_2 = -\delta_S, \quad \theta_{i1} = P' - C, \qquad (3.83)$$
$$\theta_{i2} = P'' - C, \quad \theta_{o1} = A' - \tfrac{1}{2}\pi, \quad \theta_{o2} = A'' - \tfrac{1}{2}\pi.$$

T_C and $-\delta_C$ are the compensator slow-to-fast relative attenuation and

---

[15] Note that $T_S e^{j\delta_S}$ is defined as the ratio of the x to y eigenvalues of the optical system [eqs. (3.34) and (3.44)], while $T_C e^{j\delta_C}$ is the ratio of the y to x (slow to fast) eigenvalues of the compensator.

retardation, respectively, and C is its azimuth measured from the x linear eigenpolarization of the optical system. T_S and δ_S define the ratio of magnitudes and difference in phases of the x and y eigenvalues of the optical system, respectively [eq. (3.44)]. (P', A') and (P'', A'') are the nulling polarizer-analyzer azimuth-angle pairs in the two nonequivalent zones. Substituting eqs. (3.83) into eqs. (3.80) and (3.81), we get

$$\tan(P' - C)\tan(P'' - C) = 1/T_C^2, \tag{3.84}$$
$$\tan A' \tan A'' = -T_S^2, \tag{3.85}$$

and

$$T_C\, e^{j\delta_C} = \frac{1}{\tan(P - C)} \frac{T_S\, e^{j\delta_S} + \tan C \tan A}{T_S\, e^{j\delta_S} \tan C - \tan A}, \tag{3.86}$$

$$T_S\, e^{j\delta_S} = \tan A \frac{\tan C + T_C\, e^{j\delta_C} \tan(P - C)}{T_C\, e^{j\delta_C} \tan C \tan(P - C) - 1}, \tag{3.87}$$

respectively, where

$$(P, A) = (P', A') \text{ or } (P'', A''). \tag{3.88}$$

For the PSCA ellipsometer arrangement, the quantities that appear in eqs. (3.72)–(3.82) have the following meanings

$$\begin{aligned}
T_1 &= 1/T_S, & \delta_1 &= -\delta_S, & \alpha &= +C, \\
T_2 &= T_C, & \delta_2 &= \delta_C, & \theta_{i1} &= P', \\
\theta_{i2} &= P'', & \theta_{o1} &= A' - C - \tfrac{1}{2}\pi, & \theta_{o2} &= A'' - C - \tfrac{1}{2}\pi.
\end{aligned} \tag{3.89}$$

Upon substitution of eqs. (3.89) into eqs. (3.80) and (3.81), we obtain

$$\tan P' \tan P'' = -T_S^2, \tag{3.90}$$
$$\tan(A' - C)\tan(A'' - C) = 1/T_C^2, \tag{3.91}$$

and

$$T_S\, e^{j\delta_S} = \tan P \frac{\tan C + T_C\, e^{j\delta_C} \tan(A - C)}{T_C\, e^{j\delta_C} \tan C \tan(A - C) - 1}, \tag{3.92}$$

$$T_C\, e^{j\delta_C} = \frac{1}{\tan(A - C)} \frac{T_S\, e^{j\delta_S} + \tan C \tan P}{T_S\, e^{j\delta_S} \tan C - \tan P}, \tag{3.93}$$

respectively, where

$$(P, A) = (P', A') \text{ or } (P'', A''),$$

as before.[16]

[16] Equations (3.87) and (3.92) are not new. They are the same as eqs. (3.33) and (3.55) respectively, where $T_S\, e^{j\delta_S} = \rho_S$ and $T_C\, e^{j\delta_C} = \rho_C$.

Equations (3.87) and (3.92) give the ratio ρ_S of the eigenvalues of an optical system with linear eigenpolarizations, as measured by the PCSA and PSCA ellipsometer arrangements, respectively. This requires that one set of nulling angles (P, C, A) be measured and that the slow-to-fast relative complex-amplitude transmittance ρ_C be known. Conversely, eqs. (3.86) and (3.93) give the slow-to-fast relative complex-amplitude transmittance ρ_C of the compensator if the ratio ρ_S of the eigenvalues of the optical system is known.

Equations (3.84), (3.85), (3.90) and (3.95) express the relations that exist between the azimuth-angle settings (P', A'), (P'', A'') that correspond to the two nonequivalent nulls in the PCSA and PSCA ellipsometer arrangements, when the compensator is set at a fixed azimuth C. When $T_C = 1$ (*i.e.*, when the component vibrations of the electric field incident on the compensator parallel to the fast and the slow axes experience equal attenuation), eq. (3.84) shows that P' and P'' become orthogonal in the PCSA sequence, while eq. (3.91) shows that A' and A'' become orthogonal in the PSCA sequence, as we have already found before.

3.5. Ellipsometric measurement of the ratio of eigenvalues of an optical system with orthogonal circular eigenpolarizations

The ellipsometer arrangements under consideration are PCSA and PSCA shown in figs. 3.3 and 3.5, respectively. The optical system S is presently assumed to have the right- and left-circular polarizations as its eigenpolarizations, instead of the x and y orthogonal linear eigenpolarizations previously considered in §3.4. We also assume that the instrument is operated in the nulling mode, so that the azimuth angles P, C and A of the polarizer, compensator, and analyzer, as well as the compensator relative retardation δ_C, are all such that the detected light flux is zero.[17] The quantity to be measured is the ratio ρ_S of the eigenvalues of the optical system that are associated with the left- and right-circular eigenpolarizations.

The nature of the eigenpolarizations of the optical system suggests that the orthogonal circular polarizations be used as basis states for the description of the polarization of the light beam and the optical devices that it encounters. A detailed treatment that utilizes circular Jones vectors for the light beam and circular Jones matrices for the optical devices can be carried out but is not required. Instead, use can be made of results already obtained in §3.4.

The transformation of the *circular* Jones vector between the input (I) and output (O) of the optical system S under measurement is given by

$$\boldsymbol{E}_{\mathrm{SO}}^{\ell r} = \boldsymbol{T}_{\mathrm{S}}^{\ell r} \boldsymbol{E}_{\mathrm{SI}}^{\ell r}, \tag{3.94}$$

[17] See footnote 10 on page 166.

where the superscript ℓr indicates that the *left-* and *right-*circular polarizations are being used as basis states. In expanded form, eq. (3.94) reads

$$\begin{bmatrix} E_{SO,\ell} \\ E_{SO,r} \end{bmatrix} = \begin{bmatrix} V_{e\ell} & 0 \\ 0 & V_{er} \end{bmatrix} \begin{bmatrix} E_{SI,\ell} \\ E_{SI,r} \end{bmatrix}, \qquad (3.95)$$

where $V_{e\ell}$ and V_{er} are the (generally) complex eigenvalues associated with the left- and right-circular eigenpolarizations, respectively. $E_{SI,\ell}$ and $E_{SI,r}$ are the projections of the electric-field vibration incident on the optical system along the left- and right-circular basis states, respectively, with similar meanings for the components $E_{SO,\ell}$ and $E_{SO,r}$ of the circular Jones vector $\boldsymbol{E}_{SO}^{\ell r}$ at the output of the system (see the discussion associated with fig. 1.11 of ch. 1). From eq. (3.95), we find that

$$\chi_{SO}^{\ell r} = \frac{V_{er}}{V_{e\ell}} \chi_{SI}^{\ell r}, \qquad (3.96)$$

where

$$\chi_{SI}^{\ell r} = \frac{E_{SI,r}}{E_{SI,\ell}}, \qquad \chi_{SO}^{\ell r} = \frac{E_{SO,r}}{E_{SO,\ell}}, \qquad (3.97)$$

are the complex variables that describe the ellipses of polarization at the input and output of the optical system, respectively in the circular complex-plane representation of polarization [eq. (1.92) of §1.7]. From eq. (3.96), the ratio ρ_S of the eigenvalues is given by

$$\rho_S = \frac{V_{e\ell}}{V_{er}} = \frac{\chi_{SI}^{\ell r}}{\chi_{SO}^{\ell r}}. \qquad (3.98)$$

Equations (3.94), (3.96) and (3.98) are analogous to eqs. (3.15), (3.17) and (3.35), with ℓ and r replacing x and y, respectively.

According to eq. (3.98), to determine the ratio ρ_S of the eigenvalues $V_{e\ell}$ and V_{er} of the optical system S, we need only find the circular complex polarization variables $\chi_{SI}^{\ell r}$ and $\chi_{SO}^{\ell r}$ that describe the light wave at the input and the output of S. $\chi_{SI}^{\ell r}$ is determined by the polarizing section of the ellipsometer, whereas $\chi_{SO}^{\ell r}$ is determined by its analyzing section. In §3.4, the Cartesian complex polarization variables χ_{SI}^{xy} and χ_{SO}^{xy} have been obtained, so that the corresponding circular complex polarization variables $\chi_{SI}^{\ell r}$ and $\chi_{SO}^{\ell r}$ can be found by use of the following transformations [eq. (1.104) of §1.7]

$$\chi_{SI}^{\ell r} = \frac{\chi_{SI}^{xy} + j}{-\chi_{SI}^{xy} + j}, \qquad \chi_{SO}^{\ell r} = \frac{\chi_{SO}^{xy} + j}{-\chi_{SO}^{xy} + j}. \qquad (3.99)$$

For the PCSA ellipsometer arrangement,

$$\chi_{SI}^{xy} = \frac{\tan C + \rho_C \tan(P-C)}{1 - \rho_C \tan C \tan(P-C)}, \qquad \chi_{SO}^{xy} = -\cot A, \qquad (3.100)$$

[eqs. (3.13) and (3.36)], which we substitute into eq. (3.99) to obtain

$$\chi_{SI}^{\ell r} = -\frac{(\tan C + j) + \rho_C(1 - j\tan C)\tan(P-C)}{(\tan C - j) + \rho_C(1 + j\tan C)\tan(P-C)}, \qquad (3.101)$$

$$\chi_{SO}^{\ell r} = \exp[-j2(A - \tfrac{1}{2}\pi)] = -\exp[-j2A]. \qquad (3.102)$$

If $\chi_{SI}^{\ell r}$ and $\chi_{SO}^{\ell r}$ are substituted from eqs. (3.101) and (3.102) into eq. (3.98), we get

$$\rho_S = \frac{V_{e\ell}}{V_{er}} = e^{j2A} \left[\frac{(\tan C + j) + \rho_C(1 - j\tan C)\tan(P-C)}{(\tan C - j) + \rho_C(1 + j\tan C)\tan(P-C)} \right]. \qquad (3.103)$$

For the PSCA ellipsometer arrangement,

$$\chi_{SI}^{xy} = \tan P, \qquad \chi_{SO}^{xy} = -\frac{1 - \rho_C \tan C \tan(A-C)}{\tan C + \rho_C \tan(A-C)}, \qquad (3.104)$$

[eq. (3.56)], which we use in eq. (3.99) to get

$$\chi_{SI}^{\ell r} = e^{-j2P}, \qquad (3.105)$$

$$\chi_{SO}^{\ell r} = \frac{(\tan C + j) + \rho_C(1 - j\tan C)\tan(A-C)}{(\tan C - j) + \rho_C(1 + j\tan C)\tan(A-C)}. \qquad (3.106)$$

Substitution of eqs. (3.105) and (3.106) into eq. (3.98) yields

$$\rho_S = \frac{V_{e\ell}}{V_{er}} = e^{-j2P} \left[\frac{(\tan C - j) + \rho_C(1 + j\tan C)\tan(A-C)}{(\tan C + j) + \rho_C(1 - j\tan C)\tan(A-C)} \right]. \qquad (3.107)$$

Equations (3.103) and (3.107) show that the ratio ρ_S of the eigenvalues of an optical system with right- and left-circular eigenpolarization can be measured from a set of nulling angles (P, C, A) in the PCSA and PSCA ellipsometer arrangements, respectively.

When the compensator acts as an ideal quarter-wave retarder ($\rho_C = -j$), and is set at $C = \pm\tfrac{1}{4}\pi$ azimuth, eqs. (3.103) and (3.107) simplify considerably. Thus, for the PCSA sequence, eq. (3.103) becomes

$$\rho_S = j\cot P\, e^{j2A}, \qquad \text{when } C = +\tfrac{1}{4}\pi, \qquad (3.108)$$

$$\rho_S = j\tan P\, e^{j2A}, \qquad \text{when } C = -\tfrac{1}{4}\pi. \qquad (3.109)$$

Likewise, for the PSCA sequence, eq. (3.107) becomes

$$\rho_S = -j\tan A\, e^{-j2P}, \qquad \text{when } C = +\tfrac{1}{4}\pi, \qquad (3.110)$$

$$\rho_S = -j \cot A \ e^{-j2P}, \quad \text{when } C = -\tfrac{1}{4}\pi. \tag{3.111}$$

In eqs. (3.108)–(3.111), (P, A) represent the nulling azimuth angles of the polarizer and analyzer that produce zero detected light flux.

According to eq. (3.108), if (P_2, A_2) denotes one set of nulling angles in the PCSA ellipsometer arrangement, (obtained when the compensator is set at $C = +\tfrac{1}{4}\pi$ and the polarizer and analyzer are adjusted for null), then (P_4, A_4) is another nonequivalent set, where

$$P_4 = \pi - P_2, \quad A_4 = A_2 - \tfrac{1}{2}\pi. \tag{3.112a}$$

Similarly, according to eq. (3.109), two nonequivalent nulls (P_1, A_1) and (P_3, A_3) are available when $C = -\tfrac{1}{4}\pi$ that are interrelated by

$$P_3 = \pi - P_1, \quad A_3 = A_1 - \tfrac{1}{2}\pi. \tag{3.112b}$$

From eqs. (3.110) and (3.111), the relations that correspond to eqs. (3.112a) and (3.112b) in the PSCA ellipsometer arrangement are

$$P_4 = P_2 - \tfrac{1}{2}\pi, \quad A_4 = \pi - A_2, \tag{3.113a}$$

$$P_3 = P_1 - \tfrac{1}{2}\pi, \quad A_3 = \pi - A_1. \tag{3.113b}$$

Notice that in the PCSA ellipsometer sequence [eq. (3.112)], the two conjugate positions of the analyzer at null (A_2, A_4) [or (A_1, A_3)] are orthogonal; the associated conjugate positions of the polarizer at null (P_2, P_4) [or (P_1, P_3)] are complementary. In the PSCA ellipsometer sequence [eq. (3.113)], the above relations continue to apply with the polarizer and analyzer azimuths interchanged. Similar relations have been found in §3.4, when the optical system under measurement had orthogonal linear (instead of circular) eigenpolarizations [see eqs. (3.42), (3.43), (3.59), and (3.60)].

Materials that exhibit optical rotation (OR) and/or circular dichroism (CD), either natural or induced, constitute the most common examples of systems with orthogonal circular eigenpolarizations. A sample is ordinarily tested in transmission. In this case, the eigenvalues $V_{e\ell}$ and V_{er} can be written as [see eq. (2.45)]

$$V_{e\ell} = \exp\left[-j\frac{2\pi d}{\lambda}(n_\ell - jk_\ell)\right], \quad V_{er} = \exp\left[-j\frac{2\pi d}{\lambda}(n_r - jk_r)\right], \tag{3.114}$$

where $(n_\ell - jk_\ell)$ and $(n_r - jk_r)$ are the complex indices of refraction "seen" by left- and right-circularly polarized waves of vacuum wavelength λ when they travel (a distance d) through a circularly-birefringent, circularly-dichroic medium. The ratio of eigenvalues measured by the ellipsometer is obtained from eq. (3.114) as

$$\rho_S = \frac{V_{e\ell}}{V_{er}} = T_S \, e^{j\delta_S}, \tag{3.115}$$

where

$$T_S = \exp\left[-\frac{2\pi d}{\lambda}(k_r - k_\ell)\right], \qquad \delta_S = -\frac{2\pi d}{\lambda}(n_r - n_\ell). \tag{3.116}$$

The quantities $(n_r - n_\ell)$ and $(k_r - k_\ell)$ represent the circular birefringence and circular dichroism of the medium, respectively.

3.6. Ellipsometric measurement of the ratio of eigenvalues of an optical system with elliptic eigenpolarizations

In §3.4 and §3.5, we have shown how the ratio of eigenvalues of an optical system S with linear or circular eigenpolarizations can be measured by an ellipsometer in two different arrangements. The above results will now be extended to the more general case of an optical system whose eigenpolarizations are two *known* elliptic states u and v, that may or may not be orthogonal [69]. Again, the instrument is adjusted so as to achieve the null condition of zero (minimum) detected light flux.

The procedure to be followed runs along the same lines as explained in §3.5. In particular, starting with eq. (3.94), we may write this equation, and all subsequent equations that lead to eq. (3.98), in terms of u and v (the eigenpolarizations of the optical system) as basis states, instead of ℓ and r. Therefore, the ratio ρ_S of the eigenvalues V_{eu} and V_{ev} that are associated with the elliptic eigenpolarizations u and v of the optical system S is given by

$$\rho_S = \frac{V_{eu}}{V_{ev}} = \frac{\chi_{SI}^{uv}}{\chi_{SO}^{uv}}, \tag{3.117}$$

where χ_{SI}^{uv} and χ_{SO}^{uv} are the complex variables that describe the ellipses of polarization at the input and output of S, respectively, in the *generalized* complex-plane representation of polarization whose basis states are u and v [eq. (1.97) of §1.7].

The generalized complex polarization variables χ_{SI}^{uv} and χ_{SO}^{uv} can be obtained from the corresponding Cartesian complex polarization variables χ_{SI}^{xy} and χ_{SO}^{xy} by use of the transformation that links the two, namely [eq. (1.99) of §1.7],

$$\chi_{SI}^{uv} = \frac{f_{11}\chi_{SI}^{xy} - f_{21}}{-f_{12}\chi_{SI}^{xy} + f_{22}}, \qquad \chi_{SO}^{uv} = \frac{f_{11}\chi_{SO}^{xy} - f_{21}}{-f_{12}\chi_{SO}^{xy} + f_{22}}. \tag{3.118}$$

The elements f_{ij} of the bilinear transformations that appear in eq. (3.118) are determined by the Cartesian Jones vectors $\mathcal{E}_u$ and $\mathcal{E}_v$.

$$\mathcal{E}_u = \begin{bmatrix} f_{11} \\ f_{21} \end{bmatrix}, \qquad \mathcal{E}_v = \begin{bmatrix} f_{12} \\ f_{22} \end{bmatrix}, \tag{3.119}$$

that describe the eigenpolarizations (and basis states) u and v, when these are referenced to the x and y orthogonal linear polarizations [eq. (1.61), §1.6].

For the PCSA ellipsometer arrangement, χ_{SI}^{xy} and χ_{SO}^{xy} are given by eqs. (3.100), which when substituted into eq. (3.118) lead to

$$\chi_{\text{SI}}^{uv} = -\left[\frac{(-f_{21} + f_{11}\tan C) + \rho_C(f_{11} + f_{21}\tan C)\tan(P-C)}{(-f_{22} + f_{12}\tan C) + \rho_C(f_{12} + f_{22}\tan C)\tan(P-C)}\right], \quad (3.120)$$

$$\chi_{\text{SO}}^{uv} = -\left(\frac{f_{11} + f_{21}\tan A}{f_{12} + f_{22}\tan A}\right). \quad (3.121)$$

When χ_{SI}^{uv} and χ_{SO}^{uv} are substituted from eqs. (3.120) and (3.121) into eq. (3.117), we obtain

$$\rho_S = \frac{V_{eu}}{V_{ev}} = \left(\frac{f_{12} + f_{22}\tan A}{f_{11} + f_{21}\tan A}\right)$$
$$\times \left[\frac{(-f_{21} + f_{11}\tan C) + \rho_C(f_{11} + f_{21}\tan C)\tan(P-C)}{(-f_{22} + f_{12}\tan C) + \rho_C(f_{12} + f_{22}\tan C)\tan(P-C)}\right]. \quad (3.122)$$

For the PSCA ellipsometer arrangement, χ_{SI}^{xy} and χ_{SO}^{xy} are given by eq. (3.104), which when substituted into eq. (3.118) give

$$\chi_{\text{SI}}^{uv} = -\left(\frac{f_{21} - f_{11}\tan P}{f_{22} - f_{12}\tan P}\right), \quad (3.123)$$

$$\chi_{\text{SO}}^{uv} = -\left[\frac{(f_{11} + f_{21}\tan C) + \rho_C(f_{21} - f_{11}\tan C)\tan(A-C)}{(f_{12} + f_{22}\tan C) + \rho_C(f_{22} - f_{12}\tan C)\tan(A-C)}\right]. \quad (3.124)$$

When eqs. (3.123) and (3.124) are substituted into eq. (3.117), the result takes the form

$$\rho_S = \frac{V_{eu}}{V_{ev}} = \left(\frac{f_{21} - f_{11}\tan P}{f_{22} - f_{12}\tan P}\right)$$
$$\times \left[\frac{(f_{12} + f_{22}\tan C) + \rho_C(f_{22} - f_{12}\tan C)\tan(A-C)}{(f_{11} + f_{21}\tan C) + \rho_C(f_{21} - f_{11}\tan C)\tan(A-C)}\right]. \quad (3.125)$$

Equations (3.122) and (3.125) show that the ratio of the eigenvalues of an optical system with any *known* elliptic eigenpolarizations u and v can be measured from a set of nulling angles (P, C, A) in the PCSA and PSCA ellipsometer arrangements, respectively. In addition to the compensator relative transmittance ρ_C, eqs. (3.122) and (3.125) are explicitly dependent on the parameters f_{ij} which define the eigenpolarizations u and v through eqs. (3.119).

There is no loss of generality if we assume the Jones vectors $\mathscr{E}_u$ and $\mathscr{E}_v$ of eq. (3.119) to be normal (*i.e.* of unit intensity) and of zero phase, so that f_{ij}

can be expressed as follows

$$f_{11} = (1 + \mu\mu^*)^{-\frac{1}{2}},$$
$$f_{21} = \mu(1 + \mu\mu^*)^{-\frac{1}{2}},$$
$$f_{12} = (1 + \nu\nu^*)^{-\frac{1}{2}},$$
$$f_{22} = \nu(1 + \nu\nu^*)^{-\frac{1}{2}}.$$
(3.126)

In eq. (3.126),

$$\mu = \chi^{xy}(u), \qquad \nu = \chi^{xy}(v), \tag{3.127}$$

are the Cartesian complex polarization variables that describe the eigenpolarizations u, v of the optical system under measurement, and x and y are two orthogonal axes fixed with respect to the system (fig. 3.3). Rewritten in terms of μ and ν, eqs. (3.122) and (3.125) become

$$\rho_S = \frac{V_{eu}}{V_{ev}} = \left(\frac{1 + \nu \tan A}{1 + \mu \tan A}\right)$$
$$\times \left[\frac{(-\mu + \tan C) + \rho_C(1 + \mu \tan C) \tan(P - C)}{(-\nu + \tan C) + \rho_C(1 + \nu \tan C) \tan(P - C)}\right], \tag{3.128}$$

$$\rho_S = \frac{V_{eu}}{V_{ev}} = \left(\frac{\mu - \tan P}{\nu - \tan P}\right)$$
$$\times \left[\frac{(1 + \nu \tan C) + \rho_C(\nu - \tan C) \tan(P - C)}{(1 + \mu \tan C) + \rho_C(\mu - \tan C) \tan(P - C)}\right]. \tag{3.129}$$

Equations (3.128) and (3.129) give the ratio of eigenvalues of an optical system with elliptic eigenpolarizations described in the Cartesian complex plane representation of polarization [§1.7] by (the points) μ and ν, with (P, C, A) representing one set of nulling angles in either the PCSA or PSCA ellipsometer arrangement, respectively.

The reader can verify that eqs. (3.128) and (3.129) reduce to eqs. (3.33) and eq. (3.55) of §3.4, respectively, when $\nu = 0$, $\mu = \infty$, corresponding to an optical system with x and y linear eigenpolarizations. Similarly, eqs. (3.128) and (3.129) reduce to (3.103) and (3.107) of §3.5, respectively, when $\mu = -j$ and $\nu = +j$, corresponding to an optical system whose eigenpolarizations are the left- and right-circular polarizations.

3.7. Effect of azimuth-angle errors and component imperfections on the ellipsometric measurement of ρ_S

In null ellipsometry, the ratio ρ_S of the eigenvalues of an optical system under measurement with known eigenpolarizations is a function of (1) the azimuth angles of the polarizer, compensator and analyzer that reduce the light flux reaching the photodetector to a minimum, and (2) the properties of all optical devices encountered by the light beam.

In previous sections, it was assumed that the azimuth angles are free of error and that the optical devices perform in an ideal manner. In a real situation, the azimuth angles as read on the instrument are susceptible to errors and deviations from ideal behavior by the optical devices are unavoidable.

Examples of imperfect optical components include linear polarizers that transmit elliptically polarized light with small ellipticity, compensators that have actual retardation different from their nominal retardation (*e.g.*, quarter wave) and cell windows that perturb the state of polarization of the ellipsometer light beam. The latter source of error arises in reflection and transmission ellipsometry on samples that are enclosed in test cells such that the ellipsometer light beam senses the sample through entrance and exit windows.

Various aspects of the problem of the measurement of ρ_S in the presence of errors in the instrument readings (azimuth angles) and component imperfections have been considered by several investigators [70–101]. Until recently, this has been carried out in piecemeal fashion dealing with one imperfection or the other at a time.

We present in this section a comprehensive error analysis that employs a Jones-matrix formalism. The results of the present section will be used in ch. 5, where we also consider sources of error other than azimuth-angle errors and component imperfections.

The Jones matrix of an optical device k with small imperfections can be written as

$$T_k = T_k^0 + \delta T_k,$$
$$\delta T_k = (\delta T_{ijk}), \quad |\delta T_{ijk}| \ll 1, \quad (3.130)$$

where δT_k is a 2×2 complex *imperfection matrix* representing the deviation from the ideal behavior described by T_k^0. Similarly, the azimuth angle of the kth device can be written as

$$Z_k = Z_k^0 + \delta Z_k, \quad |\delta Z_k| \ll 1, \quad (3.131)$$

where Z_k^0 is the azimuth angle as read on the ellipsometer and δZ_k is the correction to be added to this azimuth to give the true azimuth Z_k.

The ratio of eigenvalues ρ_S of the optical system under measurement is a function of the nulling azimuth angles and the optical properties of the ellipsometer components

$$\rho_S = f(Z_k, T_{ijk}). \quad (3.132)$$

The arguments Z_k and T_{ijk} of the function f represent, respectively, all the azimuth angles and the elements of the Jones matrices of all optical devices encountered by the light beam (k is a subscript that identifies an optical component). The linear Taylor expansion of eq. (3.132) around Z_k^0, T_{ijk}^0 yields

$$\rho_S = f(Z_k^0, T_{ijk}^0) + \sum_k \gamma_k' \delta Z_k + \sum_{ijk} \gamma_{ijk} \delta T_{ijk}, \tag{3.133}$$

or

$$\rho_S = \rho_S^0 + \delta\rho_S, \tag{3.134}$$

where

$$\rho_S^0 = f(Z_k^0, T_{ijk}^0), \tag{3.135}$$

$$\delta\rho_S = \sum_k \gamma_k' \delta Z_k + \sum_{ijk} \gamma_{ijk} \delta T_{ijk}. \tag{3.136}$$

In eqs. (3.133) and (3.136), the summation over ij includes the four positions of the 2×2 Jones matrix ($i, j = 1, 2$) and k runs over all the optical components encountered by the light beam. ρ_S^0 is the zero-order approximation to ρ_S, when azimuth-angle errors and component imperfections are neglected. $\delta\rho_S$ is the first-order correction term to be added to ρ_S^0.

The quantities γ_k' and γ_{ijk} that appear in eq. (3.136) are *coupling coefficients* [95] that determine the extent to which a small azimuth-angle error δZ_k or a small component imperfection δT_{ijk} in the kth optical element *couple* into an error of ρ_S, respectively. In terms of the function f of eq. (3.132), these coupling coefficients are given by

$$\gamma_k' = \left.\frac{\partial f}{\partial Z_k}\right|_{Z_k = Z_k^0}, \tag{3.137}$$

$$\gamma_{ijk} = \left.\frac{\partial f}{\partial T_{ijk}}\right|_{T_{ijk} = T_{ijk}^0}. \tag{3.138}$$

The effect of the 2×2 imperfection Jones matrix

$$\delta T_k = \begin{bmatrix} \delta T_{11k} & \delta T_{12k} \\ \delta T_{21k} & \delta T_{22k} \end{bmatrix}, \tag{3.139}$$

of the kth optical component alone is determined by a 2×2 *array* of coupling coefficients of the form

$$\Gamma_k = \begin{bmatrix} \gamma_{11k} & \gamma_{12k} \\ \gamma_{21k} & \gamma_{22k} \end{bmatrix}, \tag{3.140}$$

§3.7]   AZIMUTH-ANGLE ERRORS AND COMPONENT IMPERFECTIONS   189

such that

$$\delta\rho_S = \sum_{ij} \gamma_{ijk} \delta T_{ijk}, \quad (3.141)$$

gives the error in the ratio of eigenvalues ρ_S of the optical system under measurement due to δT_k.

In reflection ellipsometry, it is often desirable to determine the errors in the ellipsometric angles ψ and Δ (where $\rho_S = \tan\psi \, e^{j\Delta}$) separately, due to azimuth-angle errors and component imperfections. Taking the differential of the logarithm of both sides of $\rho_S = \tan\psi \, e^{j\Delta}$ [eq. (3.67)], we obtain

$$\frac{\delta\rho_S}{\rho_S} = \frac{\delta(\tan\psi)}{\tan\psi} + j\delta\Delta. \quad (3.142)$$

Equating the real and imaginary parts of both sides of eq. (3.142), we get

$$\delta\psi = \frac{1}{2}\sin 2\psi \, \mathrm{Re}\left(\frac{\delta\rho_S}{\rho_S}\right), \qquad \delta\Delta = \mathrm{Im}\left(\frac{\delta\rho_S}{\rho_S}\right). \quad (3.143)$$

According to eq. (3.143), the errors $\delta\psi$ and $\delta\Delta$ in the ellipsometric angles are directly determined by the fractional error $\delta\rho_S/\rho_S$ due to azimuth errors and component imperfections. From eq. (3.136), $\delta\rho_S/\rho_S$ is given by

$$\frac{\delta\rho_S}{\rho_S} = \sum_k \left(\frac{\gamma'_k}{\rho_S}\right)\delta Z_k + \sum_{ijk}\left(\frac{\gamma_{ijk}}{\rho_S}\right)\delta T_{ijk}. \quad (3.144)$$

In eq. (3.144), the quantities γ'_k/ρ_S and γ_{ijk}/ρ_S are *modified* coupling coefficients that relate a small azimuth error δZ_k and small component imperfection δT_{ijk} in the kth optical component to a fractional error $\delta\rho_S/\rho_S$ in the ratio of eigenvalues ρ_S of the optical system under measurement.

The objective of the ensuing analysis is to determine the coupling coefficients associated with azimuth errors and component imperfections in conventional ellipsometer arrangements. Before this, we examine the generally useful concept of the imperfection plate of a non-ideal optical device and the transposition of two optical elements.

3.7.1. *The imperfection plate of a non-ideal optical component*

Except for depolarization, the effect of a *real* optical component on the state of polarization is completely described by a 2×2 Jones matrix. This matrix assumes an *ideal* form in the absence of imperfections. For example, an ideal perfectly smooth isotropic surface has a diagonal Jones matrix in the p-s frame of reference, whereas a real surface may exhibit off-diagonal reflection coefficients due to, for example, preferential surface roughness or optical activity.

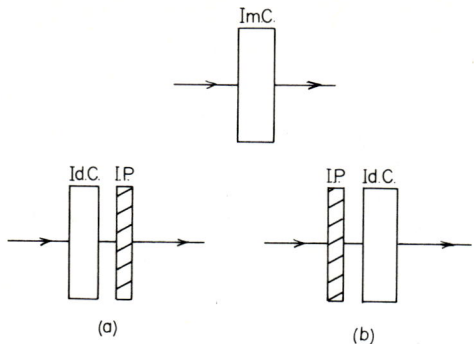

Fig. 3.8. An imperfect optical component (Im.C.) can be replaced by its ideal counterpart (Id.C.) followed (a), or preceded (b), by an imperfection plate (I.P.).

A convenient way of handling a real component with small imperfections is to replace it by a combination of two elements: its ideal counterpart plus an imperfection plate, fig. 3.8a,b. Such a plate has a Jones matrix that is approximately equal to the identity matrix. The imperfection plate can be chosen either to follow or precede the ideal element. When imperfections are assumed to be important to first order only, superposition is applicable and imperfections in a number of individual components can be lumped together into a composite imperfection plate. The Jones matrix of an imperfect optical component can be written as

$$\begin{aligned} T_{real} &= T_{real} T_{ideal}^{-1} T_{ideal} \\ &= [T_{real} T_{ideal}^{-1}] T_{ideal} \\ &= T_F^a T_{ideal}. \end{aligned} \tag{3.145}$$

According to eq. (3.145), the imperfection properties of an optical component can be isolated in an imperfection plate T_F^a that follows its ideal counterpart, where

$$T_F^a = T_{real} T_{ideal}^{-1}. \tag{3.146}$$

Similarly, we can write

$$\begin{aligned} T_{real} &= T_{ideal} T_{ideal}^{-1} T_{real} \\ &= T_{ideal} (T_{ideal}^{-1} T_{real}) \\ &= T_{ideal} T_F^b. \end{aligned} \tag{3.147}$$

In this case, the imperfection plate precedes the ideal element and its Jones matrix is given by

$$T_F^b = T_{ideal}^{-1} T_{real}. \tag{3.148}$$

§3.7]   AZIMUTH-ANGLE ERRORS AND COMPONENT IMPERFECTIONS   191

For small imperfection ($T_{\text{real}} \simeq T_{\text{ideal}}$), T_F^a and T_F^b deviate only by a small amount from the identity 2×2 matrix. In order to apply the concept of an imperfection plate, it is necessary that T_{ideal}^{-1} exists so that T_{ideal} must be non-singular. In an ellipsometer, this condition excludes the polarizer and analyzer but applies to all elements interposed between them. The determination of T_F^a or T_F^b is usually facilitated by the fact that T_{ideal} is diagonal.

When a group of imperfection plates in succession are encountered, it is convenient to find a resultant plate for the group. The Jones matrix of the kth imperfection plate takes the form

$$T_{Fk} = \begin{bmatrix} 1 + \alpha_{11k} & \alpha_{12k} \\ \alpha_{21k} & 1 + \alpha_{22k} \end{bmatrix}, \quad |\alpha_{ijk}| \ll 1; \quad i, j = 1, 2. \tag{3.149}$$

For N such plates in succession, the group matrix is

$$T_{Fg} = \prod_{k=1}^{N} T_{Fk}, \tag{3.150}$$

which upon substituting from eq. (3.149) becomes

$$T_{Fg} = \begin{bmatrix} 1 + \sum_{k=1}^{N} \alpha_{11k} & \sum_{k=1}^{N} \alpha_{12k} \\ \sum_{k=1}^{N} \alpha_{21k} & 1 + \sum_{k=1}^{N} \alpha_{22k} \end{bmatrix} = \begin{bmatrix} 1 + \alpha_{11g} & \alpha_{12g} \\ \alpha_{21g} & 1 + \alpha_{22g} \end{bmatrix}, \tag{3.151}$$

where

$$\alpha_{ijg} = \sum_{k=1}^{N} \alpha_{ijk}; \quad i, j = 1, 2. \tag{3.152}$$

Products of α_{ijk} are neglected in a first-order analysis. Equation (3.152) is an expression of the linear superposition of small imperfections to first order.

3.7.2. Transposition of two optical components

Consider two optical elements (A, B) placed in the path of a light beam such that B is after A along the direction of propagation. The effect of this pair of elements on the state of polarization of the beam is described by the combined Jones matrix

$$T = T_B T_A.$$

Post-multiplication of the right-hand side of the above equation by $T_B^{-1} T_B$ ($\equiv 2 \times 2$ identity matrix) gives

$$T = (T_B T_A T_B^{-1}) T_B,$$

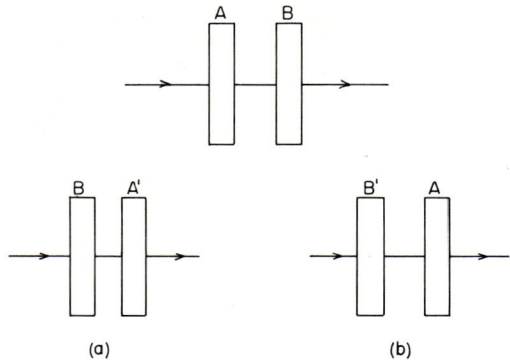

Fig. 3.9. Transposition of two optical elements A, B (above) in the path of a light beam. The sequence A, B can be replaced by either one of two equivalent sequences (a) B, A′ or (b) B′, A. The Jones matrices of A′ and B′ in terms of those of A and B are given by eqs. (3.154) and (3.156), respectively.

or

$$T = T_{A'}T_B = T_B T_A, \tag{3.153}$$

where

$$T_{A'} = T_B T_A T_B^{-1}. \tag{3.154}$$

Equation (3.153) shows that A can be moved from a position where it is placed *before* B to a position where it is placed *after* B, provided that its Jones matrix is modified as in eq. (3.154). This (A, B) → (B, A′) transposition is shown schematically in fig. 3.9. (A′ may be called the mirror image of A with respect to B.) The movement of A from a position before B to a position after B is allowed only if T_B is non-singular, *i.e.*, if B does not act as an ideal polarizer. Another transposition, (A, B) → (B′, A), such that

$$T = T_B T_A = T_A T_{B'}, \tag{3.155}$$

is also permissible, if T_B is non-singular and

$$T_{B'} = T_A^{-1} T_B T_A. \tag{3.156}$$

3.7.3. *The PCWSW′A ellipsometer arrangement*

Consider the PCWSW′A ellipsometer arrangement, fig. 3.10, in which the optical system S under measurement is assumed to be mounted in a test cell with entrance and exit optical windows W and W′, respectively. The polarizing section of the instrument consists of the linear polarizer P and the compensator C, while the analyzing section consists of the linear analyzer A. The optical system S under measurement is supposed to have orthogonal *x*

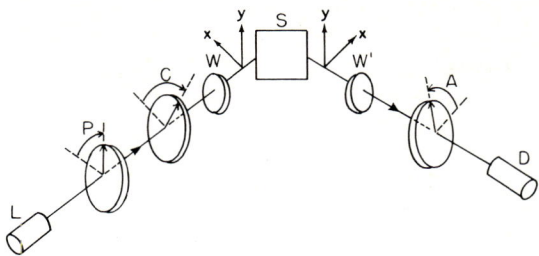

Fig. 3.10. The PCWSW'A ellipsometer arrangement. The optical system S (surface) under measurement is assumed to be mounted in a test cell with entrance and exit windows W and W', respectively. The polarizing section consists of a linear polarizer P and a compensator C, while the analyzing section consists of a linear analyzer A alone. L and D are the light source and photodetector, respectively.

and y linear eigenpolarizations and it is required to measure the ratio ρ_S of their associated eigenvalues V_{ex} and V_{ey} in the presence of azimuth-angle errors and component imperfections.

With azimuth-angle errors and component imperfections assumed small, it is possible, in a first-order analysis, to examine their effects on ρ_S separately. Furthermore, when component imperfections are treated, we can assume that only a number of components are imperfect at a time.

3.7.3.1. Azimuth-angle errors

In the imperfection-free PCWSW'A ellipsometer arrangement, the ratio of eigenvalues ρ_S of the optical system S is given by

$$\rho_S = -\tan A \, \frac{\tan C + \rho_C \tan(P-C)}{1 - \rho_C \tan C \tan(P-C)}, \tag{3.33}$$

where (P, C, A) is a set of nulling angles and ρ_C is the slow-to-fast relative complex-amplitude transmittance of the compensator. According to eq. (3.137), if we take the first partial derivatives of the above expression of ρ_S with respect to P, C and A, we obtain

$$\gamma'_P = -\frac{\rho_C \tan A \, \sec^2 C \, \sec^2(P-C)}{[1 - \rho_C \tan C \tan(P-C)]^2}, \tag{3.157}$$

$$\gamma'_C = -\frac{\tan A \, \sec^2 C [1 - \rho_C \sec^2(P-C) - \rho_C^2 \tan^2(P-C)]}{[1 - \rho_C \tan C \tan(P-C)]^2}, \tag{3.158}$$

$$\gamma'_A = -\frac{\sec^2 A [\tan C + \rho_C \tan(P-C)]}{[1 - \rho_C \tan C \tan(P-C)]}, \tag{3.159}$$

which represent the coupling coefficients for azimuth-angle errors δP, δC and δA in the polarizer, compensator, and analyzer readings, respectively.

Table 3.1. Coupling coefficients for azimuth-angle errors*

	$\gamma'_k(P, A, C, \rho_C)$	$\gamma'_k(P, A, +\tfrac{1}{4}\pi, -j)$	$\gamma'_k(P, A, -\tfrac{1}{4}\pi, -j)$
P	$\gamma'_P = -\dfrac{\rho_C \tan A \sec^2 C \sec^2(P-C)}{[1 - \rho_C \tan C \tan(P-C)]^2}$	$-2\tan A\, e^{-j2P}$	$-2\tan A\, e^{j2P}$
C	$\gamma'_C = -\dfrac{\tan A \sec^2 C[1 - \rho_C \sec^2(P-C) - \rho_C^2 \tan^2(P-C)]}{[1 - \rho_C \tan C \tan(P-C)]^2}$	$-2\tan A(-1 + j\sin 2P)e^{-j2P}$	$2\tan A(1 + j\sin 2P)e^{j2P}$
A	$\gamma'_A = -\dfrac{\sec^2 A[\tan C + \rho_C \tan(P-C)]}{[1 - \rho_C \tan C \tan(P-C)]}$	$-j\sec^2 A\, e^{-j2P}$	$j\sec^2 A\, e^{j2P}$

*P, C, and A are the polarizer, compensator, and analyzer azimuth angles at null, respectively, in the PCWSW A ellipsometer arrangement. ρ_C is the slow-to-fast relative complex-amplitude transmittance of the compensator. An azimuth error δZ couples into an error in the ratio ρ_3 of eigenvalues to be measured given by $\delta\rho_3 - \gamma' \delta Z$ ($Z = P, C, A$).

For ease of reference, table 3.1 summarizes these azimuth-angle coupling coefficients, both in general form (column 2), and in the special cases when $\rho_C = -j$, $C = +\frac{1}{4}\pi$ (column 3), and when $\rho_C = -j$, $C = -\frac{1}{4}\pi$ (column 4).

3.7.3.2. Component imperfections

The procedure that we choose to account for component imperfections uses the idea of imperfection plates introduced earlier in this section. This is based on the fact that if we know the effect on ρ_S of inserting an imperfection plate (an element with a Jones matrix approximately the same as the 2×2 identity matrix) into the ellipsometer, the effect of other component imperfections can be deduced. This requires the replacement of an imperfect optical device by its ideal counterpart followed or preceded by an imperfection plate. However, as mentioned before, this procedure does not apply to the polarizer or the analyzer but applies to the compensator, entrance window, optical system, and exit window.

Compensator, entrance-window, optical system and exit-window imperfections

Consider the ellipsometer arrangement PCSFA which is obtained from the imperfection-free PCSA ellipsometer arrangement by inserting the imperfection plate F between the optical system S and the analyzer A. The Jones matrix of F is of the form

$$T_F = \begin{bmatrix} 1 + \alpha_{11F} & \alpha_{12F} \\ \alpha_{21F} & 1 + \alpha_{22F} \end{bmatrix}, \quad |\alpha_{ijF}| \ll 1; \quad i, j = 1, 2. \tag{3.149}$$

Using the notation employed in §3.4, let χ_{AI}^{xy} represent the complex variable that describes the ellipse of polarization incident on the analyzer A, in the Cartesian representation whose basis states are the x and y linear eigenpolarizations of the optical system S under measurement. Assuming that the null condition has been reached, $\chi_{AI}^{xy} = -\cot A$ [eq. (3.36)], and because the light outgoing from F is incident on A, $\chi_{AI}^{xy} = \chi_{FO}^{xy}$. The relationship between the ellipses of polarization χ_{FI}^{xy} and χ_{FO}^{xy} at the input and output of the imperfection plate F is given by the (inverse) bilinear transformation of eq. (2.52), §2.4. Therefore, we have

$$\chi_{FI}^{xy} = \frac{(1 + \alpha_{11F})\chi_{FO}^{xy} - \alpha_{21F}}{-\alpha_{12F}\chi_{FO}^{xy} + (1 + \alpha_{22F})},$$

or

$$\chi_{FI}^{xy} = \chi_{SO}^{xy} = -\cot A \, \frac{1 + \alpha_{11F} + \alpha_{21F} \tan A}{1 + \alpha_{22F} + \alpha_{12F} \cot A}, \tag{3.160}$$

which is obtained after substituting $\chi_{\text{FO}}^{xy} = -\cot A$. The ratio ρ_S of the eigenvalues of the optical system S under measurement is equal to the ratio ($\chi_{\text{SI}}^{xy}/\chi_{\text{SO}}^{xy}$) of the complex variables χ_{SI}^{xy} and χ_{SO}^{xy} that describe the ellipses of polarization at the input and output of S, respectively [eq. (3.35), §3.4]. Because the polarizing section of the instrument is unaffected by F, χ_{SI}^{xy} ($= \chi_{\text{CO}}^{xy}$) continues to be given by eq. (3.13). Substituting χ_{SI}^{xy} ($= \chi_{\text{CO}}^{xy}$) from eq. (3.13) and χ_{SO}^{xy} from eq. (3.160) into eq. (3.35), we obtain

$$\rho_S(\text{PCSFA}) = \rho_S^0 \frac{1 + \alpha_{22F} + \alpha_{12F} \cot A}{1 + \alpha_{11F} + \alpha_{21F} \tan A}, \qquad (3.161)$$

where ρ_S^0 is the expression for ρ_S in the imperfection-free PCSA ellipsometer arrangement given by eq. (3.33). Because $|\alpha_{ijF}| \ll 1$, we can take a linear expansion of eq. (3.161) as follows[18]

$$\rho_S(\text{PCSFA}) = \rho_S^0 (1 + \alpha_{22F} + \alpha_{12F} \cot A - \alpha_{11F} - \alpha_{21F} \tan A),$$

from which the array of coupling coefficients $\Gamma_F(\gamma_{ijF})$ that relate the imperfection parameters α_{ijF} of F to an error in ρ_S [eq. (3.136)] is given by

$$\Gamma_F = \begin{bmatrix} -\rho_S^0 & \rho_S^0 \cot A \\ -\rho_S^0 \tan A & \rho_S^0 \end{bmatrix}. \qquad (3.162)$$

In the PCWSW'A ellipsometer arrangement, the effect of an imperfect exit window W' is identical to that of the imperfection plate F considered above, so that the coupling coefficients of eq. (3.162) apply to W' directly, i.e.,

$$\Gamma_{W'} = \begin{bmatrix} -\rho_S^0 & \rho_S^0 \cot A \\ -\rho_S^0 \tan A & \rho_S^0 \end{bmatrix}. \qquad (3.163)$$

To determine the effect of the entrance window W, we replace it by an equivalent element after S whose Jones matrix is given by eq. (3.154).

---

[18] More correctly, the condition $|\alpha_{21F} \tan A| \ll 1$ has to be satisfied. If $|\tan A| > 1$, the imperfection parameter α_{21F} is amplified by $|\tan A|$ (where A is the analyzer-azimuth angle at null) and second- and higher-order corrections may be necessary. Similarly, if $|\cot A| \gg 1$, the imperfection parameter α_{12F} is amplified by $|\cot A|$ and large correction factors might again be required. Because A is largely (exactly, when $\rho_C = -j$, $C = \pm\frac{1}{4}\pi$) determined by the magnitude $|\rho_S|$ of the ratio of the eigenvalues to be measured [see eq. (3.33)], large anomalous errors may be expected from the imperfection plate F in both limits of $|\rho_S| \ll 1$, and $|\rho_S| \gg 1$.

Substituting

$$T_S = \begin{bmatrix} \rho_S^0 & 0 \\ 0 & 1 \end{bmatrix}, \quad T_S^{-1} = \begin{bmatrix} 1/\rho_S^0 & 0 \\ 0 & 1 \end{bmatrix},$$

$$T_W = \begin{bmatrix} 1+\alpha_{11W} & \alpha_{12W} \\ \alpha_{21W} & 1+\alpha_{22W} \end{bmatrix}, \tag{3.164}$$

into $T_S T_W T_S^{-1}$, we find that

$$T_S T_W T_S^{-1} = \begin{bmatrix} 1+\alpha_{11W} & \alpha_{12W}\rho_S^0 \\ \alpha_{21W}/\rho_S^0 & 1+\alpha_{22W} \end{bmatrix}. \tag{3.165}$$

Because the element equivalent to W after S now has the same position as F in the PCSFA arrangement, the array of coupling coefficients Γ_W for entrance-window imperfections is obtained from Γ_F of F [eq. (3.162)] by multiplication of its off-diagonal elements γ_{12F} and γ_{21F} by ρ_S^0 and $1/\rho_S^0$, respectively [see eq. (3.165)],

$$\Gamma_W = \begin{bmatrix} -\rho_S^0 & (\rho_S^0)^2 \cot A \\ -\tan A & \rho_S^0 \end{bmatrix}. \tag{3.166}$$

The optical system under measurement is considered imperfect when its eigenpolarizations differ, to a small extent, from what they are known (presumed) to be. Thus, for a reflector exhibiting small amounts of preferential roughness, stress, or optical activity, the p and s linear polarizations are *not* its eigenpolarizations; the p-s reflection Jones matrix is non-diagonal. In general, it is assumed that the optical system S under measurement has a non-diagonal Cartesian Jones matrix in the representation whose basis states are the *presumed* x and y linear eigenpolarizations. We replace S by its ideal counterpart *followed* by an imperfection plate F_S. Substituting

$$T_{S\,\text{real}} = \begin{bmatrix} \rho_S^0 + \alpha_{11S} & \alpha_{12S} \\ \alpha_{21S} & 1+\alpha_{22S} \end{bmatrix}, \quad T_{S\,\text{ideal}}^{-1} = \begin{bmatrix} 1/\rho_S^0 & 0 \\ 0 & 1 \end{bmatrix},$$

$$|\alpha_{ijS}| \ll 1, \quad i,j = 1,2 \tag{3.167}$$

into eq. (3.146), the Jones matrix of F_S is obtained

$$T(F_S) = \begin{bmatrix} \rho_S^0 + \alpha_{11S} & \alpha_{12S} \\ \alpha_{21S} & 1 + \alpha_{22S} \end{bmatrix} \begin{bmatrix} 1/\rho_S^0 & 0 \\ 0 & 1 \end{bmatrix},$$

$$= \begin{bmatrix} 1 + (\alpha_{11S}/\rho_S^0) & \alpha_{12S} \\ \alpha_{21S}/\rho_S^0 & 1 + \alpha_{21S} \end{bmatrix}. \tag{3.168}$$

F_S has the same location as F in the PCSFA ellipsometer arrangement so that the array of coupling coefficients for the imperfection parameters α_{ijS} of S [eq. (3.167)] is obtained from eq. (3.162) by multiplication of γ_{11F} and γ_{21F} by $1/\rho_S^0$ [see eq. (3.168)]

$$\Gamma_S = \begin{bmatrix} -1 & \rho_S^0 \cot A \\ -\tan A & \rho_S^0 \end{bmatrix}. \tag{3.169}$$

The arrays of coupling coefficients for imperfections in the entrance window, optical system and exit window are summarized in table 3.2, both in general form (column 2), and for the special cases when $\rho_C = -j$, $C = +\frac{1}{4}\pi$ (column 3), and $\rho_C = -j$, $C = -\frac{1}{4}\pi$ (column 4).

The effect of a non-ideal compensator C can be treated by isolating its imperfections in an imperfection plate F_C that *follows* the ideal compensator. The imperfection plate F_C acts like the entrance window W as a source of error. We leave it as an exercise to the reader to prove that the effect of the 2×2 imperfection matrix of the compensator is described by a 2×2 array of coupling coefficients whose elements are given by

$$\begin{aligned}
\gamma_{11C} &= \frac{\rho_C \tan A \sec^2 C \tan(P - C)}{[1 - \rho_C \tan C \tan(P - C)]^2}, \\
\gamma_{12C} &= \frac{\rho_C \tan A \sec^2 C \tan^2(P - C)}{[1 - \rho_C \tan C \tan(P - C)]^2}, \\
\gamma_{21C} &= \frac{-\tan A \sec^2 C}{[1 - \rho_C \tan C \tan(P - C)]^2}, \\
\gamma_{22C} &= \frac{-\tan A \sec^2 C \tan(P - C)}{[1 - \rho_C \tan C \tan(P - C)]^2}.
\end{aligned} \tag{3.170}$$

The four coupling coefficients for imperfections in the compensator C are given in table 3.3, both in general form, column 2 [as functions of any set of nulling angles (P, C, A) and the compensator relative transmittance ρ_C], and for the special cases when $\rho_C = -j$, $C = +\frac{1}{4}\pi$ (column 3), and $\rho_C = -j$, $C = -\frac{1}{4}\pi$ (column 4).

Table 3.2. Arrays of coupling coefficients for entrance-window (W), optical system (S), and exit-window (W') imperfections*

	$\Gamma_k(P, A, C, \rho_C)$	$\Gamma_k(P, A, +\tfrac{1}{4}\pi, -\mathrm{j})$	$\Gamma_k(P, A, -\tfrac{1}{4}\pi, -\mathrm{j})$
W	$\Gamma_{\mathrm{W}} = \begin{bmatrix} -\rho_s & \rho_s^2 \cot A \\ -\tan A & \rho_s \end{bmatrix}$	$\begin{bmatrix} \mathrm{j}\tan A\, e^{-\mathrm{j}2P} & -\tan A\, e^{-\mathrm{j}4P} \\ -\tan A & -\mathrm{j}\tan A\, e^{-\mathrm{j}2P} \end{bmatrix}$	$\begin{bmatrix} -\mathrm{j}\tan A\, e^{\mathrm{j}2P} & -\tan A\, e^{\mathrm{j}4P} \\ -\tan A & \mathrm{j}\tan A\, e^{\mathrm{j}2P} \end{bmatrix}$
S	$\Gamma_{\mathrm{S}} = \begin{bmatrix} -1 & \rho_s \cot A \\ -\tan A & \rho_s \end{bmatrix}$	$\begin{bmatrix} -1 & -\mathrm{j}e^{-\mathrm{j}2P} \\ -\tan A & -\mathrm{j}\tan A\, e^{-\mathrm{j}2P} \end{bmatrix}$	$\begin{bmatrix} -1 & \mathrm{j}e^{\mathrm{j}2P} \\ -\tan A & \mathrm{j}\tan A\, e^{\mathrm{j}2P} \end{bmatrix}$
W'	$\Gamma_{\mathrm{W'}} = \begin{bmatrix} -\rho_s & \rho_s \cot A \\ -\rho_s \tan A & \rho_s \end{bmatrix}$	$\begin{bmatrix} \mathrm{j}\tan A\, e^{-\mathrm{j}2P} & -\mathrm{j}e^{-\mathrm{j}2P} \\ \mathrm{j}\tan^2 A\, e^{-\mathrm{j}2P} & -\mathrm{j}\tan A\, e^{-\mathrm{j}2P} \end{bmatrix}$	$\begin{bmatrix} -\mathrm{j}\tan A\, e^{\mathrm{j}2P} & \mathrm{j}e^{\mathrm{j}2P} \\ -\mathrm{j}\tan^2 A\, e^{\mathrm{j}2P} & \mathrm{j}\tan A\, e^{\mathrm{j}2P} \end{bmatrix}$

$$\rho_S = -\tan A\, \frac{[\tan C + \rho_C \tan(P-C)]}{[1 - \rho_C \tan C \tan(P-C)]}$$

*The Jones matrix of an imperfect component can be written as $T_k = T_k^0 + \delta T_k$, where $\delta T_k = (\delta T_{ijk})$ is a 2×2 complex imperfection matrix representing the deviation from the ideal behavior described by T_k^0. The array of coupling coefficients $\Gamma_k = (\gamma_{ijk})$ determines the error in the measured ratio of eigenvalues ρ_S due to the imperfection matrix δT_k according to the relation $\delta\rho_S = \Sigma_{ij}\, \gamma_{ijk}\delta T_{ijk}$. P, C, and A are the polarizer, compensator and analyzer azimuth angles at null, respectively; ρ_C is the slow-to-fast relative complex-amplitude transmittance of the compensator.

Table 3.3. Elements of the array of coupling coefficients for compensator (C) imperfections*

	$\gamma_{ijC}(P, A, C, \rho_C)$	$\gamma_{ijC}(P, A, +\frac{1}{4}\pi, -j)$	$\gamma_{ijC}(P, A, -\frac{1}{4}\pi, -j)$
$\gamma_{11C} =$	$\dfrac{\rho_C \tan A \sec^2 C \tan(P-C)}{[1-\rho_C \tan C \tan(P-C)]^2}$	$-\tan A \cos 2P\, e^{-j2P}$	$\tan A \cos 2P\, e^{j2P}$
$\gamma_{12C} =$	$\dfrac{\rho_C \tan A \sec^2 C \tan^2(P-C)}{[1-\rho_C \tan C \tan(P-C)]^2}$	$\tan A(1-\sin 2P)\, e^{-j2P}$	$\tan A(1+\sin 2P)\, e^{j2P}$
$\gamma_{21C} =$	$\dfrac{-\tan A \sec^2 C}{[1-\rho_C \tan C \tan(P-C)]^2}$	$-j\tan A(1+\sin 2P)\, e^{-j2P}$	$-j\tan A(1-\sin 2P)\, e^{j2P}$
$\gamma_{22C} =$	$\dfrac{-\tan A \sec^2 C \tan(P-C)}{[1-\rho_C \tan C \tan(P-C)]^2}$	$j\tan A \cos 2P\, e^{-j2P}$	$-j\tan A \cos 2P\, e^{j2P}$

*See footnote to table 3.2.

Polarizer and analyzer imperfections

To complete the discussion of component imperfections, we consider non-idealities of the polarizer P and analyzer A. An imperfect linear polarizer is assumed to transmit elliptically polarized light. Referred to a coordinate system parallel to its transmission and extinction axes, the Jones vector and complex variable that describe the output light from an imperfect polarizer P are given by[19]

$$\boldsymbol{E}_{PO}^{te} = A_c' \begin{bmatrix} 1 \\ \alpha_P \end{bmatrix}, \qquad \chi_{PO}^{te} = \alpha_P. \tag{3.171}$$

If we substitute $\boldsymbol{E}_{PO}^{te}$ from eq. (3.171) into the right-hand side of eq. (3.29), carry out the matrix multiplications involved, and equate the result to zero, we obtain

$$\rho_S = \tan A \, \frac{[\tan C + \rho_C \tan(P-C)] + \alpha_P[\rho_C - \tan C \tan(P-C)]}{[\rho_C \tan C \tan(P-C) - 1] + \alpha_P[\rho_C \tan C + \tan(P-C)]}. \tag{3.172}$$

Equation (3.172) gives the ratio of eigenvalues ρ_S of an optical system S with linear eigenpolarizations, as measured by a PCSA ellipsometer arrangement in which P acts as an elliptic polarizer. α_P represents the transmitted eigenpolarization of such an elliptic polarizer and its associated azimuth θ_P and ellipticity angle ϵ_P can be found using eqs. (1.86) and (1.87) of §1.7. In the event $\alpha_P = 0$, the polarizer becomes an ideal linear polarizer and eq. (3.172) reduces to eq. (3.33). When $|\alpha_P| \ll 1$, the linear expansion of eq. (3.172) can be

---

[19] When $\alpha_P = 0$, eq. (3.171) reduces to eq. (3.4).

taken. The coefficient of α_P in such an expansion gives the coupling coefficient[20]

$$\gamma_P = -\frac{\rho_C \tan A \sec^2 C \sec^2 (P - C)}{[1 - \rho_C \tan C \tan (P - C)]^2}, \tag{3.173}$$

that relates α_P ($|\alpha_P| \ll 1$) to an error in ρ_S. The coupling coefficient for a polarizer imperfection, eq. (3.173), is the *same* as the coupling coefficient for a polarizer azimuth-angle error, eq. (3.157).

An imperfect linear analyzer is assumed to have a non-diagonal Jones matrix of the form

$$T_A^{te} = K_A \begin{bmatrix} 1 + \alpha_{11A} & \alpha_{12A} \\ \alpha_{21A} & \alpha_{22A} \end{bmatrix}, \quad |\alpha_{ijA}| \ll 1, \quad i, j = 1, 2. \tag{3.174}$$

which reduces to eq. (3.21) when $\alpha_{ijA} = 0$. The relationship between the incident (E_{AI}^{te}) and the outgoing (E_{AO}^{te}) fields on and from the analyzer, respectively, is given by, $E_{AO}^{te} = T_A^{te} E_{AI}^{te}$ [eq. (3.22)], whose expansion yields

$$\begin{aligned} E_{AO,t} &= K_A[(1 + \alpha_{11A})E_{AI,t} + \alpha_{12A}E_{AI,e}], \\ E_{AO,e} &= K_A[\alpha_{21A}E_{AI,t} + \alpha_{22A}E_{AI,e}]. \end{aligned} \tag{3.175}$$

The transmitted (and detected) light intensity is given by

$$I_D = E_{AO,t}E_{AO,t}^* + E_{AO,e}E_{AO,e}^*, \tag{3.176}$$

which reaches a minimum as the polarizer and analyzer are adjusted, when

$$\frac{\partial I_D}{\partial P} = \frac{\partial I_D}{\partial A} = 0. \tag{3.177}$$

According to eqs. (3.175) and (3.176), the imperfection parameters α_{21A} and α_{22A} appear in the expression for the detected intensity only to second order, hence do not contribute first-order errors to ρ_S. Furthermore, when the second term on the right-hand side of eq. (3.176) is neglected, eq. (3.177) reduces to

$$E_{AO,t} = 0 = (1 + \alpha_{11A})E_{AI,t} + \alpha_{12A}E_{AI,e}. \tag{3.178}$$

Substituting $E_{AI,t}$ and $E_{AI,e}$ from eq. (3.20) into eq. (3.178), we get

$$\frac{E_{AI,x}}{E_{AI,y}} = \frac{(1 + \alpha_{11A}) \tan A + \alpha_{12A}}{\alpha_{12A} \tan A - (1 + \alpha_{11A})}. \tag{3.179}$$

Subsequently, we replace $E_{AI,x}$ and $E_{AI,y}$ from eq. (3.18) ($E_{SO}^{xy} = E_{AI}^{xy}$) into eq.

---

[20] This is also obtained directly by taking the first partial derivative of ρ_S in eq. (3.172) with respect to α_P [eq. (3.138)].

(3.179) to obtain

$$\rho_S = \frac{V_{ex}}{V_{ey}} = \rho_S^0 \frac{1 + \alpha_{11A} + \alpha_{12A} \cot A}{\alpha_{12A} \tan A - (1 + \alpha_{11A})}, \qquad (3.180)$$

where ρ_S^0 is the value of ρ_S obtained from the imperfection-free PCSA ellipsometer arrangement. The coefficients of α_{11A} and α_{12A} in the linear expansion of eq. (3.180) give the coupling coefficients γ_{11A} and γ_{12A}. Therefore, the elements of the array of coupling coefficients for the 2×2 analyzer imperfection matrix are

$$\gamma_{11A} = \gamma_{21A} = \gamma_{22A} = 0,$$

$$\gamma_{12A} = -\frac{\sec^2 A \, (\tan C + \rho_C \tan (P - C))}{(1 - \rho_C \tan C \tan (P - C))}. \qquad (3.181)$$

Note that only one imperfection parameter [α_{12A}, eq. (3.174)] for the analyzer is important, and that its associated coupling coefficient [γ_{12A}, eq. (3.181)] is the same as that for an analyzer azimuth-angle error [γ'_A, eq. (3.159)].

Table 3.4 gives the coupling coefficients for polarizer and analyzer imperfections both in general form (column 2), and for the special cases when $\rho_C = -j$, $C = +\frac{1}{4}\pi$ (column 3), and $\rho_C = -j$, $C = -\frac{1}{4}\pi$ (column 4).

Table 3.4. Coupling coefficients for polarizer (P) and analyzer (A) imperfections*

	$\gamma(P, A, C, \rho_C)$	$\gamma(P, A, +\frac{1}{4}\pi, -j)$	$\gamma(P, A, -\frac{1}{4}\pi, -j)$
P	$\gamma_P = -\dfrac{\rho_C \tan A \sec^2 C \sec^2 (P - C)}{[1 - \rho_C \tan C \tan (P - C)]^2}$	$-2 \tan A \, e^{-j2P}$	$-2 \tan A \, e^{j2P}$
A	$\gamma_{12A} = -\dfrac{\sec^2 A [\tan C + \rho_C \tan (P - C)]}{[1 - \rho_C \tan C \tan (P - C)]}$	$-j \sec^2 A \, e^{-j2P}$	$j \sec^2 A \, e^{j2P}$
	$\gamma_{11A} = \gamma_{21A} = \gamma_{22A} = 0$		

*An imperfect polarizer is assumed to transmit elliptically polarized light described by eq. (3.171). γ_P relates the polarizer imperfection parameter α_P to an error in ρ_S equal to $\gamma_P \alpha_P$. An imperfect analyzer is assumed to have a Jones matrix as in eq. (3.174). The effect of the analyzer imperfection matrix $\delta T_A = (\alpha_{ijA})$ on ρ_S is described by the array of coupling coefficients $\Gamma_A = (\gamma_{ijA})$ all of whose elements except γ_{12A} are zeros.

3.7.3.3. Errors in the ellipsometric angles ψ and Δ [21]

The ellipsometric angles ψ and Δ, corrected to first order to account for azimuth-angle errors and component imperfections, can be written as

---

[21] Although ψ and Δ have been defined with reference to reflecting surfaces (§3.4.4), the results of this subsection apply to any optical system with linear eigenpolarizations whose ratio of eigenvalues ρ_S is expressed as $\rho_S = \tan \psi \, e^{j\Delta}$.

§3.7] AZIMUTH-ANGLE ERRORS AND COMPONENT IMPERFECTIONS 203

$$\psi = \psi^0 + \delta\psi,$$
$$\Delta = \psi^0 + \delta\Delta,$$
(3.182)

which is simply an expansion of eq. (3.134) using eq. (3.67). ψ^0 and Δ^0 are the zero-order approximations of ψ and Δ, respectively, that are obtained when azimuth angles as read on the ellipsometer are substituted into the imperfection-free ellipsometry equation [eq. (3.33)]. $\delta\psi$ and $\delta\Delta$ are first-order error terms that account for azimuth-angle errors and component imperfections.

From the coupling coefficients of tables 3.1–3.4, $\delta\psi$ and $\delta\Delta$ can be obtained by use of eqs. (3.143) and (3.144). The procedure involves division of the coupling coefficient for a given azimuth error or component imperfection by ρ_S^0 to obtain the modified coupling coefficient. The latter is multiplied by the azimuth error or component imperfection in question to give the fractional error $\delta\rho_S/\rho_S$. Direct substitution of the fractional error $\delta\rho_S/\rho_S$ into eq. (3.143) yields $\delta\psi$ and $\delta\Delta$.

When the compensator acts as a quarter-wave retarder and its azimuth is set at either $C = \frac{1}{4}\pi$ or $-\frac{1}{4}\pi$, the expressions for the coupling coefficients, for ρ_S^0, and hence for $\delta\psi$ and $\delta\Delta$, are simplified. As an example, consider the errors $\delta\psi_P$ and $\delta\Delta_P$ due to the polarizer imperfection [eq. (3.171)]

$$\alpha_P = t_{1P} + jt_{2P}, \quad |t_{1P}|, \quad |t_{2P}| \ll 1. \tag{3.183}$$

When $\rho_C = -j$ and $C = \pm\frac{1}{4}\pi$, we have

$$\gamma_P = -2 \tan A \; e^{\mp j2P}, \qquad \rho_S^0 = \mp j \tan A \; e^{\mp j2P},$$

[see table 3.4 and eqs. (3.38) and (3.41)], so that the fractional error in ρ_S is given by

$$\left(\frac{\delta\rho_S}{\rho_S}\right)_P = \frac{\gamma_P}{\rho_S^0} \alpha_P$$
$$= \mp 2j(t_{1P} + jt_{2P}). \tag{3.184}$$

Substitution of eq. (3.184) into eq. (3.143) gives

$$\delta\psi_P = \pm t_{2P} \sin 2\psi, \qquad \delta\Delta_P = \mp 2t_{1P}. \tag{3.185}$$

In the above equations, the upper and lower signs correspond to $C = +\frac{1}{4}\pi$ and $C = -\frac{1}{4}\pi$, respectively. Note that t_{1P} and t_{2P} represent the small azimuth and small ellipticity of the elliptic vibration exiting from the polarizer, referenced to its transmission-extinction coordinate system.[22]

[22] For small azimuth θ and small ellipticity ϵ the Cartesian complex polarization variable χ of eq. (1.79), §1.7, simplifies to $\chi = \theta + j\epsilon$, to first order.

In terms of the nulling angles in the four zones, $\delta\psi_P$ and $\delta\Delta_P$ are given by

$$\begin{aligned}
\delta\psi_P(1) &= -t_{2P}\sin 2A_1, & \delta\Delta_P(1) &= +2t_{1P}, \\
\delta\psi_P(2) &= +t_{2P}\sin 2A_2, & \delta\Delta_P(2) &= -2t_{1P}, \\
\delta\psi_P(3) &= +t_{2P}\sin 2A_3, & \delta\Delta_P(3) &= +2t_{1P}, \\
\delta\psi_P(4) &= -t_{2P}\sin 2A_4, & \delta\Delta_P(4) &= -2t_{1P},
\end{aligned} \quad (3.186)$$

where, in going from eq. (3.185) to eq. (3.186), use has been made of the ideal zone relations summarized in table 3.5.

Table 3.5. The ideal zone relations*

Zone			
1 $C = -\frac{1}{4}\pi$	$P_1 = \frac{1}{2}\Delta - \frac{1}{4}\pi$ $A_1 = \psi$	$\Delta = 2P_1 + \frac{1}{2}\pi$ $\psi = A_1$	
2 $C = +\frac{1}{4}\pi$	$P_2 = -\frac{1}{2}\Delta - \frac{1}{4}\pi$ $A_2 = \psi$	$\Delta = -2P_2 - \frac{1}{2}\pi$ $\psi = A_2$	
3 $C = -\frac{1}{4}\pi$	$P_3 = \frac{1}{2}\Delta + \frac{1}{4}\pi$ $A_3 = -\psi$	$\Delta = 2P_3 - \frac{1}{2}\pi$ $\psi = -A_3$	
4 $C = +\frac{1}{4}\pi$	$P_4 = -\frac{1}{2}\Delta + \frac{1}{4}\pi$ $A_4 = -\psi$	$\Delta = -2P_4 + \frac{1}{2}\pi$ $\psi = -A_4$	

*(P_2, A_2) and (P_4, A_4) are the polarizer-analyzer sets of nulling angles that are available when the compensator is set at $C = +\frac{1}{4}\pi$, and the polarizer and analyzer are adjusted for null. (P_1, A_1) and (P_3, A_3) are the corresponding sets when $C = -\frac{1}{4}\pi$. The ellipsometric angles ψ and Δ are such that $\rho_S = \tan\psi\, e^{i\Delta}$ gives the ratio of the eigenvalues that correspond to the linear eigenpolarizations of the optical system under measurement. See §3.4.2.

For a different example, consider the effect of the imperfection plate F in the PCSFA ellipsometer arrangement on ψ and Δ. The fractional error in ρ_S is given by

$$\left(\frac{\delta\rho_S}{\rho_S}\right)_F = \sum_{ij}(\gamma_{ijF}/\rho_S^0)\alpha_{ijF}, \quad (3.187)$$

where α_{ijF} are the imperfection parameters of F [eq. (3.149)] and γ_{ijF} are the elements of the array of coupling coefficients [eq. (3.162)]. Substituting from eq. (3.162) into eq. (3.187), we obtain

$$\left(\frac{\delta\rho_S}{\rho_S}\right)_F = -\alpha_{11F} + \alpha_{12F}\cot A - \alpha_{21F}\tan A + \alpha_{22F}. \quad (3.188)$$

Writing the complex imperfection parameters α_{ijF} in eq. (3.188) in terms of their real and imaginary parts,

$$\alpha_{ijF} = t^r_{ijF} + jt^i_{ijF}, \quad i,j = 1, 2, \tag{3.189}$$

and subsequent use of eq. (3.143), lead to

$$\begin{aligned}\delta\psi_F &= \tfrac{1}{2}\sin 2\psi[-\alpha^r_{11F} + \alpha^r_{12F}\cot A - \alpha^r_{21F}\tan A + \alpha^r_{22F}], \\ \delta\Delta_F &= -\alpha^i_{11F} + \alpha^i_{12F}\cot A - \alpha^i_{21F}\tan A + \alpha^i_{22F}.\end{aligned} \tag{3.190}$$

Equations (3.188) and (3.190) are valid for any setting of the compensator azimuth and thus apply to the special azimuths $C = +\tfrac{1}{4}\pi$ and $C = -\tfrac{1}{4}\pi$. In terms of the nulling angles in the four zones, $\delta\psi_F$ and $\delta\Delta_F$ are summarized in table 3.6. As discussed earlier, from knowledge of the effect of the

Table 3.6. Errors in ψ and Δ caused by the imperfection plate F in the PCSFA ellipsometer arrangement*

Zone	
1, 2	$\delta\psi = \tfrac{1}{2}(\alpha^r_{12F} - \alpha^r_{21F}) + \tfrac{1}{2}(\alpha^r_{12F} + \alpha^r_{21F})\cos 2\psi + \tfrac{1}{2}(\alpha^r_{22F} - \alpha^r_{11F})\sin 2\psi$
3, 4	$\delta\psi = -\tfrac{1}{2}(\alpha^r_{12F} - \alpha^r_{21F}) - \tfrac{1}{2}(\alpha^r_{12F} + \alpha^r_{21F})\cos 2\psi + \tfrac{1}{2}(\alpha^r_{22F} - \alpha^r_{11F})\sin 2\psi$
1, 2	$\delta\Delta = (\alpha^i_{22F} - \alpha^i_{11F}) - \alpha^i_{21F}\tan\psi + \alpha^i_{12F}\cot\psi$
3, 4	$\delta\Delta = (\alpha^i_{22F} - \alpha^i_{11F}) + \alpha^i_{21F}\tan\psi - \alpha^i_{12F}\cot\psi$

*Obtained from eq. (3.190), noting that $A = \psi$ in zones 1 and 2, and $A = -\psi$ in zones 3 and 4 [see table 3.5].

imperfection plate F, the effect of imperfections in the compensator, entrance window, system and exit window on ψ and Δ can all be found.

For each imperfect component k between the polarizer and analyzer, eight real parameters (the real and imaginary parts of α_{ijk}, $i, j = 1, 2$) may contribute error terms to ψ and Δ in each of the four ellipsometry zones. Lengthy tables would therefore be required to display each and every error term. Instead, we show in tables 3.7 and 3.8 partial listings that include the more important of the imperfection parameters. In particular, off-diagonal elements in the Jones matrices of the compensator, optical system and analyzer are assumed zeros. Furthermore, the cell windows W and W' are supposed to act as weakly birefringent plates whose azimuths (direction of the principal fast axis of birefringence) are denoted by W and W', and whose respective small retardations are denoted t_{2W} and $t_{2W'}$, respectively.[23] t_{1C} +

---

[23] The principal-frame Jones matrix of a weakly birefringent window is diagonal, with diagonal elements 1 and $1 - jt_{2W}$, where t_{2W} is a small differential retardation. Pre- and post-multiplying this matrix by the counter-rotation and rotation matrices $R(-W)$ and $R(W)$, respectively (W is the azimuth of the fast axis of birefringence), we get a Jones matrix in the form of T_W of eq. (3.164), where $\alpha_{11W} = -\alpha_{22W} = \tfrac{1}{2}jt_{2W}\cos 2W$, $\alpha_{12W} = \alpha_{21W} = \tfrac{1}{2}jt_{2W}\sin 2W$.

Table 3.7. Ellipsometric angles ψ and Δ corrected to first order for azimuth-angle errors and component imperfections

Zone	
1	$\psi = A_1 - t_{2P} \sin 2A_1 - \frac{1}{2}t_{1C} \sin 2A_1 \cos 2P_1 + \sin 2A_1 \sin 2P_1 \delta C + \delta A + \frac{1}{2}t_{2W} \sin 2W \cos 2P_1 \sin 2A_1$
2	$\psi = A_2 + t_{2P} \sin 2A_2 - \frac{1}{2}t_{1C} \sin 2A_2 \cos 2P_2 + \sin 2A_2 \sin 2P_2 \delta C + \delta A - \frac{1}{2}t_{2W} \sin 2W \cos 2P_2 \sin 2A_2$
3	$\psi = -A_3 + t_{2P} \sin 2A_3 + \frac{1}{2}t_{1C} \sin 2A_3 \cos 2P_3 - \sin 2A_3 \sin 2P_3 \delta C - \delta A - \frac{1}{2}t_{2W} \sin 2W \cos 2P_3 \sin 2A_3$
4	$\psi = -A_4 - t_{2P} \sin 2A_4 + \frac{1}{2}t_{1C} \sin 2A_4 \cos 2P_4 - \sin 2A_4 \sin 2P_4 \delta C - \delta A + \frac{1}{2}t_{2W} \sin 2W \cos 2P_4 \sin 2A_4$
1	$\Delta = 2P_1 + \frac{1}{2}\pi + 2t_{1P} + 2\delta P - t_{2C} \cos 2P_1 - 2\delta C + t_{2W} \cos 2W + t_{2W'} \cos 2W' - t_{2W'} \sin 2W' \cot 2A_1$
2	$\Delta = -2P_2 - \frac{1}{2}\pi - 2t_{1P} - 2\delta P - t_{2C} \cos 2P_2 + 2\delta C + t_{2W} \cos 2W + t_{2W'} \cos 2W' - t_{2W'} \sin 2W' \cot 2A_2$
3	$\Delta = 2P_3 - \frac{1}{2}\pi + 2t_{1P} + 2\delta P - t_{2C} \cos 2P_3 - 2\delta C + t_{2W} \cos 2W + t_{2W'} \cos 2W' - t_{2W'} \sin 2W' \cot 2A_3$
4	$\Delta = -2P_4 + \frac{1}{2}\pi - 2t_{1P} - 2\delta P - t_{2C} \cos 2P_4 + 2\delta C + t_{2W} \cos 2W + t_{2W'} \cos 2W' - t_{2W'} \sin 2W' \cot 2A_4$

Table 3.8. Nulling azimuth angles as functions of ψ and Δ, azimuth-angle errors and component imperfections

Zone 1		
1	$A_1 = \psi + t_{2P}\sin 2\psi + \tfrac{1}{2}t_{1C}\sin 2\psi \sin\Delta + \sin 2\psi \cos\Delta\,\delta C - \delta A - \tfrac{1}{2}t_{2W}\sin 2W \sin\Delta \sin 2\psi$	
2	$A_2 = \psi - t_{2P}\sin 2\psi - \tfrac{1}{2}t_{1C}\sin 2\psi \sin\Delta + \sin 2\psi \cos\Delta\,\delta C - \delta A - \tfrac{1}{2}t_{2W}\sin 2W \sin\Delta \sin 2\psi$	
3	$A_3 = -\psi - t_{2P}\sin 2\psi + \tfrac{1}{2}t_{1C}\sin 2\psi \sin\Delta + \sin 2\psi \cos\Delta\,\delta C - \delta A - \tfrac{1}{2}t_{2W}\sin 2W \sin\Delta \sin 2\psi$	
4	$A_4 = -\psi + t_{2P}\sin 2\psi - \tfrac{1}{2}t_{1C}\sin 2\psi \sin\Delta + \sin 2\psi \cos\Delta\,\delta C - \delta A - \tfrac{1}{2}t_{2W}\sin 2W \sin\Delta \sin 2\psi$	
1	$P_1 = \tfrac{1}{2}\Delta - \tfrac{1}{4}\pi - t_{1P} - \delta P + \tfrac{1}{2}t_{2C}\sin\Delta + \delta C - \tfrac{1}{2}t_{2W}\cos 2W + \tfrac{1}{2}t_{2W}\sin 2W' \cot 2\psi$	
2	$P_2 = -\tfrac{1}{2}\Delta - \tfrac{1}{4}\pi - t_{1P} - \delta P + \tfrac{1}{2}t_{2C}\sin\Delta + \delta C + \tfrac{1}{2}t_{2W}\cos 2W - \tfrac{1}{2}t_{2W}\sin 2W' \cot 2\psi$	
3	$P_3 = \tfrac{1}{2}\Delta + \tfrac{1}{4}\pi - t_{1P} - \delta P - \tfrac{1}{2}t_{2C}\sin\Delta + \delta C - \tfrac{1}{2}t_{2W}\cos 2W - \tfrac{1}{2}t_{2W}\sin 2W' \cot 2\psi$	
4	$P_4 = -\tfrac{1}{2}\Delta + \tfrac{1}{4}\pi - t_{1P} - \delta P - \tfrac{1}{2}t_{2C}\sin\Delta + \delta C + \tfrac{1}{2}t_{2W}\cos 2W + \tfrac{1}{2}t_{2W}\sin 2W' \cot 2\psi$	

jt_{2C} represents the deviation from $-j$ of the slow-to-fast relative complex-amplitude transmittance of the compensator,[24] whereas $t_{1P} + jt_{2P}$ represents the polarizer imperfection as in eq. (3.183). Table 3.7 gives ψ and Δ in terms of (1) the polarizer-analyzer azimuth-angle pair at null (P_i, A_i) in the ith zone; (2) the polarizer (δP), compensator (δC), and analyzer (δA) azimuth-angle errors; (3) the polarizer $t_{1P} + jt_{2P}$ and compensator $t_{1C} + jt_{2C}$ imperfections; and (4) the entrance- and exit-window retardations t_{2W} and $t_{2W'}$ and azimuth angles W and W'. All expressions take the form $\psi = \psi^0 + \delta\psi$ and $\Delta = \Delta^0 + \delta\Delta$.

Table 3.8 gives the nulling angles in the four zones in terms of ψ and Δ and the azimuth-angle errors and component imperfections. It is obtained by inversion of the relations that appear in table 3.7, using the ideal zone relations of table 3.5.

3.7.3.4. Two-zone averages

Two-zone averaging refers to the utilization of the two conjugate pairs of nulling angles (P_2, A_2) and (P_4, A_4) or (P_1, A_1) and (P_3, A_3) that correspond to $C = +\frac{1}{4}\pi$ or $C = -\frac{1}{4}\pi$ respectively, to determine values for ψ and Δ that are free of most of the sources of error discussed. Apart from some additive multiple of $\frac{1}{2}\pi$, the values of ψ and Δ obtained from two-zone averages, assuming an ideal ellipsometer, are given by

$$\psi^0_{2,4} = -\tfrac{1}{2}(A_4 - A_2), \quad \psi^0_{1,3} = -\tfrac{1}{2}(A_3 - A_1),$$
$$\Delta^0_{2,4} = -(P_2 + P_4), \quad \Delta^0_{1,3} = (P_1 + P_3), \quad (3.191)$$

where the 2, 4 and 1, 3 subscripts correspond to the $+\frac{1}{4}\pi$ and $-\frac{1}{4}\pi$ compensator azimuths, respectively. Equation (3.191) is arrived at by use of the ideal zone relations in table 3.5. In the presence of imperfections, the values of ψ and Δ from two-zone averages become

$$\psi_{2,4} = \psi^0_{2,4} + \xi_{2,4}, \quad \psi_{1,3} = \psi^0_{1,3} + \xi_{1,3},$$
$$\Delta_{2,4} = \Delta^0_{2,4} + \eta_{2,4}, \quad \Delta_{1,3} = \Delta^0_{1,3} + \eta_{1,3}, \quad (3.192)$$

where ψ^0 and Δ^0 are the ideal averages of eq. (3.191), and ξ and η are two-zone, first-order, correction factors for ψ and Δ, respectively.

For simplicity, let us consider the sources of errors that appear in table 3.7. By addition of the ψ and Δ values in zones 2 and 4 and in zones 1 and 3,

---

[24] If we write, $\rho_C = T_C e^{j\delta_C} = -j + (t_{1C} + jt_{2C}) \simeq (1-t_{2C})\exp[-j(\tfrac{1}{2}\pi - t_{1C})]$, then $T_C = (1-t_{2C})$, $\delta_C = -\tfrac{1}{2}\pi + t_{1C}$. Thus t_{1C} represents the deviation from 90° of the compensator relative retardation, while t_{2C} represents the deviation from unity of the compensator relative attenuation.

we obtain $\psi_{2,4}$, $\Delta_{2,4}$ and $\psi_{1,3}$, $\Delta_{1,3}$ as in eq. (3.192), where

$$\xi_{2,4} = -\xi_{1,3} = t_{2P} \sin 2\psi, \tag{3.193a}$$

$$\eta_{2,4} = -2t_{1P} - 2\delta P + 2\delta C + t_{2W} \cos 2W + t_{2W'} \cos 2W',$$
$$\eta_{1,3} = 2t_{1P} + 2\delta P - 2\delta C + t_{2W} \cos 2W + t_{2W'} \cos 2W'. \tag{3.193b}$$

Equation (3.193a) shows that ψ from a two-zone average is free of all the sources of error considered in table 3.7, to first order, with the exception of the polarizer ellipticity t_{2P}. The latter leads to different values for ψ when uncorrected two-zone averages [eq. (3.191)] are taken.

Equation (3.193b) shows that Δ from a two-zone average is affected by the polarizer-imperfection parameter t_{1P}, polarizer-azimuth error δP, compensator-azimuth error δC, and entrance- and exit-window retardations t_{2W} and $t_{2W'}$.

With the help of table 3.6, which shows the effect of the imperfection plate F on ψ and Δ, the effect of more detailed imperfections in the entrance window, optical system, and exit window on two-zone averages can be determined. Thus, from table 3.6 it can be easily seen that the off-diagonal elements α_{12F} and α_{21F} in the Jones matrix of the imperfection plate F disappear when two-zone averages are taken for ψ and Δ. It immediately follows that off-diagonal elements in the Jones matrices of the entrance and exit windows, as well as the off-diagonal elements in the Cartesian Jones matrix of the optical system under measurement, all cancel to first order in two-zone averages. On the other hand, diagonal elements in F, hence in W, S, and W', contribute errors to the two-zone averages; they affect ρ_S directly and inseparably.

For the compensator, the diagonal imperfection parameters in the fast-slow Jones matrix cancel in two-zone averages, for both ψ and Δ, whereas the off-diagonal elements contribute error terms that survive two-zone averaging. The one significant off-diagonal term in the analyzer Jones matrix (α_{12A}) cancels in two-zone averaging.

A combined imperfection parameter for the two cell windows can be defined as

$$m_{W,W'} = (\alpha_{22W} - \alpha_{11W}) + (\alpha_{22W'} - \alpha_{11W'})$$
$$= m^r_{W,W'} + jm^i_{W,W'}. \tag{3.194}$$

The imaginary part $m^i_{W,W'}$ represents the combined birefringent properties of the two windows whose effect has already been described [last two terms in eq. (3.193b)]. The real part $m^r_{W,W'}$ represents the *dichroic* properties of the two windows and leads to a contribution to ξ, the two-zone ψ-correction factor, which is equal to

$$\xi_{2,4} = \xi_{1,3} = \tfrac{1}{2} m^r_{W,W'} \sin 2\psi. \tag{3.195}$$

Similarly, a single imperfection parameter may be defined to describe the combined effect of the two off-diagonal elements α_{12C} and α_{21C} in the compensator fast-slow Jones matrix

$$n_C = \alpha_{21C} + j\alpha_{12C},$$
$$= n_C^r + jn_C^i. \qquad (3.196)$$

The contributions to the two-zone correction factors ξ and η that arise from the compensator imperfection parameter n_C are given by (using the associated coupling coefficients in table 3.3)

$$\xi_{2,4} = -\xi_{1,3} = \tfrac{1}{2}n_C^r \sin 2\psi,$$
$$\eta_{2,4} = -\eta_{1,3} = n_C^i. \qquad (3.197)$$

3.7.3.5. Four-zone averages

If the four sets of nulling angles, (P_i, A_i), $i = 1, 2, 3$, and 4, that are available for the two compensator settings $C = \pm\tfrac{1}{4}\pi$ are utilized, the values of ψ and Δ so obtained are given by

$$\psi_{1-4}^0 = \tfrac{1}{2}(\psi_{2,4}^0 + \psi_{1,3}^0) = -\tfrac{1}{4}(A_3 - A_1 + A_4 - A_2),$$
$$\Delta_{1-4}^0 = \tfrac{1}{2}(\Delta_{2,4}^0 + \Delta_{1,3}^0) = \tfrac{1}{2}(P_1 + P_3 - P_2 - P_4), \qquad (3.198)$$

assuming an imperfection-free ellipsometer. Equation (3.198) is readily obtained from eq. (3.191). In the presence of imperfections, the values of ψ and Δ from four-zone averages are given by

$$\psi_{1-4} = \psi_{1-4}^0 + \tfrac{1}{2}(\xi_{2,4} + \xi_{1,3}),$$
$$\Delta_{1-4} = \Delta_{1-4}^0 + \tfrac{1}{2}(\eta_{2,4} + \eta_{1,3}), \qquad (3.199)$$

where ψ_{1-4}^0, Δ_{1-4}^0 are given by eq. (3.198), and the quantities $\tfrac{1}{2}(\xi_{2,4} + \xi_{1,3})$ and $\tfrac{1}{2}(\eta_{2,4} + \eta_{1,3})$ represent the ψ and Δ four-zone correction factors, respectively. When the sources of error of table 3.7 are considered, eq. (3.193) shows that

$$\tfrac{1}{2}(\xi_{2,4} + \xi_{1,3}) = 0,$$
$$\tfrac{1}{2}(\eta_{2,4} + \eta_{1,3}) = t_{2W} \cos 2W + t_{2W'} \cos 2W'. \qquad (3.200)$$

Thus, entrance- and exit-window birefringence represent the only imperfections that survive four-zone averaging, to cause an error in Δ. According to eq. (3.195), window dichroism, likewise, survives four-zone averaging, to cause an error in ψ. From eq. (3.197), the effect of off-diagonal elements in the fast-slow Jones matrix of the compensator cancel in four-zone averages, both for ψ and Δ.

3.7.3.6. Choice of compensator azimuth and position

Ellipsometric measurements are commonly taken with the compensator set at a fixed azimuth and the polarizer and analyzer adjusted for null. It is important to consider the optimization of the choice of the compensator azimuth from the viewpoint of minimizing the effect of ellipsometer imperfections.[25] An imperfection α_i causes an error $\delta\rho_S$ in the ratio of eigenvalues ρ_S which is given by

$$\delta\rho_S = \gamma_i \alpha_i,$$

where γ_i is the appropriate coupling coefficient. Expressions for the γ's have already been derived to account for the various sources of error in the ellipsometer. For a specific imperfection or azimuth-angle error, it is required to find the value of C which reduces the magnitude of the corresponding coupling coefficient to a minimum. We search for the minimum of

$$|\gamma| = |\gamma(P, C, A, \rho_C)|, \tag{3.201}$$

subject to the constraint that

$$\tan \psi \, e^{j\Delta} = \tan A \, \frac{\tan C + \rho_C \tan(P - C)}{\rho_C \tan C \tan(P - C) - 1}, \tag{3.202}$$

In eqs. (3.201) and (3.202), (P, C, A) represents a set of ellipsometer nulling angles, and ρ_C is the complex relative transmittance of the compensator.

The above problem was programmed for a numerical solution[26] on a computer assuming an ideal quarter-wave compensator (setting $\rho_C = -j$). With $\rho_S = \tan \psi \, e^{j\Delta}$ as a parameter, the sets of nulling angles satisfying the constraint of eq. (3.202) were substituted in eq. (3.201) in search for a minimum. The results are given below with reference to one element at a time [100].

(1) *Polarizer:* The magnitude of the coupling coefficient for the polarizer imperfection α_P (which is the same as that for an azimuth error δP) has a minimum value of $2 \tan \psi$ at $C = \pm \frac{1}{4}\pi$.

(2) *Compensator:* The magnitude of the coupling coefficient for an azimuth error δC assumes a minimum value of $2 \tan \psi (1 + \cos^2 \Delta)^{\frac{1}{2}}$ at

---

[25] Schmidt [88] studied the problem of optimization of the choice of the compensator azimuth C to minimize the errors in ψ and Δ due to the limited sensitivity of the photodetector.

[26] An analytical solution can be developed using the two relations $\tan \psi \sin \Delta = \tan A \tan(P - C) \sec^2 C / [1 + \tan^2 C \tan^2(P - C)]$, and $\sin \Delta = -\tan 2(P - C)/\sin 2C$, which are obtained from eq. (3.202) after equating the real and imaginary parts of both sides. The above equations together with the explicit expression of the coupling coefficient for a particular imperfection lead to the same conclusions derived numerically.

$C = \pm\frac{1}{4}\pi$. For the different elements of the imperfection Jones matrix, $|\gamma_{21C}|$ and $|\gamma_{22C}|$ remain constant at $\tan\psi \sin\Delta$ with the variation of C, $|\gamma_{21C}|$ and $|\gamma_{12C}|$ reach a minimum of $\tan\psi(1-\cos\Delta)$ and a maximum of $\tan\psi(1+\cos\Delta)$, respectively, at $C = \pm\frac{1}{4}\pi$. Ordinarily, the off-diagonal imperfection elements due to optical activity and birefringence are equal but of opposite signs. Therefore, the quantity $|\gamma_{21C} - \gamma_{12C}|$ was studied and found to go through a minimum of $\tan\psi(2 + 2\cos^2\Delta)^{\frac{1}{2}}$ at $C = \pm\frac{1}{4}\pi$.

(3) Analyzer: Here, only the (1, 2) element of the imperfection Jones matrix is important. The corresponding coupling coefficient (which is the same as that for an azimuth error δA) has a minimum whose position (the value of C at which it occurs) and magnitude vary with the system under measurements.

From the above, we conclude that the $\pm\frac{1}{4}\pi$ azimuths are optimum in the sense that they minimize the effect of the imperfections and azimuth errors of the polarizer and the compensator. The effect of analyzer imperfection and azimuth errors (which are not as important as those of the polarizer and compensator) is not minimized by this choice.

So far we have only considered the ellipsometer arrangement in which the compensator is placed before the optical system S. In this case, the sequence of elements encountered by the light beam is PCWSW'A where the system S is assumed to be mounted in a test cell with entrance and exit windows W and W', respectively. Assuming that the polarizing elements P and A are ideal, a null is obtained when the linear vibration E_1 emitted by P is mapped by the sequence CWSW' onto a linear vibration E_2 falling on the crossed analyzer A. The principle of reciprocity means that extinction is maintained if the detector and light source are interchanged. (This is opposite to what we stated in [100] and in agreement with [159]. The condition for reciprocity is absence of magnetic optical rotation [3]. Then, the Jones matrix of a system for backward propagation is the *transpose* of that for forward propagation. Because the null is independent of the light-source polarization, the Jones matrix of the PCWSW'A optical train is identically zero, *and so is its transpose.*) Let the sequences PCWSW'A and PWSW'CA be denoted by I and II, respectively. Sequence II is the conjugate ellipsometer arrangement with the compensator set after the optical system. In arrangement I, the ratio of eigenvalues ρ_S is a function of the nulling angles P, C, and A as well as the optical properties C, W, and W'. This functional expression also defines the relation that should be satisfied by the azimuths of E_1 and E_2 at null, and should be the same irrespective of the direction of propagation of the light beam. The inversion of sequence I, AW'SWCP is the same as sequence II except for the interchange of P and A, W and W'. Therefore, the expression of ρ_S for arrangement II is obtained from that for arrangement I using the simple transformations P↔A and W↔W'. It follows that each source of error in II corresponds to an equivalent source of error in I, both causing the same effect on ρ_S. The effect of a polarizer-azimuth error in I is the same as that of an analyzer-azimuth error in II and vice versa. The effect of a compensator-azimuth error remains the same. Imperfections in the entrance and exit windows exchange roles, whereas those of the compensator and optical system have the same effect on both arrangements.

The discussion above applies to the case of ideal polarizer and analyzer and continues to hold if small imperfections are introduced into these elements. The effect of such imperfections has been studied earlier in this section. For each of these two elements only one complex imperfection parameter (α_P for the polarizer and α_{12A} for the analyzer) proved important, and the appropriate coupling coefficients were determined. The latter were found to be the same as those pertaining to the respective azimuth-angle errors. In arrangement I, the effect of polarizer imperfection (or azimuth error) survives two-zone averaging, whereas that due to analyzer imperfection (or azimuth error) disappears. For arrangement II, these imperfections exchange roles so that the effect of the ellipticity of the polarizer light output disappears when a two-zone average is taken whereas the effect of the analyzer imperfection survives such averaging. If a circularly-polarized source beam is assumed, it becomes a simple exercise to prove that the (2, 1) and (2, 2) elements of the polarizer imperfection Jones matrix determine the ellipticity of the outgoing light. For the analyzer, it is only the element in the (1, 2) matrix location that is important. Thus if we compare I and II, we note that arrangement II has the advantage over arrangement I of cancelling the effect of the (2, 2) diagonal element of the imperfection Jones matrix of the polarizer. Apart from this difference, both arrangements are equally susceptible to the various imperfections in the ellipsometer.

3.8. Ellipsometry with imperfect components including incoherent effects

Consider an ellipsometer that consists of a light source, a polarizer, a succession of N optical devices including the optical system under measurement, an analyzer, and photodetector. The light beam and the optical devices it encounters are characterized by their Stokes vectors and Mueller matrices, respectively [99]. This description is general enough to account for such effects as a finite bandwidth and component depolarization.[27] The Mueller

---

[27] See §1.9 of ch. 1 and §2.12 of ch. 2.

matrix of the kth optical component can be written as

$$T_k = T_k^0 + \delta T_k, \qquad \delta T_k = (t_{ij})_k, \qquad |t_{ij}| \ll 1, \tag{3.203}$$

where T_k^0 describes the ideal behavior of the kth element and δT_k represents the matrix whose 16 elements carry information about various imperfections in the device. Because of δT_k, there is a noise-like Stokes vector perturbation δS_{kO} generated at the output of the kth device which is given by

$$\delta S_{kO} = \delta T_k S_{kI}^0, \tag{3.204}$$

where S_{kI}^0 is the Stokes vector of light incident on the kth element in an ideal ellipsometer. This perturbation propagates through the remaining part of the ellipsometer after the kth element, denoted by $\mathscr{S}_k$, to the photodetector. The contribution to the light flux reaching the detector arising from this perturbation is

$$\delta \mathscr{I}_{Dk} = \boldsymbol{\Gamma}_k^0 \delta T_k S_{kI}^0, \tag{3.205}$$

where $\boldsymbol{\Gamma}_k^0$ is a row vector whose elements are those of the first row of the transfer Mueller matrix of $\mathscr{S}_k^0$, the ideal version of $\mathscr{S}_k$.

Because of a possible deviation, δS_{PI}, of the source-beam Stokes vector from an assumed ideal form, S_{PI}^0, there is an additional contribution to the flux to the detector given by

$$\delta \mathscr{I}_{Ds.b.} = \boldsymbol{\Gamma}_P^0 T_P^0 \delta S_{PI}. \tag{3.206}$$

Summing the contributions from all component imperfections δT_k ($k = P$; $1, 2, \ldots, N$; A), and from the source-beam Stokes vector deviation δS_{PI}, we get, to first order,

$$\delta \mathscr{I}_D = \boldsymbol{\Gamma}_P^0 T_P^0 \delta S_{PI} + \sum_k \boldsymbol{\Gamma}_k^0 \delta T_k S_{kI}^0. \tag{3.207}$$

The flux to the detector in the absence of all imperfection is given by

$$\mathscr{I}_D^0 = \boldsymbol{\Gamma}_P^0 T_P^0 S_{PI}^0 = \boldsymbol{\Gamma}_P^0 S_{PO}^0, \tag{3.208}$$

S_{PO}^0 is the Stokes vector of light outgoing from the polarizer in an ideal ellipsometer. The detected flux in an ellipsometer with imperfections is obtained by adding eqs. (3.207) and (3.208)

$$\mathscr{I}_D = \mathscr{I}_D^0 + \delta \mathscr{I}_D$$

$$= \boldsymbol{\Gamma}_P^0 S_{PO}^0 + \boldsymbol{\Gamma}_P^0 T_P^0 \delta S_{PI} + \sum_k \boldsymbol{\Gamma}_k^0 \delta T_k S_{kI}^0 \tag{3.209}$$

$$= \boldsymbol{\Gamma}_P^0 S_{PO}^0 + \boldsymbol{\Gamma}_P^0 T_P^0 \delta S_{PI} + \boldsymbol{\Gamma}_P^0 \delta T_P S_{PI}^0 + \sum_{k=1}^N \boldsymbol{\Gamma}_k^0 \delta T_k S_{kI}^0 + \boldsymbol{\Gamma}_A^0 \delta T_A S_{AI}^0.$$

§3.8] ELLIPSOMETRY WITH IMPERFECT COMPONENTS 215

From eq. (3.209), we can proceed to consider the conditions for flux minimum using any one of the possible nulling schemes. Here we assume the commonly used method of obtaining the minimum by adjustment of the polarizer and the analyzer with other elements remaining fixed. For any other case, the procedure is similar to that which follows. For definiteness, we take all of the vectors and matrices appearing in eq. (3.209) to be in the x-y frame of reference of the optical system under measurement. Then, we have

$$\frac{\partial \mathscr{I}_D}{\partial P} = 0 = \boldsymbol{\Gamma}_P^0 \frac{\partial \boldsymbol{S}_{PO}^0}{\partial P} + \boldsymbol{\Gamma}_P^0 \frac{\partial \boldsymbol{T}_P^0}{\partial P} \delta \boldsymbol{S}_{PI}$$
$$+ \boldsymbol{\Gamma}_P^0 \frac{\partial (\delta \boldsymbol{T}_P)}{\partial P} \boldsymbol{S}_{PI}^0 + \sum_{k=1}^{N} \boldsymbol{\Gamma}_k^0 \delta T_k \frac{\partial \boldsymbol{S}_{kI}^0}{\partial P} + \boldsymbol{\Gamma}_A^0 \delta T_A \frac{\partial \boldsymbol{S}_{AI}^0}{\partial P},$$
(3.210)

$$\frac{\partial \mathscr{I}_D}{\partial A} = 0 = \frac{\partial \boldsymbol{\Gamma}_P^0}{\partial A} \boldsymbol{S}_{PO}^0 + \frac{\partial \boldsymbol{\Gamma}_P^0}{\partial A} \boldsymbol{T}_P^0 \delta \boldsymbol{S}_{PI}$$
$$+ \frac{\partial \boldsymbol{\Gamma}_P^0}{\partial A} \delta T_P \boldsymbol{S}_{PI}^0 + \sum_{k=1}^{N} \frac{\partial \boldsymbol{\Gamma}_k^0}{\partial A} \delta T_k \boldsymbol{S}_{kI}^0 + \boldsymbol{\Gamma}_A^0 \frac{\partial (\delta T_A)}{\partial A} \boldsymbol{S}_{AI}^0.$$

Note that, by definition, $\boldsymbol{S}_{kI}^0$ ($k = 1, 2, \ldots, N$; A) is a function of P only, $\boldsymbol{S}_{PI}^0$ is constant, $\boldsymbol{\Gamma}_k^0$ ($k = P; 1, 2, \ldots, N$) is a function of A only, $\boldsymbol{\Gamma}_A^0$ is a constant, δT_k ($k = 1, 2, \ldots, N$) is a constant, δT_P is a function of P only, and δT_A is a function of A only.

Let (P_n^0, A_n^0) denote the polarizer–analyzer azimuth-angle pair for flux minimum when the ellipsometer is ideal, then

$$\left(\boldsymbol{\Gamma}_P^0 \frac{\partial \boldsymbol{T}_P^0}{\partial P} \boldsymbol{S}_{PI}^0\right)_{P_n^0, A_n^0} = \left(\boldsymbol{\Gamma}_P^0 \frac{\partial \boldsymbol{S}_{PO}^0}{\partial P}\right)_{P_n^0, A_n^0} = 0,$$
$$\left(\frac{\partial \boldsymbol{\Gamma}_P^0}{\partial A} \boldsymbol{T}_P^0 \boldsymbol{S}_{PI}^0\right)_{P_n^0, A_n^0} = \left(\frac{\partial \boldsymbol{\Gamma}_P^0}{\partial A} \boldsymbol{S}_{PO}^0\right)_{P_n^0, A_n^0} = 0.$$
(3.211)

The corresponding pair of nulling angles when the ellipsometer has imperfections can be written as $(P_n^0 + \delta P, A_n^0 + \delta A)$, where δP and δA are the small-angle changes required to restore the condition of flux minimum after imperfections have been introduced into the ideal ellipsometer. Substituting the latter pair into eq. (3.210), and making use of eq. (3.211), we get

$$\delta P \left(-\boldsymbol{\Gamma}_P^0 \frac{\partial^2 \boldsymbol{S}_{PO}^0}{\partial P^2}\right) + \delta A \left(-\frac{\partial \boldsymbol{\Gamma}_P^0}{\partial A} \frac{\partial \boldsymbol{S}_{PO}^0}{\partial P}\right) = \boldsymbol{\Gamma}_P^0 \frac{\partial (\delta T_P)}{\partial P} \boldsymbol{S}_{PI}^0$$
$$+ \sum_{k=1}^{N} \boldsymbol{\Gamma}_k^0 \delta T_k \frac{\partial \boldsymbol{S}_{kI}^0}{\partial P} + \boldsymbol{\Gamma}_A^0 \delta T_A \frac{\partial \boldsymbol{S}_{AI}^0}{\partial P},$$

$$\delta P\left(-\frac{\partial \boldsymbol{\Gamma}_\mathrm{P}^0}{\partial A}\frac{\partial \boldsymbol{S}_\mathrm{PO}^0}{\partial P}\right) + \delta A\left(-\frac{\partial^2 \boldsymbol{\Gamma}_\mathrm{P}^0}{\partial A^2}\boldsymbol{S}_\mathrm{PO}^0\right) = \frac{\partial \boldsymbol{\Gamma}_\mathrm{P}^0}{\partial A}\delta T_\mathrm{P}\boldsymbol{S}_\mathrm{PI}^0$$
$$+ \sum_{k=1}^{N}\frac{\partial \boldsymbol{\Gamma}_k^0}{\partial A}\delta T_k \boldsymbol{S}_{k\mathrm{I}}^0 + \boldsymbol{\Gamma}_\mathrm{A}^0 \frac{\partial(\delta T_\mathrm{A})}{\partial A}\boldsymbol{S}_\mathrm{AI}^0, \quad (3.212)$$

to first order. All partial derivatives in eq. (3.212) are evaluated at (P_n^0, A_n^0). Thus eq. (3.212) can be solved for δP and δA which are subsequently related to $\delta \psi$ and $\delta \Delta$ by use of the ideal ellipsometry equations, which can be found in eqs. (3.33) and (3.55), or by solving eq. (3.211). Note that by virtue of eq. (3.211), the effect of $\delta \boldsymbol{S}_\mathrm{PI}$, the deviation from an assumed ideal form of the source-beam's Stokes vector, disappears to first order.

3.8.1. Application to the PCWSW'A ellipsometer arrangement

In this subsection we will apply the above results to the commonly used PCWSW'A ellipsometer arrangement in which the optical system S is mounted in a test cell with entrance and exit windows W and W', respectively. The compensator C is set at $\pm\frac{1}{4}\pi$ azimuth to give near $\frac{1}{2}\pi$ retardation, and the flux minimum is obtained by adjustment of the polarizer P and the analyzer A.

The source beam is assumed to be essentially isotropic, that is, it exhibits weak linear preferences only. This condition can always be met in practice by placing a circular polarizer in the source beam oriented for maximum transmitted flux. Thus the normalized Stokes vector of the source beam incident on the polarizer can be written as

$$\boldsymbol{S}_\mathrm{PI} = \{1, \zeta_1, \zeta_2, \mu\}, \qquad |\zeta_1|, |\zeta_2| \ll 1, \quad (3.213)$$

where ζ_1 and ζ_2 are the residual horizontal and $+45°$ linear preferences, and μ represents any degree of circular polarization. In eq. (3.213), and throughout, the 4×1 Stokes column vectors are written horizontally between curly brackets { } and the elements are separated by commas to save space. If ζ_1 and ζ_2 are neglected in eq. (3.212), we get

$$\boldsymbol{S}_\mathrm{PI}^0 = \{1, 0, 0, \mu\}, \quad (3.214)$$

which is the idealized source-beam Stokes vector. Due to ζ_1, and ζ_2 it can be easily seen that the output flux from the polarizer has a sinusoidal ripple component of the form $\zeta \sin(2P + \alpha)$, where $\zeta = (\zeta_1^2 + \zeta_2^2)^{\frac{1}{2}}$ and $\alpha = \tan^{-1}\zeta_1/\zeta_2$. However, as discussed in going from eq. (3.210) to eq. (3.212), this ripple does not affect the nulling angles to first order.

To determine δP and δA using eq. (3.212), $\boldsymbol{S}_{k\mathrm{I}}^0$ and $\boldsymbol{\Gamma}_k^0$, together with their partial derivatives, must be first evaluated assuming an ideal ellipsometer. Using the idealized source-beam Stokes vector and the matrices characteris-

[§3.8]   ELLIPSOMETRY WITH IMPERFECT COMPONENTS   217

tic of an ideal ellipsometer (appendix A) we calculate S_{kI}^0, $\partial S_{kI}^0/\partial P$ and Γ_k^0, $\partial \Gamma_k^0/\partial A$ as functions of P and A, respectively. The results are given in the 2nd columns of tables 3.9 and 3.10. The values assumed by these quantities when

$$P = P_n^0(\nu) = (-)^{\nu+1} \tfrac{1}{2}\Delta \mp \tfrac{1}{4}\pi, \qquad A = A_n^0(\nu) = \pm \psi, \tag{3.215}$$

are given in the 3rd and 4th columns of the same tables. Equation (3.215) gives the ideal nulling angles, where ν represents a zone number and the upper and lower signs on the right apply to $\nu = 1, 2$ and $\nu = 3, 4$, respectively. Also, recall that zones 1, 3 and zones 2, 4 correspond to compensator azimuths of $-\tfrac{1}{4}\pi$ and $+\tfrac{1}{4}\pi$, respectively. To determine the coefficients of δP and δA on the left of eq. (3.212), we note that $S_{PO}^0 = S_{CI}^0$, and thus S_{PO}^0, $\partial S_{PO}^0/\partial P$, Γ_P^0, and $\partial \Gamma_P^0/\partial A$ are directly obtainable from the 2nd columns of tables 3.9 and 3.10. Differentiating, we evaluate the second derivatives $\partial^2 S_{FO}^0/\partial P^2$ and $\partial^2 \Gamma_P^0/\partial A^2$. Direct substitution of these results and

Table 3.9. S_{kI}^0 and $\partial S_{kI}^0/\partial P$*

Element k	$S_{kI}^0(C = \mp\tfrac{1}{4}\pi)$	$S_{kI}^0(1, 3)$	$S_{kI}^0(2, 4)$
P	$\{1, 0, 0, \mu\}$	$\{1, 0, 0, \mu\}$	$\{1, 0, 0, \mu\}$
C	$\{1, c_{2P}, s_{2P}, 0\}$	$\{1, \pm s_{1\Delta}, \mp c_{1\Delta}, 0\}$	$\{1, \mp s_{1\Delta}, \mp c_{1\Delta}, 0\}$
W, S	$\{1, 0, s_{2P}, \mp c_{2P}\}$	$\{1, 0, \mp c_{1\Delta}, \mp s_{1\Delta}\}$	$\{1, 0, \mp c_{1\Delta}, \mp s_{1\Delta}\}$
W', A	$\{1, -c_{2\psi}, s_{2\psi}s_{2P\mp\Delta}, \mp s_{2\psi}c_{2P\mp\Delta}\}$	$\{1, -c_{2\psi}, \mp s_{2\psi}, 0\}$	$\{1, -c_{2\psi}, \mp s_{2\psi}, 0\}$

Element k	$\partial S_{kI}^0/\partial P(C = \mp\tfrac{1}{4}\pi)$	$\partial S_{kI}^0/\partial P(1, 3)$	$\partial S_{kI}^0/\partial P(2, 4)$
P	0	0	0
C	$\{0, -2s_{2P}, 2c_{2P}, 0\}$	$\{0, \pm 2c_{1\Delta}, \pm 2s_{1\Delta}, 0\}$	$\{0, \pm 2c_{1\Delta}, \mp 2s_{1\Delta}, 0\}$
W, S	$\{0, 0, 2c_{2P}, \pm 2s_{2P}\}$	$\{0, 0, \pm 2s_{1\Delta}, \mp 2c_{1\Delta}\}$	$\{0, 0, \mp 2s_{1\Delta}, \pm 2c_{1\Delta}\}$
W', A	$\{0, 0, 2s_{2\psi}c_{2P\mp\Delta}, \pm 2s_{2\psi}s_{2P\mp\Delta}\}$	$\{0, 0, 0, \mp 2s_{2\psi}\}$	$\{0, 0, 0, \pm 2s_{2\psi}\}$

*S_{kI}^0 represents the Stokes vector of light incident on the kth element in an ideal ellipsometer. Starting with $S_{PI}^0 = \{1, 0, 0, \mu\}$ and using the Mueller matrices of the ideal elements in appendix A we proceed to determine the Stokes vectors at the input of the successive elements encountered by the beam along its path by making use of the relation $S_{(k+1)I}^0 = T_k^0 S_{kI}^0$. Simple differentiation gives $\partial S_{kI}^0/\partial P$ and use of the ideal nulling angles of eq. (3.215) provides the third and fourth columns of this table. Upper signs are for zones 1, 2; lower for 3, 4, in the 3rd and 4th columns.

Table 3.10. Γ_k^0 and $\partial \Gamma_k^0/\partial A$*

Element k	$\Gamma_k^0 (C = \mp \tfrac{1}{4}\pi)$	$\Gamma_k^0(1, 3)$	$\Gamma_k^0(2, 4)$
A	$[1, 0, 0, 0]$	$[1, 0, 0, 0]$	$[1, 0, 0, 0]$
W′, S	$[1, c_{2A}, s_{2A}, 0]$	$[1, c_{2\psi}, \pm s_{2\psi}, 0]$	$[1, c_{2\psi}, \pm s_{2\psi}, 0]$
W, C	$[(1 - c_{2\psi}c_{2A}), (c_{2A} - c_{2\psi}), s_{2\psi} c_{1A} s_{2A}, s_{2\psi} s_{1A} s_{2A}]$	$s_{2\psi}^2[1, 0, \pm c_{1\Delta}, \pm s_{1\Delta}]$	$s_{2\psi}^2[1, 0, \pm c_{1\Delta}, \pm s_{1\Delta}]$
P	$[(1 - c_{2\psi}c_{2A}), \mp s_{2\psi} s_{1A} s_{2A}, s_{2\psi} c_{1A} s_{2A}, \mp(c_{2\psi} - c_{2A})]$	$s_{2\psi}^2[1, \mp s_{1\Delta}, \pm c_{1\Delta}, 0]$	$s_{2\psi}^2[1, \pm s_{1\Delta}, \pm c_{1\Delta}, 0]$

Element k	$\partial\Gamma_k^0/\partial A\ (C = \mp \tfrac{1}{4}\pi)$	$\partial\Gamma_k^0/\partial A(1, 3)$	$\partial\Gamma_k^0/\partial A(2, 4)$
A	0	0	0
W′, S	$[0, -2s_{2A}, 2c_{2A}, 0]$	$[0, \mp 2s_{2\psi}, 2c_{2\psi}, 0]$	$[0, \mp 2s_{2\psi}, 2c_{2\psi}, 0]$
W, C	$[2c_{2\psi} s_{2A}, -2s_{2A}, 2s_{2\psi} c_{1A} c_{2A}, 2s_{2\psi} s_{1A} c_{2A}]$	$2s_{2\psi}[\pm c_{2\psi}, \mp 1, c_{2\psi} c_{1\Delta}, c_{2\psi} s_{1\Delta}]$	$2s_{2\psi}[\pm c_{2\psi}, \mp 1, c_{2\psi} c_{1\Delta}, c_{2\psi} s_{1\Delta}]$
P	$[2c_{2\psi} s_{2A}, \mp 2s_{2\psi} s_{1A} c_{2A}, 2s_{2\psi} c_{1A} c_{2A}, \mp 2s_{2A}]$	$2s_{2\psi}[\pm c_{2\psi}, -c_{2\psi} s_{1\Delta}, c_{2\psi} c_{1\Delta}, \mp 1]$	$2s_{2\psi}[\pm c_{2\psi}, c_{2\psi} s_{1\Delta}, c_{2\psi} c_{1\Delta}, \pm 1]$

*Γ_k^0 is a 1×4 row vector that determines the contribution to the detected light flux arising from the Stokes-vector perturbation generated at the output of the kth device due to its imperfection, eqs. (3.203)–(3.205). Starting with $\Gamma_A^0 = [1, 0, 0, 0]$, and using the Mueller matrices of the ideal elements in appendix A, we work backwards to determine Γ_k^0 ($k = $ W′, S, W, C and P) using the relation $\Gamma_{k-1}^0 = \Gamma_k^0 T_k^0$ (e.g., $\Gamma_{W'}^0 = \Gamma_A^0 T_A^0$) which follows from simple rules of matrix algebra. Differentiation gives $\partial\Gamma_k^0/\partial A$, and use of the ideal nulling angles in eq. (3.215) provides the third and fourth columns of this table. Upper signs are for zones 1, 2; lower for 3, 4, in the 3rd and 4th columns.

subsequent use of the ideal nulling angles of eq. (3.215) yield

$$\frac{\partial \Gamma_P^0}{\partial A}\frac{\partial S_{PO}^0}{\partial P} = 4s_{2\psi} c_{2A} c_{2P\mp\Delta} = 0,$$

$$\Gamma_P^0 \frac{\partial^2 S_{PO}^0}{\partial P^2} = -4s_{2\psi} s_{2A} s_{2P\mp\Delta} = 4s_{2\psi}^2, \qquad (3.216)$$

$$\frac{\partial^2 \Gamma_P^0}{\partial A^2} S_{PO}^0 = 4c_{2\psi} c_{2A} - 4s_{2\psi} s_{2A} s_{2P\mp\Delta} = 4.$$

To save space, we use the notation $c_{2\theta}$ and $s_{2\theta}$ to mean $\cos 2\theta$ and $\sin 2\theta$, respectively, in eq. (3.216) and throughout. Substituting from eq. (3.216) into eq. (3.212), we get

$$-4s_{2\psi}^2 \delta P = \Gamma_P^0 \frac{\partial(\delta T_P)}{\partial P} S_{PI}^0 + \sum_{k=1}^{N} \Gamma_k^0 \delta T_k \frac{\partial S_{kI}^0}{\partial P} + \Gamma_A^0 \delta T_A \frac{\partial S_{AI}^0}{\partial P},$$

$$-4\delta A = \frac{\partial \Gamma_P^0}{\partial A}\delta T_P S_{PI}^0 + \sum_{k=1}^{N} \frac{\partial \Gamma_k^0}{\partial A}\delta T_k S_{kI}^0 + \Gamma_A^0 \frac{\partial(\delta T_A)}{\partial A}S_{AI}^0. \qquad (3.217)$$

Equation (3.217) gives an explicit solution for δP and δA in terms of the imperfections in the various components. Note that we now have $N = 4$, and $k = 1, 2, 3$ and 4 represent the compensator, entrance window, system, and exit window, respectively.

We are now in a position to determine the effect of the 4×4 Mueller imperfection matrices (t_{ij}) of all of the optical elements on the nulling angles. If one element is taken at a time, the results can be put in the form

$$\delta P = \sum_{i,j} \beta_{ij} t_{ij}, \qquad \delta A = \sum_{i,j} \alpha_{ij} t_{ij}, \qquad (3.218)$$

where (β_{ij}) and (α_{ij}) represent 4×4 arrays of coupling coefficients. Each array element determines the extent to which the corresponding imperfection-matrix element couples to cause an error of the nulling angles. In what follows, we will determine these arrays for each component imperfection.

3.8.1.1. Polarizer and analyzer imperfections
(i) *Polarizer imperfection*
From eq. (3.217) it can be seen that δP and δA introduced by the polarizer imperfection are given by

$$-4s_{2\psi}^2 \delta P = \Gamma_P^0 \frac{\partial(\delta T_P)}{\partial P} S_{PI}^0, \qquad -4\delta A = \frac{\partial \Gamma_P^0}{\partial A}\delta T_P S_{PI}^0. \qquad (3.219)$$

Let $(t'_{ij})_P$ denote the imperfection matrix of the polarizer in its principal frame,

then
$$\delta T_P = \boldsymbol{R}(-P)(t'_{ij})_P \boldsymbol{R}(P), \qquad (3.220)$$

where $\boldsymbol{R}(P)$ is the 4×4 rotation matrix (table A.4 of appendix A). Using eq. (3.220) in eq. (3.219), and noting that $\boldsymbol{R}(P)\boldsymbol{S}^0_{PI} = \boldsymbol{S}^0_{PI}$ because $\boldsymbol{S}^0_{PI} = \{1, 0, 0, \mu\}$, we find that

$$-4s^2_{2\psi}\delta P = \boldsymbol{\Gamma}^0_P \frac{d}{dP}\boldsymbol{R}(-P) \times \{\epsilon_0, \epsilon_1, \epsilon_2, \epsilon_3\},$$
$$-4\delta A = \frac{\partial \boldsymbol{\Gamma}^0_P}{\partial A}\boldsymbol{R}(-P) \times \{\epsilon_0, \epsilon_1, \epsilon_2, \epsilon_3\}, \qquad (3.221)$$

where
$$\{\epsilon_0, \epsilon_1, \epsilon_2, \epsilon_3\} = (t'_{ij})_P \boldsymbol{S}^0_{PI}$$
$$= \{(t'_{11} + \mu t'_{14}), (t'_{21} + \mu t'_{24}), (t'_{31} + \mu t'_{34}), (t'_{41} + \mu t'_{44})\}. \qquad (3.222)$$

Equation (3.222) represents the Stokes-vector perturbation generated at the output of the polarizer due to its own imperfection, evaluated in its principal frame. Substituting for the rotation matrix and its derivative from table A.4 and eq. (A.9) of the appendix and for $\boldsymbol{\Gamma}^0_P$ and $\partial \boldsymbol{\Gamma}^0_P/\partial A$ from table 3.10, we obtain

$$\boldsymbol{\Gamma}^0_P \frac{d}{dP}\boldsymbol{R}(-P) = s^2_{2\psi}[0, 0, 2, 0], \quad \text{for all zones,}$$

$$\frac{\partial \boldsymbol{\Gamma}^0_P}{\partial A}\boldsymbol{R}(-P) = [\pm 2s_{2\psi}c_{2\psi}, \mp 2s_{2\psi}c_{2\psi}, 0, \mp 2s_{2\psi}], \quad \text{for zones 1 and 3,}$$
$$\qquad (3.223)$$
$$= [\mp 2s_{2\psi}c_{2\psi}, \mp 2s_{2\psi}c_{2\psi}, 0, \pm 2s_{2\psi}], \quad \text{for zones 2 and 4.}$$

From eqs. (3.221) and (3.223), we have

$$\delta P = -\tfrac{1}{2}\epsilon_2, \quad \text{for all zones,}$$
$$\delta A = \mp\tfrac{1}{2}(\epsilon_0 - \epsilon_1)s_{2\psi}c_{2\psi} \pm \tfrac{1}{2}s_{2\psi}\epsilon_3, \quad \text{for zones 1 and 3,} \qquad (3.224)$$
$$= \mp\tfrac{1}{2}(\epsilon_0 - \epsilon_1)s_{2\psi}c_{2\psi} \mp \tfrac{1}{2}s_{2\psi}\epsilon_3, \quad \text{for zones 2 and 4.}$$

Substituting the values of ϵ_0, ϵ_1, ϵ_2 and ϵ_3 from eq. (3.222) into eq. (3.224), we get expressions of the form of eq. (3.218). The arrays of coupling coefficients are given in table 3.11. These determine the effect of the principal-frame imperfection matrix $(t'_{ij})_P$ on the nulling angles and hence on ψ and Δ.

Now we examine the degree of polarization of light leaving the polarizer. For a Stokes vector $\boldsymbol{S} = \{S_0, S_1, S_2, S_3\}$, the degree of polarization [eq. (1.127), §1.9] is

$$\mathcal{P} = \frac{(S^2_1 + S^2_2 + S^2_3)^{\frac{1}{2}}}{S_0}. \qquad (3.225)$$

Table 3.11. Arrays of coupling coefficients for the four zones*

	Polarizer (P)	Compensator (C)	Analyzer (A)
(β_{ij})	$\dfrac{1}{2}\begin{bmatrix} 0 & 0 & 0 & 0 \\ 0 & 0 & 0 & 0 \\ -1 & 0 & 0 & -\mu \\ 0 & 0 & 0 & 0 \end{bmatrix}$	$\dfrac{1}{2}\begin{bmatrix} 0 & \pm s_{1\Delta} & \mp c_{1\Delta} & 0 \\ 0 & -s_{1\Delta}c_{1\Delta} & c_{1\Delta}^{2} & 0 \\ 0 & 0 & 0 & 0 \\ 0 & s_{1\Delta}^{2} & -s_{1\Delta}c_{1\Delta} & 0 \end{bmatrix}$	$\dfrac{1}{2}s_{2\psi}\begin{bmatrix} 0 & 0 & 0 & \pm 1 \\ 0 & 0 & 0 & 0 \\ 0 & 0 & 0 & 0 \\ 0 & 0 & 0 & 0 \end{bmatrix}$
(α_{ij})	$\dfrac{s_{2\psi}}{2}\begin{bmatrix} \mp c_{2\psi} & 0 & 0 & \mp \mu c_{2\psi} \\ \pm c_{2\psi} & 0 & 0 & \pm \mu c_{2\psi} \\ 0 & 0 & 0 & 0 \\ \mp 1 & 0 & 0 & \mp \mu \end{bmatrix}$	$\dfrac{s_{2\psi}}{2}\begin{bmatrix} \mp c_{2\psi} & -c_{2\psi}c_{1\Delta} & -c_{2\psi}s_{1\Delta} & 0 \\ c_{2\psi}c_{1\Delta} & \pm c_{2\psi}c_{1\Delta}^{2} & \pm c_{2\psi}s_{1\Delta}c_{1\Delta} & 0 \\ \pm 1 & c_{1\Delta} & s_{1\Delta} & 0 \\ -c_{2\psi}s_{1\Delta} & \mp c_{2\psi}s_{1\Delta}c_{1\Delta} & \mp c_{2\psi}s_{1\Delta}^{2} & 0 \end{bmatrix}$	$\dfrac{1}{2}\begin{bmatrix} 0 & 0 & -1 & 0 \\ 0 & 0 & 0 & 0 \\ 0 & 0 & 0 & 0 \\ 0 & 0 & 0 & 0 \end{bmatrix}$

Table 3.11. (*continued*)

Entrance Window (W)

$$(\beta_{ij}) \quad \frac{1}{2}\begin{bmatrix} 0 & 0 & \mp s_{1\Delta} & \pm c_{1\Delta} \\ 0 & 0 & 0 & 0 \\ 0 & 0 & -s_{1\Delta}c_{1\Delta} & c^2_{1\Delta} \\ 0 & 0 & -s^2_{1\Delta} & s_{1\Delta}c_{1\Delta} \end{bmatrix}$$

$$(\alpha_{ij}) \quad \frac{s_{2\psi}}{2}\begin{bmatrix} \mp c_{2\psi} & 0 & c_{2\psi}c_{1\Delta} & c_{2\psi}s_{1\Delta} \\ \pm 1 & 0 & -c_{1\Delta} & -s_{1\Delta} \\ -c_{2\psi}c_{1\Delta} & 0 & \pm c_{2\psi}c^2_{1\Delta} & \pm c_{2\psi}c_{1\Delta}s_{1\Delta} \\ -c_{2\psi}s_{1\Delta} & 0 & \pm c_{2\psi}s_{1\Delta}c_{1\Delta} & \pm c_{2\psi}s^2_{1\Delta} \end{bmatrix}$$

Optical System (S)

$$(\beta_{ij}) \quad \frac{1}{2s^2_{2\psi}}\begin{bmatrix} 0 & 0 & \mp s_{1\Delta} & \pm c_{1\Delta} \\ 0 & 0 & \mp c_{2\psi}s_{1\Delta} & \pm c_{2\psi} \\ 0 & 0 & -s_{2\psi}s_{1\Delta} & s_{2\psi}c_{1\Delta} \\ 0 & 0 & 0 & 0 \end{bmatrix}$$

$$(\alpha_{ij}) \quad \frac{1}{2}\begin{bmatrix} 0 & 0 & 0 & 0 \\ \pm s_{2\psi} & 0 & -s_{2\psi}c_{1\Delta} & -s_{2\psi}s_{1\Delta} \\ c_{2\psi} & 0 & \pm c_{2\psi}c_{1\Delta} & \pm c_{2\psi}s_{1\Delta} \\ 0 & 0 & 0 & 0 \end{bmatrix}$$

Exit Window (W′)

$$(\beta_{ij}) \quad \frac{1}{2s_{2\psi}}\begin{bmatrix} 0 & 0 & 0 & \pm 1 \\ 0 & 0 & 0 & \pm c_{2\psi} \\ 0 & 0 & 0 & s_{2\psi} \\ 0 & 0 & 0 & 0 \end{bmatrix}$$

$$(\alpha_{ij}) \quad \frac{1}{2}\begin{bmatrix} 0 & 0 & 0 & 0 \\ \pm s_{2\psi} & \mp s_{2\psi}c_{2\psi} & -s^2_{2\psi} & 0 \\ -c_{2\psi} & c^2_{2\psi} & \pm s_{2\psi}c_{2\psi} & 0 \\ 0 & 0 & 0 & 0 \end{bmatrix}$$

*These arrays determine the effect on the nulling angles (and hence on ψ and Δ) in zones 1 (upper signs) and 3 (lower signs) of the imperfection matrices (t_{ij}) of the respective elements according to the relations $\delta P = \Sigma_{ij}\beta_{ij}t_{ij}$ and $\delta A = \Sigma_{ij}\alpha_{ij}t_{ij}$. For the polarizer, compensator, analyzer, and system, the imperfection matrices are referenced to their principal frames; whereas for the windows they are given in the x–y frame. To get the arrays appropriate to zones 2 and 4, multiply the elements having a bar on the top by -1. The multiplier on the left multiplies each of the elements of the array.

§3.8]   ELLIPSOMETRY WITH IMPERFECT COMPONENTS   223

The perturbed Stokes vector at the output of the polarizer is

$$S_{PO} = S_{PO}^0 + \delta S_{PO} = \{1 + \epsilon_0, 1 + \epsilon_1, \epsilon_2, \epsilon_3\}, \tag{3.226}$$

in its principal frame. Using eq. (3.225) we find that the degree of polarization of this light is

$$\begin{aligned}\mathcal{P} &= 1 + \epsilon_1 - \epsilon_0 + \tfrac{1}{2}(2\epsilon_0^2 + \epsilon_1^2 + \epsilon_2^2 + \epsilon_3^2) + \cdots \\ &= 1 + \epsilon_1 - \epsilon_0, \quad \text{to first order.}\end{aligned} \tag{3.227}$$

From eqs. (3.224) and (3.227), the polarizer output is seen to be depolarized by an amount $(\epsilon_0 - \epsilon_1)$; this depolarization affects the analyzer nulling angle A and hence ψ, whereas P and Δ are unaffected.

If we set $\epsilon_0 = \epsilon_1$ in eq. (3.226) we can identify the resulting Stokes vector with a completely polarized elliptic vibration with the small azimuth $\tfrac{1}{2}\epsilon_2$, and the small ellipticity $\tfrac{1}{2}\epsilon_3$, referenced to the polarizer's principal frame. The corresponding δP and δA, obtained from eq. (3.224), are the same as we found before for this special case [eq. (3.185) and table 3.5].

From the above, we conclude that besides the effects expected when an elliptic vibration with small ellipticity is assumed there is also a contribution caused by depolarization. This small depolarization affects only the value of ψ but not Δ and, as will be seen later, is retained when two- and four-zone averages are taken.

(ii) *Analyzer imperfection*

For the analyzer imperfection alone, we have

$$-4s_{2\psi}^2 \delta P = \Gamma_A^0 \delta T_A \frac{\partial S_{AI}^0}{\partial P}, \qquad -4\delta A = \Gamma_A^0 \frac{\partial (\delta T_A)}{\partial A} S_{AI}^0, \tag{3.228}$$

where

$$\delta T_A = R(-A)(t'_{ij})_A R(A), \tag{3.229}$$

and $(t'_{ij})_A$ represents the analyzer's principal frame imperfection matrix. Because $\Gamma_A^0 = [1, 0, 0, 0]$, eq. (3.228) simplifies to

$$\begin{aligned}-4s_{2\psi}^2 \delta P &= [t'_{11}, t'_{12}, t'_{13}, t'_{14}] \times R(A) \frac{\partial S_{AI}^0}{\partial P}, \\ -4\delta A &= [t'_{11}, t'_{12}, t'_{13}, t'_{14}] \times \frac{dR(A)}{dA} S_{AI}^0.\end{aligned} \tag{3.230}$$

Using the rotation matrix and its derivative from table A.4 and eq. (A.10) of appendix A, and S^0_{AI}, $\partial S^0_{AI}/\partial P$ from table 3.9, we find that

$$R(A)\frac{\partial S^0_{AI}}{\partial P} = [0, 0, 0, \mp 2s_{2\psi}], \quad \text{for zones 1 and 3,}$$
$$= [0, 0, 0, \pm 2s_{2\psi}], \quad \text{for zones 2 and 4,} \quad (3.231)$$

$$\frac{dR(A)}{dA}S^0_{AI} = [0, 0, 2, 0], \quad \text{for all zones.}$$

Substituting from eq. (3.231) into eq. (3.230), we obtain

$$\delta P = \pm\frac{t'_{14}}{2s_{2\psi}}, \quad \text{for zones 1 and 3,}$$
$$= \mp\frac{t'_{14}}{2s_{2\psi}}, \quad \text{for zones 2 and 4,} \quad (3.232)$$
$$\delta A = -\tfrac{1}{2}t'_{13}, \quad \text{for all zones.}$$

The arrays of coupling coefficients (β_{ij}) and (α_{ij}) assume the simple forms given in table 3.11 with only one non-zero element in each 4×4 array.

In §3.7 we have treated the effect of an imperfect analyzer assuming a non-diagonal 2×2 Jones matrix of the form of eq. (3.174). We proved that α_{12A} is the only significant imperfection element and its effect was determined. This result is in agreement with that in eq. (3.232), as could be proved by finding the 4×4 Mueller matrix equivalent of the 2×2 Jones matrix. Such a transformation [eq. (2.243), §2.12] yields elements in the 1, 3 and 1, 4 matrix locations which are simply the real and imaginary parts of α_{12A}, respectively. This shows, as expected, that the results of an analysis that uses the Mueller-matrix formalism should include the results of a corresponding analysis using the Jones-matrix formalism as a special case.

3.8.1.2. Compensator imperfection

The effect of a compensator imperfection on the nulling angles is given by

$$-4s^2_{2\psi}\delta P = \boldsymbol{\Gamma}^0_C \delta \boldsymbol{T}_C \frac{\partial S^0_{CI}}{\partial P}, \quad -4\delta A = \frac{\partial \boldsymbol{\Gamma}^0_C}{\partial A} \delta \boldsymbol{T}_C S^0_{CI}, \quad (3.233)$$

where

$$\delta \boldsymbol{T}_C = \boldsymbol{R}(-C)(t'_{ij})_C \boldsymbol{R}(C), \quad C = \mp\tfrac{1}{4}\pi. \quad (3.234)$$

$(t'_{ij})_C$ is the fast-slow imperfection matrix. Substituting the rotation matrix from table A.4 of appendix A and S^0_{CI}, $\partial S^0_{CI}/\partial P$, $\boldsymbol{\Gamma}^0_C$ and $\partial \boldsymbol{\Gamma}^0_C/\partial A$ from tables

3.9, 3.10, we find that eq. (3.233) becomes

$$\delta P = \tfrac{1}{2}[1, \mp c_{1\Delta}, 0, \pm s_{1\Delta}](t'_{ij})_C\{0, \pm s_{1\Delta}, \mp c_{1\Delta}, 0\}, \quad \text{for zones 1 and 3,}$$
$$= \tfrac{1}{2}[1, \pm c_{1\Delta}, 0, \pm s_{1\Delta}](t'_{ij})_C\{0, \pm s_{1\Delta}, \pm c_{1\Delta}, 0\}, \quad \text{for zones 2 and 4,}$$
$$\delta A = \tfrac{1}{2}s_{2\psi}[\mp c_{2\psi}, c_{2\psi}c_{1\Delta}, \pm 1, -c_{2\psi}s_{1\Delta}](t'_{ij})_C\{1, \pm c_{1\Delta}, s_{1\Delta}, 0\},$$
$$\text{for zones 1 and 3,}$$
$$= \tfrac{1}{2}s_{2\psi}[\mp c_{2\psi}, -c_{2\psi}c_{1\Delta}, \mp 1, -c_{2\psi}s_{1\Delta}](t'_{ij})_C\{1, \mp c_{1\Delta}, \pm s_{1\Delta}, 0\},$$
$$\text{for zones 2 and 4.}$$
(3.235)

Direct expansion of eq. (3.235) yields the arrays of coupling coefficients that are given in table 3.11.

3.8.1.3. Entrance-window, system, and exit-window imperfections

With the imperfection matrices δT_k ($k = $ W, S and W') all expressed in the x-y frame of reference, their effect on the nulling angles becomes

$$-4s_{2\psi}^2 \delta P = \Gamma_k^0 \delta T_k \frac{\partial S_{k1}^0}{\partial P}, \quad -4\delta A = \frac{\partial \Gamma_k^0}{\partial A}\delta T_k S_{k1}^0; \quad k = \text{W, S and W'},$$
(3.236)

where $\delta T_k = (t_{ij})_k$ and S_{k1}^0, $\partial S_{k1}^0/\partial P$; Γ_k^0 and $\partial \Gamma_k^0/\partial A$ ($k = $ W, S and W') are given by tables 3.9 and 3.10. Substituting these into eq. (3.236), we get for the entrance window,

$$\delta P = \tfrac{1}{2}[1, 0, \pm c_{1\Delta}, \pm s_{1\Delta}](t_{ij})_W\{0, 0, \mp s_{1\Delta}, \pm c_{1\Delta}\}, \quad \text{for zones 1 and 3,}$$
$$= \tfrac{1}{2}[1, 0, \pm c_{1\Delta}, \pm s_{1\Delta}](t_{ij})_W\{0, 0, \pm s_{1\Delta}, \mp c_{1\Delta}\}, \quad \text{for zones 2 and 4,}$$
$$\delta A = \tfrac{1}{2}s_{2\psi}[\mp c_{2\psi}, \pm 1, -c_{2\psi}c_{1\Delta}, -c_{2\psi}s_{1\Delta}](t_{ij})_W\{1, 0, \mp c_{1\Delta}, \mp s_{1\Delta}\},$$
$$\text{for zones 1 and 3, and zones 2 and 4;}$$
(3.237)

for the system,

$$\delta P = \frac{1}{2s_{2\psi}^2}[1, c_{2\psi}, \pm s_{2\psi}, 0](t_{ij})_S\{0, 0, \mp s_{1\Delta}, \pm c_{1\Delta}\}, \quad \text{for zones 1 and 3,}$$
$$= \frac{1}{2s_{2\psi}^2}[1, c_{2\psi}, \pm s_{2\psi}, 0](t_{ij})_S\{0, 0, \pm s_{1\Delta}, \mp c_{1\Delta}\}, \quad \text{for zones 2 and 4,}$$
$$\delta A = \tfrac{1}{2}[0, \pm s_{2\psi}, -c_{2\psi}, 0](t_{ij})_S\{1, 0, \mp c_{1\Delta}, \mp s_{1\Delta}\}, \quad \text{for zones 1 and 3,}$$
$$\text{and zones 2 and 4;}$$
(3.238)

and for the exit window,

$$\delta P = \frac{1}{2s_{2\psi}^2}[1, c_{2\psi}, \pm s_{2\psi}, 0](t_{ij})_W\{0, 0, 0, \pm s_{2\psi}\}, \quad \text{for zones 1 and 3,}$$

$$= \frac{1}{2s_{2\psi}^2}[1, c_{2\psi}, \pm s_{2\psi}, 0](t_{ij})_W\{0, 0, 0, \mp s_{2\psi}\}, \quad \text{for zones 2 and 4,}$$

$$\delta A = \tfrac{1}{2}[0, \pm s_{2\psi}, -c_{2\psi}, 0](t_{ij})_W\{1, -c_{2\psi}, \mp s_{2\psi}, 0\}, \quad \text{for zones 1 and 3;} \\ \text{and zones 2 and 4.}$$

(3.239)

Expansion of eqs. (3.237)–(3.239) provides the required arrays of coupling coefficients found in table 3.11.

3.8.2. Zone averaging

(i) Two zones

The small-angle changes $\delta P(\nu)$ and $\delta A(\nu)$ required to restore the condition of flux minimum in the νth zone after imperfections have been introduced into an initially ideal ellipsometer are given by

$$\delta P(\nu) = \sum_{i,j;k} \beta_{ij}(k, \nu) t_{ij}(k), \qquad \delta A(\nu) = \sum_{i,j;k} \alpha_{ij}(k, \nu) t_{ij}(k), \qquad (3.240)$$

where $[\beta_{ij}(k, \nu)]$ and $[\alpha_{ij}(k, \nu)]$ represent the two arrays of coupling coefficients associated with the kth-element imperfection in the νth zone [e.g., $[\beta_{ij}(C, 3)]$ represents the (β_{ij})-array of the compensator in zone 3]. These are given in table 3.11 for $k = P, C, W, S, W'$ and A and $\nu = 1, 3$ corresponding to $C = -\tfrac{1}{4}\pi$. The arrays appropriate to $\nu = 2, 4$ corresponding to $C = \tfrac{1}{4}\pi$ are obtained by multiplying the elements having a bar on the top by -1. Thus, the measured nulling angles in the νth zone in terms of the system's ψ and Δ and the various imperfections are

$$P_n(\nu) = P_n^0(\nu) + \sum_{i,j;k} \beta_{ij}(k, \nu) t_{ij}(k),$$

$$A_n(\nu) = A_n^0(\nu) + \sum_{i,j;k} \alpha_{ij}(k, \nu) t_{ij}(k),$$

(3.241)

where $P_n^0(\nu)$, $A_n^0(\nu)$ are given by eq. (3.215), and $[\beta_{ij}(k, \nu)]$, $[\alpha_{ij}(k, \nu)]$ by table 3.11. From eqs. (3.241) and (3.215) the values of ψ and Δ from two-zone averages are

$$\Delta = [P_n(1) + P_n(3)] - \eta'_{1,3}, \qquad \Delta = -[P_n(2) + P_n(4)] + \eta'_{2,4},$$

$$\psi = \tfrac{1}{2}[A_n(1) - A_n(3)] - \xi'_{1,3}, \qquad \psi = \tfrac{1}{2}[A_n(2) - A_n(4)] - \xi'_{2,4},$$

(3.242)

where η' and ξ' represent the two-zone correction factors and are given by

$$\eta'_{1,3} = \sum_{i,j;k} [\beta_{ij}(k, 1) + \beta_{ij}(k, 3)] t_{ij}(k),$$
$$2\xi'_{1,3} = \sum_{i,j;k} [\alpha_{ij}(k, 1) - \alpha_{ij}(k, 3)] t_{ij}(k),$$
(3.243)

with similar expressions for $\nu = 2, 4$. From table 3.11, we note that

$$\beta_{ij}(k, 1) = \pm \beta_{ij}(k, 3), \qquad \alpha_{ij}(k, 1) = \pm \alpha_{ij}(k, 3),$$
(3.244)

and similarly for zones 2 and 4. Using eq. (3.244), eq. (3.243) simplifies to

$$\eta'_{1,3} = 2 \sum_{i,j;k}{}' \beta_{ij}(k, 1) t_{ij}(k), \qquad \xi'_{1,3} = \sum_{i,j;k}{}'' \alpha_{ij}(k, 1) t_{ij}(k).$$
(3.245)

Σ' indicates a summation only over those elements in the (β_{ij})-array that are preceded by a single sign and not by $\pm$ or $\mp$ signs, whereas Σ'' indicates a summation only over those elements in the (α_{ij})-array that are preceded by the $\pm$ or $\mp$ signs. The location of the elements in the imperfection matrices of the different optical components that are effective when two-zone averages are taken are schematically indicated by X (or $\bar{X}$) in table 3.12. The zeros (0) indicate that the effect of the corresponding imperfection-matrix elements disappears in a two-zone average. According to eq. (3.245), the contribution to $\eta'_{1,3}$ and $\xi'_{1,3}$ from elements with locations marked $\bar{X}$ are specified by $2\beta_{ij}(k, 1)$ and $\alpha_{ij}(k, 1)$, respectively. The bar over X is carried over from table 3.11 and its significance will be clear shortly.

(ii) *Four zones*
The values of ψ and Δ from four-zone averages are given by

$$\Delta = \tfrac{1}{2}[P_n(1) + P_n(3) - P_n(2) - P_n(4)] - \tfrac{1}{2}\eta'_{1,3} + \tfrac{1}{2}\eta'_{2,4},$$
$$\psi = \tfrac{1}{4}[A_n(1) - A_n(3) + A_n(2) - A_n(4)] - \tfrac{1}{2}\xi'_{1,3} - \tfrac{1}{2}\xi'_{2,4},$$
(3.246)

which follow directly from eq. (3.242). From eq. (3.246), the four-zone correction factors of Δ and ψ are $\tfrac{1}{2}(-\eta'_{1,3} + \eta'_{2,4})$ and $-\tfrac{1}{2}(\xi'_{1,3} + \xi'_{2,4})$, respectively. Because the elements of the (β_{ij}) and (α_{ij}) arrays applicable for zones 2 and 4 are either the same as those for zones 1 and 3 or differ by a sign only, the following simple rule applies. In table 3.12(a), the effect on Δ of the imperfection-matrix elements with locations indicated by X disappears in a four-zone average. In table 3.12(b), the effect on ψ disappears for elements whose locations are indicated by $\bar{X}$. This leads to table 3.13 in which the locations of the imperfection-matrix elements which affect four-zone averages are indicated by X. The contributions to the four-zone correction factors for Δ and ψ from the remaining elements in tables 3.13(a) and 3.13(b) are specified by $2\beta_{ij}(k, 1)$ and $\alpha_{ij}(k, 1)$, respectively.

We examined earlier the degree of polarization of the light from the

Table 3.12. Location of effective elements–two-zone averages*

	P				C				W				S				W'				A			
(a)	0 0 0 0				0 0 0 0				0 0 0 0				0 0 0 0				0 0 0 0				0 0 0 0			
	0 0 0 0				0 X̄ X 0				0 0 0 0				0 0 0 0				0 0 0 0				0 0 0 0			
	X 0 X 0				0 X̄ 0 0				0 X̄ X̄ 0				0 X̄ X̄ 0				0 0 0 X̄				0 0 0 0			
	0 0 0 0				0 0 0 0				0 X̄ X̄ 0				0 0 0 0				0 0 0 X̄				0 0 0 0			
(b)	X 0 0 0				0 0 0 0				0 0 0 0				0 0 0 0				0 0 0 0				0 0 0 0			
	X 0 0 0				0 X̄ 0 0				0 0 0 0				0 X 0 0				0 0 0 0				0 0 0 0			
	0 0 X̄ 0				0 0 0 0				0 0 X X				0 0 X X				0 0 X 0				0 0 0 0			
	0 X̄ 0 0				0 X̄ X X				0 0 X X				0 0 0 0				0 0 0 X				0 0 0 0			

*This tables gives the location of those elements of the Mueller imperfection matrices of the different ellipsometer components that affect (a) Δ, (b) ψ; when a two-zone average is taken. X (or X̄) and 0 denote that the corresponding matrix elements are effective and not effective, respectively. The bar over X has the meaning explained in the text.

Table 3.13. Location of effective elements–four-zone averages*

	P				C				W				S				W'				A			
(a)	0 0 0 0				0 0 0 0				0 0 0 0				0 0 0 0				0 0 0 0				0 0 0 0			
	0 0 0 0				0 X 0 0				0 0 0 0				0 0 0 0				0 0 0 0				0 0 0 0			
	0 0 0 X				0 0 0 X				0 0 X X				0 0 X X				0 0 0 X				0 0 0 0			
	0 0 0 0				0 0 0 0				0 0 X X				0 0 0 0				0 0 0 X				0 0 0 0			
(b)	X 0 0 X				X 0 0 0				X 0 0 0				0 X 0 0				0 0 0 0				0 0 0 0			
	X 0 0 X				0 X 0 0				X 0 0 0				0 X 0 0				0 0 0 0				0 0 0 0			
	0 0 0 0				0 0 0 0				0 0 X X				0 0 X X				0 0 X 0				0 0 0 0			
	0 0 0 0				0 0 X 0				0 0 X X				0 0 0 0				0 0 0 0				0 0 0 0			

*Same as in table 3.12 for four-zone averages.

polarizer and how it is related to the nulling angles and hence to ψ and Δ. When the light beam passes through the remainder of the ellipsometer, its degree of polarization decreases due to the incoherent effects in the optical devices. Though small, this depolarization is of interest; in the following we will determine the elements in the imperfection matrix of a device that contribute to this quantity.

3.8.3. Component depolarization

If the degree of polarization defined by eq. (2.225) is calculated just before and just after a given device, the difference will give the amount of depolarization. This is a measure of the strength of the incoherent processes within the device. Obviously, this quantity is a function of the device Mueller matrix and the incident-light Stokes vector. In a near-ideal ellipsometer, and with the elements set to give minimum detected flux, the state of polarization at the different points along the ellipsometer is known. The perturbed Stokes vector at the output of the kth device is

$$S_{kO} = S_{kO}^0 + \delta S_{kO} = S_{kO}^0 + \delta T_k S_{kI}^0, \tag{3.247}$$

where δT_k is the imperfection matrix of the device. Substituting the ideal values of S_{kI}^0 and S_{kO}^0 into eq. (3.247), we can determine the degree of polarization of S_{kO}, and hence the amount of depolarization introduced by the kth element. For example, to consider the system depolarization, we take S_{SI}^0 and $S_{SO}^0 (= S_{WI}^0)$ from table 3.9, so that

$$\begin{aligned} S_{SO} &= \{1, -c_{2\psi}, \mp s_{2\psi}, 0\} + (t_{ij})_S\{1, 0, \mp c_{1\Delta}, \mp s_{1\Delta}\}, \\ &= \{1 + \delta S_0, -c_{2\psi} + \delta S_1, \mp s_{2\psi} + \delta S_2, \delta S_3\}, \end{aligned} \tag{3.248a}$$

where

$$\{\delta S_0, \delta S_1, \delta S_2, \delta S_3\} = (t_{ij})_S\{1, 0, \mp c_{1\Delta}, \mp s_{1\Delta}\}, \tag{3.248b}$$

for both zone pairs. Substituting from eq. (3.248a) into eq. (3.225), we get

$$\begin{aligned} \mathcal{P} &= 1 - \delta S_0 - c_{2\psi}\,\delta S_1 \mp s_{2\psi}\,\delta S, \\ 1 - \mathcal{P} &= \delta S_0 + c_{2\psi}\,\delta S_1 \pm s_{2\psi}\,\delta S_2. \end{aligned} \tag{3.249}$$

Then, from eqs. (3.248b) and (3.249), we obtain the degree of depolarization, $(1 - \mathcal{P})$, as a function of the system's ψ and Δ and the system-imperfection matrix $(t_{ij})_S$. Although the effect of $(t_{ij})_S$ on depolarization can be displayed in detail by an array of coupling coefficients, we indicate here only the locations of those elements contributing to the depolarization. This, together with the results obtained for the other components, by use of the procedure explained above, are shown in table 3.14 with the effective elements indicated by X.

Table 3.14. Location of the elements contributing to the depolarization*

	P				C				W				S				W'				A			
	X	0	X	0	X	X	X	0	X	0	X	X	X	0	X	X	X	X	X	0	X	X	0	0
	X	0	X	0	X	X	X	0	0	0	0	X	0	X	X	X	X	X	X	0	X	X	0	0
	0	0	0	0	0	0	0	0	X	0	X	X	X	0	X	X	X	X	X	0	X	X	0	0
	0	0	0	0	X	X	X	0	X	0	X	X	0	0	0	0	0	0	0	0	X	X	0	0

*The above table indicates the location of the elements in the Mueller imperfection matrices of the different optical components that contribute to the depolarization of light when the ellipsometer is set for minimum detected flux.

3.8.4. Discussion

The results summarized in tables 3.11–3.14 are very useful in analyzing the effect of detailed imperfections in the optical components, including the optical system under measurement, on the measured ellipsometric angles ψ and Δ. They display in an elegant and most direct fashion the effects of the Mueller imperfection matrices of the elements on the nulling angles and hence on ψ and Δ. Many important conclusions, which would be difficult to prove otherwise, can be obtained readily. The following is a non-exhaustive list:

(1) Polarizer

In §3.7 a polarizer imperfection was assumed to be such that the light leaving the polarizer is totally polarized with small ellipticity. In addition to the error terms found based on this assumption, we see from eqs. (3.224)–(3.227) that if the polarizer output is depolarized by a small amount, $1 - \mathcal{P}$, this will cause an error in ψ equal to $\frac{1}{4}(1 - \mathcal{P}) \sin 4\psi$. For a system with $\psi = 22.5°$ this means that an error of 0.15° results from a degree of depolarization of 1%. The principal frame imperfection-matrix elements responsible for this depolarization are $t'_{11}, t'_{21}, t'_{14}$ and t'_{24} [table 3.14, or eqs. (3.222)]. According to tables 3.12 and 3.13, the effect of these elements survives two- and four-zone averaging and hence can be important if the depolarization is appreciable. (Note that in the PCA straight-through position, the nulling angles are unaffected by slight depolarization, to first order ($\sin 4\psi = \sin \pi = 0$). If $t'_{11} = t'_{21}$ and $t'_{14} = t'_{24}$, the depolarization disappears for all values of μ, the degree of circular polarization of the source beam.)

(2) Analyzer

It is useful to introduce the number-pair (n_Δ, n_ψ) to represent the number of elements in the principal-frame imperfection matrix of a device that affect Δ and ψ, respectively. From table 3.11, we see that for the analyzer in one zone, $(n_\Delta, n_\psi) = (1, 1)$. Two- and four-zone averaging reduce (n_Δ, n_ψ) to $(0, 0)$, that is, the effect of any small but general imperfection in the analyzer disappears to first order, tables 3.12 and 3.13. Thus ψ and Δ are more susceptible to errors caused by polarizer imperfection than analyzer imperfection, and the better element should be used in the polarizer position.

(3) Compensator

From table 3.11, (n_Δ, n_ψ) for the compensator are (6, 12), where the six elements affecting Δ are among the twelve affecting ψ. Two- and four-zone averaging reduce (n_Δ, n_ψ) to (4, 6) and (2, 3), respectively, as shown by tables 3.12 and 3.13. So far, we have assumed that the 16 elements of the imperfection Mueller matrix are different. However, we expect that by invoking both physical and symmetry considerations, the number can be reduced considerably. For example, if incoherent effects are absent in the compensator, only 6 different elements completely specify the ellipsometric effect of the compensator imperfection. These 6 elements can be obtained most directly by finding the Mueller matrix corresponding to a Jones matrix of the form $\begin{bmatrix} 1 & \epsilon_3 \\ \epsilon_4 & -j+\epsilon_2 \end{bmatrix}$. ϵ_2 arises from a possible deviation of the retardation from 90° and from multiple reflections within a thin-plate compensator which cause the magnitude of its relative transmittance to deviate from unity. ϵ_3 and ϵ_4 represent small amounts of birefringence and optical activity. Applying the transformation of eq. (2.243) we obtain the imperfection Mueller matrix of the compensator

$$(t'_{ij})_C = \begin{bmatrix} -v & v & x & z \\ v & -v & w & y \\ z & -y & u & -v \\ -x & w & v & u \end{bmatrix}, \tag{3.250a}$$

where

$$u + jv = \epsilon_2, \quad x - jy = \epsilon_3 + j\epsilon_4, \quad z + jw = \epsilon_4 + j\epsilon_3. \tag{3.250b}$$

It is convenient to put Mueller imperfection matrices, such as that in eq. (3.250a), in the form $\begin{bmatrix} H_1 & H_3 \\ H_4 & H_2 \end{bmatrix}$, where H_1, H_2, H_3 and H_4 are 2×2 submatrices. Note that off-diagonal elements ϵ_3 and ϵ_4 in the Jones matrix have contributed elements only in the H_3 and H_4 off-diagonal submatrices, to first order. Substituting $(t'_{ij})_C$ from eq. (3.250a), and $(\beta_{ij})_C$, $(\alpha_{ij})_C$ from table 3.11, into eq. (3.240), we get

$$\begin{aligned} \delta P(1,3) &= \pm \tfrac{1}{2} v \, s_{1\Delta} \mp \tfrac{1}{2} x \, c_{1\Delta} + \tfrac{1}{2} w, \\ \delta A(1,3) &= \tfrac{1}{2} u s_{2\psi} \, s_{1\Delta} \pm \tfrac{1}{2} z s_{2\psi} - \tfrac{1}{2} y s_{2\psi} \, c_{1\Delta}, \end{aligned} \tag{3.251}$$

in zones 1 and 3. Upon averaging over two zones, the effect of ϵ_2 disappears whereas the effect of optical activity and birefringence remains through w and z. The effect of w is similar to that caused by a polarizer azimuth-angle error and that of z to a polarizer ellipticity. Similar results were also obtained using the Jones-matrix formulation in §3.7.

The above example has been carried out in some detail to demonstrate the use of the arrays of coupling coefficients in table 3.11.

The effect of the imperfection elements in the off-diagonal submatrices H_3 and H_4 disappears in a four zone average. This can be seen from table 3.13 and proves that both coherent and incoherent cross-scattering do not affect four-zone averages. By cross-scattering we mean that for an incident linear vibration along the fast (slow) axis there is a component (either coherent or incoherent) of the emergent light along the slow (fast) axis.

(4) Cell windows
From table 3.11, we find that the values of (n_Δ, n_ψ) are (6, 12) and (3, 6) for the entrance and exit windows, respectively. Upon averaging over two zones (table 3.12), these become (4, 6) and (1, 3) for W and W', respectively. Four-zone averaging does not provide any extra advantage, table 3.13. Also, from tables 3.11 and 3.14, we see that exit window depolarization does not affect Δ although it might affect ψ. In view of the above, we conclude that the better window should be used as the entrance window.

The main window errors are expected to arise from stress birefringence. These can be determined by following a procedure similar to that explained in connection with the compensator imperfection. The results found this way are the same as derived in §3.7. Note that the elements in the 2×2 off-diagonal submatrices H_3 and H_4 cancel in two- and four-zone averaging. This includes both coherent and incoherent cross-scattering and follows from tables 3.12 and 3.13 which show zero off-diagonal 2×2 submatrices.

(5) System
The imperfection-matrix elements in the main diagonal submatrices H_1 and H_2 cause errors that are attached to and inseparable from ψ and Δ and thus do not cancel in two- or four-zone averages, tables 3.12 and 3.13. However, these same tables tell us that both coherent and incoherent $x \leftrightarrow y$ cross-scattering cancel in two- and four-zone averages. In reflection ellipsometry, cross-scattering can be due to surface optical activity and coherent and incoherent scattering by surface roughness. Thus, two-zone averaging helps determine a correct value for the ratio of the diagonal ($p \leftrightarrow p$ and $s \leftrightarrow s$) reflection coefficients in the presence of small spurious coherent or incoherent $p \leftrightarrow s$ cross-scattering. Four-zone averaging has no more to offer compared to two-zone averaging, table 3.13.

3.9. Generalized ellipsometry

Conventional ellipsometry can be readily generalized to determine the normalized Jones matrix, hence the polarization transfer function (PTF), $\chi_o = f(\chi_i)$ [eq. (3.1)], of an optical system [66]. Recall that χ_i and χ_o are complex numbers that represent the polarization states of light incident on and emergent from the optical system. No information about the internal structure of the optical system or its eigenpolarizations is required except that it does not depolarize the light passing through it and neither frequency mixing nor multiplication effects are present. The transfer function maps the incident to outgoing polarization states between two specified coordinate systems (x_i, y_i) and (x_o, y_o) at the input and output of the system, respectively.

The experimental set-up is such that the optical system S under measurement is placed between a linear polarizer P and a linear analyzer A. A compensator C is positioned either between the polarizer and the system or between the system and the analyzer. First, we consider the polarizer-compensator-system-analyzer (PCSA) ellipsometer arrangement.

The main features of the present analysis are: (1) interpretation of the measurements from the viewpoint of the bilinear-mapping property of optical systems, and (2) placing ellipsometry in a wider perspective in its application to arbitrary polarizing optical systems.

It should be noted that the polarization transfer function involves the input and output polarization *forms* only, without any account of overall amplitude or phase transmission factors (§2.3).

The ellipsometer arrangement under consideration is shown schematically in fig. 3.11. The optical system is indicated by a black box whose input and

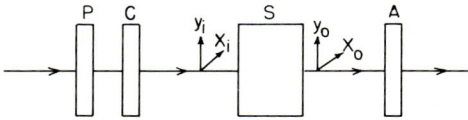

Fig. 3.11. Schematic diagram of PCSA ellipsometer arrangement. The optical system whose normalized Jones matrix is to be measured is indicated by a black box S, with input and output coordinate systems (x_i, y_i) and (x_o, y_o), respectively. P, C, and A represent the polarizer, compensator, and analyzer, respectively.

output are specified by the coordinate systems (x_i, y_i) and (x_o, y_o), respectively. The light beam enters and leaves the optical system in a direction perpendicular to these coordinate systems. The compensator is assumed to act as an ideal quarter-wave retarder and is placed between the polarizer and the optical system. The azimuths of the polarizer P and of the compensator's fast axis C are both measured from x_i, whereas the aximuth of the analyzer A is measured from x_o.

3.9.1. *The polarizer-compensator combination as a controlled polarization filter*

When an isotropically polarized light beam[28] is passed through a linear polarizer followed by a quarter-wave compensator, the polarization of the

---

[28]This means that the Stokes parameters S_1 and S_2 that represent the horizontal and $+45°$ linear preferences are zeros [see eq. (3.214)].

emergent light can be made to assume all possible states (represented by all points in the complex plane or on the Poincaré sphere) by rotating the polarizer and compensator around the beam axis. During this rotation, the intensity of the emergent light remains constant. Referring to fig. 3.12, let the azimuths P of the transmission axis of the polarizer and C of the fast axis of

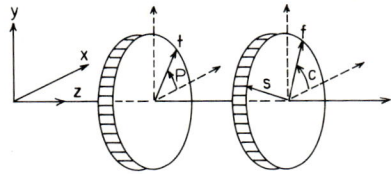

Fig. 3.12. Isotropically polarized light is incident on a combination of a linear polarizer and a quarter-wave compensator. The polarization of the emergent light can be controlled to assume all possible states by varying the azimuthal orientations P and C of the polarizer and compensator. These azimuths are measured with reference to a space-fixed orthogonal and right-handed xyz coordinate system.

the compensator be measured from the x-axis (in a counter-clockwise sense looking into the beam) of an xyz space-fixed Cartesian coordinate system whose z-axis is along the direction of propagation of the incident light beam. By use of the two linear polarizations along the x and y directions as basis states, the complex variable χ that describes the ellipse of polarization of the emergent light is given by

$$\chi(P, C) = \frac{[\tan C - j \tan (P - C)]}{[1 + j \tan C \tan (P - C)]}, \qquad (3.252)$$

as obtained from χ_{CO}^{xy} of eq. (3.13) after setting $\rho_C = -j$. Because the Cartesian complex polarization variable χ that describes an elliptic polarization of azimuth θ and ellipticity $\tan \epsilon$ is given by [eq. (1.79), §1.7]

$$\chi = \frac{\tan \theta + j \tan \epsilon}{1 - j \tan \theta \tan \epsilon},$$

eq. (3.252) represents an elliptic vibration of azimuth C and ellipticity $-\tan (P - C)$.[29]

[29]The components of the linearly vibrating electric vector incident on the compensator parallel to its fast and slow axes are obviously given by $\cos (P - C)$ and $\sin (P - C)$, apart from a constant multiplier. Upon passing through the quarter-wave compensator, the component parallel to the slow axis is retarded in phase by $\frac{1}{2}\pi$ relative to that to the fast axis. The superposition of these components at the output of the compensator gives rise to an ellipse whose major and minor axes are aligned with the fast and slow axes of the compensator and whose axial ratio (ellipticity) is $\tan (P - C)$. This ellipticity is negative because the slow component *lags* the fast component by $\frac{1}{2}\pi$.

Simultaneous and uncorrelated rotations of both the polarizer and the compensator cause $\chi(P, C)$ [eq. (3.252)] to describe a complicated trajectory in the complex plane. Three special cases are of interest. (1) The polarizer and compensator are rotated together as one unit so that $(P - C) = $ constant; (2) the compensator is set at a fixed azimuth, $C = $ constant, and the polarizer is rotated; and (3) the polarizer is set at a fixed azimuth, $P = $ constant, and the compensator is rotated.

Case 1, $(P - C) = $ constant: The locus $\chi(P, C)$ in this case is an equi-ellipticity circle that encloses either the point $L(0, -1)$ or $R(0, +1)$, representing the left- and right-circular polarizations, respectively. [See fig. 1.17, ch. 1]. The ellipticity assumes all possible values from $+1$ to -1 as $(P - C)$ is allowed to take all values in the range

$$-\tfrac{1}{4}\pi \leq (P - C) \leq +\tfrac{1}{4}\pi, \tag{3.253}$$

and all members of the equi-ellipticity family of circles are generated.

Case 2, $C = $ constant: If the compensator is set at a fixed azimuth C and the polarizer is rotated through one quadrant such that $(P - C)$ varies from $-\tfrac{1}{4}\pi$ to $+\tfrac{1}{4}\pi$, the emergent vibration will stay at constant azimuth C while its ellipticity varies from $+1$ to -1. The representative point $\chi(P, C)$ describes an equi-azimuth circle arc joining the point R and L. [See fig. 1.16, ch. 1]. All equi-azimuth circle arcs are generated if C is allowed to scan the interval

$$-\tfrac{1}{2}\pi \leq C < \tfrac{1}{2}\pi. \tag{3.254}$$

Case 3, $P = $ constant: The case when the polarizer is set at a fixed azimuth and the quarter-wave compensator is rotated leads to an interesting contour of a different nature.[30] The parametric equations of this contour can be obtained by taking the real and imaginary parts of the equation for χ, expanding $\tan(P - C)$, and considering $\tan C$ as the parameter and $\tan P$ as a constant. Alternatively, the equation can be used directly to computer plot the contour of $\chi(P, C)$. Figure 3.13 shows one contour obtained in this manner when $P = 60°$ and C is varied over a range of π from $-\tfrac{1}{2}\pi$ to $+\tfrac{1}{2}\pi$. (A 2π rotation of the compensator results in two traversals of the contour.)

The special cases when $P = 0°$ and $90°$ are shown in fig. 3.14. The $P = 0$ contour[31] is symmetrical with respect to both the real and imaginary axes and cuts the imaginary axis at right angles and the real axis at $+45°$ and $-45°$. The $P = 90°$ contour is a hyperbola whose asymptotes are the bisectors of the coordinate axes. The contour for $P = +45°$ (and also that for $P = -45°$) lies entirely on one side of the imaginary axis touching it at R and L. Figure

---

[30] See ref. 102. The only previous mention of this contour can be found in ref. 103 where it is only sketched on the Poincaré sphere.

[31] This contour can be identified with the Lemniscate of Bernoulli (two-leaved rose).

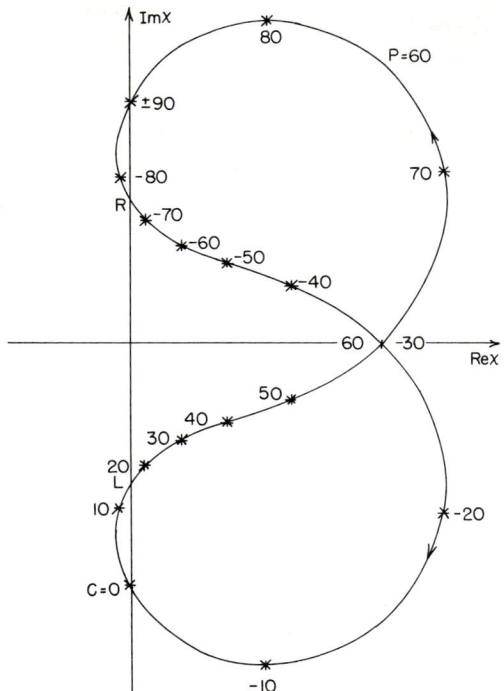

Fig. 3.13. When the polarizer is set at a fixed azimuth ($P = 60°$) and the compensator is rotated from $C = -90°$ to $C = +90°$, the polarization of the emergent light follows a double-lobed contour in the complex plane whose basis states are the x and y linear polarizations.

3.15 shows a collective picture of these contours drawn for various values of P.

The constant-P-variable-C contours have the following properties.

(1) Each contour consists of two lobes that are mirror images of one another in the real axis.

(2) The contour for $-P$ is obtained from that for $+P$ by mirror reflection in the imaginary axis.

(3) Each contour intersects the real axis twice at the same point $(\tan P, 0)$ and intersects the imaginary axis four times at R(0, 1), L(0, −1), (0, tan P) and (0, −tan P).

(4) In addition to R and L which are common to all contours, any two contours P_1 and P_2 intersect at two more points.

Because of this last property a π rotation of C, for every value of P from $-\frac{1}{2}\pi$ to $+\frac{1}{2}\pi$, results in generating every polarization state in the complex plane twice.

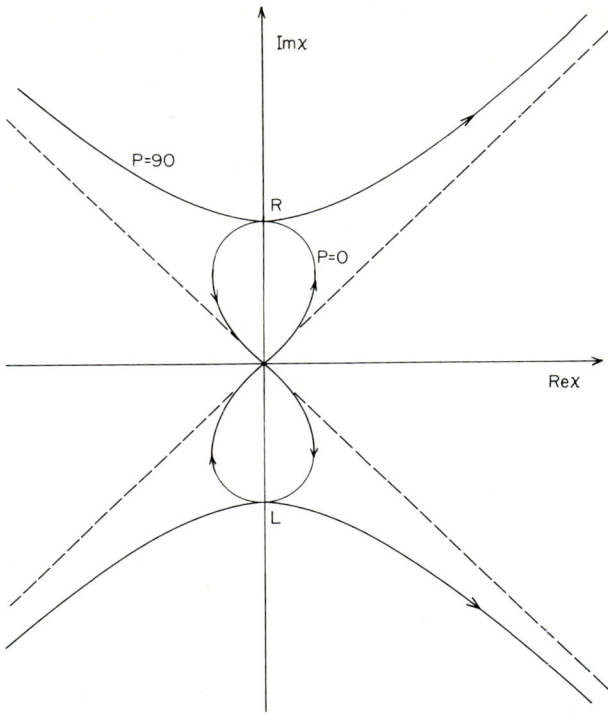

Fig. 3.14. Same as in fig. 3.13 when the polarizer is set at 0° and 90° instead of 60°. The $P = 90°$ contour is hyperbola whose asymptotes are the bisectors of the coordinate axes.

3.9.2. Nulling schemes and conditions of compensation

In §3.2, we presented the basis of generalized ellipsometry. In particular, we have shown that the normalized Jones matrix (hence the PTF) of an optical system can be measured from the mapping of three input ellipses of polarization into the corresponding three output ellipses of polarization. This requires that a minimum of three different null measurements be carried out using, for example, the polarizer-compensator-system-analyzer (PCSA) ellipsometer arrangement.

A number of nulling schemes are available for obtaining the multiple nulls. If we exclude the compensator relative retardation δ_C as a possible adjustable parameter in search for the null, we are left with the three azimuth angles P, C, and A of the polarizer, compensator and analyzer, respectively. From a practical point of view, it is desirable to adjust the minimum number of parameters to reach the null condition and, in ellipsometry, this number is

§3.9]   GENERALIZED ELLIPSOMETRY   239

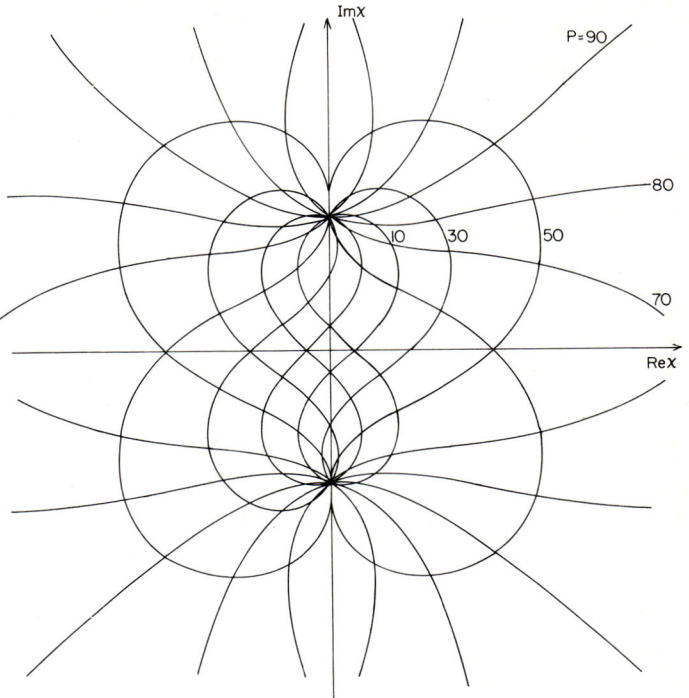

Fig. 3.15. A collective view of the constant-P–variable-C contours.

in general two. Thus we can distinguish between three nulling schemes obtained by setting one of the three elements (polarizer, compensator or analyzer) at a fixed azimuth and adjusting the other two for null.

As will be shown shortly, when one element is set at a fixed azimuth, a null (extinction of the light flux that reaches the photodetector) may or *may not* be found by the adjustment of the other two elements. *The conditions of compensation* (existence of a null) must therefore be investigated.

3.9.2.1. The fixed-analyzer nulling scheme
When the analyzer is set at any fixed azimuth, and the polarizer and compensator azimuths are adjusted, nulling is always possible. This is because a polarizer-compensator azimuth-angle pair (P, C) can be found that leads to a state of polarization at the input of the optical system which is mapped by the system to a linear polarization,

$$\chi_o = \tan(A - \tfrac{1}{2}\pi), \tag{3.255}$$

at its output crossed with the transmission axis of the linear analyzer A. Note that there is one-to-one correspondence between the polarization states at the input and output of the optical system under measurements, hence only *one* null can always be obtained for *every* setting of A, as P and C are adjusted [19]. The range of compensation (range of analyzer azimuth) is unlimited (*i.e.* equal to 2π).

3.9.2.2. The fixed-compensator nulling scheme
When the compensator is set at a fixed azimuth C and the polarizer is rotated, the state of polarization χ_i at the input of the optical system describes a complete circle in the complex polarization χ_i-plane through the points R(0, 1) and L(0, −1) that represent the right- and left-circular polarizations. Because of the bilinear-mapping relationship between input and output polarizations, the state of polarization χ_o of the light emergent from the system will describe a circle through the points R' and L' in the χ_o-plane where R' and L' represent the system's response to incident right (R)- and left (L)-handed circular polarizations, respectively. For another value of C and for a π-rotation of P, χ_i will describe another circle in the χ_i-plane through the same points R and L, and χ_o will follow the image circle through R' and L' in the χ_o-plane. Both of the two families of circles obtained are shown in fig. 3.16. For compensation, the circle of χ_o should intersect the real axis of the χ_o-plane [66]. In this case the combination of the compensator and optical system is such that two linear vibrations incident on the compensator will emerge from the system as two linear vibrations χ_{o1} and χ_{o2}. Therefore, two measurable non-equivalent nulls are physically realizable.

From fig. 3.16, the conditions of compensation can be conveniently stated with reference to the system's response to incident circularly polarized light of both handedness. Thus with R' and L' denoting the polarization states of light emergent from the optical system when the incident light is right (R)- and left (L)-circularly polarized, respectively, the different possibilities of compensation can be summarized as follows.

(1) R' and L' have opposite handedness (hence lie on opposite sides of the real axis of the χ_o-plane). A distinct pair of two non-equivalent nulls is available for *every* setting C of the compensator azimuth.

(2) R' and L' have the same handedness (hence lie on the same side of, above or below, the real axis of the χ_o-plane). A distinct pair of two non-equivalent nulls is available for every compensator azimuth setting C in a *limited range* $C_1 < C < C_2$. When $C = C_1$ or C_2, the two nulls coincide. (C_1 and C_2 are determined by the two circles in the χ_o-plane through R' and L' that touch the real axis.)

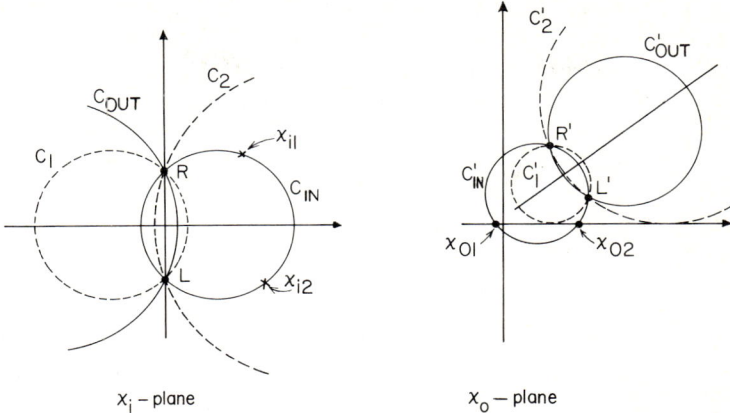

Fig. 3.16. The complex χ_i and χ_o planes, where χ_i and χ_o are the polarization states at the input and output of the optical system S. R' and L' represent the system's response to right (R)- and left-(L) circularly polarized incident light, respectively. For fixed compensator azimuth, the locus of χ_i is a circle through R and L as the polarizer is rotated. The locus of χ_o is the image circle through R' and L'. If the circle of χ_o intersects the real axis, two nonequivalent nulls are obtainable in the PCSA ellipsometer arrangement in which the compensator is fixed. In the case shown, R' and L' lie on the same side of the real axis and there is a limited range of C values $C_1 \le C \le C_2$ for which the image circle intersects the real axis. C_1 and C_2 define the range of compensation (the image circles C_1' and C_2' touch the real axis). C_{IN} is inside the range of compensation with two nonequivalent nulls corresponding to (χ_{i1}, χ_{o1}) and (χ_{i2}, χ_{o2}). C_{OUT} lies outside the range of compensation.

(3) Either R' or L' is linearly polarized (hence lie on the real axis of the χ_o-plane). Only one new null is available for every compensator setting *after the first*, for which two different nulls are obtained.

(4) Both R' and L' are linearly polarized (hence lie on the real axis of the χ_o-plane). For this type of system an infinite number of nulls can be obtained at one particular setting (C_{part}) of the compensator. For every other setting, the same two nulls are always obtained.[32]

From the above, we conclude that for N different settings of the compensator azimuth C, the number of non-equivalent nulls that are attainable in cases (1), (2), (3), and (4) are, respectively: (1) $2N$; (2) $2N_1$, $(2N_1 - 1)$, or $(2N_1 - 2)$ where N_1 (which is less than or equal to N) is the number of

---

[32] The circle locus 𝒞 of incident polarization states that become linearly polarized after passing through the optical system is shown in fig. 3.18, for each of the above four cases. In case (4), 𝒞 coincides with a constant-C–variable-P circle through R and L and one distinct null becomes available for *every* setting of P (obviously, the analyzer has to be rotated to maintain the null condition.)

settings within the range of compensation, one or two of which may coincide with the limiting azimuths C_1 and C_2; (3) $N + 1$; and (4) infinite or two depending on whether one or none of the N settings coincide with C_{part}.

A different approach to the conditions of compensation that are summarized above is given in ref. [19] and is based on the use of the circular (instead of the Cartesian) complex-plane representation of polarization [§1.7, ch. 1]. In ref. [19], it is shown that the two azimuthal settings P' and P'' of the polarizer at the two nulls that are available for the same compensator azimuth C are related by

$$\tan(P' - C - \tfrac{1}{4}\pi) \tan(P'' - C - \tfrac{1}{4}\pi) = K, \qquad (3.256)$$

where K is a constant independent of C which is determined only by the optical system under measurement. ($K = -1$ when the optical system has orthogonal linear eigenpolarizations; and eq. (3.256) reduces to eq. (3.84).) Furthermore, it is shown in ref. [19] that K is an indicator of the range of compensation and of the crowding of the nulls.

3.9.2.3. The fixed-polarizer nulling scheme

The null condition occurs when incident light polarized in the state χ_i emerges from the optical system S linearly polarized in a state χ_o. The latter linear polarization χ_o can be crossed by the analyzer A to give extinction or null. The locus of the incident polarization states χ_i that produce the null condition is a circle $\mathscr{C}$ in the complex χ_i-plane.

When the polarizer is set at a fixed azimuth and the compensator is rotated through a range of π, the polarization state χ_i incident on the optical system traces a double-lobed contour of the type shown in fig. 3.13. In general, this contour intersects the circle locus $\mathscr{C}$ of the optical system S in four points leading to four non-equivalent nulls, fig. 3.17.

Compared with the nulling scheme in which either A, C, or $(P-C)$ is fixed, the fixed-polarizer scheme [102] provides the largest number of nulls per setting of the fixed element.

The various conditions of compensation and the counting of nulls are based on the different ways in which the circle locus $\mathscr{C}$ of the optical system cuts the family of constant P contours of fig. 3.15. Because $\mathscr{C}$ in the χ_i-plane maps into the real axis of the χ_o-plane, the interior and exterior domains of $\mathscr{C}$ are mapped into the two half-planes above and below the real axis of the χ_o-plane. If R' and L' denote the system's response to the right (R)- and left (L)-circular polarizations (as before), the different possibilities of compensation can be stated as follows. [See figs. 3.15 and 3.18.]

(1) R' and L' have opposite handedness. R' and L' lie on opposite sides of the real axis of the χ_o-plane so that R and L lie on opposite sides of $\mathscr{C}$. In

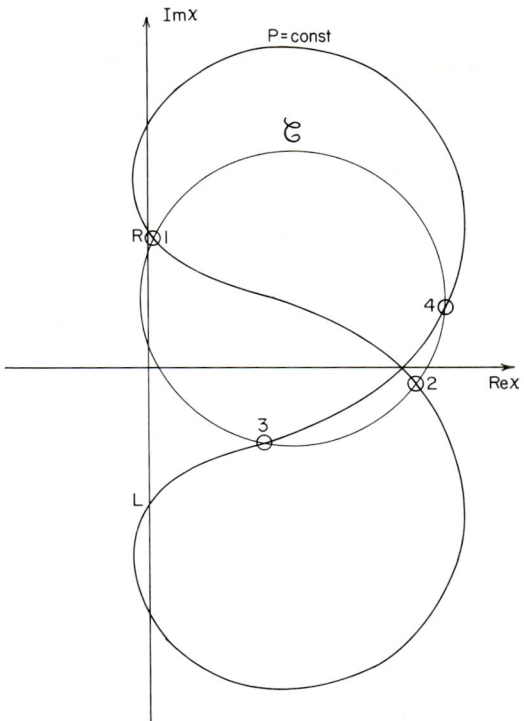

Fig. 3.17. This figure illustrates the availability of up to four nulls per setting of the polarizer in the fixed-polarizer nulling scheme. The circle 𝒞 represents locus of the polarization states which, when incident on the optical system S, become linearly polarized at its output leading to the null condition. The double-lobed contour is the locus of the polarization state emergent from the compensator incident on the optical system as the compensator is rotated while the polarizer is held fixed. Polarization matching, hence nulling, occurs at the points of intersection 1, 2, 3, and 4.

other words, the circle 𝒞 encloses either R or L and cuts all of the constant P contours in fig. 3.15. From one to four nulls are available for every setting of the polarizer azimuth depending on the value of P and whether 𝒞 lies entirely above or below, touches or cuts the real axis of the χ_i-plane.

(2) R' and L' have the same handedness. R' and L' lie on the same side of the real axis of the χ_o-plane, hence both R and L lie either inside or outside of 𝒞. From one to four nulls are available for every polarizer setting P in a limited range $P_1 < P < P_2$. The limiting contours at P_1 and P_2 touch 𝒞. Outside this range nulling is impossible.

(3) Either R' or L' is linearly polarized. 𝒞 passes through R or L and nulling is possible for all values of P. For all settings of the polarizer after the first, the number of *new* nulls varies from zero to a maximum of three.

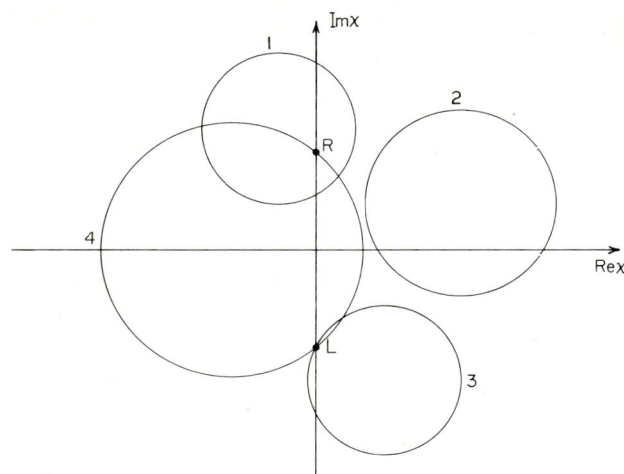

Fig. 3.18. The four different possibilities of compensation (1, 2, 3, and 4) depend upon the relationship between the circle locus $\mathscr{C}$ of the system and the points R and L which represent the right- and left-circular polarizations.

(4) Both R' and L' are linearly polarized. $\mathscr{C}$ passes through R and L and nulling is possible for all values of P. For all settings of the polarizer after the first the number of *new* nulls that can be obtained varies from zero to two.

The different possibilities for the circle locus $\mathscr{C}$ that lead to cases (1)–(4) above are shown in fig. 3.18.

For N different settings of the polarizer, the number of available non-equivalent nulls depends on how these settings are chosen and the position of $\mathscr{C}$, as explained above. Because of the wide range of possibilities, counting of the nulls in these individual cases will not be attempted.

It is interesting to compare the fixed-compensator and fixed-polarizer nulling schemes when measurements are taken on a system for which R' and L' are linearly polarized (case 4). (Such a system may be composed of an arbitrary optical rotator [including half-wave plates] preceded or followed by a quarter-wave retarder.) For the fixed-compensator scheme, two nulls are readily obtained at the first setting of compensator (those for which $\chi_i =$ R and L). Any trial to obtain new nulls at another compensator azimuth will fail since the *a priori* probability of setting the compensator at C_{part} is zero. On the other hand, in the fixed-polarizer scheme a polarizer setting can be easily found which produces four non-equivalent nulls including those produced by R and L. This number exceeds by one that required to determine the PTF of the optical system.

As another example, consider the use of the fixed-polarizer nulling scheme in conventional ellipsometry where measurements are made to determine ψ and Δ of a surface with zero p↔s cross reflection coefficients. The locus $\mathscr{C}$ becomes a straight line through the origin at an angle with the real axis equal to Δ. Excluding trivial (unwanted) nulls obtained when the incident light is either p- or s-polarized, the following conclusions (which the reader could verify for himself) can be reached. (1) If $|\Delta| \leq \frac{1}{4}\pi$ or $|\Delta| \geq \frac{3}{4}\pi$, *two* non-equivalent nulls are available for *every* setting of P except for $P = 0°$ and $90°$; and (2) if $\frac{1}{4}\pi < |\Delta| < \frac{3}{4}\pi$, a range of P values exists inside which *four* non-equivalent nulls are available for *every* value of P. This four-null range of P depends on Δ. Recall that *only two* nulls are *always* available for *every* setting of the compensator in the fixed-compensator nulling scheme.

3.9.3. Multiple-null measurements

The polarization transfer function is determined if the system's response to three different incident polarizations is known. Consequently, only three null measurements are needed to determine the function completely. With more measurements the unknown parameters are overdetermined. Therefore, an accurate solution can be obtained in the presence of various sources of error in the ellipsometer.[33]

Let $\{P_n(\nu), C_n(\nu), A_n(\nu)\}$ denote the νth set of nulling angles, where $\nu = 1, 2, \ldots N$ and $N \geq 3$. From the polarizer and compensator azimuths $P_n(\nu)$ and $C_n(\nu)$ at null, the polarization state of the light incident on the optical system is given by

$$\chi_i(\nu) = \frac{\tan[C_n(\nu)] - j \tan[P_n(\nu) - C_n(\nu)]}{1 + j \tan[C_n(\nu)] \tan[P_n(\nu) - C_n(\nu)]}. \tag{3.257}$$

Because the light emergent from the optical system is extinguished by the analyzer, it must be linearly polarized with a polarization state

$$\chi_o(\nu) = \tan[A_n(\nu) - \tfrac{1}{2}\pi]. \tag{3.258}$$

Except for a slight change of notation, eqs. (3.257) and (3.258) are the same as eqs. (3.252) and (3.255), respectively.

First, consider three different sets of nulling angles ($\nu = 1, 2,$ and 3). From eqs. (3.257) and (3.258) the corresponding polarization states at the input and

---

[33] In conventional ellipsometry on systems with known eigenpolarizations, a single null or one-zone measurement is adequate, in theory, to determine the ratio of the associated eigenvalues. However, it is common practice in reflection ellipsometry to make multiple null (two- and four-zone) measurements to reduce or cancel the effect of systematic errors [§3.7 and §3.8]. In generalized ellipsometry, where three complex quantities are to be measured instead of one, as many as six to twelve nulls may be desirable for error minimization.

output of the optical system are determined. Direct substitution into eq. (3.3) provides the unnormalized elements of the Jones matrix of the optical system under measurement, hence the bilinear transformation that defines the polarization transfer between its input and output [eq. (3.1)]. Next, consider N different sets of nulling angles. Because $\chi_o(\nu)$ is confined to the real axis of the complex χ_o-plane, and the mapping $\chi_i \to \chi_o$ is bilinear, $\chi_i(\nu)$ ($\nu = 1, 2, \ldots N$) should correspond to points on a circle in the complex χ_i-plane. The polarization transfer function is determined by the mapping of any three points in the χ_i-plane onto the image points on the real axis of the χ_o-plane. Thus the normalized Jones matrix is overdetermined and some best-fit criterion can be utilized. Because of experimental errors, the points $\chi_i(\nu)$ are expected to be scattered slightly around the circumference of the best-fit circle. Note that the use of eq. (3.257) to obtain $\chi_i(\nu)$ is subject to errors arising from polarizer and compensator imperfections and azimuth-angle errors. (The former include the polarizer ellipticity and deviations from ideal quarter-wave compensator retardation.) On the other hand, the use of eq. (3.258) to obtain $\chi_o(\nu)$ is susceptible to analyzer imperfection and azimuth-angle errors.

The search for the normalized Jones matrix (T_{ij}) should be directed to finding such a matrix that transforms the experimentally measured incident polarizations $\chi_i^{\text{meas}}(\nu)$ [eq. (3.257)] into *computed* output polarizations $\chi_o^{\text{comp}}(\nu)$ that are as *near* as possible to the *measured* output polarizations $\chi_o^{\text{meas}}(\nu)$ [eq. (3.258)]. Thus we need to consider the *collective* nearness function [47b]

$$\mathcal{N}_{\text{coll}}(T_{ij}) = \sum_{\nu=1}^{N} \mathcal{N}[\chi_o^{\text{comp}}(\nu), \chi_o^{\text{meas}}(\nu)], \qquad (3.259)$$

where $\mathcal{N}(\chi_1, \chi_2)$ is the nearness of the two polarizations represented by the complex numbers χ_1 and χ_2, as defined by eqs. (2.149), (2.151) or (2.153) of §2.8. When the nearness function $\mathcal{N}$ of eq. (2.149) is used, the collective nearness function $\mathcal{N}_{\text{coll}}(T_{ij})$ of eq. (3.259) has to be maximized. On the other hand, if $\mathcal{N}$ of either eq. (2.151) or eq. (2.153) is used, $\mathcal{N}_{\text{coll}}(T_{ij})$ has to be minimized. Best-fit calculations of this nature are typically carried out by a digital computer.

3.9.4. Alternate PSCA ellipsometer arrangement

In the introduction to this section, we have indicated that the quarter-wave compensator C can be placed either before or after the optical system S under measurement, leading to the two well-known PCSA and PSCA ellipsometer arrangements, respectively. Although the above discussion has been confined to the PCSA arrangement, the conclusions can be readily extended to apply to the alternate PSCA arrangement. In particular, it can be shown that the conditions of compensation and the counting of the available nulls for the fixed-polarizer, fixed-compensator, and fixed-analyzer nulling schemes in the PSCA ellipsometer arrangement are exactly the same as those for the fixed-analyzer, fixed-compensator, and fixed-polarizer nulling schemes, respectively, already stated for the PCSA ellipsometer arrangement. In other words, the roles of the polarizer and analyzer are simply interchanged while that of the compensator stays the same when a switch is made between the PCSA and the PSCA ellipsometer arrangements.

The equations that replace eqs. (3.257) and (3.258), applicable to the PSCA ellipsometer arrangement, are given by [see eq. (3.56)]

$$\chi_i(\nu) = \tan [P_n(\nu)], \tag{3.260}$$

$$\chi_o(\nu) = -\frac{1 + j \tan [C_n(\nu)] \tan [A_n(\nu) - C_n(\nu)]}{\tan [C_n(\nu)] - j \tan [A_n(\nu) - C_n(\nu)]}, \tag{3.261}$$

respectively. The normalized Jones matrix can be determined from three sets of nulling angles $[P_n(\nu), C_n(\nu), A_n(\nu)]$, $\nu = 1, 2$, and 3 in the PSCA ellipsometer arrangement by substituting $\chi_i(\nu)$ and $\chi_o(\nu)$ from eqs. (3.260) and (3.261) into eq. (3.3). For multiple nulls in excess of three, the criterion given in connection with eq. (3.259) is still applicable.

3.10. Other ellipsometer arrangements

In earlier sections of this chapter, we have considered the commonly used polarizer-compensator-system(surface)-analyzer (PCSA) and polarizer-system(surface)-compensator-analyzer (PSCA) ellipsometer arrangements that are operated in the nulling mode. Here we discuss other null, photometric and interferometric ellipsometers that have recently been introduced. Because the basic principles are the same, we do not consider as essentially different those variants of the PCSA or PSCA ellipsometer arrangements that incorporate additional modulated electro-optic elements to slightly perturb the state of polarization of the ellipsometer light beam for higher precision in locating the settings of the optical elements at zero (minimum) detected dc light flux. Also not considered as fundamentally different are the all-electronic null ellipsometers in which adjustments of component azimuth angles or retardations are effected electro-optically instead of mechanically. These alternative arrangements differ from one another in instrumental detail and will be later discussed in chapter 5.

3.10.1. Null ellipsometers that do not employ compensators

3.10.1.1. An ellipsometer based on the detection of the azimuth of the polarization ellipse only, with no measurement of ellipticity

Monin and Boutry [104] have recently proposed an ellipsometer arrangement that dispenses with the compensator (or quarter-wave plate) but employs instead a modulated optical rotator [*e.g.*, a Faraday cell excited by an alternating current (ac)] for the sensitive detection of the azimuth of the ellipse of polarization. Figure 3.19 represents a schematic of this ellipsometer in which L is the light source, P is a linear polarizer, S is the optical

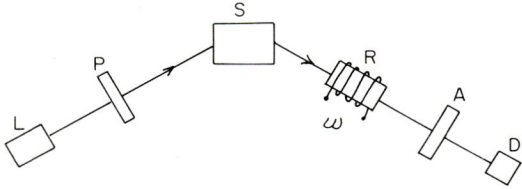

Fig. 3.19. Schematic diagram of Monin and Boutry's ellipsometer which consists of a light source L, polarizer P, optical system under measurement S, optical rotator R fed by an alternating current of frequency ω, analyzer A, and photodetector D. After ref. 104.

system under measurement (*e.g.*, a reflecting surface), R is the optical rotator which is driven by an ac current of frequency ω, A is a linear analyzer, and D is the photoelectric detector. Notice that this PSRA ellipsometer arrangement differs from the PSCA ellipsometer arrangement previously discussed in §3.4 only in that the compensator C has now been replaced by the modulated optical rotator R.

Monin and Boutry's ellipsometer (MBE) is based on the fact that when the transmission axis of the linear analyzer A is aligned parallel to either the major or minor axis of the ellipse of polarization of light at the output of the system S, alternating-current modulation of the optical rotator R produces no first-harmonic (modulator frequency) light-flux component after the analyzer A. Thus, with the linear polarizer P set at a fixed azimuth, the linear analyzer A can be rotated until the component of the detected photoelectric current at the modulator frequency ω is nulled. Alternatively, the analyzer may be set at a fixed azimuth and the polarizer is rotated in the incident beam until the null condition is reached. In both cases, the orientation of the analyzer A at which the fundamental component of the photoelectric current is nulled determines the orientation of the ellipse of polarization at the output of the system S.

The main features of the MBE are that no measurement of ellipticity is carried out (no compensator is required) and that measurement of azimuth alone provides enough information[34] about the system (surface) S, as explained below.

To prove the validity of the method of azimuth detection in MBE, we use the expression for the intensity transmittance τ of an ideal analyzer given by eq. (2.144), §2.7. If θ, ϵ and I denote the azimuth, ellipticity angle and intensity of the totally polarized light to be analyzed, the expression for the detected signal after the linear analyzer A can be written as

$$\mathcal{I}_D = K_D I\tau,$$
$$= \bar{\mathcal{I}}_D[1 + Q \cos 2(\theta - A)], \qquad (3.262)$$

where

$$\bar{\mathcal{I}}_D = \tfrac{1}{2} K_D I\tau_{max}, \qquad Q = \cos 2\epsilon. \qquad (3.263)$$

τ_{max} is the maximum intensity transmittance of the analyzer for light linearly polarized parallel to its transmission axis at azimuth A and K_D is a detector sensitivity factor. If the azimuth θ is modulated such that

$$\theta = \theta_0 + \alpha_m \cos \omega t, \qquad |\alpha_m| \ll 1, \qquad (3.264)$$

the cosine term in eq. (3.262) becomes

$$\begin{aligned}
\cos 2(\theta - A) &= \cos [2(\theta_0 - A) + 2\alpha_m \cos \omega t], \\
&= \cos 2(\theta_0 - A) \cos (2\alpha_m \cos \omega t) \\
&\quad - \sin 2(\theta_0 - A) \sin (2\alpha_m \cos \omega t), \\
&\cong \cos 2(\theta_0 - A)[1 - \tfrac{1}{2}(2\alpha_m \cos \omega t)^2] \\
&\quad - \sin 2(\theta_0 - A)[2\alpha_m \cos \omega t], \qquad (3.265)
\end{aligned}$$

to second order in α_m. Substitution of eq. (3.265) into eq. (3.262) leads to the following harmonic analysis of the photoelectric current

$$\mathcal{I}_D(0) = \bar{\mathcal{I}}_D[1 + Q(1 - \alpha_m^2) \cos 2(\theta_0 - A)], \qquad (3.266)$$
$$\mathcal{I}_D(\omega) = 2\bar{\mathcal{I}}_D Q \alpha_m \sin 2(\theta_0 - A) \cos (\omega t + \pi), \qquad (3.267)$$
$$\mathcal{I}_D(2\omega) = \bar{\mathcal{I}}_D Q \alpha_m^2 \cos 2(\theta_0 - A) \cos (2\omega t + \pi). \qquad (3.268)$$

From eq. (3.267), it is evident that

$$\mathcal{I}_D(\omega) = 0 \qquad (3.269)$$

when

$$A = \theta_0, \qquad \text{or} \qquad A = \theta_0 \pm \tfrac{1}{2}\pi. \qquad (3.270)$$

---

[34] The same principle has also been reported on by Som and Chowdhury [105] who used a static photometric method for azimuth detection.

Therefore, the fundamental component of the photoelectric current vanishes when the analyzer transmission axis is aligned either parallel to the major axis ($A = \theta_0$) or minor axis ($A = \theta_0 \pm \frac{1}{2}\pi$) of the ellipse of polarization under measurement. From eq. (3.267), it can also be shown that the amplitude of the fundamental component has equal sensitivity with respect to small rotations of the analyzer around the settings $A = \theta_0$ and $A = \theta_0 \pm \frac{1}{2}\pi$. However, the dc light flux levels are different at these two orthogonal positions; their ratio is equal to the square of the ellipticity, $\tan^2 \epsilon = b^2/a^2$, where b and a are the semi-minor and semi-major axes of the ellipse of polarization, respectively. If the detection process is shot-noise limited, measurement of the orientation of the ellipse of polarization using the above scheme would be less susceptible to noise (hence is more precise) with the analyzer aligned parallel to the minor axis of the ellipse rather than with it aligned parallel to the major axis.

If χ_i and χ_o denote the complex polarization numbers at the input and output of the optical system S, which is assumed to have orthogonal linear eigenpolarizations (*e.g.*, an isotropic reflecting surface), then [see eq. (3.35)]

$$\chi_o = \chi_i/(\tan \psi \, e^{j\Delta}),$$

where $\tan \psi \, e^{j\Delta}$ is the ratio of eigenvalues (p and s reflection coefficients) to be measured. In MBE, the incident polarization is linear and $\chi_i = \tan P$, where P is the polarizer azimuth so that

$$\chi_o = \tan P \cot \psi \, e^{-j\Delta}. \tag{3.271}$$

The azimuth θ_o of the polarization state χ_o is given by eq. (1.86) of §1.7,

$$\tan 2\theta_o = 2\text{Re}\,(\chi_o)/(1 - |\chi_o|^2), \tag{3.272}$$

and is also equal to the analyzer azimuth A or $A + \frac{1}{2}\pi$, when the component of the photoelectric current at the modulator frequency is nullified. Therefore, from eqs. (3.271) and (3.272), we obtain

$$\tan 2A = \frac{2 \tan P \cot \psi \cos \Delta}{1 - \tan^2 P \cot^2 \psi}$$

$$= \frac{2 \tan P \tan \psi \cos \Delta}{\tan^2 \psi - \tan^2 P}. \tag{3.273}$$

By taking two ac-null measurements (P_1, A_1) and (P_2, A_2), we get two equations from eq. (3.273) that can be solved for ψ and Δ

$$\tan \psi = (\tan P_1 \tan P_2)^{\frac{1}{2}} \frac{\tan P_1 \tan 2A_1 - \tan P_2 \tan 2A_2}{\tan P_2 \tan 2A_1 - \tan P_1 \tan 2A_2},$$

$$\cos \Delta = \tan 2A_1 \frac{\tan^2 \psi - \tan^2 P_1}{2 \tan \psi \tan P_1}. \tag{3.274}$$

Based upon a study of the sensitivity of the fundamental component of the detected photoelectric current to small rotations of the polarizer and analyzer around their null settings, Monin and Boutry [104] proposed that the two measurements be taken by setting the analyzer at A_1 and $A_2 = 90° - A_1$ and adjusting the polarizer until $\mathcal{I}_D(\omega) = 0$. The two ac-null settings of the polarizer are denoted by P_1 and P_2, as before. This choice of $A_1 + A_2 = 90°$ simplifies eqs. (3.274)

$$\tan \psi = (\tan P_1 \tan P_2)^{\frac{1}{2}},$$

$$\cos \Delta = \frac{\tan 2A_1 (\tan P_2 - \tan P_1)}{(2 \tan P_1 \tan P_2)^{\frac{1}{2}}}.$$

(3.275)

A discussion of the different sources of error in this ellipsometer arrangement and their effect on the accuracy of ψ and Δ is given in ref. 104. The main advantages of MBE are that no compensator (hence no measurement of ellipticity) is required and that the null condition for the component of the photoelectric current at the modulator frequency can be precisely detected.

3.10.1.2. *Self-compensated ellipsometers based on the angle-of-incidence tunability of the reflection phase difference* Δ

In reflection ellipsometry on isotropic surfaces, the ratio $\tan \psi \, e^{j\Delta}$ of the complex amplitude-reflection coefficients for light polarized parallel (p) and perpendicular (s) to the plane of incidence is a function of the angle of incidence ϕ. The possibility therefore exists for a null ellipsometer without a compensator that uses the inherent tunability of the reflection phase difference Δ as a function of the angle of incidence ϕ. The latter is a convenient and easily controlled experimental parameter.

For simplicity, let us first consider the case of a film-free dielectric substrate. In the absence of absorption and surface films, the reflection phase difference Δ is either π or zero dependent on the angle of incidence being smaller or larger than the Brewster angle ϕ_B, respectively (see fig. 4.4 of §4.2). When the incident light is linearly polarized at an arbitrary azimuth from the plane incidence, the reflected light is also linearly polarized at a different azimuth, hence can be extinguished by a linear analyzer. Thus a polarizer-surface-analyzer (PSA) ellipsometer arrangement can be used, in which ψ at any angle of incidence ϕ is obtained from one *null* measurement (P, A) by

$$\tan \psi = |\tan P / \tan (A - \tfrac{1}{2}\pi)|,$$

(3.276)

as can be easily proven by use of eq. (3.35). If the substrate is absorbing or a *thin* film is present, the reflection phase difference Δ decreases *monotonically* and continuously from π to zero as the angle of incidence ϕ is increased from zero (normal incidence) to $\tfrac{1}{2}\pi$ (grazing incidence) [see figs. 4.7 and 4.10, §4.2]. Therefore, at all angles of incidence except zero and $\tfrac{1}{2}\pi$, the reflected light will be elliptically polarized when the incident light is linearly polarized, hence cannot be extinguished by a linear analyzer alone without the use of a compensator. This is true, however, because only one reflection is employed. At an angle of incidence ϕ_M such that Δ is equal to an integral fraction of π, π/M, the cumulative phase shift after M reflections between *identical parallel surfaces* will be $(\pi/M) \times M = \pi$. Under this condition, incident linearly polarized light emerges after M reflections also linearly polarized and can be extinguished by a linear analyzer. This is the basis of a $PS_M A$ null ellipsometer, fig. 3.20, that does not employ a compensator. In this arrange-

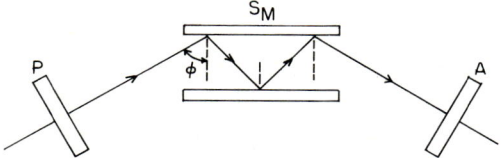

Fig. 3.20. A $PS_M A$ null ellipsometer without a compensator. S_M is a composite two-mirror system that allows M multiple reflections (only 3 are shown here for simplicity). A null can be obtained by the polarizer P and analyzer A when the angle of incidence ϕ is such that the cummulative phase difference Δ after M reflection equals π.

ment, S_M identifies a composite two-mirror system that allows M multiple reflections to take place. A null measurement can be obtained by setting the polarizer P (or the analyzer A) at some suitable azimuth from the plane of incidence (neither zero nor $\frac{1}{2}\pi$) and adjusting both the angle of incidence ϕ and the analyzer (or polarizer P) for extinction. From this measurement, the values of Δ and ψ characteristic of a single reflection are given by

$$\Delta = \pi/M,$$
$$\tan \psi = |\tan P/\tan (A - \tfrac{1}{2}\pi)|^{1/M}. \tag{3.277}$$

Notice that the above principle continues to apply in case of M reflections between nonparallel surfaces[35] at different incidence angles ϕ_i ($i = 1, 2, \ldots M$) provided that the cumulative phase difference Δ is equal to π, i.e.,

$$\Delta = \sum_{i=1}^{M} \Delta_i = \pi. \tag{3.278}$$

Also, only a single mirror need be the one under investigation, the other can act as a reference mirror if it has *known* properties. Because the intensity reflectance $\mathcal{R}_p$ for the p polarization is smaller than the intensity reflectance $\mathcal{R}_s$ for the s polarization when reflection takes place at a bare or thin-film-

---

[35] This assumes that the multiply reflected (zig-zag) beam is confined to one plane so that the incident beam must be perpendicular to the line of intersection of the planes of the two mirrors.

covered surface (*i.e.*, $\tan \psi < 1$), a small number of multiple reflections is desirable to avoid the near extinction of the p polarization after multiple reflections between the mirrors.

The "self-compensation" inherent in a system that employs multiple reflections and the angle-of-incidence tunability of the reflection phase difference Δ was first observed by Brewster [106] in 1830. Over a century later (1936) O'Bryan [107] devised an elegant null ellipsometer based on a novel modification of Brewster's scheme. A schematic of such an ellipsometer is shown in fig. 3.21 where B is a beam splitter, P a linear polarizer, S the

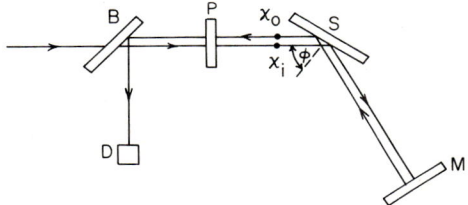

Fig. 3.21. O'Bryan's folded-path ellipsometer consists of a beam splitter B, polarizer P, surface S under measurement, retroreflecting mirror M and detector D. χ_i and χ_o represent the polarization states of the light beam before and after a complete round trip between the surface S and the mirror M. After ref. [107].

specimen surface under measurement, and M is a mirror that reflects the light beam back onto itself to give a signal at the detector D. In this folded-path ellipsometer, *two* reflections occur at the surface S at the same angle of incidence ϕ, each producing a reflection phase difference Δ. The normal incidence on the optically isotropic auxiliary mirror M does not change the ellipse of polarization. If χ_i and χ_o are the complex polarization numbers that describe the polarization states of the ellipsometer light beam before and after a complete round trip between the surface S and the mirror M (fig. 3.21), then

$$(\tan \psi \; e^{j\Delta})^2 = \chi_i/\chi_o,$$

or

$$\chi_i \chi_o = \chi_i^2/(\tan^2 \psi \; e^{j2\Delta}). \tag{3.279}$$

χ_i is determined solely by the azimuth of the linear polarizer P, $\chi_i = \tan P$. Using this substitution, eq. (3.279) becomes

$$\chi_i \chi_o = (\tan P/\tan \psi)^2 \; e^{-j2\Delta}. \tag{3.280}$$

If the state χ_o can be made orthogonal to χ_i, null at the detector D will be obtained. In this case, the return beam is extinguished by the polarizer P (now acting as an analyzer), with no light reaching the detector D. From eq. (3.280), the condition of orthogonality of χ_i and χ_o [eq. (1.81)],

$$\chi_i \chi_o = -1,$$

is satisfied when

$$\Delta = \pm \tfrac{1}{2}\pi, \tag{3.281a}$$

$$P = \pm \psi. \tag{3.281b}$$

The reflection phase difference Δ for a film-free or film-covered surface can always be adjusted to $\tfrac{1}{2}\pi$ [eq. (3.281a)] by varying the angle of incidence ϕ until it coincides with the *principal angle* ϕ_P. Simultaneously, the polarizer can be rotated around the beam axis so that $P = \pm \psi$ [eq. (3.281b)]. Therefore, by controlling the angle of incidence ϕ and the polarizer azimuth P, two distinct nulls can be reached using this simple ellipsometer.

The advantages of O'Bryan's ellipsometer are [107, 108]: 1) the use of only *one* polarizing optical component, namely the linear polarizer P which simultaneously acts both as a polarizer and as an analyzer. This is in contrast with conventional ellipsometers that require a compensator and an additional polarizer. 2) Because of the minimum number of optical elements employed (one), additional systematic errors due to component imperfections (especially those from the compensator) or azimuth errors are avoided in this ellipsometer. 3) The instrument can be operated over a wider bandwidth (wavelength range) because no compensator is required and only one polarizer is used. 4) Increased sensitivity due to the presence of two reflections. 5) Because of its simple construction, the essential optical parts of this ellipsometer can be enclosed in a test chamber (*e.g.*, under vacuum) for *in situ* measurements without incurring errors that would otherwise arise from stress-induced cell-window birefringence.

The above discussion has been primarily concerned with surfaces which are either film-free or covered with very thin films ($\lesssim \tfrac{1}{20}$ of a wavelength). In these cases, the reflection phase difference Δ is a monotonically decreasing function of the angle of incidence ϕ with a maximum value of π at normal incidence and zero at grazing incidence. For film-substrate systems in which the film thickness is comparable to or a significant fraction of a wavelength ($\gtrsim \tfrac{1}{10}$), the reflection phase difference Δ may exhibit a non-monotonic (oscillatory) variation with angle of incidence ϕ between $\phi = 0$ and $\phi = \tfrac{1}{2}\pi$ [see fig. 4.23 of §4.4]. More importantly, if the film thickness is greater than a certain minimum value (which is a function of the film-substrate system under investigation), the reflection phase difference Δ can become equal to π at one or more angles of incidence *between* $\phi = 0$ and $\phi = \tfrac{1}{2}\pi$. A single reflection, polarizer-surface-analyzer (PSA), ellipsometer arrangement can then be used [109] in which the null condition is reached by adjustment of the angle of incidence ϕ together with the azimuth setting around the beam axis of either the polarizer P or the analyzer A (P, $A \neq 0$ or $\frac{1}{2}\pi$). If the number of angles of incidence that make $\Delta = \pi$ (excluding $\phi = 0$, $\phi = \frac{1}{2}\pi$) is equal to or greater than three, enough information is obtained by making null measurements at these angles to determine the five pertinent optical parameters of the film-substrate system.

3.10.2. Photometric ellipsometers

In null ellipsometry information about the optical system under measurement is contained in those values of the azimuthal settings (P, C, A) of the optical elements, the relative phase retardation of the compensator (δ_C) and, in case of measurements on surfaces, the angle of incidence (ϕ) that reduce the dc or an ac-component of the detected light flux to zero. Photometric ellipsometry, on the other hand, is based on utilization of the variation of the detected light flux as a function of one or more of the above parameters (azimuth angle, phase retardation, or angle of incidence). The raw data from a photometric ellipsometer includes intensity (light flux) signals that are obtained at prescribed conditions.

For simplicity, we consider measurement of the ratio of eigenvalues V_ex and V_ey of an optical system with orthogonal x and y linear eigenpolarizations by use of the polarizer-compensator-system(surface)-analyzer (PCSA) ellipsometer arrangement. By writing $V_\text{ex}/V_\text{ey} = \tan \psi \, e^{j\Delta}$, the expression for the signal output of the photodetector D of eqs. (3.26) and (3.27) becomes

$$\mathcal{I}_\text{D} = G|V_\text{ey}|^2|\tan \psi \, e^{j\Delta} \cos A[\cos C \cos(P-C) \\ - T_\text{C} e^{j\delta_\text{C}} \sin C \sin(P-C)] \\ + \sin A[\sin C \cos(P-C) + T_\text{C} e^{j\delta_\text{C}} \cos C \sin(P-C)]|^2, \quad (3.282)$$

where $|X|$ indicates the absolute value of the complex number X, $T_\text{C} e^{j\delta_\text{C}}$ is the slow-to-fast relative transmittance of the compensator C, and G is a constant factor that depends on the source-beam intensity, detector sensitivity, and the transmittances of the optical elements P, C and A [see eq. (3.28)]. For the alternate PSCA arrangement, the detected signal $\mathcal{I}_\text{D}$ is also given by eq. (3.282), with P and A interchanged.

We distinguish between static and dynamic photometric ellipsometers. In a static photometric ellipsometer, the detected signal $\mathcal{I}_\text{D}$ (usually dc unless a chopper is used to interrupt the source beam) is recorded at predetermined fixed settings of the ellipsometer components, i.e., at specific values of P, C, A, and δ_C. In a dynamic photometric ellipsometer, one or more of the parameters P, C, A, and δ_C is periodically varied with time and the detected signal $\mathcal{I}_\text{D}$ is Fourier-analyzed.

3.10.2.1. Static photometric ellipsometers

In principle, no compensator is required for the determination of the ellipsometric parameters ψ and Δ by photometric methods. It is, therefore, sufficient to consider the polarizer-system(surface)-analyzer (PSA) ellipsometer arrangement. The detected signal $\mathscr{I}_D$ is a function of the azimuthal angles P of the polarizer and A of the analyzer that can be readily obtained by setting $C = 0$ and $T_C e^{j\delta_c} = 1$ in eq. (3.282)

$$\mathscr{I}_D(P, A) = G|V_{ey}|^2|\tan \psi \, e^{j\Delta} \cos A \cos P + \sin A \sin P|^2. \tag{3.283}$$

After some manipulations, eq. (3.283) becomes

$$\mathscr{I}_D(P, A) = F'[1 - \cos 2\psi \,(\cos 2A + \cos 2P) + \cos 2A \cos 2P \\ + \sin 2\psi \cos \Delta \sin 2A \sin 2P]. \tag{3.284}$$

Let $\mathscr{I}_{D1}$, $\mathscr{I}_{D2}$, and $\mathscr{I}_{D3}$ represent the detected signals at three different sets (P_1, A_1), (P_2, A_2), and (P_3, A_3) of polarizer-analyzer azimuth angles. Substitution of this data into eq. (3.284) produces three equations in F', ψ, and Δ. Division of two of these equations by the third eliminates F' and gives two equations that can be solved for the two ellipsometric parameters ψ and Δ.

From eq. (3.284), it is evident that a minimum of three data sets $\mathscr{I}_{Di}, (P_i, A_i)$, $i = 1, 2, 3$ is required. Additional measurements would be redundant, but may be necessary for accurate determination of ψ and Δ in the presence of imperfections in the ellipsometer.

We have outlined above a general routine of making measurements of ψ and Δ from photometric data using a PSA ellipsometer. The choice of settings (P, A) for the polarizer and analyzer azimuth angles has been left arbitrary. One choice that simplifies data reduction is to set the polarizer at a fixed azimuth of $+\tfrac{1}{4}\pi$ and record the signal $\mathscr{I}_D$ at the three analyzer settings $-\tfrac{1}{4}\pi, 0$, and $\tfrac{1}{4}\pi$. From eq. (3.284), we obtain

$$\mathscr{I}_{D1} = \mathscr{I}_D(\tfrac{1}{4}\pi, -\tfrac{1}{4}\pi) = F'(1 - \sin 2\psi \cos \Delta), \tag{3.285a}$$

$$\mathscr{I}_{D2} = \mathscr{I}_D(\tfrac{1}{4}\pi, 0) = F'(1 - \cos 2\psi), \tag{3.285b}$$

$$\mathscr{I}_{D3} = \mathscr{I}_D(\tfrac{1}{4}\pi, \tfrac{1}{4}\pi) = F'(1 + \sin 2\psi \cos \Delta). \tag{3.285c}$$

Equations (3.285a) and (3.285c) can be combined to produce two simpler equations

$$\mathscr{I}_{D1} + \mathscr{I}_{D3} = 2F', \tag{3.286a}$$

$$\mathscr{I}_{D3} - \mathscr{I}_{D1} = 2F' \sin 2\psi \cos \Delta. \tag{3.286b}$$

Equations (3.285b), (3.286a), and (3.286b) are readily solved for ψ and Δ in terms of $\mathscr{I}_{D1}$, $\mathscr{I}_{D2}$, and $\mathscr{I}_{D3}$

$$\psi = \tfrac{1}{2} \arccos [(\mathscr{I}_{D1} - 2\mathscr{I}_{D2} + \mathscr{I}_{D3})/(\mathscr{I}_{D1} + \mathscr{I}_{D3})], \tag{3.287a}$$

$$\Delta = \arccos\left[(\csc 2\psi)(\mathcal{I}_{D3} - \mathcal{I}_{D1})/(\mathcal{I}_{D1} + \mathcal{I}_{D3})\right]. \tag{3.287b}$$

Other choices of the settings (P, A) may be dictated by accuracy and/or precision considerations. These considerations may also suggest that a compensator be added, in which case eq. (3.282) can be used to relate the measured photometric signal $\mathcal{I}_D$ to ψ and Δ. For further discussion of photometric methods see refs. 8, 110, and 136.

3.10.2.2. Dynamic photometric ellipsometers

In a dynamic photometric ellipsometer one or more optical parameter (azimuth angle or relative retardation) is modulated and the detector signal $\mathcal{I}_D$ is Fourier-analyzed. There is obviously a large number of possibilities dependent on which parameter, or combination of parameters is chosen for modulation. We limit ourselves to two such ellipsometer systems that have received more attention recently. These are the rotating-analyzer ellipsometer (RAE) [111–115] and the polarization-modulated ellipsometer (PME) [116, 117].

(1) The Rotating-Analyzer Ellipsometer (RAE)

The arrangements of the optical components of the rotating-analyzer ellipsometer are the same as in a null ellipsometer. Either the PCSA (polarizer-compensator-system-analyzer) or PSCA (polarizer-system-compensator-analyzer) sequences can be used. The compensator C is not essential, in this case, so that the PSA sequence may also be employed. With the other elements (P, C) set at fixed azimuths, the analyzer A is synchronously rotated around the beam axis at constant angular speed ω and the detector signal $\mathcal{I}_D$ is Fourier analyzed. For definiteness, we consider in more detail the PCSA RAE shown in fig. 3.22.

Let χ_{A0} be the complex number that represents the polarization state passed by the analyzer A at some reference-azimuth setting around the beam axis. For generality, χ_{A0} is assumed to be arbitrary so that the analyzer may be elliptic. When the analyzer is rotated from its reference position by an angle A in a counter-clockwise sense looking into the beam, the transmitted

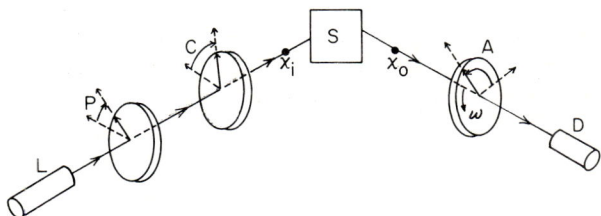

Fig. 3.22. A rotating-analyzer ellipsometer (RAE) consists of the light source L, polarizer P, compensator C, system S, analyzer A rotating at angular speed ω, and detector D.

polarization χ_A in this new orientation is related to χ_{A0} by

$$\chi_A = \frac{\tan A + \chi_{A0}}{1 - \tan A \chi_{A0}}, \qquad (3.288)$$

which is a bilinear transformation that expresses the effect of the rotation A [see eq. (1.105), §1.7].

If I_o denotes the intensity of the light beam at the output of the optical system S that is polarized in the state χ_o, the proportion I_D of I_o passed by the analyzer A and detected by the photodetector D is a function of χ_o and χ_A given by [see eq. (2.149), §2.8]

$$I_D = kI_o \frac{\chi_o \chi_o^* \chi_A \chi_A^* + \chi_o \chi_A^* + \chi_o^* \chi_A + 1}{\chi_o \chi_o^* \chi_A \chi_A^* + \chi_o \chi_o^* + \chi_A \chi_A^* + 1}, \qquad (3.289)$$

where the superscript asterisk indicates the complex conjugate and k represents the maximum transmittance of the analyzer when $\chi_o = \chi_A$.

In the rotating-analyzer ellipsometer (RAE), χ_o is determined from the variation with azimuth of the detected signal $\mathscr{I}_D$ after the rotating analyzer A. If we substitute χ_A from eq. (3.288) into eq. (3.289), we obtain

$$\mathscr{I}_D = \bar{\mathscr{I}}[1 + \alpha \cos 2A + \beta \sin 2A], \qquad (3.290)$$

where

$$\bar{\mathscr{I}} = \tfrac{1}{2} k' I_o \frac{(1 + |\chi_o|^2)(1 + |\chi_{A0}|^2) + 4\mathrm{Im}\,(\chi_o)\,\mathrm{Im}\,(\chi_{A0})}{(1 + |\chi_o|^2)(1 + |\chi_{A0}|^2)} \qquad (3.291)$$

and

$$\alpha = \frac{(1 - |\chi_o|^2)(1 - |\chi_{A0}|^2) + 4\mathrm{Re}\,(\chi_o)\,\mathrm{Re}\,(\chi_{A0})}{(1 + |\chi_o|^2)(1 + |\chi_{A0}|^2) + 4\mathrm{Im}\,(\chi_o)\,\mathrm{Im}\,(\chi_{A0})} \qquad (3.292a)$$

$$\beta = \frac{2\mathrm{Re}\,(\chi_o)(1 - |\chi_{A0}|^2) - 2\mathrm{Re}\,(\chi_{A0})(1 - |\chi_o|^2)}{(1 + |\chi_o|^2)(1 + |\chi_{A0}|^2) + 4\mathrm{Im}\,(\chi_o)\,\mathrm{Im}\,(\chi_{A0})}. \qquad (3.292b)$$

In eqs. (3.290)–(3.292), $\bar{\mathscr{I}}$ represents the average detected signal over one full (or half) rotation of the analyzer, k' is a constant factor, and α and β represent the normalized cosine and sine Fourier coefficients of the signal $\mathscr{I}_D(A)$. By solving eqs. (3.292a) and (3.292b), χ_o can be reconstructed from the normalized Fourier coefficients α and β as well as the polarization state χ_{A0} transmitted by the analyzer in its reference position.

For simplicity, consider the case when the analyzer is linear, as often it is. If we take the reference position of the analyzer to be that in which its transmission axis is parallel to the x linear eigenpolarization of the optical system under measurement, then

$$\chi_{A0} = 0. \qquad (3.293)$$

Substitution of eq. (3.293) into eqs. (3.292a) and (3.292b) simplifies them considerably

$$\alpha = \frac{1 - |\chi_o|^2}{1 + |\chi_o|^2}, \qquad (3.294a)$$

$$\beta = \frac{2\text{Re}(\chi_o)}{1 + |\chi_o|^2}. \qquad (3.294b)$$

Equations (3.294a) and (3.294b) can be solved for $|\chi_o|$ and $\text{Re}(\chi_o)$

$$|\chi_o| = \left(\frac{1-\alpha}{1+\alpha}\right)^{\frac{1}{2}}, \qquad (3.295a)$$

$$\text{Re}(\chi_o) = \frac{\beta}{1+\alpha}. \qquad (3.295b)$$

Finally, from $|\chi_o|$ and $\text{Re}(\chi_o)$, the output polarization χ_o is given by the identity

$$\chi_o = \text{Re}(\chi_o) \pm j\{|\chi_o|^2 - [\text{Re}(\chi_o)]^2\}^{\frac{1}{2}},$$

so that

$$\chi_o = \frac{1}{1+\alpha}[\beta \pm j(1 - \alpha^2 - \beta^2)^{\frac{1}{2}}], \qquad (3.296)$$

if eqs. (3.295) are used.

Equation (3.296) shows that two output polarizations χ_o and χ_o^* lead to the same normalized Fourier coefficients α and β. If (θ_o, ϵ_o) represent the azimuth and ellipticity angle of χ_o, then $(\theta_o, -\epsilon_o)$ are the azimuth and ellipticity angle of χ_o^* (see fig. 1.19, ch. 1). This means that the handedness (sign of ϵ) is indeterminate by this polarization-detection technique that employs a rotating linear analyzer, as may be intuitively expected.

The state of polarization χ_i at the input of the optical system S is determined by the fixed azimuth settings P of the polarizer and C of the compensator (and the latter's relative transmittance ρ_C) as

$$\chi_i = \frac{\tan C + \rho_C \tan(P - C)}{1 - \rho_C \tan C \tan(P - C)}. \qquad (3.13)$$

The ratio $\tan \psi \, e^{j\Delta}$ of the eigenvalues of the optical system S (the p and s reflection coefficients of a surface) is given by

$$\tan \psi \, e^{j\Delta} = \chi_i / \chi_o,$$

or

$$\tan \psi \, e^{j\Delta} = \frac{(1+\alpha)}{[\beta \pm j(1-\alpha^2-\beta^2)^{\frac{1}{2}}]} \times \frac{\tan C + \rho_C \tan(P-C)}{1 - \rho_C \tan C \tan(P-C)} \quad (3.297)$$

if eqs. (3.296) and (3.13) are substituted.

Equation (3.297) is the main working equation of the rotating analyzer ellipsometer (RAE). It gives the ratio of the complex eigenvalues associated with the orthogonal x and y linear eigenpolarizations of the optical system S under measurement in terms of (1) the azimuth angles P and C of the polarizer and compensator, (2) the slow-to-fast complex relative transmittance ρ_C of the compensator, and (3) the normalized Fourier coefficients α and β of the detected photoelectric current after the rotating analyzer A. Detailed discussion of the accuracy and precision of rotating-analyzer ellipsometers are adequately covered in several recent publications [115, 118–120].

(2) The Polarization-Modulated Ellipsometer (PME)

A polarization-modulated ellipsometer (PME) is one in which the state of polarization of the ellipsometer light beam at a suitable point along its path is modulated in a prescribed fashion so that information on the optical system under measurement is retrievable from harmonic analysis of the resulting time-varying detected photoelectric current. Again, several possibilities exist, dependent on where and how the modulation is applied. A convenient arrangement is that proposed by Jasperson et al. [116, 117]. The sequence of optical components, Fig. 3.23, includes a linear polarizer P, a modulator M,

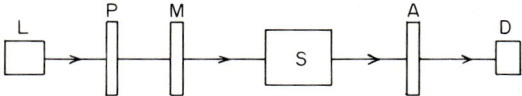

Fig. 3.23. Jasperson et al.'s polarization modulated ellipsometer (PME) consists of a light source L, polarizer P, polarization modulator M, optical system under measurement S, analyzer A, and photodetector D. After refs. 116, 117.

the optical system under measurement S and a linear analyzer A. This PMSA ellipsometer arrangement may be considered the same as a conventional PCSA arrangement in which the compensator's relative retardation δ_C is periodically modulated as a function of time. In a PME all optical components remain stationary. This freedom of mechanical movement is a distinct advantage and allows very high speed of measurement.

The output signal of the photodetector for any general orientation of the optical components of the PME is given by eq. (3.282). In this equation, C

represents the azimuth of the modulator's fast axis[36] and $T_C e^{j\delta_C}$ is its slow-to-fast complex relative transmittance. When the relative phase shift δ_C is sinusoidally modulated at a frequency ω, we may write

$$\delta_C = \mathscr{A} \sin \omega t, \tag{3.298}$$

where the amplitude $\mathscr{A}$ is a function of the modulator's driving voltage and the wavelength of light.

Because the orientation of the different optical components of a PME may be chosen at will, such a choice can be made to lead to simple relationships between ψ and Δ and the Fourier components of the photoelectric current. For example, Jasperson et al. use an arrangement in which $P = 90°$, $C = +45°$, and $A = -45°$, where all azimuths are measured from the direction of the x linear eigenpolarization of the optical system S under measurement. Substitution of the above values of P, C, and A, together with $T_C = 1$ (a condition that is very closely met in practice), into eq. (3.282) gives

$$\mathscr{I}_D = \tfrac{1}{8} F |\tan \psi \, e^{j\Delta} (1 - e^{j\delta_C}) - (1 + e^{j\delta_C})|^2,$$

which, after some manipulations, reduces to

$$\mathscr{I}_D = \tfrac{1}{4} F[(1 + \tan^2 \psi) + (1 - \tan^2 \psi) \cos \delta_C - 2 \tan \psi \sin \Delta \sin \delta_C]. \tag{3.299}$$

When δ_C varies sinusoidally with time, $\delta_C = \mathscr{A} \cos \omega t$ [eq. (3.298)], the cosine and sine functions in eq. (3.299) can be expanded in their Bessel-function series [121] as follows

$$\begin{aligned}
\cos \delta_C &= \cos [\mathscr{A} \sin (\omega t)] \\
&= J_0(\mathscr{A}) + 2 \sum_{m=1}^{\infty} J_{2m}(\mathscr{A}) \cos (2m\omega t), \\
\sin \delta_C &= \sin [\mathscr{A} \sin (\omega t)] \\
&= 2 \sum_{m=0}^{\infty} J_{2m+1}(\mathscr{A}) \sin [2(m+1)\omega t].
\end{aligned} \tag{3.300}$$

In eqs. (3.300), $J_m(\mathscr{A})$ is the Bessel function of the first kind of order m and argument $\mathscr{A}$. If the voltage applied to the modulator is adjusted such that

$$J_0(\mathscr{A}) = 0,$$

the detected signal $\mathscr{I}_D$, eq. (3.299), becomes

$$\mathscr{I}_D = \mathscr{I}_{dc} - \mathscr{I}_\omega \sin \omega t + \mathscr{I}_{2\omega} \cos \omega t + \cdots \tag{3.301}$$

---

[36]Because the relative retardation alternates sinusoidally with time, the fast axis coincides with the stress direction during the negative half-cycles of the relative retardation δ_C and becomes orthogonal to it during the positive half-cycles. For simplicity, we retain the notation used in static ellipsometry and call the stress axis the fast axis independent of time. This does not lead to any problem as far as the mathematical development is concerned.

where

$$\mathcal{I}_{dc} = \tfrac{1}{4}F(1 + \tan^2 \psi),$$
$$\mathcal{I}_{\omega} = \tfrac{1}{4}F(2 \tan \psi \sin \Delta)[2J_1(\mathcal{A})], \quad (3.302)$$
$$\mathcal{I}_{2\omega} = \tfrac{1}{4}F(1 - \tan^2 \psi)[2J_2(\mathcal{A})].$$

The fundamental (ω) and second-harmonic (2ω) components of the detected photoelectric current can be isolated and measured, and their ratios to the dc component

$$R_{\omega} = \mathcal{I}_{\omega}/\mathcal{I}_{dc} = 2 \sin 2\psi \sin \Delta J_1(\mathcal{A}),$$
$$R_{2\omega} = \mathcal{I}_{2\omega}/\mathcal{I}_{dc} = 2 \cos 2\psi J_2(\mathcal{A}), \quad (3.303)$$

can be determined.

Jasperson et al. [117] suggest a simple calibration procedure to avoid the need for calculating the Bessel functions. By placing a circular analyzer between the system S and modulator M (see fig. 3.23), it can be shown that the ratio of the fundamental component to the dc component becomes

$$R_{\omega}(\text{cal}) = 2J_1(\mathcal{A}). \quad (3.304a)$$

If a linear analyzer instead of a circular analyzer is placed between S and M, with its transmission axis parallel to the x or y linear eigenpolarization of S, the ratio of the second harmonic and dc components becomes

$$R_{2\omega}(\text{cal}) = 2J_2(\mathcal{A}). \quad (3.304b)$$

From eqs. (3.303) and (3.304), we obtain

$$R_{\omega}/R_{\omega}(\text{cal}) = \sin 2\psi \sin \Delta, \quad (3.305a)$$
$$R_{2\omega}/R_{2\omega}(\text{cal}) = \cos 2\psi. \quad (3.305b)$$

Equations (3.305) show how ψ and Δ can be measured by a polarization-modulated ellipsometer (PME) when it is operated in the mode suggested by Jasperson et al. [117]. Notice that ψ is first obtained from eq. (3.305b), then Δ from eq. (3.305a).

A study of the effect of systematic errors on the operation of the above PME has been made by O'Handley [122]. However, no detailed study of the precision of such an instrument has been conducted.

3.10.3. Interferometric Ellipsometers (IE)

Hazebroek and Holscher [123] have recently proposed an interferometric technique for the determination of the ratio of the p and s complex-amplitude reflection coefficients of a surface. Following ref. [123], the operation of such an interferometric ellipsometer is explained with the help

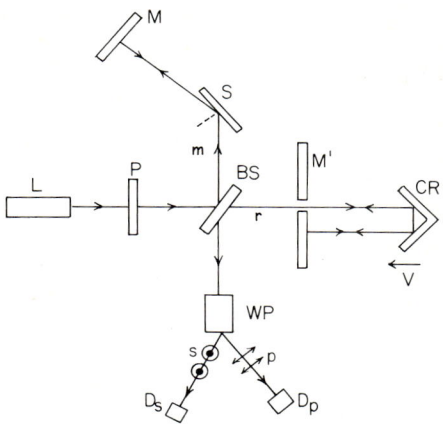

Fig. 3.24. Hazebroek and Holscher's interferometric ellipsometer consists of a light source L, polarizer P, beam splitter BS, surface S under measurement, Wollaston prism WP, return mirror M and retroreflecting cavity composed of the corner reflector CR and mirror M'. D_p and D_s are detectors that measure the intensity of the component polarizations parallel (p) and perpendicular (s) to the plane of incidence on the surface S that are separated by Wollaston prism WP. The corner reflector CR is scanned at constant linear speed v. After ref. 123.

of fig. 3.24. The arrangement is essentially that of a Michelson interferometer. L is a collimated light source (preferably a laser), P is a linear polarizer oriented at an oblique azimuth with respect to the common plane of incidence on the beam splitter BS and the test surface S. CR is a corner reflector, M and M' are return mirrors, and D_p and D_s are photodetectors that measure the intensities of the parallel (p) and perpendicular (s) components of light after they are separated by the Wollaston prism WP.

Let E_{0p} and E_{0s} be the p and s components of the electric vector of the light beam at the output of the linear polarizer P. The effect of the beam splitter (BS) on these two polarizations can be described by two complex factors $r_{BS}^p \exp(j\delta_{BS}^p)$ and $r_{BS}^s \exp(j\delta_{BS}^s)$, where $r_{BS}^{p,s}$ and $\delta_{BS}^{p,s}$ represent the p, s amplitude attenuations and phase shifts. After the beam splitter BS, the light beam is divided into two beams along each of the two arms of the interferometer, a measuring beam m which strikes the test surface S under measurement, and a reference beam r which is retroreflected by the corner reflector (CR)–mirror (M') combination. By choosing the polarization of the forward-travelling reference beam to coincide with one of the two eigenpolarizations of the CR–M' cavity, the backward-travelling (retroreflected) reference beam will have the same polarization as the forward-travelling reference beam. There is only an isotropic change (same for p and s) in amplitude and phase caused by the CR–M' cavity which can be described by

$r_{CRM'} \exp(j\delta_{CRM'})$. The effect of a single reflection at the surface S can be described by the complex amplitude-reflection coefficients $r_S^p \exp(j\delta_S^p)$ and $r_S^s \exp(j\delta_S^s)$. Normal incidence on the mirror M causes isotropic changes in amplitude and phase that are described by $r_M \exp(j\delta_M)$. Upon recombination by the beam splitter BS, the complex amplitudes of the measuring (m) and reference (r) component beams parallel (p) and perpendicular (s) to the plane of incidence become

$$\begin{aligned} E_{mp} &= E_{0p} r_{BS}^p (r_S^p)^2 r_M \exp[j(\delta_{BS}^p + 2\delta_S^p + \delta_M)], \\ E_{ms} &= E_{0s} r_{BS}^s (r_S^s)^2 r_M \exp[j(\delta_{BS}^s + 2\delta_S^s + \delta_M)], \\ E_{rp} &= E_{0p} r_{BS}^p r_{CRM'} \exp[j(\delta_{BS}^p + \delta_{CRM'} + \delta)], \\ E_{rs} &= E_{0s} r_{BS}^s r_{CRM'} \exp[j(\delta_{BS}^s + \delta_{CRM'} + \delta)], \end{aligned} \quad (3.306)$$

where δ is the difference in phase that corresponds to the difference in the paths traversed by the reference and measuring beams.

Because the Wollaston prism WP separates the p and s components, the observable signals at the detectors D_p and D_s are

$$\begin{aligned} \mathcal{I}_{Dp} &= K_p |E_{mp} + E_{rp}|^2, \\ \mathcal{I}_{Ds} &= K_s |E_{ms} + E_{rs}|^2, \end{aligned} \quad (3.307)$$

where K_p and K_s are constants. For simplicity, we assume that the effect of the mirror M and of the CR–M' cavity can be made identical, so that

$$r_M = r_{CRM'}, \qquad \delta_M = \delta_{CRM'}. \quad (3.308)$$

Substituting eqs. (3.306) into eqs. (3.307) and making use of the simplifying assumption of eqs. (3.308), we obtain

$$\begin{aligned} \mathcal{I}_{Dp} &= C_p \{\tfrac{1}{2}[1 + (r_S^p)^4] + (r_S^p)^2 \cos(2\delta_S^p - \delta)\}, \\ \mathcal{I}_{Ds} &= C_s \{\tfrac{1}{2}[1 + (r_S^s)^4] + (r_S^s)^2 \cos(2\delta_S^s - \delta)\} \end{aligned} \quad (3.309)$$

where C_p and C_s are constants.

If the corner reflector CR is translated at a constant linear speed v, the path phase difference δ that appears in eqs. (3.309) can be written as

$$\begin{aligned} \delta &= 2\left(\frac{vt}{\lambda_a}\right) \times 2\pi, \\ &= 4\omega_0 vt \frac{n_a}{c}, \end{aligned} \quad (3.310)$$

where λ_a is the wavelength of light in air, n_a is the air refractive index, ω_0 and c are the angular frequency and free-space speed of electromagnetic radiation, respectively. For simplicity, we rewrite eq. (3.310) as

$$\delta = \omega t, \quad (3.311)$$

where

$$\omega = 4n_a \left(\frac{v}{c}\right) \omega_0. \tag{3.312}$$

By substituting $\delta = \omega t$ in eqs. (3.309), it becomes evident that the translation of the corner reflector generates ac signals of frequency ω in the two photodetectors D_p and D_s.

$$\begin{aligned}\mathcal{I}_{\omega D_p} &= C_p(r_S^p)^2 \cos[\omega t - 2\delta_S^p], \\ \mathcal{I}_{\omega D_s} &= C_s(r_S^s)^2 \cos[\omega t - 2\delta_S^s],\end{aligned} \tag{3.313}$$

whose phase difference gives the ellipsometric parameter Δ,

$$2(\delta_S^p - \delta_S^s) = 2\Delta, \tag{3.314}$$

and whose amplitude ratio gives the ellipsometric parameter ψ

$$(C_p/C_s)[(r_S^p)^2/(r_S^s)^2] = (C_p/C_s)\tan^2\psi. \tag{3.315}$$

The factor C_p/C_s that appears in eq. (3.315) can be made equal to one or can be measured accurately by blocking the path to S using a mirror placed for normal incidence to the measuring beam so that $\tan\psi = 1$, and adjusting the electronics associated with the detectors D_p and D_s.

By designing a suitable driver for the corner reflector, Hazebroek and Holscher managed to successfully operate such an interferometric ellipsometer. Feeding the signals $\mathcal{I}_{\omega D_p}$ and $\mathcal{I}_{\omega D_s}$ [eqs. (3.313)] to the X and Y deflecting plates of a cathode-ray tube, they provided a direct display of the ellipse of polarization. This direct display of the ellipse of polarization is rather interesting and quite useful. In addition, the interferometric ellipsometer (IE) has the advantage of not requiring a compensator, hence wide-band operation is possible. By placing M and CR–M' inside an ultrahigh vacuum system, it is possible to avoid the errors caused by window birefringence. The speed of measurement of the IE is higher than that realizable with automatic mechanically-nulled ellipsometers.

3.11. Modulated ellipsometry

The term "modulated ellipsometry" was coined by Buckman and Bashara [124–126] to denote the use of an ordinary ellipsometer to measure small changes in the optical parameters of a surface that are induced by the application of an external field (electrical, mechanical, etc). Here we assume a surface S with linear eigenpolarizations parallel (p) and perpendicular (s) to the plane of incidence, with associated eigenvalues (reflection coefficients) R_p and R_s. (The discussion generally applies to any optical system with orthogonal linear eigenpolarizations.) Under the action of a modulating field, small changes $\delta\psi$ of ψ and $\delta\Delta$ of Δ (where $\tan\psi\ e^{j\Delta} = R_p/R_s$) take place and the objective of modulated ellipsometry is to measure these changes. In principle, if $\delta\psi$ and $\delta\Delta$ are transient (aperiodic), they can be followed by a fast automatically nulled ellipsometer, a rotating-analyzer ellipsometer, or a polarization-modulated ellipsometer (§3.10). However, the applications of modulated ellipsometry (band-structure studies of solids and thin films) usually involve periodic changes $\delta\psi$ and $\delta\Delta$ that are induced by periodic modulating fields. An ellipsometer with fixed (non-moving) optical elements is used in which $\delta\psi$ and $\delta\Delta$ are retrieved from the periodic photodetector signal. The usual PCSA (polarizer-compensator-surface-analyzer) or PSCA ellipsometer arrangements can be used. Notice that because $\delta\psi$ and $\delta\Delta$ are derived from photometric signals, rather than from component null settings, the compensator is not essential and can be removed which gives rise to the PSA arrangement.

3.11.1. *PSA arrangement*

We first consider the simple PSA ellipsometer arrangement. The expression for the detected signal in this case is given by eq. (3.284). By further assuming that the polarizer is oriented with its transmission axis at 45° to the plane of incidence, we can substitute

$$\cos 2P = 0, \quad \sin 2P = 1,$$

into eq. (3.284); this gives

$$\mathscr{I}_D = \text{const } \mathscr{R}_s \sec^2 \psi [1 - \cos 2\psi \cos 2A + \sin 2\psi \cos \Delta \sin 2A], \quad (3.316)$$

where

$$\mathscr{R}_s = |R_s|^2 = |V_{ey}|^2,$$

is the reflectance of the surface for s-polarized light. At a given analyzer azimuth A, with the surface perturbation (modulation) applied, the fractional change $\delta\mathscr{I}_D/\mathscr{I}_D$ of the detector signal $\mathscr{I}_D$ can be related to the induced changes $\delta\mathscr{R}_s/\mathscr{R}_s$, $\delta\psi$ and $\delta\Delta$ in the surface parameters $\mathscr{R}_s$, ψ and Δ, respectively. Thus, by taking the logarithmic differential of eq. (3.316), we obtain

$$\delta\mathscr{I}_D/\mathscr{I}_D = \delta\mathscr{R}_s/\mathscr{R}_s + \alpha_1\delta\psi + \alpha_2\delta\Delta \quad (3.317)$$

where the coefficients α_1 and α_2 are given by

$$\alpha_1 = \frac{2[\tan\bar{\psi}(1+\cos 2A)+\cos\bar{\Delta}\sin 2A]}{[1-\cos 2\bar{\psi}\cos 2A+\sin 2\bar{\psi}\cos\bar{\Delta}\sin 2A]},$$

$$\alpha_2 = \frac{-\sin 2\bar{\psi}\sin\bar{\Delta}\sin 2A}{[1-\cos 2\bar{\psi}\cos 2A+\sin 2\bar{\psi}\cos\bar{\Delta}\sin 2A]}. \quad (3.318)$$

By setting the analyzer A at N different azimuths $A_1, A_2, A_3 \ldots A_N$ and measuring $\delta\mathcal{I}_D/\mathcal{I}_D$ at each azimuth, we get a set of N different equations from eq. (3.317) where α_1 and α_2 are calculated from eqs. (3.318). In eqs. (3.318), $\bar{\psi}$ and $\bar{\Delta}$ represent the unmodulated quiescent values of ψ and Δ. If the analyzer is set at three different azimuths ($N = 3$), three equations are generated by eq. (3.317) that can be solved for the three independent changes of interest $\delta\mathcal{R}_s/\mathcal{R}_s$, $\delta\psi$ and $\delta\Delta$. More equations ($N > 3$) lead to an overdetermined set of equations which may be desirable for higher precision in the determination of $\delta\mathcal{R}_s/\mathcal{R}_s$, $\delta\psi$ and $\delta\Delta$ in the presence of error in the measurement scheme.

Because $\mathcal{R}_s$, the s reflectance, $\mathcal{R}_p$, the p reflectance, and ψ are interrelated by

$$\tan\psi = (\mathcal{R}_p/\mathcal{R}_s)^{\frac{1}{2}}, \tag{3.319}$$

$\delta\mathcal{R}_s/\mathcal{R}_s$, $\delta\mathcal{R}_p/\mathcal{R}_p$ and $\delta\psi$ are interrelated by

$$\frac{\delta\mathcal{R}_p}{\mathcal{R}_p} = \frac{\delta\mathcal{R}_s}{\mathcal{R}_s} + \frac{4\delta\psi}{\sin 2\psi}, \tag{3.320}$$

as is obtained from the logarithmic differential of eq. (3.319). Equation (3.320) shows that $\delta\mathcal{R}_p/\mathcal{R}_p$ is determined once $\delta\mathcal{R}_s/\mathcal{R}_s$ and $\delta\psi$ are known.

Table 3.15. Gives the coefficients α_1, α_2 and $\delta\mathcal{I}_D/\mathcal{I}_D$ [eq. (3.317)] for special settings of the analyzer azimuth A in PSA modulated ellipsometry

α_1, α_2 \ $A \rightarrow$	0	$\pm\frac{1}{4}\pi$	$\frac{1}{2}\pi$
α_1	$\dfrac{4\tan\bar{\psi}}{(1-\cos 2\bar{\psi})}$	$\dfrac{2(\tan\bar{\psi} \pm \cos\bar{\Delta})}{(1 \pm \sin 2\bar{\psi}\cos\bar{\Delta})}$	0
α_2	0	$\dfrac{\mp\sin 2\bar{\psi}\sin\bar{\Delta}}{(1 \pm \sin 2\bar{\psi}\cos\bar{\Delta})}$	0
$\delta\mathcal{I}_D/\mathcal{I}_D$	$\delta\mathcal{R}_s/\mathcal{R}_s + \alpha_1\delta\psi$	$\delta\mathcal{R}_s/\mathcal{R}_s + \alpha_1\delta\psi + \alpha_2\delta\Delta$	$\delta\mathcal{R}_s/\mathcal{R}_s$

Table 3.15 gives expressions for the coefficients α_1, α_2, and $\delta\mathcal{I}_D/\mathcal{I}_D$ [see eq. (3.317)] as a function of the analyzer azimuth A, when the latter assumes the special values 0, $\pm\frac{1}{4}\pi$ and $\pm\frac{1}{2}\pi$.

3.11.2. PCSA arrangement

We assume that the compensator C acts as an exact quarter-wave retarder and that the azimuth of its fast axis is $+\frac{1}{4}\pi$ from the plane of incidence. In such case, the detected signal is given by eq. (3.48). The latter equation can be rewritten as

$$\mathcal{I}_D = \text{const } \mathcal{R}_s \sec^2 \psi [1 - \cos 2\psi \cos 2A + \sin 2\psi \sin (2P + \Delta) \sin 2A], \tag{3.321}$$

if the substitutions

$$T_S e^{j\delta_S} = \tan \psi \, e^{j\Delta}, \qquad |V_{ey}|^2 = \mathcal{R}_s,$$

are made.

With the polarizer P and analyzer A set at some arbitrary but fixed azimuths, application of the modulation to the surface leads to changes $\delta\psi$, $\delta\Delta$ and $\delta\mathcal{R}_s$ and a signal $\delta\mathcal{I}_D$ is detected by the photodetector. From the logarithmic differential of eq. (3.321), we obtain an expression of the form of eq. (3.317), where the coefficients α_1 and α_2 are now given by

$$\begin{aligned}
\alpha_1 &= \frac{2[\tan \bar{\psi}(1 + \cos 2A) - \sin 2A \sin (2P + \bar{\Delta})]}{[1 - \cos 2\bar{\psi} \cos 2A + \sin 2\bar{\psi} \sin (2P + \bar{\Delta}) \sin 2A]}, \\
\alpha_2 &= \frac{-\sin 2\bar{\psi} \cos (2P + \bar{\Delta}) \sin 2A}{[1 - \cos 2\bar{\psi} \cos 2A + \sin 2\bar{\psi} \sin (2P + \bar{\Delta}) \sin 2A]}.
\end{aligned} \tag{3.322}$$

In eqs. (3.222), $\bar{\psi}$ and $\bar{\Delta}$ are the unperturbed (quiescent) values of ψ and Δ and P and A are the polarizer and analyzer azimuths respectively.

By recording $\delta\mathcal{I}_D/\mathcal{I}_D$ at three different pairs (P_1, A_1), (P_2, A_2) and (P_3, A_3) of the polarizer and analyzer azimuths, we get three equations from eq. (3.317) [and (3.322)] that can be solved for the changes $\delta\mathcal{R}_s/\mathcal{R}_s$, $\delta\psi$ and $\delta\Delta$ in the surface properties due to the modulation. For higher precision, more than three measurements may be used.

The effect of using an inexact quarter-wave compensator has been studied [127] but a complete analysis of systematic errors has not been carried out. Choice of the azimuth-angle pair (P, A) leading to signals $\delta\mathcal{I}_D/\mathcal{I}_D$ that are (almost) proportional to either $\delta\Delta$ or $\delta\psi$ only have been investigated [128]. Because $\delta\mathcal{I}_D/\mathcal{I}_D$ is related to $\delta\psi$ and $\delta\Delta$ via the coefficients α_1 and α_2 [eq. (3.317)], which are themselves functions of P and A [eq. (3.322)], the extent to which given changes $\delta\psi$ and $\delta\Delta$ couple to a signal $\delta\mathcal{I}_D/\mathcal{I}_D$ is controlled by P and A. In general, it can be proven that higher sensitivity is realized when P and A are set near their null settings [128].

CHAPTER 4

Reflection and Transmission of Polarized Light by Stratified Planar Structures

4.1. Introduction

In order to interpret ellipsometer data taken when polarized light is reflected from or transmitted by bare or filmed substrates, the electromagnetic theory of light should be used to derive expressions for the complex amplitude reflection and transmission coefficients in terms of the macroscopic optical properties that characterize the specific structure under measurement.

A considerable amount of theoretical work has been done on this topic, a detailed discussion of which can be found in a number of books [129–131] and review articles [132–135]. Our primary interest here is to present results that are essential for the analysis of ellipsometric measurements on bare and filmed substrates.[1] We will consider the reflection and transmission of polarized light by structures of increasing degree of complexity. We start with the simple case of a planar interface between two homogeneous optically isotropic media (§4.2) and progress to systems that involve one (§4.3–§4.5) or more (§4.6) films. Subsequently, we consider the more-complicated case of stratified anisotropic structures (§4.7) that have been

---

[1] Determination of the optical properties of materials in bulk or thin-film phase from ellipsometric measurements as a function of wavelength (spectroscopic ellipsometry) usually requires that such data (dispersion spectra) be subsequently interpreted quantum mechanically (*e.g.*, by the use of the band theory of solids). This subject of considerable interest is not treated here. The reader may consult a number of excellent books, refs. [136–140], for an adequate coverage of this related topic.

treated recently using 4×4-matrix methods. We conclude with a brief review (§4.8) of the theory of light reflection from surfaces covered with discontinuous films and surfaces with rough boundaries.

The notation recommended by Muller and other participants at the 1968 International Conference on Ellipsometry held at the University of Nebraska [1, 2] will be used. In particular, *we choose the time dependence of all harmonic fields to be according to* $e^{j\omega t}$. The propagation of plane waves in an isotropic absorbing medium is described by the *complex index of refraction N* (also called the complex refractive index) which is written as

$$N = n - jk \tag{4.1}$$

where n is called the *index of refraction* and k the *extinction coefficient* of the medium.[2] (n and k may also be simply referred to as the real and imaginary parts of the complex index of refraction N.) The full expression for the electric vector of an optical plane wave travelling in the positive direction of the z-axis in an isotropic absorbing medium such that the planes of constant phase and those of constant amplitude of the wave are parallel is given by

$$\begin{aligned} E &= E_0 \, e^{j(\omega t + \delta)} \, e^{-j\omega N z/c} \\ &= E_0 \, e^{j(\omega t + \delta)} \, e^{-j\omega n z/c} \, e^{-\omega k z/c}, \end{aligned} \tag{4.2}$$

where δ is a constant phase angle, c is the free-space wave velocity, and E_0, which is in general complex, defines both the amplitude and polarization of the wave. From eq. (4.2), it is obvious that the wave velocity in the medium is c/n, and that the wave amplitude decays exponentially along the direction of propagation at a rate of $\omega k/c$ or $2\pi k/\lambda$ nepers/meter,[3] where λ is the free-space wavelength of light.

4.2. Reflection and refraction at the planar interface between two isotropic media

With reference to fig. 4.1, consider the oblique reflection and transmission of an optical plane wave at the planar interface between two semi-infinite homogeneous optically isotropic media 0 and 1 with complex indices of

---

[2] The complex index of refraction N is sometimes written as $N = n(1 - j\kappa)$, where κ (kappa) is called the *absorption index* [2]. Also, in discussing the oblique reflection and refraction of light by absorbing media, an angle-of-incidence-dependent index of refraction and extinction coefficient can be defined [see, for example, refs. 2, 131, 136]. Such notations and definitions will be avoided.

[3] The neper is a unit of attenuation commonly used in electrical communications engineering. An attenuation of one neper corresponds to a reduction of amplitude (of a travelling wave) by $1/e$.

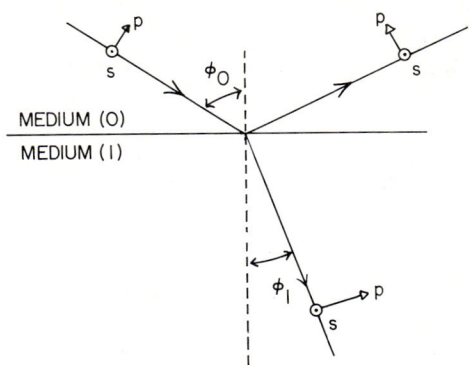

Fig. 4.1. Oblique reflection and transmission of a plane wave at the planar interface between two semi-infinite media 0 and 1. ϕ_0 and ϕ_1 are the angles of incidence and refraction, respectively. p and s are axes parallel and perpendicular to the plane of incidence, respectively. The vector product $\hat{p} \times \hat{s}$ of unit vectors along these axes is parallel to the direction of propagation.

refraction N_0 and N_1, respectively. It is assumed that the change in the index of refraction is abrupt (a step function) across the interface.

The wave incident from medium 0 gives rise to a reflected wave in the same medium and a transmitted (or refracted) wave in medium 1. The angle of incidence ϕ_0 and the angle of refraction ϕ_1 are both measured from the direction of the normal to the interface. The total fields inside media 0 and 1 obey Maxwell's equations and the boundary conditions (BC) at the interface. For the BC to be satisfied, (1) the directions of propagation of the incident, reflected and transmitted waves must all lie in one plane called the *plane of incidence*, perpendicular to the interface (this is the plane of the paper in fig. 4.1), (2) the angle of incidence must equal the angle of reflection, and (3) the angles of incidence ϕ_0 and refraction ϕ_1 must be related by

$$N_0 \sin \phi_0 = N_1 \sin \phi_1, \qquad (4.3)$$

which is *Snell's law*.

If media 0 and 1 are transparent, so that N_0 and N_1 are both real numbers, the angles ϕ_0 and ϕ_1 are also real and the above picture of how a plane wave is reflected and refracted at the interface between two media is simple. However, when either one or both media is absorbing, the angles ϕ_0 and ϕ_1 become, in general, complex and the discussion continues to hold only formally but the physical picture of the fields becomes complicated [141].

For a given amplitude and polarization of the incident wave, the amplitude and polarization of the reflected and transmitted waves can be determined from the continuity of the tangential components of the electric (E) and magnetic (H) field vectors across the interface. It can be shown that when the incident wave is linearly polarized with the electric vector vibrating parallel (p) to the plane of incidence, the reflected and transmitted waves are similarly polarized with their electric vectors also vibrating parallel to the plane of incidence. Likewise, when the incident wave is linearly polarized perpendicular (s) to the plane of incidence,[4] the reflected and transmitted waves are also linearly polarized perpendicular to the same plane. In other words, the *eigenpolarizations* of reflection and refraction are the linear vibrations parallel (p) and perpendicular (s) to the plane of incidence. It should be remembered, however, that this condition is true only when the two media are optically isotropic, as is presently assumed.

It is adequate to determine the amplitudes of the reflected and transmitted waves in terms of those of the incident wave for the p and s polarizations only. This is because an arbitrarily polarized incident wave can be resolved into its p and s components, each then is treated separately and the results are later combined.

Let (E_{ip}, E_{is}), (E_{rp}, E_{rs}) and (E_{tp}, E_{ts}) represent the complex amplitudes of the components of the electric vectors of the incident, reflected and transmitted waves, respectively, at opposite points immediately above and below the interface. Matching the tangential E and H fields across the interface leads to

$$\frac{E_{rp}}{E_{ip}} = r_p = \frac{N_1 \cos \phi_0 - N_0 \cos \phi_1}{N_1 \cos \phi_0 + N_0 \cos \phi_1}, \tag{4.4}$$

$$\frac{E_{rs}}{E_{is}} = r_s = \frac{N_0 \cos \phi_0 - N_1 \cos \phi_1}{N_0 \cos \phi_0 + N_1 \cos \phi_1}, \tag{4.5}$$

$$\frac{E_{tp}}{E_{ip}} = t_p = \frac{2 N_0 \cos \phi_0}{N_1 \cos \phi_0 + N_0 \cos \phi_1}, \tag{4.6}$$

$$\frac{E_{ts}}{E_{is}} = t_s = \frac{2 N_0 \cos \phi_0}{N_0 \cos \phi_0 + N_1 \cos \phi_1}, \tag{4.7}$$

which are the Fresnel complex-amplitude reflection (r) and transmission (t) coefficients for the p and s polarizations. Snell's law [eq. (4.3)] can be used to recast eqs. (4.4)–(4.7) in a form that depends only on the angles of incidence ϕ_0 and refraction ϕ_1[5]

$$r_p = \frac{\tan(\phi_0 - \phi_1)}{\tan(\phi_0 + \phi_1)}, \tag{4.8}$$

---

[4] The use of the letter s to indicate a component vibration perpendicular to the plane of incidence has its origin in the German word *senkrecht* meaning perpendicular.

[5] Alternatively, because $N_1 \cos \phi_1 = (N_1^2 - N_0^2 \sin^2 \phi_0)^{\frac{1}{2}}$ from Snell's law, the angle of refraction ϕ_1 may be deleted from eqs. (4.4)–(4.7) leading to expressions for the interface reflection and transmission coefficients that are functions of the angle of incidence ϕ_0 and the indices of refraction N_0 and N_1 only.

$$r_s = \frac{-\sin(\phi_0 - \phi_1)}{\sin(\phi_0 + \phi_1)}, \tag{4.9}$$

$$t_p = \frac{2 \sin\phi_1 \cos\phi_0}{\sin(\phi_0 + \phi_1)\cos(\phi_0 - \phi_1)}, \tag{4.10}$$

$$t_s = \frac{2 \sin\phi_1 \cos\phi_0}{\sin(\phi_0 + \phi_1)}. \tag{4.11}$$

It should be mentioned that eqs. (4.4)–(4.11) are based on choosing the directions of p and s so as to form *right-handed* Cartesian coordinate systems with the direction of propagation of the incident, reflected, and transmitted waves, fig. 4.1.

In order to examine the effect of reflection and refraction (or transmission) on the amplitude and phase of the wave *separately*, we write the complex Fresnel coefficients as

$$r_p = |r_p|\, e^{j\delta_{rp}}, \tag{4.12} \qquad t_p = |t_p|\, e^{j\delta_{tp}}, \tag{4.14}$$

$$r_s = |r_s|\, e^{j\delta_{rs}}, \tag{4.13} \qquad t_s = |t_s|\, e^{j\delta_{ts}}. \tag{4.15}$$

$|r_p|$ and $|t_p|$ give the ratios of the amplitudes of the (sinusoidal) vibrations of the electric vectors of the reflected and transmitted waves, respectively, to that of the incident wave when the latter is polarized parallel to the plane of incidence, with similar meanings for $|r_s|$ and $|t_s|$. δ_{rp} and δ_{tp} give the phase shifts upon reflection and refraction, respectively, experienced by the electric vibration parallel to the plane of incidence, with similar meanings for δ_{rs} and δ_{ts}.

When the incident wave is polarized in an arbitrary state (other than the p- or s-linear states), the polarization of the reflected and transmitted waves will be different from that of the incident wave. This is an immediate consequence of the fact that the Fresnel coefficients for the p and s components are different, so that the relative amplitude and phase relationships between these two components are changed upon reflection and refraction. It is obvious that this change of polarization will take place when either the absolute values or the angles (or both) of the Fresnel coefficients of reflection (r_p and r_s) or refraction (t_p or t_s) are different.

Reflection ellipsometry is a technique based on measurements of the states of polarization of the incident and reflected waves, leading to the determination of the ratio ρ of the complex Fresnel reflection coefficients for the p and s polarizations,

$$\rho = \frac{r_p}{r_s}. \tag{4.16}$$

It is often convenient to write ρ in the form

$$\rho = \tan \psi \, e^{j\Delta}. \tag{4.17}$$

From eqs. (4.16) and (4.17), it is readily seen that

$$\tan \psi = \frac{|r_p|}{|r_s|}, \tag{4.18}$$

$$\Delta = \delta_{rp} - \delta_{rs}. \tag{4.19}$$

Thus ψ and Δ determine the differential changes in amplitude and phase, respectively, experienced upon reflection by the component vibrations of the electric vector parallel and perpendicular to the plane of incidence.

If we substitute for r_p and r_s in eq. (4.16) their values from eqs. (4.4) and (4.5) and make use of Snell's law [eq. (4.3)], we get an equation that can be solved for N_1/N_0 in terms of ρ and ϕ_0 only, as follows

$$\frac{N_1}{N_0} = \sin \phi_0 \left[1 + \left(\frac{1-\rho}{1+\rho} \right)^2 \tan^2 \phi_0 \right]^{\frac{1}{2}} \tag{4.20a}$$

or in a slightly different but equivalent form

$$N_1 = N_0 \tan \phi_0 \left[1 - \frac{4\rho}{(1+\rho)^2} \sin^2 \phi_0 \right]^{\frac{1}{2}}. \tag{4.20b}$$

Equations (4.20) show that the complex index of refraction of medium 1 can be determined if the index of refraction of medium 0 (the medium of incidence) is known and the ellipsometric ratio ρ is measured at one angle of incidence ϕ_0.

In addition to polarization, other measurable quantities include the reflectances $\mathcal{R}_p$ and $\mathcal{R}_s$

$$\mathcal{R}_p = |r_p|^2, \tag{4.21}$$
$$\mathcal{R}_s = |r_s|^2, \tag{4.22}$$

which give the fraction of the total intensity of an incident plane wave that appears in the reflected wave for the p and s polarizations.[6]

Figures 4.2–4.10 show the variation with angle of incidence $\phi (= \phi_0)$ of the phase shifts δ_{rp} and δ_{rs} and the reflectances $\mathcal{R}_p$ and $\mathcal{R}_s$ when light is incident

---

[6]When medium 1 is sufficiently transparent (*e.g.*, a dielectric) and the refracted wave is accessible for measurements, the ratio of the complex-amplitude transmission coefficients $\rho = t_p/t_s$ can be measured ellipsometrically, while the intensity transmittances, $\tau_p = (n_1 \cos \phi_1 / n_0 \cos \phi_0)|t_p|^2$ and $\tau_s = (n_1 \cos \phi_1 / n_0 \cos \phi_0)|t_s|^2$, can be measured photometrically. The prefactors of $|t_p|^2$ and $|t_s|^2$ in the above expressions for the measurable transmittances τ_p and τ_s arise from the change in the cross-section of a finite beam upon refraction and from the difference of refractive index between the two media (see, for example, ref. 131, pp. 38–41).

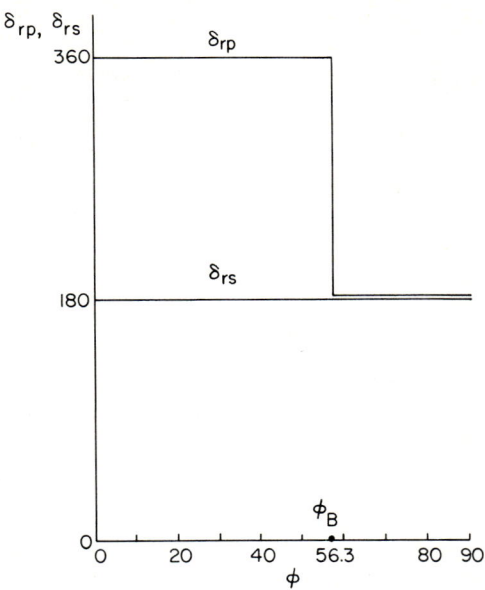

Fig. 4.2. The reflection phase shifts δ_{rp} and δ_{rs} for the p and s polarizations as functions of the angle of incidence ϕ (degrees) for an air/glass interface; $\lambda = 5461$ Å, $N_{\text{glass}} = 1.50$.

from vacuum (or air) onto the surface of a transparent dielectric, semiconductor, and metal. Also shown are the variations of the ellipsometric angles Δ and ψ with the angle of incidence under the same conditions.

When a p-polarized light wave is incident on the interface between two transparent media the reflected wave entirely disappears at a particular angle of incidence called the *Brewster angle* ϕ_B, and the incident wave is *totally refracted* into the second medium. It can be readily proved from eqs. (4.4) and (4.3) that ϕ_B at which $r_p = 0$ is given by

$$\tan \phi_B = (n_1/n_0), \tag{4.23}$$

where n_0 and n_1 are the indices of refraction of the two media. Figure 4.3 shows that the reflectance $\mathcal{R}_p$ for p-polarized light reaches a minimum value of zero at the Brewster angle, while fig. 4.2 shows that the phase shift δ_{rp} experiences an abrupt jump from $2\pi(\phi_0 < \phi_B)$ to $\pi(\phi_0 > \phi_B)$. If the incident plane wave is s-polarized (instead of p-polarized), the corresponding reflectance $\mathcal{R}_s$ increases monotonically (without going through a minimum at the Brewster angle), fig. 4.3; the phase shift δ_{rs} remains constant at π for all

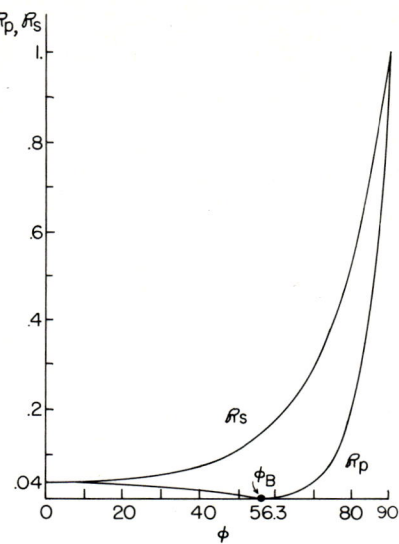

Fig. 4.3. The intensity reflectances $\mathscr{R}_p$ and $\mathscr{R}_s$ for the p and s polarizations as functions of the angle of incidence ϕ (degrees) for an air/glass interface; $\lambda = 5461$ Å, $N_{\text{glass}} = 1.50$.

angles of incidence,[7] fig. 4.2. The accompanying changes of the ellipsometric angles Δ and ψ are shown in fig. 4.4. In particular, note that Δ ($\delta_{rp} - \delta_{rs}$) drops discontinuously from a value of π just below the Brewster angle ϕ_B to zero just above it, while ψ reaches a minimum value of zero at ϕ_B. If unpolarized or partially polarized light is incident at the Brewster angle, the reflected light is *totally* s-polarized (perpendicular to the plane of incidence). For this reason, the term *polarizing angle* is sometimes used instead of the Brewster angle. The phenomenon of total polarization upon reflection at the Brewster angle is the basis of a class of infrared reflection polarizers (see, for example, ref. 142).

If an arbitrarily polarized optical plane wave is incident on a two-transparent-media interface from the side of high index of refraction, *total internal reflection* takes place at angles of incidence larger than a *critical*

---

[7] At normal incidence, the p and s polarizations are physically indistinguishable. However, with $\phi_0 = \phi_1 = 0$, and $n_1 > n_0$, eqs. (4.4) and (4.5) indicate that r_p and r_s have equal positive and negative real values respectively. Thus we get $\delta_{rp} = 2\pi$, and $\delta_{rs} = \pi$. This apparent contradiction is quickly resolved when it is recognized that at normal incidence the p directions in fig. 4.1 for the incident and reflected waves are opposite to one another, while the s directions are parallel.

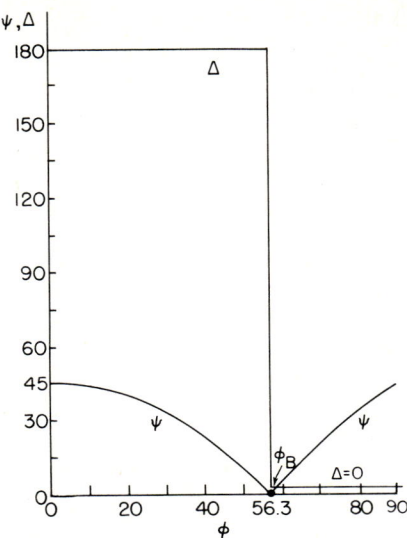

Fig. 4.4. The ellipsometric angles ψ ($= \tan^{-1}(\mathcal{R}_p/\mathcal{R}_s)^{\frac{1}{2}}$) and Δ ($= \delta_{rp} - \delta_{rs}$) as functions of the angle of incidence ϕ (degrees) for reflection at an air/glass interface; $\lambda = 5461$ Å, $N_{glass} = 1.50$.

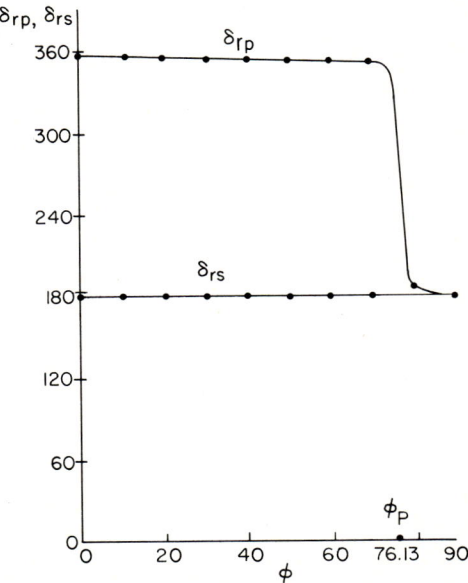

Fig. 4.5. The reflection phase shifts δ_{rp} and δ_{rs} for the p and s polarizations as functions of the angle of incidence ϕ (degrees) for an air/silicon interface; $\lambda = 5461$ Å, $N_{Si} = 4.05 - j0.028$.

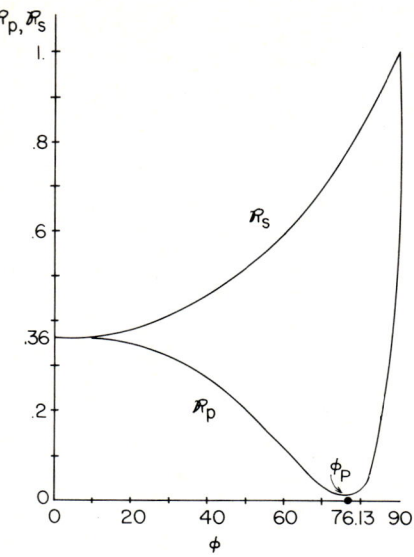

Fig. 4.6. The intensity reflectances $\mathcal{R}_p$ and $\mathcal{R}_s$ for the p and s polarizations as functions of the angle of incidence ϕ (degrees) for an air/silicon interface; $\lambda = 5461$ Å, $N_{Si} = 4.05 - j0.028$.

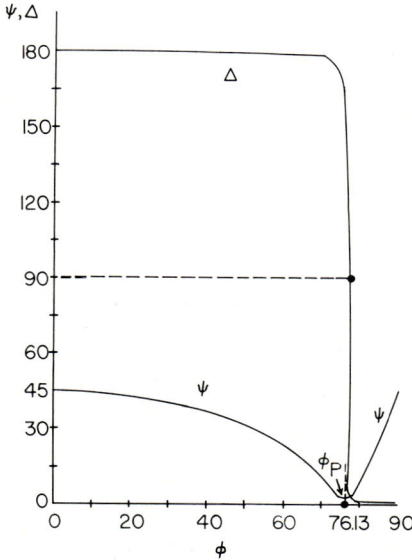

Fig. 4.7. The ellipsometric angles ψ ($= \tan^{-1}(\mathcal{R}_p/\mathcal{R}_s)^{\frac{1}{2}}$) and Δ ($= \delta_{rp} - \delta_{rs}$) as functions of the angle of incidence ϕ (degrees) for reflection at an air/silicon interface; $\lambda = 5461$ Å, $N_{Si} = 4.05 - j0.028$.

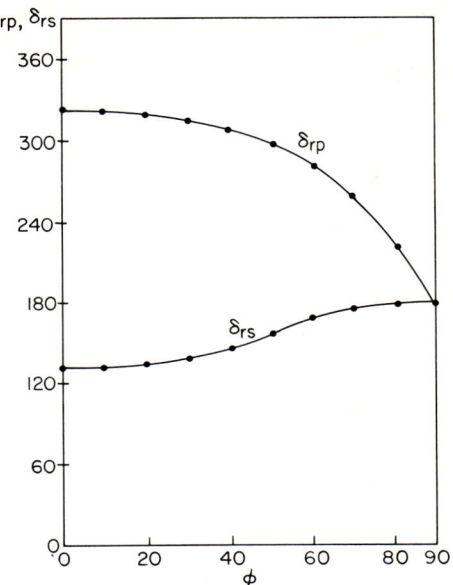

Fig. 4.8. The reflection phase shifts δ_{rp} and δ_{rs} for the p and s polarizations as functions of the angle of incidence ϕ (degrees) for an air/gold interface; $\lambda = 5461$ Å, $N_{Au} = 0.35 - j2.45$.

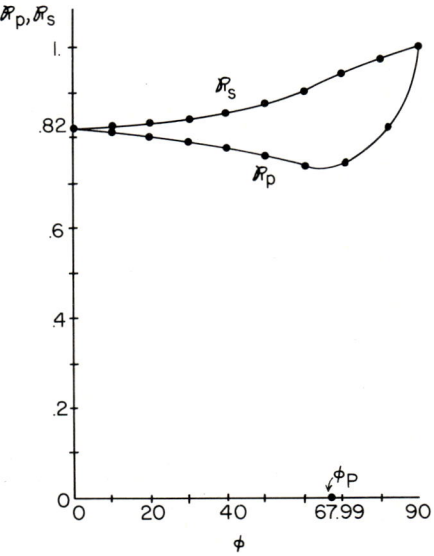

Fig. 4.9. The intensity reflectances $\mathcal{R}_p$ and $\mathcal{R}_s$ for the p and s polarizations as functions of the angle of incidence ϕ (degrees) for an air/gold interface; $\lambda = 5461$ Å, $N_{Au} = 0.35 - j2.45$.

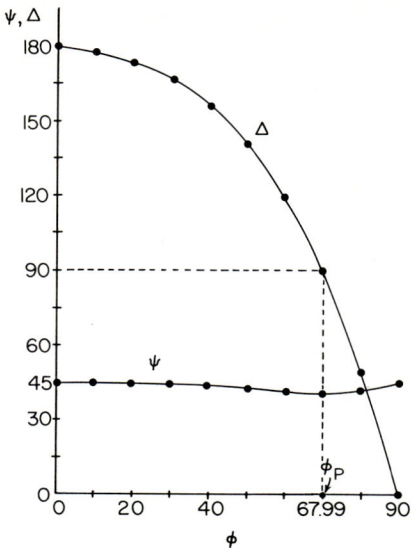

Fig. 4.10. The ellipsometric angles ψ $(=\tan^{-1}(\mathcal{R}_p/\mathcal{R}_s)^{\frac{1}{2}})$ and Δ $(=\delta_{rp}-\delta_{rs})$ as functions of the angle of incidence ϕ (degrees) for reflection at an air/gold interface; $\lambda = 5461$ Å, $N_{Au} = 0.35 - j2.45$.

angle ϕ_c given by

$$\sin \phi_c = n_1/n_0 \qquad (4.24)$$

where n_0 and n_1 are the indices of refraction of the two media, as before, and $n_0 > n_1$. Under conditions of total internal reflection, the absolute values of the Fresnel reflection coefficients for both the p and s polarizations are unity

$$|r_p| = |r_s| = 1, \quad \text{for } \phi_0 \geq \phi_c, \qquad (4.25)$$

while their associated phase angles δ_{rp} and δ_{rs} are given by[8]

$$\tan \tfrac{1}{2}\delta_{rp} = (n_0/n_1)[(n_0/n_1)^2 \sin^2 \phi_0 - 1]^{\frac{1}{2}}/\cos \phi_0, \qquad (4.26)$$
$$\tan \tfrac{1}{2}\delta_{rs} = (n_1/n_0)[(n_0/n_1)^2 \sin^2 \phi_0 - 1]^{\frac{1}{2}}/\cos \phi_0. \qquad (4.27)$$

The ellipsometric angle $\Delta(\delta_{rp} - \delta_{rs})$ is determined from

$$\tan \tfrac{1}{2}\Delta = (n_1/n_0)[(n_0/n_1)^2 \sin^2 \phi_0 - 1]^{\frac{1}{2}}/\sin \phi_0 \tan \phi_0. \qquad (4.28)$$

Equations (4.25)–(4.28) are derived from eqs. (4.4) and (4.5) for the Fresnel coefficients. Figures 4.11 and 4.12 show the variation with angle of incidence

---

[8] The fact that the p and s linear polarizations are reflected unattenuated though differentially phase retarded is the basis of the Fresnel-rhomb retarder.

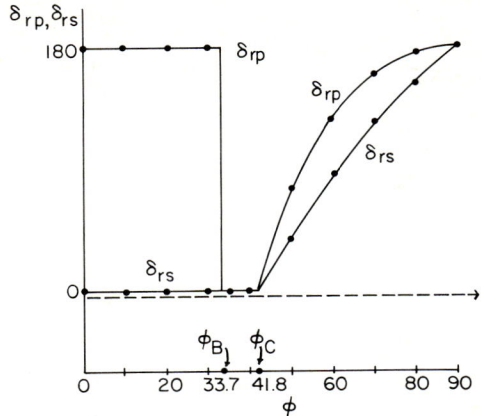

Fig. 4.11. The internal-reflection phase shifts δ_{rp} and δ_{rs} for the p and s polarizations as functions of the angle of incidence ϕ (degrees) for a glass/air interface; $\lambda = 5461$ Å, $N_{glass} = 1.50$.

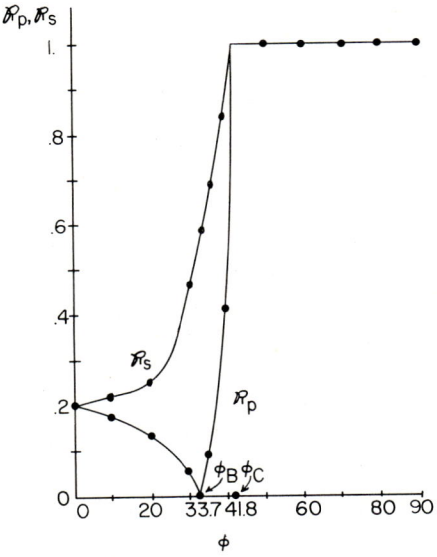

Fig. 4.12. The internal intensity reflectances $\mathcal{R}_p$ and $\mathcal{R}_s$ for the p and s polarizations as functions of the angle of incidence ϕ (degrees) for a glass/air interface; $\lambda = 5461$ Å, $N_{glass} = 1.50$.

of the phase shifts δ_{rp} and δ_{rs} and the reflectances $\mathcal{R}_p$ and $\mathcal{R}_s$ when light is incident on a glass/air interface. Figure 4.13 shows the corresponding variations of the ellipsometric parameters Δ and $\psi = \tan^{-1}(\mathcal{R}_p/\mathcal{R}_s)^{\frac{1}{2}}$. Note that for $\phi_0 \geq \phi_c$, ψ is equal to $\frac{1}{4}\pi$.

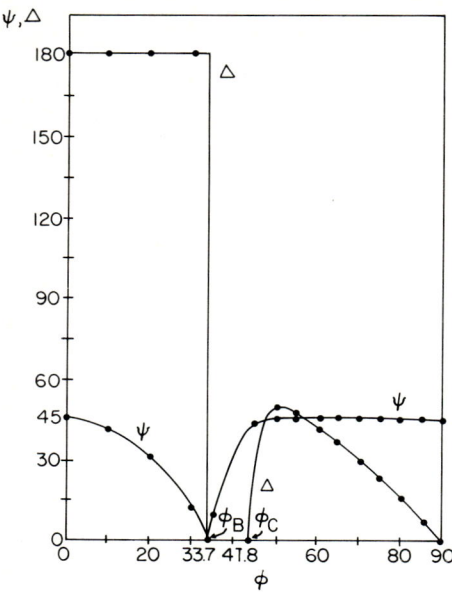

Fig. 4.13. The ellipsometric angles ψ $(=\tan^{-1}(\mathcal{R}_p/\mathcal{R}_s)^{\frac{1}{2}})$ and Δ $(=\delta_{rp}-\delta_{rs})$ as functions of the angle of incidence ϕ (degrees) for internal reflection at a glass/air interface; $\lambda = 5461$ Å, $N_{\text{glass}} = 1.50$.

If we go back to figs. 4.6 and 4.9, we see that for an absorbing material (semiconductor or metal) the reflectance $\mathcal{R}_p$ for p-polarized light does not reach zero as the angle of incidence is varied, but rather exhibits a minimum whose value depends on the extinction coefficient k. The angle of incidence at which $\mathcal{R}_p$ is minimum is called the *pseudo-Brewster angle*[9] $\phi_{B'}$. Another important angle of incidence is the *principal angle* ϕ_P at which the difference Δ between the phase shifts δ_{rp} and δ_{rs} experienced on reflection by the p and s linear vibrations is $\frac{1}{2}\pi$. The difference between the pseudo-Brewster angle and the principal angle is usually small (less than 1° in the visible) and tends to zero as the extinction coefficient k approaches zero in a perfect dielectric [133, 143]. Note the gradual change in Δ around the principal angle for the case of a semiconductor, fig. 4.7, or a metal, fig. 4.10, in contrast with the discontinuous transition displayed in fig. 4.4 for a dielectric. The smaller the value of the extinction coefficient, the more sharp the change in Δ

---

[9]This is not the same angle as that at which the *ratio* of reflectances $\mathcal{R}_p/\mathcal{R}_s$ is minimum (an angle which is sometimes called the *second Brewster angle*). There is some confusion concerning these angles in the literature and our definitions follow those adopted by Bennett and Bennett in the Handbook of Optics [10].

around the principal angle. At the principal angle, incident linearly polarized light at an azimuth from the plane of incidence which is neither zero nor $\frac{1}{2}\pi$ is reflected elliptically polarized with the major and minor axes of the ellipse of polarization aligned parallel and perpendicular to the plane of incidence. This follows because the oscillating p and s components of the electric vector which are in phase before reflection acquire a phase difference of $\frac{1}{2}\pi$ after reflection ($\Delta = \frac{1}{2}\pi$).

Before concluding this section, we note that if the direction of propagation of the refracted wave in medium 1 is reversed (see fig. 4.1), the reflection and transmission coefficients r_{10} and t_{10} at medium 1–medium 0 interface are related to the corresponding coefficients r_{01} and t_{01} at medium 0–medium 1 interface by

$$r_{10} = -r_{01}, \tag{4.29}$$

$$t_{10} = (1 - r_{01}^2)/t_{01}, \tag{4.30}$$

as can be proved by interchanging N_0 and N_1 in eqs. (4.4)–(4.7). Equations (4.29) and (4.30) are applicable to both linear polarizations parallel and perpendicular to the plane of incidence.

4.3. Reflection and transmission by an ambient-film-substrate system

A case of considerable importance in ellipsometry is that in which polarized light is reflected from, or transmitted by, a substrate covered by a single film. As shown in fig. 4.14, we assume that the film has parallel-plane boundaries of separation (film thickness) d_1 and is sandwiched between semi-infinite ambient (immersion) and substrate media. The ambient (medium 0), the film (medium 1), and the substrate (medium 2) are all homogeneous and optically

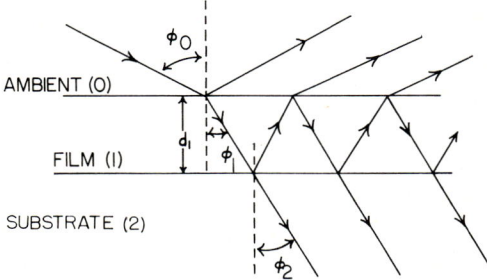

Fig. 4.14. Oblique reflection and transmission of a plane wave by an ambient(0)-film(1)-substrate(2) system with parallel-plane boundaries. d_1 is the film thickness. ϕ_0 is the angle of incidence in the ambient and ϕ_1, ϕ_2 are the angles of refraction in the film and substrate, respectively.

isotropic, with complex indices of refraction N_0, N_1, and N_2, respectively. In most instances, the medium of incidence is transparent and N_0 is real.

A plane wave incident in medium 0 (at an angle ϕ_0) will give rise to a *resultant* reflected wave in the same medium and to a *resultant* transmitted wave (at the angle ϕ_2) in medium 2 (the substrate). Our objective is to relate the complex amplitudes of the resultant reflected and transmitted waves to the amplitude of the incident wave, when the latter is linearly polarized parallel (p) and perpendicular (s) to the plane of incidence. The procedure to be followed (originally due to Airy [144]) is based on the physical picture of fig. 4.14. Specifically, we assume that when the incident wave *first* meets the 0–1 interface, part of it is reflected in medium 0 and part is refracted in the film, according to the interface Fresnel reflection and transmission coefficients previously given in §4.2. The refracted wave inside the film subsequently suffers multiple internal reflections at the 1–2 and 1–0 film-bounding interfaces which are, in general, not perfectly reflecting. Thus each time the multiply-reflected wave in the film strikes the 1–0 or the 1–2 interface, a component (partial) wave is leaked (refracted) into the semi-infinite ambient or substrate medium, respectively. If the Fresnel reflection and transmission coefficients at the 0–1 (1–0) and 1–2 interfaces are denoted by r_{01}, $t_{01}(r_{10}, t_{10})$ and r_{12}, t_{12}, respectively, the complex amplitudes of the successive *partial* plane waves that make up the resultant reflected wave in medium 0 are given by r_{01}, $t_{01}t_{10}r_{12}\,e^{-j2\beta}$, $t_{01}t_{10}r_{10}r_{12}^2\,e^{-j4\beta}$, $t_{01}t_{10}r_{10}^2r_{12}^3\,e^{-j6\beta}$, ..., while the complex amplitudes of the successive partial plane waves that make up the resultant transmitted wave in medium 2 are given by $t_{01}t_{12}\,e^{-j\beta}$, $t_{01}t_{12}r_{10}r_{12}\,e^{-j3\beta}$, $t_{01}t_{12}r_{10}^2r_{12}^2\,e^{-j5\beta}$, ..., where β is the phase change that the multiply-reflected wave inside the film experiences as it traverses the film once from one boundary to the other. In terms of the free-space wavelength λ, the film thickness d_1, the film complex index of refraction N_1 and the (complex) angle of refraction in the film ϕ_1 (the angle between the direction of propagation of the zig-zag wave in the film and the normal to the film boundaries), the phase angle β (*film phase thickness*) is given by

$$\beta = 2\pi\left(\frac{d_1}{\lambda}\right) N_1 \cos\phi_1, \tag{4.31}$$

or

$$\beta = 2\pi\left(\frac{d_1}{\lambda}\right)(N_1^2 - N_0^2 \sin^2\phi_0)^{\frac{1}{2}}, \tag{4.32}$$

if Snell's law [eq. (4.3)] is applied, where ϕ_0 is the angle of incidence (in medium 0). In the above discussion, the incident wave is assumed to be of unit amplitude and is either p- or s-polarized. (For simplicity, the appropriate subscript p or s has been dropped from all interface reflection and transmission coefficients but will be restored later.)

Addition of the partial waves leads to an infinite geometric series for the total reflected amplitude R

$$R = r_{01} + t_{01}t_{10}r_{12}\,e^{-j2\beta} + t_{01}t_{10}r_{10}r_{12}^2\,e^{-j4\beta} + t_{01}t_{10}r_{10}^2r_{12}^3\,e^{-j6\beta} + \cdots, \qquad (4.33)$$

whose summation gives

$$R = r_{01} + \frac{t_{01}t_{10}r_{12}\,e^{-j2\beta}}{1 - r_{10}r_{12}\,e^{-j2\beta}},$$

or

$$R = \frac{r_{01} + r_{12}\,e^{-j2\beta}}{1 + r_{01}r_{12}\,e^{-j2\beta}} \qquad (4.34)$$

if the substitutions $r_{10} = -r_{01}$ and $t_{01}t_{10} = 1 - r_{01}^2$ [eqs. (4.29) and (4.30)] are made. Similarly, the total transmitted amplitude T is given by the infinite geometric series

$$T = t_{01}t_{12}\,e^{-j\beta} + t_{01}t_{12}r_{10}r_{12}\,e^{-j3\beta} + t_{01}t_{12}r_{10}^2r_{12}^2\,e^{-j5\beta} + \cdots, \qquad (4.35)$$

whose summation yields

$$T = \frac{t_{01}t_{12}\,e^{-j\beta}}{1 + r_{01}r_{12}\,e^{-j2\beta}}, \qquad (4.36)$$

where, again, r_{10} has been replaced by $-r_{01}$.

Because the incident wave has unit amplitude, eqs. (4.34) and (4.36) also give the *overall* complex-amplitude reflection (R) and transmission (T) coefficients for the ambient-film-substrate system in terms of (1) the *interface* Fresnel reflection (r_{01}, r_{12}) and transmission (t_{01}, t_{12}) coefficients at the ambient-film (0–1) and film-substrate (1–2) interfaces, and (2) the phase change β [eq. (4.32)] experienced by the multiply-reflected wave inside the film on a single transversal between its boundaries. These equations are valid when the incident wave is linearly polarized either parallel (p) or perpendicular (s) to the plane of incidence. Thus, we may restore the information on polarization by adding the p and s subscripts as follows:

$$R_p = \frac{r_{01p} + r_{12p}\,e^{-j2\beta}}{1 + r_{01p}r_{12p}\,e^{-j2\beta}}, \qquad (4.37)$$

$$R_s = \frac{r_{01s} + r_{12s}\,e^{-j2\beta}}{1 + r_{01s}r_{12s}\,e^{-j2\beta}}, \qquad (4.38)$$

$$T_p = \frac{t_{01p}t_{12p}\,e^{-j\beta}}{1 + r_{01p}r_{12p}\,e^{-j2\beta}}, \qquad (4.39)$$

$$T_s = \frac{t_{01s}t_{12s}\,e^{-j\beta}}{1 + r_{01s}r_{12s}\,e^{-j2\beta}}, \qquad (4.40)$$

where β is the same for both the p and s polarizations and is given by eq. (4.31) or (4.32). The Fresnel reflection and transmission coefficients at the 0–1 and 1–2 interfaces that appear in eqs. (4.37)–(4.40) are given by

$$r_{01p} = \frac{N_1 \cos \phi_0 - N_0 \cos \phi_1}{N_1 \cos \phi_0 + N_0 \cos \phi_1}, \tag{4.41}$$

$$r_{12p} = \frac{N_2 \cos \phi_1 - N_1 \cos \phi_2}{N_2 \cos \phi_1 + N_1 \cos \phi_2}, \tag{4.42}$$

$$r_{01s} = \frac{N_0 \cos \phi_0 - N_1 \cos \phi_1}{N_0 \cos \phi_0 + N_1 \cos \phi_1}, \tag{4.43}$$

$$r_{12s} = \frac{N_1 \cos \phi_1 - N_2 \cos \phi_2}{N_1 \cos \phi_1 + N_2 \cos \phi_2}; \tag{4.44}$$

$$t_{01p} = \frac{2N_0 \cos \phi_0}{N_1 \cos \phi_0 + N_0 \cos \phi_1}, \tag{4.45}$$

$$t_{12p} = \frac{2N_1 \cos \phi_1}{N_2 \cos \phi_1 + N_1 \cos \phi_2}, \tag{4.46}$$

$$t_{01s} = \frac{2N_0 \cos \phi_0}{N_0 \cos \phi_0 + N_1 \cos \phi_1}, \tag{4.47}$$

$$t_{12s} = \frac{2N_1 \cos \phi_1}{N_1 \cos \phi_1 + N_2 \cos \phi_2}; \tag{4.48}$$

as can be readily obtained from eqs. (4.4)–(4.7) of §4.2. The three angles ϕ_0, ϕ_1 and ϕ_2 between the directions of propagation of the plane waves in media 0, 1 and 2 respectively, and the normal to the film boundaries are interrelated by Snell's law [eq. (4.3)]

$$N_0 \sin \phi_0 = N_1 \sin \phi_1 = N_2 \sin \phi_2. \tag{4.49}$$

As pointed out by Heavens [133], when applying the above theory to practical film-substrate systems, the following conditions have to be met – (1) the lateral dimension of the film must be many times its thickness so that the multiply-reflected and transmitted partial waves can be summed to infinity, (2) the source bandwidth, beam diameter, and degree of collimation, as well as film thickness must all be such that the multiply-reflected and transmitted waves combine coherently, and (3) the film material must not be amplifying.[10] These conditions are readily met in most practical applications of ellipsometry.

To examine the change of amplitude and phase separately, as a plane wave is obliquely reflected from or transmitted by a film-covered substrate, the

---

[10]Notice also that the theory assumes perfectly smooth parallel-plane interfaces and media that are homogeneous and isotropic. Surface roughness, inhomogeneity and anisotropy are therefore unaccounted for in the present model (see §4.8).

overall complex-amplitude reflection (R_p, R_s) and transmission (T_p, T_s) coefficients are written in terms of their absolute values and angles

$$R_p = |R_p| e^{j\Delta_{rp}}, \quad (4.50)$$

$$R_s = |R_s| e^{j\Delta_{rs}}, \quad (4.51)$$

$$T_p = |T_p| e^{j\Delta_{tp}}, \quad (4.52)$$

$$T_s = |T_s| e^{j\Delta_{ts}}. \quad (4.53)$$

$|R_p|$ ($|T_p|$) and $\Delta_{rp}(\Delta_{tp})$ represent the amplitude attenuation and phase shift respectively, as p-polarized light is reflected (transmitted) by the film-covered substrate, with similar meanings for $|R_s|$ ($|T_s|$) and $\Delta_{rs}(\Delta_{ts})$ in the case of s-polarization.

From eqs. (4.37)–(4.40) it is evident that because the *interface* Fresnel reflection and transmission coefficients for the p and s polarizations are different (see §4.2), the *overall* reflection and transmission coefficients of an ambient-film-substrate system for these two polarizations are also different. This is the basis of reflection and transmission ellipsometry on ambient-film-substrate systems where a change of polarization takes place on reflection and transmission due to the difference in amplitude attenuation and phase shift experienced by the p and s components. From measurements of the incident and reflected polarizations, the ratio

$$\rho_r = \frac{R_p}{R_s} \quad (4.54)$$

of the overall complex-amplitude reflection coefficients of the ambient-film-substrate system for the p and s polarizations is determined. If ellipsometric measurements are carried out on the incident and transmitted (instead of the reflected) waves, the ratio

$$\rho_t = \frac{T_p}{T_s}, \quad (4.55)$$

of the complex-amplitude transmission coefficients of the same system for the p and s polarizations can be determined. If we express ρ_r or ρ_t in terms of the ellipsometric angles ψ and Δ, as in eq. (4.17), we find that

$$\tan \psi_r = \frac{|R_p|}{|R_s|}, \quad (4.56)$$

$$\Delta_r = \Delta_{rp} - \Delta_{rs}, \quad (4.57)$$

in case of reflection ellipsometry, and

$$\tan \psi_t = \frac{|T_p|}{|T_s|}, \tag{4.58}$$

$$\Delta_t = \Delta_{tp} - \Delta_{ts}, \tag{4.59}$$

in transmission ellipsometry.

4.4. The equations of reflection and transmission ellipsometry for ambient-film-substrate systems

4.4.1. Reflection ellipsometry

For the majority of applications of reflection ellipsometry, the model of an ideal optically isotropic three-phase ambient-film-substrate system is adequate. The quantity measured by the ellipsometer is the ratio ρ of the complex-amplitude reflection coefficients R_p and R_s for the p and s polarizations, eq. (4.54). If we substitute for R_p and R_s in eq. (4.54) their values given by eqs. (4.37) and (4.38), we obtain[11]

$$\tan \psi \, e^{j\Delta} = \rho = \frac{r_{01p} + r_{12p} e^{-j2\beta}}{1 + r_{01p} r_{12p} e^{-j2\beta}} \times \frac{1 + r_{01s} r_{12s} e^{-j2\beta}}{r_{01s} + r_{12s} e^{-j2\beta}}, \tag{4.60}$$

where the ambient-film (r_{01p}, r_{01s}) and the film-substrate (r_{12p}, r_{12s}) interface Fresnel reflection coefficients are given by eqs. (4.41)–(4.44), and the film phase thickness β is given by eq. (4.32). Equation (4.60) relates the measured ellipsometric angles ψ and Δ to the optical properties of the three-phase system, namely the (complex) refractive indices of the ambient (N_0), film (N_1), substrate (N_2), as well as the film thickness (d_1), for given values of the vacuum wavelength (λ) of the ellipsometer light beam and its angle of incidence (ϕ_0) in the ambient. The functional dependence of ψ and Δ on the system parameters can be symbolically written as

$$\tan \psi \, e^{j\Delta} = \rho(N_0, N_1, N_2, d_1, \phi_0, \lambda), \tag{4.61}$$

where ρ is given by the right-hand side of eq. (4.60). Equation (4.61) may be broken into two real equations for ψ and Δ separately

$$\psi = \tan^{-1} |\rho(N_0, N_1, N_2, d_1, \phi_0, \lambda)|, \tag{4.62}$$

$$\Delta = \arg [\rho(N_0, N_1, N_2, d_1, \phi_0, \lambda)], \tag{4.63}$$

where $|\rho|$ and $\arg(\rho)$ are the absolute value and argument (angle) of the complex function ρ, respectively.

---

[11] For simplicity, the subscript r which we used in §4.3 to distinguish reflection from transmission ellipsometry is deleted from ρ, ψ, and Δ.

Although the function ρ may appear from eq. (4.60) to be deceptively simple, it is, in reality, quite complicated and can be handled satisfactorily only by a digital computer. Notice that ρ is, in general, explicitly dependent on *nine*[12] real arguments: the real and imaginary parts of the three complex refractive indices N_0, N_1, N_2, film thickness d_1, angle of incidence ϕ_0, and wavelength λ. It will require an incredibly voluminous library of computer printout (which, fortunately, is not necessary) to document the behavior of the function ρ, as each of its arguments is allowed to scan all possible values. In fact, an entire book[145] has been published that contains ellipsometric tables and curves showing the dependence of ψ and Δ (or, equivalently, ρ) on just two parameters: the thickness and (real) refractive index of the oxide film for the air–SiO_2–Si system at select mercury and He–Ne–laser spectral lines.

4.4.1.1. Constant-Angle-of-Incidence Contours (CAIC) of the ellipsometric function $\rho(\phi, d)$

Perhaps the easiest feature of the function ρ to examine analytically is its dependence on film thickness, which is very important in many film growth applications of ellipsometry. To do that [109] we cast eqs. (4.37) and (4.38) in the form

$$R_p = \frac{a + bX}{1 + abX}, \qquad R_s = \frac{c + dX}{1 + cdX}. \tag{4.64}$$

Subsequently, eq. (4.60) becomes

$$\rho = \frac{(a + bX)(1 + cdX)}{(1 + abX)(c + dX)},$$

or

$$\rho = \frac{A + BX + CX^2}{D + EX + FX^2}, \tag{4.65}$$

where

$$X = e^{-j2\beta}, \tag{4.66}$$

$$(a, b) = (r_{01p}, r_{12p}), \qquad (c, d) = (r_{01s}, r_{12s}), \tag{4.67}$$

and

$$\begin{aligned} A &= a, & B &= (b + acd), & C &= bcd, \\ D &= c, & E &= (d + abc), & F &= abd. \end{aligned} \tag{4.68}$$

---

[12] Although for all practical purposes the ambient is sufficiently transparent to make N_0 almost real, it might be useful to examine the effect of the very small (but physically nonzero) imaginary part of N_0 (the extinction coefficient).

Equations (4.64) show that each of the complex-amplitude reflection coefficients R_p and R_s is related to the complex exponential function of film thickness X, eq. (4.66), by a bilinear transformation with coefficients determined by the 0–1 and 1–2 interface Fresnel reflection coefficients,[13] eqs. (4.67). Equation (4.65) shows that the ratio of the p and s reflection coefficients ρ is a rational function of X in the form of a quadratic divided by another quadratic with coefficients determined by the interface Fresnel reflection coefficients, eqs. (4.67) and (4.68).

If we substitute the film phase thickness β from eq. (4.32) into eq. (4.66) and write $\phi_0 = \phi$, $d_1 = d$, the complex exponential function X becomes

$$X = \exp[-j4\pi(d/\lambda)(N_1^2 - N_0^2 \sin^2 \phi)^{\frac{1}{2}}]$$

or

$$X = \exp[-j2\pi(d/D_\phi)], \qquad (4.69)$$

where[14]

$$D_\phi = +\tfrac{1}{2}\lambda(N_1^2 - N_0^2 \sin^2 \phi)^{-\frac{1}{2}}. \qquad (4.70)$$

When the ambient and film media are transparent, N_0, N_1, and D_ϕ are real.[15] In this case, it can be seen from eq. (4.69) that at any angle of incidence ϕ, the representative point of the complex exponential function X moves uniformly in a clockwise direction around the unit circle in the complex X-plane, as the film thickness d is increased starting from the point $X = 1$ when $d = 0$, fig. 4.15 (left). Notice that X is a periodic function of d so that

---

[13]This relationship was first recognized by Winterbottom [71]. Notice that a similar relationship describes the evolution of the ellipse of polarization as a function of the distance travelled by a light wave in a homogeneous anisotropic medium [eq. (2.182), §2.10].

[14]From the discussion in §4.3 on the addition of multiply reflected and refracted waves to compute the overall reflection and transmission coefficients of an ambient-film-substrate system, it is obvious that β represents a propagation phase *delay*. Consequently, only the square root on the right-hand side of eq. (4.70) that has a *positive* real part is acceptable, as is emphasized by the + sign in that equation.

[15]This assumes that $N_1 > N_0 \sin \phi$, which is always satisfied when $N_1 > N_0$. However, when the ambient is of higher refractive index than the film, $N_0 > N_1$, total reflection occurs at the ambient-film interface for angles of incidence above the critical angle $\phi_C = \sin^{-1}(N_1/N_0)$ and D_ϕ becomes pure imaginary. In this case, X traces the segment of the real axis between the points $X = 1$ and $X = 0$. The ellipsometric function ρ, which is related to X by an analytic rational function [eq. (4.65)], traces an image trajectory between $\rho = \rho_{02}$ and ρ_{01} where ρ_{02} and ρ_{01} are the values of ρ for the ambient-substrate and ambient-film interfaces. For incidence at the critical angle, $N_1 = N_0 \sin \phi$ and $X = 1$ for all values of film thickness. Under this condition, it can be shown from eq. (4.60) that $\rho = \rho_{02}$ (for the ambient-substrate interface) independent of film thickness. This means that the state of polarization of light reflected at the critical angle is completely insensitive to the growth of a film phase. The constant-angle-of-incidence contour (CAIC) collapses onto a single point in the complex ρ-plane.

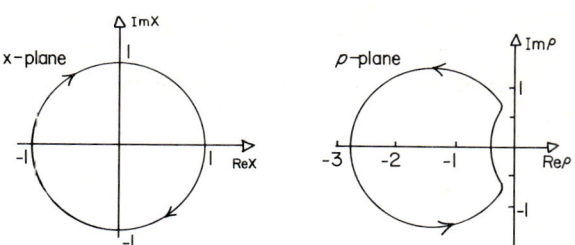

Fig. 4.15. *Left:* the unit circle in the complex X-plane. *Right:* constant-angle-of-incidence contour (CAIC) (or polar curve) of the ellipsometric function ρ at an angle of incidence $\phi = 60°$ for the air–SiO$_2$–Si system at wavelength $\lambda = 6328$ Å.

the unit circle is cyclically traversed. The first full revolution around the unit circle is completed as the film thickness sweeps the thickness interval or period

$$0 \leq d < D_\phi.$$

The thickness period D_ϕ, eq. (4.70), is a function of the angle of incidence ϕ and is completely determined by the wavelength λ and the ambient and film refractive indices N_0 and N_1. Figure 4.16 shows a plot of the thickness period D_ϕ versus the angle of incidence ϕ when $\lambda = 6328$ Å, $N_0 = 1$ and $N_1 = 1.46$ (which corresponds to a SiO$_2$ film). The minimum value of D_ϕ occurs at $\phi = 0°$ ($D_0 = 2167.13$ Å) while the maximum value occurs at $\phi = 90°$ ($D_{90} = 2974.34$ Å).

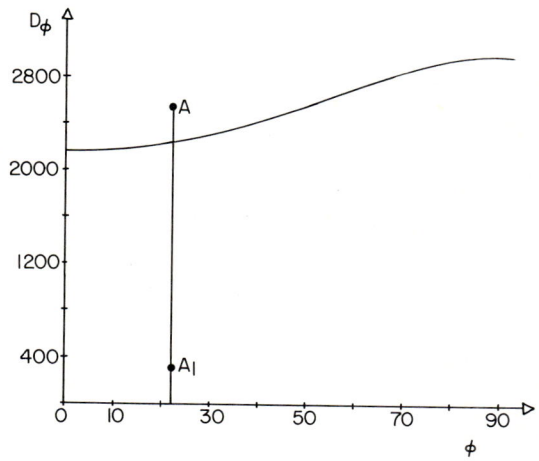

Fig. 4.16. Film-thickness period D_ϕ in angstroms, when $N_0 = 1$, $N_1 = 1.46$ (SiO$_2$) and $\lambda = 6328$ Å, as a function of the angle of incidence ϕ in degrees. The ellipsometric function ρ has the same value at points A (ϕ, d) and $A_1 = (\phi, d - D_\phi)$.

Because the ellipsometric function ρ is related to the complex exponential X by an analytic rational function of X, eq. (4.65), the representative point of ρ must trace a closed contour in the complex ρ-plane as X traces the unit circle in the complex X-plane (when the film thickness d is increased at a given angle of incidence ϕ). Figure 4.15 (right) shows such a *constant-angle-of-incidence contour* (CAIC) (also called *polar curve* [146–148]) when $\phi = 60°$ for the air–SiO$_2$–Si system at $\lambda = 6328$ Å. As ϕ scans its entire range

$$0 \le \phi \le 90°,$$

the coefficients of the rational function $\rho(X)$, and the period D_ϕ change such that a whole family of nonintersecting constant-ϕ (variable-d) contours is generated. Figure 4.17 shows such a family for the air–SiO$_2$–Si system at $\lambda = 6328$ Å. The $\phi = 0°$ (normal incidence) and $\phi = 90°$ (grazing incidence) contours collapse to the null points $\rho = -1$ and $\rho = +1$ on the real axis because, at these angles, $R_p = -R_s$ and $R_p = R_s$, respectively, for any film thickness. All CAIC's that correspond to angles of incidence in a certain range

$$0 < \phi < \phi_s,$$

enclose the point $\rho = -1$, whereas the remaining contours that correspond to

$$\phi_s < \phi < 90°,$$

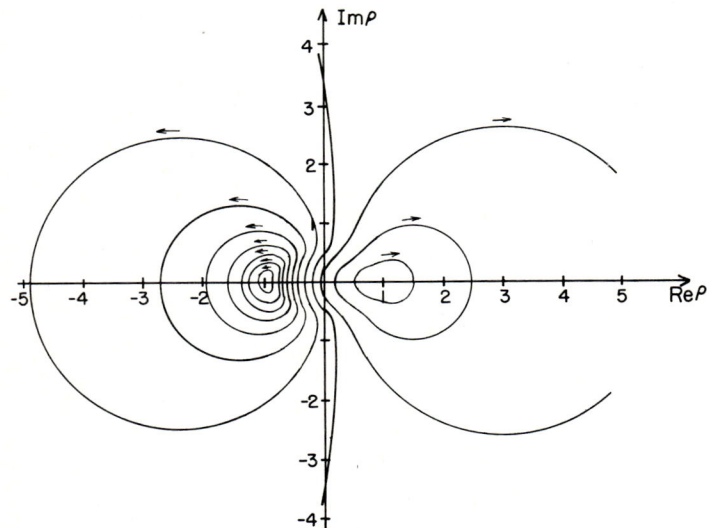

Fig. 4.17. Constant-angle-of-incidence contours (CAIC's) in the complex ρ-plane for the air–SiO$_2$–Si system at $\lambda = 6328$ Å. The arrows indicate the direction in which the film thickness increases.

enclose the point $\rho = +1$. The direction in which the thickness increases is indicated by an arrow on each contour. Arrows on all CAIC's enclosing $\rho = -1$ point the same way and opposite to the arrows on all the other contours enclosing $\rho = +1$. At the angle of incidence ϕ_s the CAIC passes through the point at infinity. This means that at this angle a value (a set of values equally spaced by the period D_{ϕ_s}) of film thickness can be found such that $\rho = \infty$. Under such conditions, the film-substrate system acts as an s-suppressing reflection polarizer. The locus of points on the CAIC's where the film thickness is zero, $d = 0$, is the same as the locus of

$$\rho = \rho_{02} = r_{02p}/r_{02s},$$

which is the ratio of the Fresnel reflection coefficients of the ambient-substrate interface. This zero-thickness contour (ZTC) for the air–SiO$_2$–Si system at $\lambda = 6328$ Å is shown on an expanded scale in fig. 4.18.

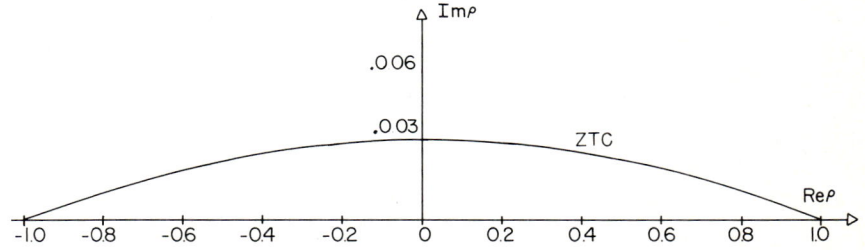

Fig. 4.18. Zero-thickness contour (ZTC) in the complex ρ-plane for the air–SiO$_2$–Si system at $\lambda = 6328$ Å.

When the film is absorbing, D_ϕ in eq. (4.70) becomes complex and the exponential function X describes a logarithmic spiral in the complex plane, fig. 4.19 (left), as the film thickness d is increased at a fixed angle of incidence ϕ. When this logarithmic spiral X is transformed by the analytic rational function $\rho(X)$, eq. (4.65), the resulting CAIC, fig. 4.19 (right), is a distorted spiral that is traced once as the film thickness is increased starting from $\rho = \rho_{02}$ (the ratio of reflection coefficients for the ambient-substrate interface) at zero film thickness and asymptotically converging to $\rho = \rho_{01}$ (the ratio of reflection coefficients for the ambient-film interface) at infinite film thickness.

4.4.1.2. *Constant-Thickness Contours (CTC) of the ellipsometric function* $\rho(\phi, d)$

The bare-substrate zero-thickness contour, fig. 4.18, is one member of a complementary family of *constant-thickness contours* (CTC) that can be drawn in the complex ρ-plane. For a given film thickness d, the associated CTC can be generated from the zero-thickness contour (ZTC) with the aid of the constant-angle-of-incidence contours (CAIC). Each point on the ZTC

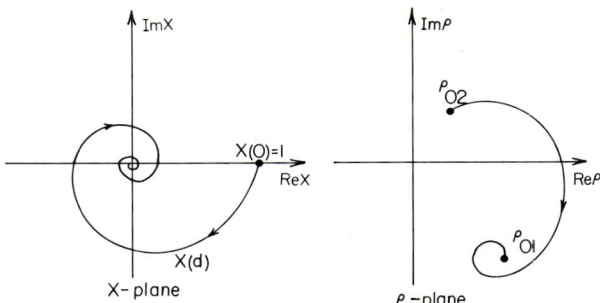

Fig. 4.19. *Left:* the logarithmic spiral of X when the film is absorbing. *Right:* the constant-angle-of-incidence contour of ρ becomes a distorted spiral whose initial and final points are ρ_{02} ($d = 0$) and ρ_{01} ($d = \infty$), respectively.

corresponds to one angle of incidence ϕ, and as ϕ is increased from $\phi = 0°$ to $\phi = 90°$, we move along the ZTC once from $\rho = -1$ to $\rho = +1$. At a given angle of incidence ϕ, a point on the ZTC can be moved around the constant ϕ contour to a new point that corresponds to the thickness d. Joining such points for the same value of film thickness d, at different values of ϕ, generates the constant-thickness contour, fig. 4.20. The points $\rho = -1$ and $\rho = +1$ stay the same for all values of d.

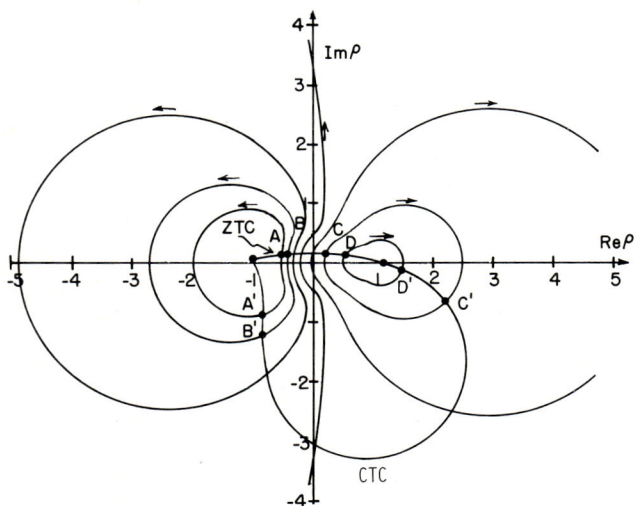

Fig. 4.20. Generation of an arbitrary constant-thickness contour (CTC) from the bare-substrate zero-thickness contour (ZTC). Each point on the ZTC (A, B, C ... etc.) is translated along the CAIC passing through that point to a new point (A', B', C', ... etc.) corresponding to the given film thickness.

Figure 4.21 shows the $d = 1.5\ \mu$m constant-thickness contour of the ellipsometric function ρ in the complex ρ-plane for the air–SiO$_2$–Si system at $\lambda = 6328$ Å. This contour is traversed as the angle of incidence ϕ is increased starting from the point N ($\rho = -1$) when $\phi = 0°$ (normal incidence) and ending at the point G ($\rho = +1$) when $\phi = 90°$ (grazing incidence). The spiralling behavior near N shown by the constant-thickness contour of fig. 4.21 is shared by other such contours for thicknesses comparable to or greater than λ. This spiralling can be attributed to the high density of the CAIC's near N (see fig. 4.17) and to the fact that the thickness period D_ϕ of the CAIC is shortest (hence the rate of rotation is maximum) at and near $\phi = 0°$ (see fig. 4.16).

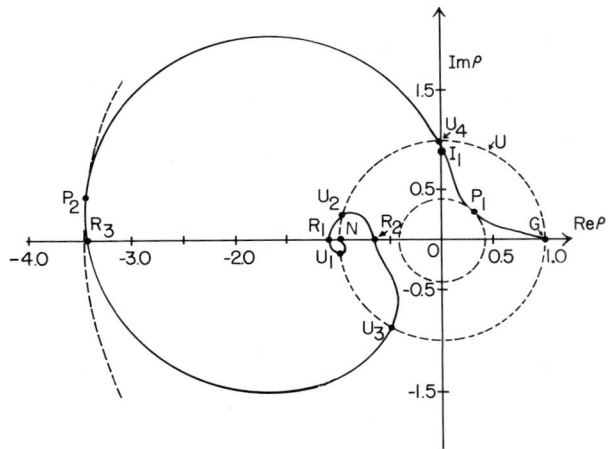

Fig. 4.21. Constant-thickness contour (CTC) of film thickness $d = 1.5\ \mu$m in the complex ρ-plane for the air–SiO$_2$–Si system at $\lambda = 6328$ Å.

The angles ψ and Δ associated with the ellipsometric function $\rho = \tan\psi\ e^{j\Delta}$ are plotted in figs. 4.22 and 4.23 versus the angle of incidence ϕ as we move along the 1.5 μm constant-thickness contour of fig. 4.21. Several significant points are marked on the CTC of fig. 4.21 and on the ψ and Δ curves of figs. 4.22 and 4.23 respectively. Excluding N and G, these significant points are grouped as follows.

(1) The intersection points $R_1, R_2, \ldots$ of the CTC with the real axis of the complex ρ-plane which occur at angles of incidence $\phi_{r1}, \phi_{r2}\ldots$: Dependent on whether an intersection R_i of the CTC with the real axis is to the left or to the right of the origin, $\Delta = \pi$ or 0, respectively. All intersections R_1, R_2, and R_3 for the $d = 1.5\ \mu$m CTC shown in fig. 4.21 correspond to $\Delta = \pi$ and occur at angles of incidence $\phi_{r1} = 30.13°$, $\phi_{r2} = 46.75°$, and $\phi_{r3} = 62.44°$, respec-

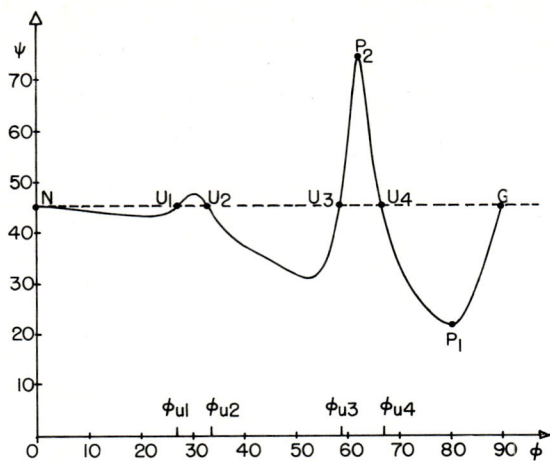

Fig. 4.22. The angle ψ as a function of the angle of incidence ϕ for the air–SiO$_2$–Si system at $\lambda = 6328$ Å and film thickness $d = 1.5$ μm; ψ and ϕ are in degrees.

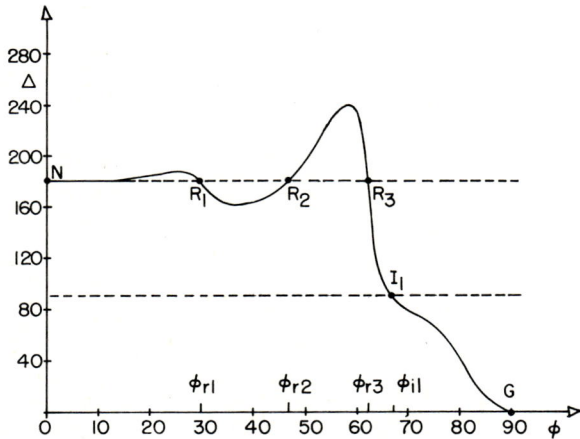

Fig. 4.23. The angle Δ as a function of the angle of incidence ϕ for the air–SiO$_2$–Si system at $\lambda = 6328$ Å and film thickness $d = 1.5$ μm; Δ and ϕ are in degrees.

tively. Such points appear in fig. 4.23 as the intersections of the $\Delta - \phi$ curve with the $\Delta = \pi$ line. When $\Delta = 0$ or π, incident light which is linearly polarized at any azimuth from the plane of incidence is linearly polarized after it is reflected from the film-coated substrate. This is the basis of the polarizer-surface-analyzer (PSA) null ellipsometer arrangement without a compensator briefly discussed in §3.10.

(2) The intersections $I_1, I_2, \ldots$ of the CTC with the imaginary axis of the complex ρ-plane which occur at angles of incidence $\phi_{i1}, \phi_{i2}, \ldots$: At such points, $\Delta = \frac{1}{2}\pi$ or $\frac{3}{2}\pi$ dependent on whether the intersection I is above or below the origin. When $\Delta = \frac{1}{2}\pi$ or $\frac{3}{2}\pi$ the major and minor axes of the ellipse of polarization of the reflected light, when the incident light is linearly polarized at an arbitrary azimuth, are aligned parallel and perpendicular to the plane of incidence. The angles of incidence $\phi_{i1}, \phi_{i2}, \ldots$ may be called the principal angles of incidence for the film-substrate system. For the $d = 1.5\ \mu$m CTC of fig. 4.21, only one principal angle is available $\phi_{i1} = 67.19°$. The I intersections appear in fig. 4.23 as the intersections of the Δ versus ϕ curve with the $\Delta = \frac{1}{2}\pi$ line.

(3) The intersection points $U_1, U_2, \ldots$ with the unit circle of the complex ρ-plane: At such intersections, we have $\tan\psi = 1$, hence $\psi = 45°$. These points are shown in fig. 4.22 as intersections of the $\psi - \phi$ curve with the $\psi = 45°$ line. When $\tan\psi = 1$, we have $|R_p| = |R_s|$ and $\rho = e^{j\Delta}$, and the film-substrate system acts as a reflection retarder. Such a retarder exhibits isotropic absorption because the absolute reflectances $\mathcal{R} = |R_p|^2 = |R_s|^2$ are necessarily less than 1. For the air–SiO$_2$–Si system with $d = 1.5\ \mu$m at $\lambda = 6328$ Å, the angles of incidence at which exact operation as a retarder is achieved are $\phi_{u1} = 27.0°$, $\phi_{u2} = 33.25°$, $\phi_{u3} = 58.75°$, and $\phi_{u4} = 66.75°$. The associated retardations are $\Delta_{u1} = 187.62°$, $\Delta_{u2} = 167.23°$, $\Delta_{u3} = 240.9°$, and $\Delta_{u4} = 92.03°$, whereas the reflectances are $\mathcal{R}_{u1} = 0.105$, $\mathcal{R}_{u2} = 0.108$, $\mathcal{R}_{u3} = 0.155$, and $\mathcal{R}_{u4} = 0.208$, respectively.

(4) The points P_1 and P_2 where the minimum-radius and maximum-radius circles, centered on the origin, touch the CTC, respectively: These two concentric circles define an annular domain in which the CTC is confined. At P_1 and P_2, $|\rho| = \tan\psi$ is minimum and maximum, respectively. These points indicate how a given film-substrate system comes close to operating as a p- or s-suppressing polarizer. When the minimum radius is zero, $\psi_{P1} = 0$, the CTC passes through the origin. At the corresponding angle of incidence ϕ_{P1} the film-substrate system acts exactly as a p-suppressing polarizer. On the other hand, when the maximum radius becomes infinitely large, $\psi_{P2} = 90°$, the CTC passes through the point at infinity. At the corresponding angle of incidence ϕ_{P2}, the film-substrate system acts exactly as an s-suppressing polarizer.

Figures 4.24 and 4.25 show two sequences of constant-thickness contours (CTC). As can be appreciated from these sequences, there is an infinite variety of such CTC's. Two CTC's may intersect one another, so that a given value of the ellipsometric function ρ can be realized at two different thicknesses at the same angle of incidence. (The thicknesses must be separated by an integral multiple of the film thickness period evaluated at the common angle of incidence.) This is in contrast with the nonintersecting

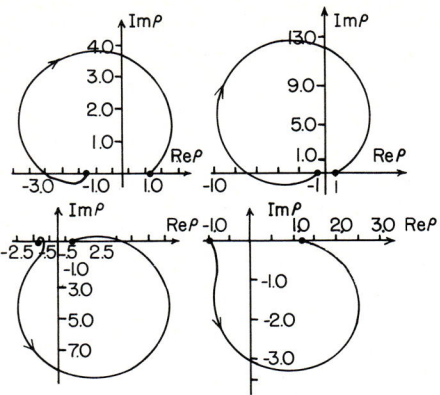

Fig. 4.24. A sequence of constant-thickness contours (CTC's) in the complex ρ-plane for the air–SiO$_2$–Si system at $\lambda = 6328$ Å for film thicknesses $d = 0.21\,\lambda$, (*upper left*), $d = 0.22\,\lambda$, (*upper right*), $d = 0.23\,\lambda$, (*lower left*), and $d = 0.24\,\lambda$, (*lower right*).

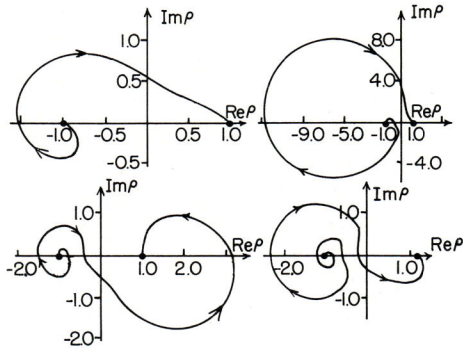

Fig. 4.25. A sequence of constant-thickness contours (CTC's) in the complex ρ-plane for the air–SiO$_2$–Si system at $\lambda = 6328$ Å for film thicknesses $d = \lambda$, (*upper left*), $d = 2\lambda$, (*upper right*), $d = 3\lambda$, (*lower left*), and $d = 4\lambda$, (*lower right*).

character of the constant-angle-of-incidence contours at different angles of incidence, fig. 4.17.

4.4.1.3. Cartesian curves

Instead of the complex-plane (polar) curves which relate tan ψ to Δ as polar coordinates, it may be convenient to plot Δ directly versus ψ in Cartesian coordinates, with the film thickness as the variable parameter along the curve [145, 149–152]. Obviously, such properties of the polar curves as closeness or spiral shape are preserved when these curves are transformed to Cartesian coordinates. When a family of such curves, computed for different values of film refractive index for a given ambient-film-substrate system are drawn together, a useful nomogram is obtained through which the experimentally measured Δ, ψ values can be related to film thickness and refractive index. One such family of curves [150] is shown in fig. 4.26 for an air-film-Si system at $\lambda = 5461$ Å. The primary advantage of the Cartesian curves over the polar curves stems from the fact that the ellipsometric angles ψ and Δ are simply obtained from the instrument readings and can be readily entered on Cartesian coordinates. On the other hand, the polar plots have the advantage of being analytically related to the exponential of film thickness $e^{-j2\beta}$.

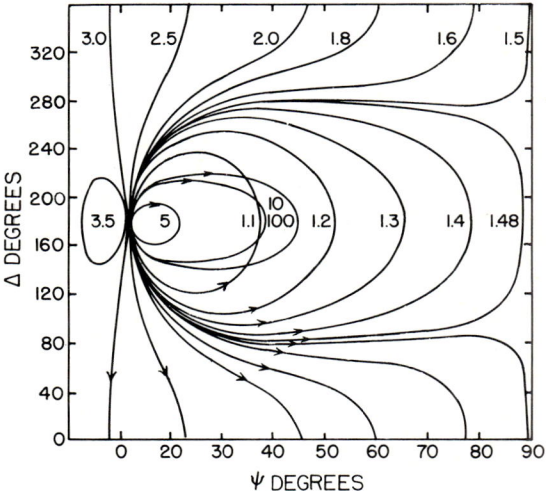

Fig. 4.26. Δ versus ψ for an air-film-silicon system at angle-of-incidence $\phi = 70°$ and wavelength $\lambda = 5461$ Å, for different values of the film refractive index (marked by each curve). The arrows indicate the direction of increase of film thickness. After Archer [150].

4.4.2. Transmission ellipsometry

Ellipsometry can be carried out on the transmitted wave and (or instead of) the reflected wave when the ambient-film-substrate system is transparent enough to make the transmitted wave accessible for measurement. Care must be taken, however, to avoid, or account for, the influence of inhomogeneity and strain in the usually solid substrate. The quantity measured by the ellipsometer is the ratio ρ of the complex-amplitude transmission coefficients T_p and T_s for the p and s polarizations, eq. (4.55). If we substitute for T_p and T_s in eq. (4.55) their values given by eqs. (4.39) and (4.40), we obtain

$$\tan\psi\, e^{j\Delta} = \rho = \left(\frac{t_{01p}}{t_{01s}}\right)\left(\frac{t_{12p}}{t_{12s}}\right) \times \frac{1 + r_{01s}r_{12s}\, e^{-j2\beta}}{1 + r_{01p}r_{12p}\, e^{-j2\beta}}. \tag{4.71}$$

In eq. (4.71) the first and second bracketed prefactors on the right-hand side are the ratios of the p to s Fresnel transmission coefficients for the 0–1 (ambient-film) and 1–2 (film-substrate) interfaces [eqs. (4.45)–(4.48)]. r_{01p}, r_{01s}, r_{12p}, r_{12s} are the interface Fresnel reflection coefficients, and β is the film phase thickness, as before. Equation (4.71) is the main equation of transmission ellipsometry on optically isotropic ambient-film-substrate systems. The functional dependence of the ratio of transmission coefficients ρ (or ψ, Δ) on the refractive indices of the three-phase system, angle of incidence and wavelength can be cast in the form of eq. (4.61), or eqs. (4.62) and (4.63).

The main equation of transmission ellipsometry [eq. (4.71)] is simpler than the corresponding equation of reflection ellipsometry [eq. (4.60)]. This simplicity is more evident when we consider the dependence of ρ on film thickness d_1, $\rho(d_1)$.

Equation (4.71) takes the form

$$\rho = a\frac{1 + bX}{1 + cX}, \tag{4.72}$$

where

$$\begin{aligned}a &= (t_{01p}/t_{01s})(t_{12p}/t_{12s}),\\ b &= r_{01s}r_{12s}, \qquad c = r_{01p}r_{12p},\end{aligned} \tag{4.73}$$

and X is the complex exponential function of eqs. (4.66) and (4.69).

Equation (4.72) shows that the ellipsometric ratio of transmission coefficients ρ is related to the complex exponential function X of film thickness by a bilinear transformation. The coefficients a, b and c of such a transformation, eq. (4.73), are determined by the ambient-film and film-substrate interface Fresnel reflection and transmission coefficients. Recall that bilinear transformations also govern the reflection of p- and s-polarized light by three-phase systems [eq. (4.64)] and the mapping of the ellipse of polarization between the input and output of a linear nondepolarizing optical system [eq. (2.50), §2.3].

From the above we conclude that the complex-plane trajectory of the ratio of transmission coefficients ρ as a function of film thickness at a fixed angle of incidence (the CAIC or polar curve in the complex ρ-plane) is an image through a bilinear transformation of the complex exponential function X. The trajectory of X when the film is transparent is the unit circle in the complex X-plane, fig. 4.27 (left). Because a bilinear transformation between two complex planes maps a circle in one plane onto a circle in the other, it follows that the CAIC (polar curve) for transmission ellipsometry on a transparent ambient-film substrate system is exactly a circle, fig. 4.27 (right). The circle loci of X and ρ are both periodically traced as the film thickness is increased where the thickness period is D_ϕ [eq. (4.70)]. Notice also that as the angle of incidence ϕ is changed, the coefficients a, b and c [eqs. (4.73)] of the bilinear transformation [eq. (4.72)] will also change and a family of circles of ρ is generated.

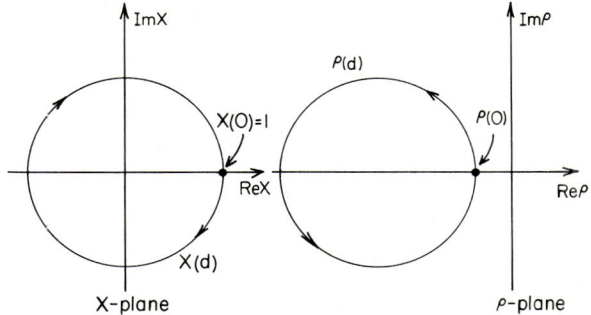

Fig. 4.27. *Left*: at a fixed angle of incidence ϕ, the complex exponential function $X(d)$ [eq. (4.69)] traces the unit circle in the complex X-plane as the thickness d of a *transparent* film is increased. *Right*: the ratio of the three-phase transmission coefficients $\rho(d)$, being related to $X(d)$ by a bilinear transformation [eq. (4.72)], describes a circular CAIC (polar curve) in the complex ρ-plane.

When the film is absorbing, the film refractive index N_1 becomes complex as does D_ϕ [eq. (4.70)]. In this case, X traces a convergent logarithmic spiral that starts at the point $X = 1$ and ends at the origin $X = 0$, fig. 4.28 (left). The bilinear transformation $X \to \rho$ does not preserve the shape of a spiral, so that the polar curve of ρ (ϕ = constant) takes the form of a distorted spiral, fig. 4.28 (right).

4.4.3. *Linear approximation of the equations of reflection and transmission ellipsometry for ambient-film-substrate systems*

4.4.3.1. *Reflection*

When the thickness of the film in an ambient-film-substrate system changes by an increment that is a small fraction of the wavelength of the ellipsometer light beam used to probe such a system, a linear approximation of the equation of reflection ellipsometry, eq. (4.60), can be utilized with reasonable

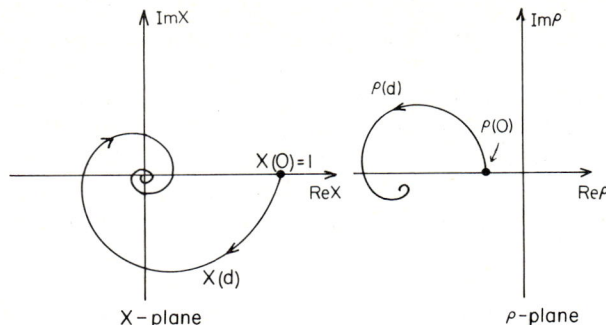

Fig. 4.28. *Left*: at a fixed angle of incidence ϕ, the complex exponential $X(d)$ [eq. (4.69)] traces a logarithmic spiral in the complex X-plane as the thickness d of an *absorbing* film is increased. *Right*: the ratio of the three-phase transmission coefficients $\rho(d)$, which is the image of $X(d)$ through a bilinear transformation [eq. (4.72)], traces a distorted spiral curve in the complex ρ-plane.

accuracy [153–160]. There are instances, however, where the error introduced by use of such a linear approximation, instead of the exact equation, may become unacceptable, even for small thickness changes. In the following, we will derive the linear approximation of eq. (4.60) which is valid for any ambient-film-substrate system with arbitrary initial film thickness.

By taking the logarithmic differential of eq. (4.54), we obtain

$$\delta\rho/\rho = (\delta R_p/R_p) - (\delta R_s/R_s), \tag{4.74}$$

which states that the fractional change $(\delta\rho/\rho)$ of the ellipsometric quantity ρ is equal to the difference between the fractional changes $(\delta R_p/R_p)$ and $(\delta R_s/R_s)$ of the complex-amplitude reflection coefficients R_p and R_s for the p and s polarizations, respectively. To determine the latter fractional changes, we write eqs. (4.37) and (4.38) as

$$R_\nu = N_\nu/D_\nu \tag{4.75}$$

where

$$N_\nu = r_{01\nu} + r_{12\nu}X, \qquad D_\nu = 1 + r_{01\nu}r_{12\nu}X, \tag{4.76}$$

and

$$X = e^{-j2\beta} = e^{\gamma d_1}, \tag{4.77}$$

$$\gamma = -j(4\pi/\lambda)(N_1^2 - N_0^2 \sin^2\phi_0)^{\frac{1}{2}}. \tag{4.78}$$

The logarithmic differential of eq. (4.75) is

$$\delta R_\nu/R_\nu = (\delta N_\nu/N_\nu) - (\delta D_\nu/D_\nu). \tag{4.79}$$

Above and throughout, all equations with ν subscripts should be considered applicable for both $\nu = p$ and $\nu = s$. When all differential changes are induced by an incremental change δX of the complex exponential function X, eq. (4.77), we can rewrite eq. (4.79) as

$$\delta R_\nu / R_\nu = [(1/N_\nu)(\partial N_\nu / \partial X) - (1/D_\nu)(\partial D_\nu / \partial X)]\delta X. \tag{4.80}$$

If, in turn, the increment δX of X results from an increment δd_1 of the film thickness d_1, then, from eq. (4.77),

$$\delta X = \gamma X \delta d_1. \tag{4.81}$$

Substituting eq. (4.81) and eqs. (4.76) and their derivatives

$$\partial N_\nu / \partial X = r_{12\nu}, \qquad \partial D_\nu / \partial X = r_{01\nu} r_{12\nu}, \tag{4.82}$$

into eq. (4.79) we find that

$$\delta R_\nu / R_\nu = \gamma d_1 X \left[\frac{r_{12\nu}}{r_{01\nu} + r_{12\nu} X} - \frac{r_{01\nu} r_{12\nu}}{1 + r_{01\nu} r_{12\nu} X} \right] (\delta d_1 / d_1)$$

$$= Q_\nu (\delta d_1 / d_1), \tag{4.83}$$

where

$$Q_\nu = \gamma d_1 X \left[\frac{r_{12\nu}}{r_{01\nu} + r_{12\nu} X} - \frac{r_{01\nu} r_{12\nu}}{1 + r_{01\nu} r_{12\nu} X} \right]. \tag{4.84}$$

Equations (4.83) and (4.84) are important because they relate a fractional change of film thickness ($\delta d_1 / d_1$) to the resulting fractional change of the overall (three-phase) complex-amplitude reflection coefficient[16] ($\delta R_\nu / R_\nu$) for the two fundamental polarizations $\nu = p$ and $\nu = s$ through complex coupling coefficients Q_ν.

In ellipsometry we are interested in the fractional change $\delta \rho / \rho$ of the ratio of reflection coefficients ρ that is caused by a fractional change $\delta d_1 / d_1$ of film thickness d_1. This is obtained by substituting from eqs. (4.83) and (4.84) into eq. (4.74)

$$\delta \rho / \rho = K (\delta d_1 / d_1), \tag{4.85}$$

where

$$K = Q_p - Q_s$$
$$= \gamma d_1 X \left[\frac{r_{12p}}{r_{01p} + r_{12p} X} - \frac{r_{01p} r_{12p}}{1 + r_{01p} r_{12p} X} - \frac{r_{12s}}{r_{01s} + r_{12s} X} + \frac{r_{01s} r_{12s}}{1 + r_{01s} r_{12s} X} \right]. \tag{4.86}$$

[16] In differential reflection spectroscopy [161] the fractional change of reflectance $\delta \mathcal{R}_\nu / \mathcal{R}_\nu$ is measured where $\mathcal{R}_\nu = R_\nu R_\nu^*$. We can write $\delta \mathcal{R}_\nu / \mathcal{R}_\nu = (\delta R_\nu / R_\nu) + (\delta R_\nu^* / R_\nu^*) = 2 \operatorname{Re}(\delta R_\nu / R_\nu) = [2 \operatorname{Re}(Q_\nu)](\delta d_1 / d_1)$, or $= [2 \operatorname{Re}(\bar{Q}_\nu)]\delta d_1$.

The quantity K, eq. (4.86), is significant because it defines a complex sensitivity factor that couples a fractional change of film thickness $\delta d_1/d_1$ to a corresponding fractional change of the ellipsometric ratio of reflection coefficients $\delta\rho/\rho$, eq. (4.85).

The above results are general in that the small changes of film thickness δd_1 are assumed to take place in the neighborhood of an arbitrary initial value d_1. Many cases of practical interest involve the growth of very thin films at the interface between an ambient and a substrate starting from zero film thickness d_1. In this case, we desire to relate $\delta R_\nu/R_\nu$ and $\delta\rho/\rho$ directly to δd_1, instead of $\delta d_1/d_1$ which becomes infinite when $d_1 = 0$. The equations that apply when the initial film thickness is zero can be readily obtained by setting $X = 1$ [from eq. (4.77), $X = 1$ when $d_1 = 0$] in eqs. (4.83)–(4.86). They are listed below

$$\delta R_\nu/R_\nu = \gamma \left[\frac{r_{12\nu}}{r_{01\nu} + r_{12\nu}} - \frac{r_{01\nu}r_{12\nu}}{1 + r_{01\nu}r_{12\nu}} \right] \delta d_1$$
$$= \bar{Q}_\nu \delta d_1, \tag{4.87}$$

$$\bar{Q}_\nu = \gamma \left[\frac{r_{12\nu}}{r_{01\nu} + r_{12\nu}} - \frac{r_{01\nu}r_{12\nu}}{1 + r_{01\nu}r_{12\nu}} \right], \tag{4.88}$$

$$\delta\rho/\rho = \bar{K}\delta d_1, \tag{4.89}$$

$$\bar{K} = \bar{Q}_p - \bar{Q}_s$$
$$= \gamma \left[\frac{r_{12p}}{r_{01p} + r_{12p}} - \frac{r_{01p}r_{12p}}{1 + r_{01p}r_{12p}} - \frac{r_{12s}}{r_{01s} + r_{12s}} + \frac{r_{01s}r_{12s}}{1 + r_{01s}r_{12s}} \right]. \tag{4.90}$$

Recall that γ appearing in the above equations is given by eq. (4.78).

4.4.3.2. Transmission

The main equation of transmission ellipsometry, eq. (4.71), can be written as

$$\rho = M(D_s/D_p), \tag{4.91}$$

where

$$M = (t_{01p}/t_{01s})(t_{12p}/t_{12s}), \tag{4.92}$$

and D_ν ($\nu = p, s$) are the same as in eqs. (4.76). By taking the logarithmic differential of eq. (4.91), we obtain

$$\delta\rho/\rho = (\delta M/M) + (\delta D_s/D_s) - (\delta D_p/D_p). \tag{4.93}$$

If all differential changes arise from changes of film thickness alone, then

and eq. (4.93) simplifies to
$$\delta\rho/\rho = (\delta D_s/D_s) - (\delta D_p/D_p)$$
$$= [(1/D_s)(\partial D_s/\partial X) - (1/D_p)(\partial D_p/\partial X)]\delta X. \qquad (4.94)$$

δX is the differential change of the complex exponential function X, eq. (4.77), that corresponds to a change of film thickness δd_1. If we substitute D_ν and their derivatives $\partial D_\nu/\partial X$ ($\nu = p, s$) from eqs. (4.76) and (4.82), respectively, and δX from eq. (4.81) into eq. (4.94) we get

$$\delta\rho/\rho = -\gamma d_1 X \left[\frac{r_{01p}r_{12p}}{1 + r_{01p}r_{12p}X} - \frac{r_{01s}r_{12s}}{1 + r_{01s}r_{12s}X} \right] (\delta d_1/d_1)$$
$$= K(\delta d_1/d_1), \qquad (4.95)$$

where

$$K = -\gamma d_1 X \left[\frac{r_{01p}r_{12p}}{1 + r_{01p}r_{12p}X} - \frac{r_{01s}r_{12s}}{1 + r_{01s}r_{12s}X} \right]. \qquad (4.96)$$

Equation (4.95) shows that K of eq. (4.96) defines a complex sensitivity factor that couples a fractional change of film thickness $\delta d_1/d_1$ to a corresponding fractional change $\delta\rho/\rho$ of the ellipsometric ratio of transmission coefficients. Equations (4.95) and (4.96) apply when film-thickness changes δd_1 occur around an arbitrary initial value d_1. For film growth starting from zero thickness, we set $X = 1$ in eq. (4.95)

$$\delta\rho/\rho = -\gamma \left[\frac{r_{01p}r_{12p}}{1 + r_{01p}r_{12p}} - \frac{r_{01s}r_{12s}}{1 + r_{01s}r_{12s}} \right] \delta d_1$$
$$= \bar{K}\delta d_1, \qquad (4.97)$$

where

$$\bar{K} = -\gamma \left[\frac{r_{01p}r_{12p}}{1 + r_{01p}r_{12p}} - \frac{r_{01s}r_{12s}}{1 + r_{01s}r_{12s}} \right], \qquad (4.98)$$

and γ is given by eq. (4.78).

4.4.3.3. ψ and Δ sensitivity factors

Instead of the fractional change $\delta\rho/\rho$ of the ratio of reflection or transmission coefficients, it is usually more convenient to deal with the corresponding changes $\delta\psi$ and $\delta\Delta$ of the ellipsometric angles ψ and Δ. By taking the logarithmic differential of eq. (4.17), we obtain

$$\delta\rho/\rho = (2\delta\psi/\sin 2\psi) + j\delta\Delta, \qquad (4.99)$$

so that

$$\delta\psi = (\tfrac{1}{2}\sin 2\psi)\,\mathrm{Re}\,(\delta\rho/\rho), \qquad (4.100a)$$

$$\delta\Delta = \text{Im}\,(\delta\rho/\rho). \tag{4.100b}$$

$\text{Re}\,(Y)$ and $\text{Im}\,(Y)$ denote the real and imaginary parts of Y. When $\delta\rho/\rho$ is caused by a fractional change of film thickness $\delta d_1/d_1$, we can rewrite eqs (4.100) as

$$\delta\psi = [(\tfrac{1}{2}\sin 2\psi)\,\text{Re}\,(K)](\delta d_1/d_1), \tag{4.101a}$$

$$\delta\Delta = [\text{Im}\,(K)](\delta d_1/d_1), \tag{4.101b}$$

for an arbitrary initial film thickness d_1, or as

$$\delta\psi = [(\tfrac{1}{2}\sin 2\bar{\psi})\,\text{Re}\,(\bar{K})]\delta d_1, \tag{4.102a}$$

$$\delta\Delta = [\text{Im}\,(\bar{K})]\delta d_1, \tag{4.102b}$$

for zero initial film thickness. Equations (4.101) and (4.102) are obtained by substituting $\delta\rho/\rho$ from eq. (4.85) [or (4.89)] and eq. (4.95) [or (4.97)] into eqs (4.100), respectively. Cast somewhat differently, eqs. (4.101) and (4.102) read

$$\psi = \psi_0 + S_\psi(\delta d_1/d_1), \tag{4.103a}$$

$$\Delta = \Delta_0 + S_\Delta(\delta d_1/d_1), \tag{4.103b}$$

and

$$\psi = \bar{\psi} + \bar{S}_\psi \delta d_1, \tag{4.104a}$$

$$\Delta = \bar{\Delta} + \bar{S}_\Delta \delta d_1, \tag{4.104b}$$

where

$$S_\psi = (\tfrac{1}{2}\sin 2\psi_0)\,\text{Re}\,(K), \tag{4.105a}$$

$$S_\Delta = \text{Im}\,(K), \tag{4.105b}$$

$$\bar{S}_\psi = (\tfrac{1}{2}\sin 2\bar{\psi})\,\text{Re}\,(\bar{K}), \tag{4.106a}$$

$$\bar{S}_\Delta = \text{Im}\,(\bar{K}). \tag{4.106b}$$

ψ_0, Δ_0 are the ellipsometric angles that correspond to an arbitrary initial film thickness d_1 and $\bar{\psi}$, $\bar{\Delta}$ are the same quantities when $d_1 = 0$ (i.e., the parameters characteristic of the film-free ambient-substrate interface). Equations (4.99)–(4.106) are the same for both reflection and transmission ellipsometry. $S_{\psi R}$, $S_{\Delta R}$ and $S_{\psi T}$, $S_{\Delta T}$ are psi and delta sensitivity factors [162, 163] that determine the extent to which small changes of thickness of the film phase influence the ellipse of polarization of the reflected and transmitted light beams. These quantities are, therefore, important and their magnitudes and behaviour with angle of incidence are of great interest. If the complex sensitivity factors K and $\bar{K}$ are expressed in terms of their magnitudes and angles

$$K = |K|e^{j\theta}, \qquad \bar{K} = |\bar{K}|e^{j\bar{\theta}}, \tag{4.107}$$

the psi and delta sensitivity factors of eqs. (4.105) and (4.106) become

$$S_\psi = (180/\pi)(\tfrac{1}{2}\sin 2\psi_0)|K|\cos\theta, \tag{4.108a}$$

$$S_\Delta = (180/\pi)|K|\sin\theta, \tag{4.108b}$$

and

$$\bar{S}_\psi = (180/\pi)(\tfrac{1}{2}\sin 2\bar{\psi})|\bar{K}|\cos\bar{\theta}, \tag{4.109a}$$

$$\bar{S}_\Delta = (180/\pi)|\bar{K}|\sin\bar{\theta}, \tag{4.109b}$$

respectively, in units of degrees per angstrom. For reflection ellipsometry, the complex sensitivity factors $|K|e^{j\theta}$, $|\bar{K}|e^{j\bar{\theta}}$ that appear in eqs. (4.108) and (4.109) are given by eqs. (4.86) and (4.90), whereas for transmission ellipsometry, they are given by eqs. (4.96) and (4.98).

The above analysis provides a unified treatment of the theory of linear approximation for reflection and transmission ellipsometry [164]. The linear approximation of the equation of reflection ellipsometry differs from previous forms [144, 153–160] in that an arbitrary initial nonzero film thickness is assumed. It is less complicated, because it starts with the fractional change of the ellipsometric ratio $\delta\rho/\rho$ and relates it to the fractional change of film thickness $\delta d_1/d_1$ through complex sensitivity factors K and $\bar{K}$. By leaving the complex sensitivity factors K and $\bar{K}$ expressed in terms of the interface Fresnel coefficients [eqs. (4.86), (4.90), (4.96), and (4.98)] rather than the optical properties of the three media, considerable complexity is avoided and, with it, the possibility of erroneous results.[17] Furthermore, we have seen that the psi and delta sensitivity factors S_ψ and S_Δ are proportional to the projections ($|K|\cos\theta$, $|K|\sin\theta$) of the complex sensitivity factor K on the real and imaginary axes respectively, eqs. (4.108) and (4.109).

In figs. 4.29–4.31, the computed complex sensitivity factor $\bar{K}$ ($=|\bar{K}|e^{j\bar{\theta}}$), and the real psi and delta sensitivity factors $\bar{S}_\psi$ and $\bar{S}_\Delta$ for the reflected and transmitted waves are plotted as functions of angle of incidence ϕ for different films at an air (vacuum)-glass interface. Both cases of external (figs. 4.29 and 4.30) and internal (fig. 4.31) incidence are represented.

In fig. 4.29 (left), notice that the magnitude of the complex sensitivity factor $\bar{K}_R$ goes through a peak of infinite height at the Brewster angle of the

---

[17] Some of the earlier results are in error, as explained by Burge and Bennett [156]. By retaining the sensitivity factor $\bar{K}$ [eq. (4.90)] in complex form, the possibility of obtaining inordinately complicated and possibly incorrect expressions is avoided.

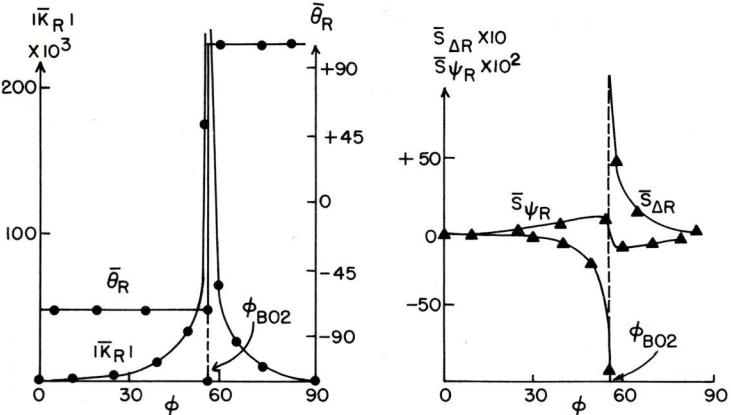

Fig. 4.29. Dependence of the magnitude $|\bar{K}_R|$ and angle $\bar{\theta}_R$ of the complex sensitivity factor $\bar{K}_R$ (left) and of the real psi and delta sensitivity factors $\bar{S}_\psi$ and $\bar{S}_\Delta$ (right) on the angle of incidence ϕ for external-reflection ellipsometry on an air-gold-glass system at $\lambda = 5461$ Å. $|\bar{K}|$ is in reciprocal angstroms, $\bar{S}_\psi$ and $\bar{S}_\Delta$ are in degrees per angstrom, and $\bar{\theta}_R$, ϕ are in degrees. $N_{\text{glass}} = 1.50$, $N_{\text{Au}} = 0.35 - j2.45$.

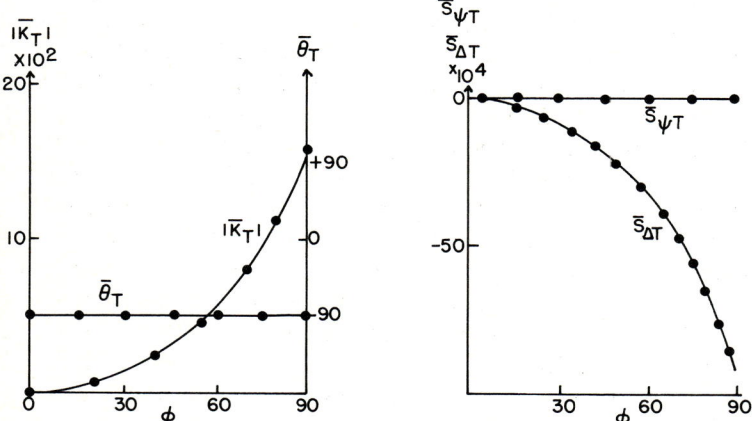

Fig. 4.30. Dependence of the magnitude $|\bar{K}_T|$ and angle $\bar{\theta}_T$ of the complex sensitivity factor $\bar{K}_T$ (left) and of the real psi and delta sensitivity factors $\bar{S}_\psi$ and $\bar{S}_\Delta$ (right) on the angle of incidence ϕ for external-transmission ellipsometry on an air-magnesium fluoride-glass system at $\lambda = 5461$ Å. $|\bar{K}|$ is in reciprocal angstroms, $\bar{S}_\psi$ and $\bar{S}_\Delta$ are in degrees per angstrom, and $\bar{\theta}_T$, ϕ are in degrees. $N_{\text{glass}} = 1.50$, $N_{\text{MgF}_2} = 1.38$.

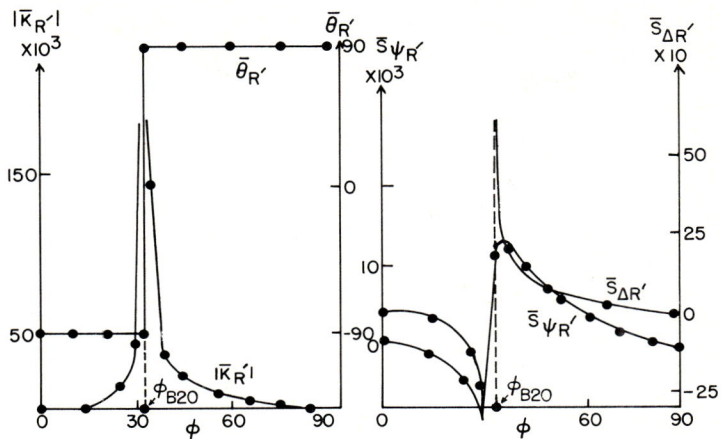

Fig. 4.31. Dependence of the magnitude $|\bar{K}_{R'}|$ and angle $\bar{\theta}_{R'}$ of the complex sensitivity factor $\bar{K}_{R'}$ (left) and of the real psi and delta sensitivity factors $\bar{S}_\psi$ and $\bar{S}_\Delta$ (right) on the angle of incidence ϕ for *internal* incidence on a glass-silicon-air system at $\lambda = 5461$ Å. $|\bar{K}_{R'}|$ is in reciprocal angstroms, $\bar{S}_\psi$ and $\bar{S}_\Delta$ are in degrees per angstrom, $\bar{\theta}_{R'}$ and ϕ are in degrees. $N_{\text{glass}} = 1.50$, $N_{\text{Si}} = 4.05 - j0.028$.

air-glass interface

$$\phi_{B02}(\text{air-glass}) = \tan^{-1}(N_{\text{glass}}/N_{\text{air}})$$
$$= \tan^{-1} 1.5 = 56.31°.$$

This is true for any film material and can be proven analytically as follows. At zero film thickness ($d_1 = 0$, $X = 1$), eqs. (4.37) and (4.38) reduce to the identity

$$R_\nu = r_{02\nu} = \frac{r_{01\nu} + r_{12\nu}}{1 + r_{01\nu} r_{12\nu}}, \qquad (\nu = p, s) \tag{4.110}$$

where $r_{02\nu}$ are the Fresnel reflection coefficients for medium 0-medium 2 interface. For parallel-polarized light ($\nu = p$) at the Brewster angle ϕ_{B02}

$$r_{02p} = 0,$$

hence, from eq. (4.110)

$$r_{01p} + r_{12p} = 0, \qquad \text{at } \phi = \phi_{B02}. \tag{4.111}$$

Substituting $r_{01p} + r_{12p} = 0$, we find that the first term in the expression of $\bar{K}_R$ [eq. (4.90)] becomes infinite.

Notice also in fig. 4.29 (left) that the angle $\bar{\theta}_R$ of the complex sensitivity factor $\bar{K}_R$ stays essentially constant near $-\frac{1}{2}\pi$ as the angle of incidence ϕ is increased from zero to ϕ_{B02}, experiences a discontinuous jump of π at ϕ_{B02}, then stays essentially constant again near $+\frac{1}{2}\pi$ between ϕ_{B02} and 90°. $\bar{\theta}_R$ is exactly constant and equal to $-\frac{1}{2}\pi$ ($0 \le \phi < \phi_{B02}$) or $+\frac{1}{2}\pi$ ($\phi_{B02} < \phi \le 90°$) only when the film is non-absorbing.[18]

The psi and delta sensitivity factors $\bar{S}_{\psi R}$ and $\bar{S}_{\Delta R}$ are obtained from the complex sensitivity factor $\bar{K}_R$ by eqs. (4.109). Because the delta sensitivity factor $\bar{S}_{\Delta R}$ is obtained by multiplying $|\bar{K}_R|$ by $\sin \bar{\theta}_R$ [eq. (4.109b)] (and a constant factor $180/\pi$ which is required to convert $\bar{S}_{\Delta R}$ to units of degrees per angstrom) and because $\bar{\theta}_R$ in fig. 4.29 (left) stays essentially constant except for a change of π at the Brewster angle ϕ_{B02}, $\bar{S}_{\Delta R}$ in fig. 4.29 (right) has the same behaviour as $-|\bar{K}_R|$ in the range $0 < \phi \le \phi_{B02}$, and as $+|\bar{K}_R|$ in the range $\phi_{B02} < \phi \le 90°$. Therefore, the delta sensitivity factor $\bar{S}_{\Delta R}$ first decreases rapidly from zero to $-\infty$ as the angle of incidence ϕ is increased from zero (normal incidence) to the Brewster angle ϕ_{B02}, then it experiences a discontinuous jump to $+\infty$ at ϕ_{B02}, and finally it decreases from $+\infty$ to zero as ϕ continues to increase from ϕ_{B02} to 90° (grazing incidence). This behaviour is the same for any film formed at the air-glass interface. Positive and negative $\bar{S}_{\Delta R}$ indicate that Δ increases or decreases, respectively, with increase of thickness of the thin-film phase, starting from the film-free substrate value $\bar{\Delta}$.

From eq. (4.109a), $\bar{S}_{\psi R}$ is obtained by multiplying $|\bar{K}_R|$ by three factors $F_1 = \cos \bar{\theta}_R$, $F_2 = \frac{1}{2} \sin 2\bar{\psi}_R$, and $F_3 = 180/\pi$,

$$\bar{S}_{\psi R} = F_1 F_2 F_3 |\bar{K}_R|.$$

The first of the three factors, $F_1 = \cos \bar{\theta}_R$, is identically zero for transparent films, because $\bar{\theta}_R = \pm \frac{1}{2}\pi$. Therefore,

$$\bar{S}_{\psi R} = 0, \quad \text{at } 0 \le \phi \le 90°,$$

for all transparent films. Thus, to first order, the ellipsometric parameter ψ_R is insensitive to the presence of a transparent thin-film phase between transparent ambient and substrate media. When the film is absorbing, $\bar{\theta}_R$ is nearly a constant which is close, but not equal, to $\pm \frac{1}{2}\pi$, fig. 4.29 (left). Therefore, $\cos \bar{\theta}_R$ is a small nonzero quantity that stays fairly unchanged with angle-of-incidence variations except that it switches sign at the Brewster angle. The second factor, $F_2 = \frac{1}{2} \sin 2\bar{\psi}_R$, is a function of the angle of incidence ϕ. It decreases from a maximum value of $\frac{1}{2}$ at $\phi = 0$ ($\bar{\psi}_R = \frac{1}{4}\pi$) to zero at $\phi = \phi_{B02}$ ($\bar{\psi}_R = 0$), then increases again to $\frac{1}{2}$ at $\phi = 90°$ ($\bar{\psi}_R = \frac{1}{4}\pi$).

---

[18] This can be seen as follows. When the three phases are transparent, all interface Fresnel reflection coefficients that appear in eq. (4.90) become real. Thus the angle of $\bar{K}_R$ is that of γ (except for an uncertainty of $\pm \pi$). From eq. (4.78), we see that γ is pure imaginary when N_1 is real, that is when the film is non-absorbing. Thus, $\bar{\theta}_R$ is either $\frac{1}{2}\pi$ or $-\frac{1}{2}\pi$. The extent to which $\bar{\theta}_R$ stays constant near $\pm \frac{1}{2}\pi$ depends on the amount of film absorption.

Notice that the product $F_2|\bar{K}_R|$ becomes $0 \times \infty$ at the Brewster angle. It can be shown that this "indeterminate" product is actually zero so that

$$\bar{S}_{\psi R} = 0, \quad \text{at } \phi = \phi_{B02},$$

for any absorbing film. Because $\bar{S}_{\psi R}$ is zero also at $\phi = 0$ and $\phi = 90°$ and changes sign at $\phi = \phi_{B02}$, it must show a minimum (maximum) in the interval $0 < \phi < \phi_{B02}$ and a maximum (minimum) in the interval $\phi_{B02} < \phi < 90°$. Because of the behaviour of $|\bar{K}_R|$, fig. 4.29 (left), the angles of incidence at which $\bar{S}_{\psi R}$ is maximum and minimum are close to the Brewster angle ϕ_{B02}, fig. 4.29 (right).

In fig. 4.30 (left) the magnitude $|\bar{K}_T|$ of the transmission complex sensitivity factor $\bar{K}_T$ increases monotonically as a function of angle of incidence with no structure in the neighbourhood of the Brewster angle of reflection ϕ_{B02}. This is true for other film materials. Maximum sensitivity occurs at grazing incidence ($\phi = 90°$). The angle $\bar{\theta}_T$ of the complex sensitivity factor $\bar{K}_T$ stays exactly constant at $-\frac{1}{2}\pi$. In contrast with $\bar{\theta}_R$ in reflection, $\bar{\theta}_T$ does not experience any discontinuity at the Brewster angle or any other angle of incidence. $\bar{\theta}_T = -\frac{1}{2}\pi$ is exactly satisfied only for transparent films.

The psi and delta sensitivity factors $\bar{S}_{\psi T}$ and $\bar{S}_{\Delta T}$ for transmission ellipsometry are obtained from $\bar{K}_T$ by eqs. (4.109). They are plotted in fig. 4.30 (right). The salient characteristics of these curves (and others obtained for absorbing films) are as follows. Because $\bar{S}_{\Delta T} \propto |\bar{K}_T| \sin \bar{\theta}_T$ [eq. (4.109b)], and $\bar{\theta}_T$ is essentially (exactly for the transparent films) constant, $\bar{S}_{\Delta T}$ has the same form as $|\bar{K}_T|$ when plotted as a function of the angle of incidence ϕ. $|\bar{S}_{\Delta T}|$ increases monotonically with ϕ to a maximum value at $\phi = 90°$ (grazing incidence). Because $\bar{S}_{\psi T} \propto |\bar{K}_T|(\sin 2\bar{\psi}_T) \cos \bar{\theta}_T$ [eq. (4.109a)], we have

$$\bar{S}_{\psi T} = 0, \quad \text{at } 0 \le \phi \le 90°,$$

when $\bar{\theta}_T = \pm 90$, i.e., for all transparent films. Therefore, the growth of very thin films in an all-transparent three-phase system is not observable, to first order, in changes of the ellipsometric parameter ψ_T. $\bar{\psi}_T$, the bare-substrate value of ψ_T, increases monotonically but gently as a function of the angle of incidence between $\phi = 0$ and $\phi = 90°$. The increase of $|\bar{K}_T|$ with ϕ, however, is more rapid, with the result that $|\bar{S}_{\psi T}|$ increases monotonically with angle of incidence in the general case of absorbing films. The maximum values of $\bar{S}_{\psi T}$ occur at grazing incidence.

Figure 4.31 shows the sensitivity functions $|\bar{K}_{R'}|$, $\bar{\theta}_{R'}$ (left) and $\bar{S}_{\psi R'}$, $\bar{S}_{\Delta R'}$ (right) for the glass-Si-air system. Other film materials lead to similar behavior, only the magnitudes are different. Notice that $|\bar{K}_{R'}|$ and $|\bar{S}_{\Delta R'}|$ become infinitely large at the Brewster angle of internal incidence

$$\phi_{B20} = \tan^{-1}(N_0/N_2) = 90° - \tan^{-1}(N_2/N_0)$$
$$= 90° - \phi_{B02} = 33.69°,$$

and that $\bar{\theta}_{R'}$ experiences a discontinuity at that angle. Very high sensitivities are attainable near ϕ_{B20} in internal incidence as they are realizable near ϕ_{B02} in external incidence.

One interesting observation that can be made from fig. 4.31 (as well as from the corresponding figures for other films) is that no structure exists in any of the three ellipsometric sensitivity factors $|\bar{K}_{R'}|$, $\bar{S}_{\psi R'}$ and $\bar{S}_{\Delta R'}$ at or near the critical angle of total internal reflection. The angle of importance in internal-reflection ellipsometry is the Brewster angle ϕ_{B20}.

4.4.3.4. Validity of the linear approximations
To examine the range of thickness values over which the linear approximations are of acceptable accuracy, we show in figs. 4.32–4.35 the computed curves for ψ and Δ as functions of film thickness in the range 0–300 Å based on the exact equations of ellipsometry. These curves are compared with straight lines drawn through the point $\bar{\psi}$, $\bar{\Delta}$ (which characterizes the film-free ambient-substrate interface) with slopes given by the psi and delta sensitivity factors $\bar{S}_{\psi}$ and $\bar{S}_{\Delta}$, respectively.

The case of external reflection near the Brewster angle for the air-MgF$_2$-glass system is shown in fig. 4.32 (left). Here $\bar{S}_{\psi R} = 0$ and ψ_R increases parabolically with thickness. This restricts the linear range for ψ to sub-angstrom thicknesses near the minimum of the parabola. The linearity of Δ_R holds for thicknesses up to 30 Å. The case of external transmission at $\phi = 75°$ (this represents a compromise between high sensitivity near grazing

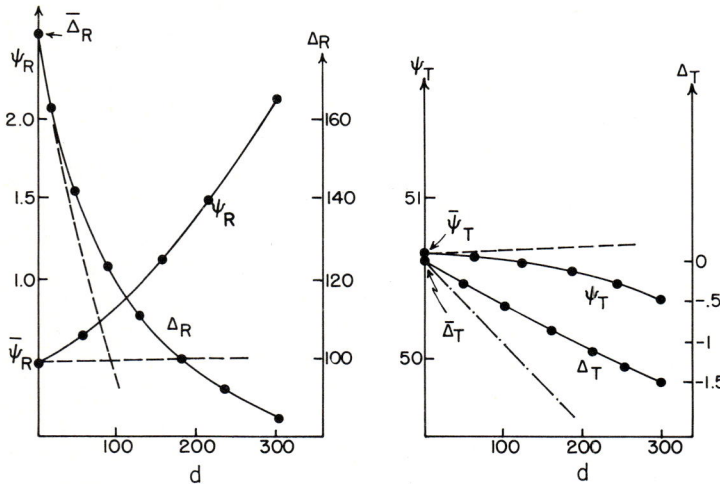

Fig. 4.32. Dependence of the ellipsometric angles ψ and Δ on film thickness d for the air-magnesium fluoride-glass system at $\lambda = 5461$ Å in case of external reflection at $\phi = 56°$ (left) and external transmission at $\phi = 75°$ (right). ψ, Δ are in degrees and d is angstroms. Notice the range of validity of the linear approximation. $N_{\text{glass}} = 1.50$, $N_{\text{MgF}_2} = 1.38$.

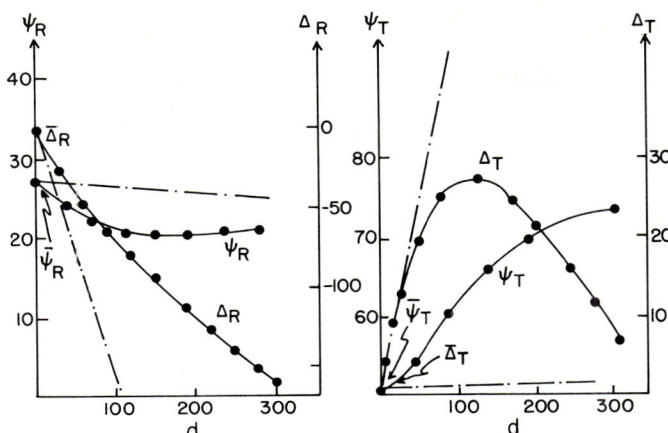

Fig. 4.33. Dependence of the ellipsometric angles ψ and Δ on film thickness d for the air-silicon-glass system at $\lambda = 5461$ Å in case of external reflection (left) and external transmission (right) at $\phi = 75°$. ψ, Δ are in degrees and d is angstroms. Notice the range of validity of the linear approximation. $N_{\text{glass}} = 1.50$, $N_{\text{Si}} = 4.05 - j0.028$.

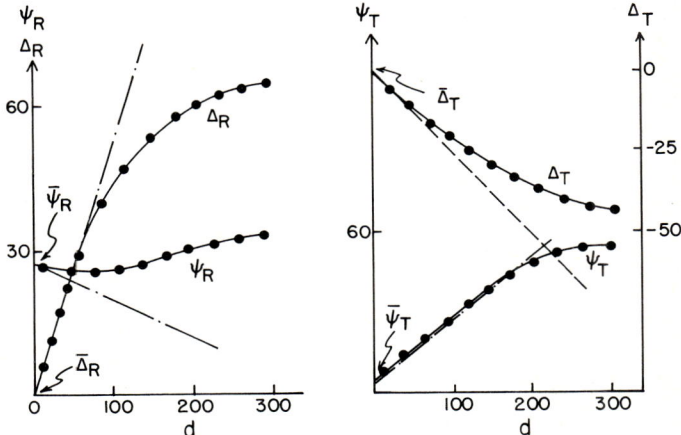

Fig. 4.34. Dependence of the ellipsometric angles ψ and Δ on film thickness d for the air-gold-glass system at $\lambda = 5461$ Å in case of external reflection (left) and external transmission (right) at $\phi = 75°$. Notice the range of validity of the linear approximation. $N_{\text{glass}} = 1.50$, $N_{\text{Au}} = 0.35 - j2.45$.

incidence and experimental convenience), for the same system, is shown in fig. 4.32 (right). Because $\bar{S}_{\psi T} = 0$ for transparent films, ψ_T is also a parabolic function of film thickness near $d = 0$. The linearity ranges for ψ_T and Δ_T are less than 5 Å in this case.

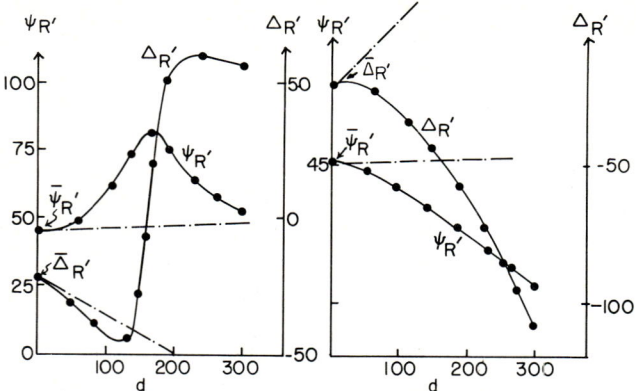

Fig. 4.35. Dependence of the ellipsometric angles ψ and Δ on film thickness d for the glass-gold-air system (left) and glass-silicon-air system (right) at $\lambda = 5461$ Å in case of total internal reflection at $\phi = 75°$. ψ, Δ are in degrees and d is in angstroms. Notice the range of validity of the linear approximation. $N_{\text{glass}} = 1.50$, $N_{\text{Au}} = 0.35 - j2.45$, $N_{\text{Si}} = 4.05 - j0.028$.

Figure 4.33 (left) shows the results for external reflection at $\phi = 75°$ for the air-Si-glass system. The linear ranges for ψ and Δ are about 1 Å and 10 Å, respectively. Transmission ellipsometry on the same system at the same angle of incidence, fig. 4.33 (right), indicates that the linear approximations of ψ_T and Δ_T are valid over thickness ranges of 2 and 20 Å, respectively.

Results for external reflection and transmission by the air-Au-glass system at $\phi = 75°$ are shown in fig. 4.34. Notice that ψ_T is accurately obtained from the linear approximation for thicknesses up to about 140 Å.

Examples of total internal reflection ellipsometry are shown in fig. 4.35 for the glass-Au-air (left) and the glass-Si-air (right) systems, respectively, at $\phi = 75°$. The thickness range for linearity of ψ and Δ is about 10 Å. In fig. 4.35 (left), it is interesting to observe the considerable structure in the curves in the thickness range 0–300 Å for Au. This is not unexpected, because of the strong influence that the film exerts on the evanescent wave which has a spatial extent of the same order of magnitude as the film thickness [165].

The above examples show that the linear approximations in external and internal reflection and transmission ellipsometry are generally applicable for the growth of very thin films (~ 10 Å) and are, sometimes, further limited to submonolayer coverage (~ 1 Å). This is a range of thickness of considerable interest for the application of ellipsometry to the detection of extremely small amounts of adsorbed atomic or molecular species at interfaces. An obvious conclusion to reach from the above examples is that the range of validity of the linear approximation is quite dependent on the particular system under consideration.

4.5. Numerical inversion of the exact equation of reflection ellipsometry

The emphasis throughout this chapter is to present analytical expressions which can be used to compute measurable ratios of reflection or transmission coefficients for a given stratified structure of known optical properties. The calculations involved in this *forward* problem are readily programmable for digital computation. The *reverse* problem in which ellipsometrically measured ratios of reflection or transmission coefficients are to be used to find some unknown optical properties (including film thicknesses) of a stratified structure under measurement is of considerable practical interest. Because of their nonlinear and transcendental nature, the equations that govern the forward problem [*e.g.*, eqs. (4.60) and (4.71)] usually defy analytical inversion for all but a few simple special cases. Fortunately, with the availability of digital computers, accurate numerical inversion of such sets of nonlinear transcendental equations is possible.[19]

In the present section, we confine our attention primarily to the inversion of the exact equation of *reflection* ellipsometry, eq. (4.60), for an ambient-film-substrate three-phase system. Several aspects of this problem are, however, quite general.

A numerical-inversion computer program that inverts sets of optical equations for a stratified ambient-*multiple film*-substrate structure (to be discussed in the subsequent section of this chapter) has been developed by Hansen [166]. This program uses a generalized secant method for solving simultaneous nonlinear equations, and accepts as input measured (intensity) reflectances and transmittances, as well as ellipsometric ratios of (complex-amplitude) reflection and transmission coefficients. Hansen, however, applied his program largely to spectrophotometric measurements of reflectance and transmittance, but not to ellipsometric data.

4.5.1. Cases that allow analytical inversion

Two cases that allow eq. (4.60) to be *analytically* inverted are (a) two-phase (ambient-substrate) system with known refractive index for one phase, and (b) three-phase (ambient-film-substrate) system with known refractive indices for all phases. In case (a), the only unknown is the refractive index of the second phase which can be computed from eq. (4.20) using the ratio ρ of the

---

[19] It is a fact that many of the advances that have been made in ellipsometry since about 1960 [1 and 64] have significantly depended on the availability of digital computers.

p-to-s reflection coefficients at a single angle of incidence ϕ_0, with the wavelengths λ being that at which the refractive index of the first phase is known. In case (b) [167], the only unknown is the film thickness d_1 which can be determined by solving eq. (4.65) for the complex exponential $X = e^{-j2\beta}$ in terms of one measured value of ρ and the calculable Fresnel interface reflection coefficients for the p and s polarizations. The solution is

$$X = \frac{-(B - \rho E) \pm [(B - \rho E)^2 - 4(C - \rho F)(A - \rho D)]^{\frac{1}{2}}}{2(C - \rho F)} \tag{4.112}$$

where A, B, C, D, E, and F are functions of the interface reflection coefficients r_{01p}, r_{12p}, r_{01s}, and r_{12s} given by eqs. (4.67) and (4.68). From the known values of the refractive indices N_0, N_1, N_2 of the three phases (ambient, film, and substrate, respectively) and the angle of incidence ϕ_0, the Fresnel coefficients r_{01p}, r_{12p}, r_{01s}, r_{12s} for the p and s polarizations at the ambient-film (0-1) and film-substrate (1-2) interfaces can be computed using eqs. (4.41)–(4.44). These coefficients, together with the ellipsometrically measured value of ρ, are next substituted into eq. (4.112) to yield two values X_1, X_2 for X (corresponding to the $+$ and $-$ signs that appear in that equation). From eq. (4.77), the film thickness d_1 is given by

$$d_1 = \frac{1}{\gamma} \log X,$$

or

$$d_1 = [j(4\pi)^{-1}(N_1^2 - N_0^2 \sin^2 \phi_0)^{-\frac{1}{2}} \log X]\lambda, \tag{4.113}$$
$$X = X_1 \quad \text{or} \quad X_2,$$

when eq. (4.78) is used.

In the absence of experimental and model errors[20], eq. (4.113) should yield a positive real value for the film thickness d_1. In fact, the appropriate root of eq. (4.112) (X_1 or X_2) should be chosen such that it leads to a positive real calculated film thickness d_1 from eq. (4.113) or, in the presence of errors, to a complex value with the smaller imaginary part and non-negative real part. In this latter case, the actual film thickness can be taken equal to the real part of the right-hand side of eq. (4.113), with a possible error of the order of magnitude of the associated imaginary part. When both the ambient and film

---

[20]Such errors arise because real ambient-film-substrate systems deviate in several subtle ways from the idealized theoretical models that are used to describe them. For example, rough boundaries, optical inhomogeneity and anisotropy are always inevitably present, to some extent. When such effects are left unaccounted for in an optical model, the resulting theoretical equations should only be considered as approximate; the corresponding inaccuracies define the model errors.

§4.5] INVERSION OF ELLIPSOMETRY EQUATION 317

phases are transparent, and d_1^0 is a solution of eq. (4.113), then

$$d_1^m = d_1^0 + mD_\phi, \quad m = 0, \pm 1, \pm 2, \ldots \tag{4.114a}$$

is also a solution, where D_ϕ is the film thickness period (the period of the cyclic complex exponential function X and the cyclic ratio of reflection coefficients ρ). Substitution of D_ϕ from eq. (4.70) into eq. (4.114a) leads to

$$d_1^m = d_1^0 + [\tfrac{1}{2}(N_1^2 - N_0^2 \sin^2 \phi_0)^{-\tfrac{1}{2}}]m\lambda, \quad m = 0, \pm 1, \pm 2, \ldots \tag{4.114b}$$

The indeterminancy of film thickness indicated by eq. (4.114b) represents no difficulty when the range of film thickness is known *a priori*, from other considerations or from additional nonellipsometric (*e.g.*, reflectance) measurements.

4.5.2. *General formulation of the numerical-inversion problem and choice of error function*

For a given ambient-film-substrate system, measurement of the ratio of reflection coefficients ρ (or, equivalently, ψ and Δ) at one wavelength λ and one angle of incidence ϕ_0 provides enough information to determine *two* real optical parameters of the system *only*, assuming that all the remaining parameters are known. Thus, for example, the complex refractive index of the film N_1 can be determined if its thickness d_1 and the refractive indices N_0 of the ambient and N_2 of the substrate are known.

When the number of unknown optical parameters of a three-phase system exceeds two, it is necessary to acquire additional experimental data to determine (or overdetermine) such parameters. In the most general case, all seven parameters: film thickness d_1 and the real and imaginary parts of the three complex refractive indices N_0, N_1, and N_2 of the three phases, may all have to be measured. McCrackin and Colson [167] investigated a number of possible ways to increase the number of independent ellipsometric measurements. These can be generally grouped into two categories: (i) multiple measurements with a deliberate change of an adjustable phase, and (ii) multiple-angles-of-incidence (MAI) measurements. In (i), it is possible, for example, to change a transparent inert fluid ambient without disturbing the properties of the film-substrate subsystem and make repeated measurements of the ratio of reflection coefficients ρ at fixed wavelength and angle of incidence for the different ambients used. The substrate may also be changed when interest is primarily in the properties of a thin film (*e.g.*, a Langmuir-Blodgett layer), and there is reason to believe that the substrate does not exert a considerable influence on the film. It is still even more common to take repeated measurements of ρ at successively increasing thickness values for a growing interphasial film between an ambient and a substrate, and interpret the data assuming that the refractive index of the film phase stays the same independent of film thickness. In (ii), measurements of the ratio of reflection coefficients ρ at an adequate number of properly chosen angles of incidence provide enough information to determine all of the optical parameters of a given ambient-film-substrate system, without having to perturb such a system in any manner. Because of its simplicity, directness, and its nondestructiveness, the use of the multiple-angle-of-incidence approach will be emphasized in this section.

Besides ellipsometric measurements in reflection, ellipsometric measurements in transmission on sufficiently transparent systems can be carried out [164]. Also, complementary photometric measurements [168–170] of intensity reflectances or transmittances may conveniently provide additional data when the number of unknown optical parameters for a three-phase system exceeds two. Measurements at several wavelengths can be useful.

Let ρ_i^m denote the ratio of reflection coefficients from the ith measurement on an ambient-film-substrate system, and ρ_i^c be the computed value of this ratio from eq. (4.60) assuming a given set of values for the unknown optical parameters. The computational problem in reflection ellipsometry consists of the search for a set of optical parameters (N_0, N_1, N_2, d_1) that characterizes the optical system under measurement such that the quantity

$$F = \sum_{i=1}^{M} |\rho_i^m - \rho_i^c(N_0, N_1, N_2, d_1, \phi_0, \lambda)|^2, \tag{4.115}$$

is zero or minimum, where M is the number of measurements. Dependent upon which procedure is used to secure additional measurements, any one or combination of the optical parameters N_0, N_2, d_1, or ϕ_0 can be different from one measurement to the next. An added subscript i may be necessary for these quantities in eq. (4.115), but is dropped for simplicity. Also, for a given procedure, the total number of unknown optical parameters can be easily determined. For example, when measurements are repeated M times at M different thickness values of a film that grows between the same ambient and substrate media, the total number of unknown real optical parameters is (at most) $M + 6$, corresponding to the M film thicknesses and the real and imaginary parts of N_0, N_1, and N_2. Because each measurement of ρ provides two real quantities ψ and Δ, the number of independent measurements is $2M$. Obviously, we must have $2M \geq M + 6$, or $M \geq 6$, if all the unknowns are to be determined. As an exercise, the reader may work out for himself the number of unknowns and (hence the number of required measurements) for the different methods that can be used to increase the optical data.

In an ideal situation where there are no experimental or model errors, it should be possible to compute values for the unknown optical parameters that would make F of eq. (4.115) identically zero. However, instrumental and model imperfections are unavoidable and F represents an error function that has to be minimized by numerical computation. The error function that may be chosen is not unique and the form of eq. (4.115) represents only one possibility. An alternate error function is based on dealing with ψ and Δ directly and not $\rho\,(=\tan\psi\,e^{j\Delta})$. Thus, an error function G can be defined as

$$G = \sum_{i=1}^{M} [(\Delta_i^c - \Delta_i^m)^2 + (\psi_i^c - \psi_i^m)^2], \qquad (4.116)$$

which would also be required to be zero or minimum. In eq. (4.116), (ψ_i^m, Δ_i^m) are the ellipsometric parameters from the ith measurement, while (ψ_i^c, Δ_i^c) are the corresponding values of these parameters computed using eq. (4.60). The computational problem is certainly dependent on the choice of the error function. The difference between the error functions F and G can be brought out clearly by noting that a given error $\delta\rho$ in ρ is related to errors $\delta\psi$ and $\delta\Delta$ in ψ and Δ, respectively, by [see eq. (4.99)]

$$\delta\rho = \tan\psi\,e^{j\Delta}(2\csc 2\psi\,\delta\psi + j\delta\Delta),$$

or

$$|\delta\rho|^2 = \tan^2\psi\,[(\delta\Delta)^2 + 4\csc^2 2\psi\,(\delta\psi)^2]. \qquad (4.117)$$

Upon substitution of the expression for $|\delta\rho|^2$ of eq. (4.117) into eq. (4.115), we may write

$$F = \sum_{i=1}^{M} |\delta\rho_i|^2 = \sum_{i=1}^{M} \tan^2\psi_i\,[(\delta\Delta)_i^2 + 4\csc^2 2\psi_i\,(\delta\psi)_i^2] \qquad (4.118)$$

Comparison between eq. (4.116) and (4.118) reveals that in the error function F the ψ-error term from the ith measurement is weighted by $4\csc^2 2\psi_i$ and the combined-error term from the ith measurement is weighted by $\tan^2\psi_i$, compared to the respective error terms in the error function G. Because the accuracies of the measurement of ψ and Δ are more or less the same, the weighting involved in the error function F is unwarranted, and it is generally preferable to use the error function G as the basis of the computational numerical-inversion approach.

Analyses of some aspects of the computational problem in reflection ellipsometry have been published [167–175]. An inversion procedure that promises to be efficient, rapidly convergent and accurate is that based on the fact that the optical constants (refractive indices and extinction coefficients) of the film and substrate can be separated from the film thickness [109]. In the case of transparent films, this follows because the optical constants of the film and the substrate should satisfy the condition $|X| = 1$ [see eqs. (4.69) and (4.70)] independently of the film thickness d. From a number of independent measurements of ρ (this depends on the number of unknowns to be determined) we write the required equations, one for each measurement, in the form of $|X| = 1$, where X is given by eq. (4.112). The numerical inversion procedure is to iterate on the optical constants of both the film and the substrate until the condition $|X| = 1$ is satisfied for all such equations. Substituting the optical constants thus obtained into eq. (4.113), we obtain the film thickness. When the film is absorbing, iteration on the optical constants is directed to zeroing or minimizing the imaginary part of the right-hand side of each of a set of equations (one for each independent measured ρ) of the form of eq. (4.113), where X in the latter equation is given by eq. (4.112).

4.5.3. Multiple-Angle-of-Incidence (MAI) ellipsometry [175]

In most cases of practical interest, the medium of incidence (the ambient) has a known refractive index N_0 and only the optical parameters of the film-substrate subsystem are to be determined. These optical parameters are the complex refractive indices N_1 and N_2 of the film and substrate respectively, and the film thickness d_1. For simplicity, a notation similar to that adopted by Ibrahim and Bashara [175] is used here. We denote the complex refractive indices N_1 of the film and N_2 of the substrate, the film thickness d_1, and the angle of incidence ϕ_0 by

$$\begin{aligned}
N_1 &= n_F - jk_F, \\
N_2 &= n_S - jk_S, \\
d_1 &= d, \\
\phi_0 &= \phi.
\end{aligned} \quad (4.119)$$

It is being assumed that the ambient, film, and substrate media are all homogeneous and optically isotropic and that the refractive index experiences a sharp (discontinuous) step-like jump across the ambient-film and film-substrate parallel plane interfaces.

Let multiple-angle-of-incidence (MAI) measurements be made at M different angles of incidence ϕ_i. From eqs. (4.60), (4.62) and (4.63), $2M$ simultaneous non-linear equations are obtained that relate the measured ellipsometric angles ψ_i^m and Δ_i^m at a given wavelength to the optical parameters $n_S, k_S, n_F, k_F,$ and d of the film-substrate subsystem as follows

$$\psi_i^m = \psi_i^c(n_S, k_S, n_F, k_F, d, \phi_i), \quad (4.120)$$

$$\Delta_i^m = \Delta_i^c(n_S, k_S, n_F, k_F, d, \phi_i), \quad (4.121)$$

where

$$i = 1, 2, \ldots, M,$$

and the superscripts m and c distinguish the measured and computed values of ψ and Δ respectively. Because of experimental and/or model errors, eqs. (4.120) and (4.121) cannot all be satisfied exactly so that a least-squares solution must be sought.

If the parameters n_S, k_S, n_F, k_F and d are represented as the components of a vector $\boldsymbol{B} = (b_1, b_2, \ldots)$, the computational part of the problem is to find a vector $\boldsymbol{B}_0$ such that the sum of the squares of the residuals $G(\boldsymbol{B})$

$$G(\boldsymbol{B}) = \sum_{i=1}^{M} \{[\Delta_i^m - \Delta_i^c(\boldsymbol{B}, \phi_i)]^2 + [\psi_i^m - \psi_i^c(\boldsymbol{B}, \phi_i)]^2\} \tag{4.122}$$

is minimum.

In certain instances, the sensitivity of ψ is poor with respect to the optical parameters (i.e., $\partial \psi / \partial b_i$ is very small) and $\psi_i^c(\boldsymbol{B}, \phi_i)$ will not change appreciably as a function of $\boldsymbol{B}$. Consequently, the solution $\boldsymbol{B}_0$ is mainly determined by minimizing the first summation in eq. (4.122)

$$G(\boldsymbol{B}) \approx \sum_{i=1}^{M} [\Delta_i^m - \Delta_i^c(\boldsymbol{B}, \phi_i)]^2. \tag{4.123}$$

4.5.3.1. Independence (or interdependence) of the MAI system of equations: The parameter correlation test

A linear Taylor-series expansion of $\psi_i^c(\boldsymbol{B}, \phi_i)$ and $\Delta_i^c(\boldsymbol{B}, \phi_i)$ around $\boldsymbol{B}_0$ yields

$$\psi_i^c(\boldsymbol{B}, \phi_i) = \psi_i^c(\boldsymbol{B}_0, \phi_i) + [\nabla \psi_i^c(\boldsymbol{B}_0, \phi_i)] \cdot \delta \boldsymbol{B}, \tag{4.124}$$

$$\Delta_i^c(\boldsymbol{B}, \phi_i) = \Delta_i^c(\boldsymbol{B}_0, \phi_i) + [\nabla \Delta_i^c(\boldsymbol{B}_0, \phi_i)] \cdot \delta \boldsymbol{B}, \tag{4.125}$$

$$i = 1, 2, \ldots M,$$

where ∇ is the gradient operator in the parameter $\boldsymbol{B}$-space, and $\delta \boldsymbol{B}$ is an increment of $\boldsymbol{B}$ around $\boldsymbol{B}_0$. Solutions of eqs. (4.120) and (4.121), $\boldsymbol{B}_0 + \delta \boldsymbol{B}$ (which are different from $\boldsymbol{B}_0$ by $\delta \boldsymbol{B}$), may be obtained with almost the same degree of fit to the ellipsometric measurements [i.e., with the sum of squares of the residuals equal to $G(\boldsymbol{B}_0)$] if the set of equations

$$[\nabla \psi_i^c(\boldsymbol{B}_0, \phi_i)] \cdot \delta \boldsymbol{B} = 0, \tag{4.126}$$

$$[\nabla \Delta_i^c(\boldsymbol{B}_0, \phi_i)] \cdot \delta \boldsymbol{B} = 0, \tag{4.127}$$

$$i = 1, 2, \ldots M,$$

have a nonzero solution for $\delta \boldsymbol{B}$. Equations (4.126) and (4.127) can be expanded to read

$$\sum_{j=1}^{K} \frac{\partial \psi_i^c}{\partial b_j} \delta b_j = 0, \qquad (4.128)$$

$$\sum_{j=1}^{K} \frac{\partial \Delta_i^c}{\partial b_j} \delta b_j = 0, \qquad (4.129)$$

$$i = 1, 2, \ldots M,$$

where K is the number of unknown optical parameters and also gives the dimensionality of the parameter $\boldsymbol{B}$-space. Equations (4.128) and (4.129) can further be rewritten explicitly, when all of the parameters of the film-substrate subsystem are to be determined, as

$$\frac{\partial \psi_i^c}{\partial n_S} \delta n_S + \frac{\partial \psi_i^c}{\partial k_S} \delta k_S + \frac{\partial \psi_i^c}{\partial n_F} \delta n_F + \frac{\partial \psi_i^c}{\partial k_F} \delta k_F + \frac{\partial \psi_i^c}{\partial d} \delta d = 0 \qquad (4.130)$$

$$\frac{\partial \Delta_i^c}{\partial n_S} \delta n_S + \frac{\partial \Delta_i^c}{\partial k_S} \delta k_S + \frac{\partial \Delta_i^c}{\partial n_F} \delta n_F + \frac{\partial \Delta_i^c}{\partial k_F} \delta k_F + \frac{\partial \Delta_i^c}{\partial d} \delta d = 0 \qquad (4.131)$$

$$i = 1, 2, \ldots M.$$

The independence or interdependence of the MAI system of equations [eqs. (4.120) and (4.121)] will be judged by whether a nonzero solution of eqs. (4.128) and (4.129) for the error vector $\delta \boldsymbol{B} = (\delta b_1, \delta b_2, \ldots \delta b_K)$ can be found or not.

Equations (4.128) and (4.129) represent a set of $2M$ simultaneous linear homogeneous equations in the K components of the error vector $\delta \boldsymbol{B}$. If all of these $2M$ equations are *linearly independent*, and $2M \geq K$, the only solution of eqs. (4.128) and (4.129) is $\delta \boldsymbol{B} = 0$, or $\delta b_1 = \delta b_2 = \ldots = \delta b_K = 0$. In this case, there will be a unique solution, $\boldsymbol{B} = \boldsymbol{B}_0$, satisfying the ellipsometric MAI system of eqs. (4.120) and (4.121) that minimizes the sum of the squares of the residuals $G(\boldsymbol{B})$ of eq. (4.122). This corresponds to the desired optimal use of the MAI technique in determining (when $2M = K$) or overdetermining (when $2M > K$) the K optical parameters of a film-substrate subsystem. As will be shown later, conditions for such optimal use of MAI measurements may be achieved even for very thin films (where the validity of the MAI approach was previously thought to be questionable [156]) by proper choice of angles of incidence and/or by use of measurements at different wavelengths.

If some of the $2M$ linear homogeneous eqs. (4.128) and (4.129) are *linearly dependent* on the others and the net number of independent equations is less than K (although $2M \geq K$), a nonzero solution or infinity of solutions for $\delta \boldsymbol{B}$ exist. Under these conditions, there will be a multiplicity of solution vectors, $\boldsymbol{B} = \boldsymbol{B}_0 + \delta \boldsymbol{B}$, that satisfy the MAI system of eqs. (4.120) and (4.121),

§4.5] INVERSION OF ELLIPSOMETRY EQUATION

all leading to the same value of the sum of squares of residuals $G(\boldsymbol{B})$. Consider only two equations of the system of eqs. (4.126) and (4.127) [these are the same as eqs. (4.128) and (4.129)][21]

$$\nabla\Delta_i \cdot \delta\boldsymbol{B} = 0,$$
$$\nabla\Delta_j \cdot \delta\boldsymbol{B} = 0,$$

corresponding to the ith and jth angles of incidence. These two equations will be linearly dependent on each other, hence are equivalent to one equation only, if

$$\nabla\Delta_i = C\nabla\Delta_j, \tag{4.132}$$

where C is a constant. From eq. (4.132), it can be seen that measurements of Δ at two angles of incidence i and j do *not* provide two independent equations if the vector gradients of Δ in the parameter $\boldsymbol{B}$-space at these two angles are parallel. The condition in eq. (4.132) can also be stated as

$$\frac{(\partial\Delta_i/\partial b_1)}{(\partial\Delta_j/\partial b_1)} = \frac{(\partial\Delta_i/\partial b_2)}{(\partial\Delta_j/\partial b_2)} = \cdots = \frac{(\partial\Delta_i/\partial b_K)}{(\partial\Delta_j/\partial b_K)}, \tag{4.133}$$

where the numerators and denominators are the components of the gradient vectors $\nabla\Delta_i$ and $\nabla\Delta_j$ respectively, along the coordinate axes $b_1, b_2, \ldots b_K$ of the K-dimensional $\boldsymbol{B}$-space. Instead of the *total* gradient vectors at the angles i and j being parallel, only the *projections* of these two vectors in a subspace of $\boldsymbol{B}$ may be parallel. Furthermore, this condition may hold for all angles of incidence. Under these circumstances, the parameters that constitute the subspace in which the projections of the total gradient vectors are parallel are said to be correlated. Alternatively, we may say that ℓ parameters $b_1, b_2, \ldots b_\ell (\ell \leq K)$ are correlated when $\partial\Delta/\partial b_1, \partial\Delta/\partial b_2, \ldots,$ and $\partial\Delta/\partial b_\ell$ show (almost) identical variation with angle of incidence. The above definition of correlation has been based on the derivatives of Δ only because often the derivatives of ψ, $\partial\psi/\partial b_j$ ($j = 1, 2, \ldots \ell$), are sufficiently small to make $\Sigma_j (\partial\psi/\partial b_j) \delta b_j$ lie within the range of experimental error ($\sim 0.1°$) even for relatively large values of $\delta b_j/b_j$ (~ 0.3) at all angles of incidence. This leads to the gradient-of-ψ equations [eqs. (4.126)] being automatically satisfied (within experimental error) for relatively large error vectors $\delta\boldsymbol{B}$ so that these equations become of little or no utility. In general, ℓ correlated optical parameters $b_1, b_2, \ldots b_\ell$, cannot be determined simultaneously from MAI measurements. However, MAI data can still be used to determine one of these correlated parameters if the remaining $(\ell - 1)$ are known, in addition to the other $(K - \ell)$ uncorrelated parameters.

---

[21] For simplicity, the superscript c that we used earlier to distinguish the computed from the measured (m) values of Δ and ψ is deleted in subsequent equations.

The relative uncertainties of the correlated parameters can be calculated from eq. (4.129). For example, the uncertainty of estimation of b_1 due to its correlation with b_2 is

$$\delta b_1 = \delta b_2 (\partial \Delta / \partial b_2)/(\partial \Delta / \partial b_1). \tag{4.134}$$

For ℓ correlated parameters, we get

$$\delta b_k = \left[\sum_{i=1}^{\ell-1} \delta b_i (\partial \Delta / \partial b_i) \right] \Big/ (\partial \Delta / \partial b_k). \tag{4.135}$$

In the test for cross correlation to determine whether independent estimates for parameters are needed and/or the extent to which MAI measurements will overdetermine the solutions, the criterion that the relative derivative remains constant with angle of incidence does not depend critically on exact values of the optical parameters. This is of considerable practical importance because rough estimates of the parameters can be used in developing an experimental plan to minimize the interdependence between parameters.

4.5.3.2. The Hessian matrix and rate of convergence

Assume that $G(\boldsymbol{B})$, the sum of the squares of the residuals in the least-squares analysis, has continuous first and second derivatives in the parameter $\boldsymbol{B}$-space. The matrix of the second derivatives, the Hessian matrix $\boldsymbol{H}(\boldsymbol{B})$, is defined by [176, 177]

$$\boldsymbol{H}(\boldsymbol{B}) = \frac{\partial^2 G(\boldsymbol{B})}{\partial b_i \partial b_j}, \quad i, j = 1, \ldots, K, \tag{4.136}$$

where K is the number of parameters. Expansion of $G(\boldsymbol{B})$ in a Taylor series in the neighbourhood of its minimum at $\boldsymbol{B}_0$ gives

$$G(\boldsymbol{B}) - G(\boldsymbol{B}_0) = \tfrac{1}{2} (\boldsymbol{B} - \boldsymbol{B}_0)^{\mathrm{T}} \boldsymbol{H}(\boldsymbol{B}_0)(\boldsymbol{B} - \boldsymbol{B}_0) + \ldots$$

$$= \frac{1}{2} \sum_{i=1}^{K} \sum_{j=1}^{K} \frac{\partial^2 G}{\partial b_i \, \partial b_j} \Delta b_i \Delta b_j + \ldots \tag{4.137}$$

where T denotes the transpose.

Cross sections of the surface $G(\boldsymbol{B})$ when all parameters are fixed except b_i are nearly parabolic in the vicinity of $\boldsymbol{B}_0$ because

$$G(\boldsymbol{B}_i) - G(\boldsymbol{B}_0) = \frac{1}{2}(b_i - b_{0i})^2 \frac{\partial^2 G(\boldsymbol{B})}{\partial b_i^2} + \ldots, \tag{4.138}$$

if higher-order terms are ignored. The diagonal elements of the Hessian matrix (which will hereafter be referred to as DEHM) determine whether the minima of $G(\boldsymbol{B})$ are deep or shallow because increments will be directly proportional to the diagonal element. Also, the DEHM determine the uncertainty δb_j in a given parameter b_j as follows

$$\delta b_j = \left(\frac{2\delta G}{\partial^2 G/\partial b_j^2}\right)^{\frac{1}{2}} \tag{4.139}$$

Computation of the error function $G(\boldsymbol{B})$ of eq. (4.122) involves guesses for the optical-parameter vector $\boldsymbol{B}$. Also, an acceptable lower limit on G is critically dependent on the shape of the G surface. For example, when minimization methods of the gradient type [176, 177] are used, the iteration sequence is

$$\boldsymbol{B}^{(n+1)} = \boldsymbol{B}^{(n)} - \alpha \nabla G(\boldsymbol{B}^{(n)}) \tag{4.140}$$

where ∇G is the gradient of G in $\boldsymbol{B}$ space. If the Hessian matrix $\boldsymbol{H}(\boldsymbol{B})$ is positive definite and bounded in norm $\|\boldsymbol{H}\|$, the sufficient condition for convergence is [176]

$$\epsilon \leq \alpha \leq \frac{2}{\|\boldsymbol{H}\|} - \epsilon, \tag{4.141}$$

where $\epsilon \geq 0$.

Figure 4.36 shows typical cross sections of the G surface in the vicinity of the minimum, evaluated as a function of the real part of the film refractive index n_F for an air–SiO$_2$–Si system. Curves 1 and 2 correspond to thick and thin oxide films, with large and small second derivatives, respectively. When $\partial^2 G/\partial n_F^2$ is large, as in curve 1 in the figure, an initial estimate n_1 slightly different from the correct value n_0 will lead to a large initial value of $G(\boldsymbol{B})$. Under these conditions, convergence will be slow, in general, and it may be difficult to minimize $G(\boldsymbol{B})$ after a reasonable number of iterations. Equation (4.141) indicates that for this case the α increments in the sequence will be small because, for convergence, $\alpha \leq (2/\|\boldsymbol{H}\|)$ and $\|\boldsymbol{H}\|$ will be large when the DEHM are large. In this case, it may be helpful to combine iteration with direct search where different values of initial estimates of the parameters are used and the interation is started only if the initial $G(\boldsymbol{B})$ is less than a preset value. When $\partial^2 G/\partial b_i^2$ is very small, the minimum will be almost flat as in curve 2. In this latter case, small final values of G may be reached rapidly and a fit to the experimental data is obtained for a wide range of initial estimates of the parameters. Nevertheless, acceptable solutions will depend critically on the initial estimates if the usual gradient-iteration sequence is followed. Typically, when such a procedure is used, the computation is terminated when a given (small) increment is achieved. However, because of the small value of the second derivative in curve 2, computation could be terminated at a value considerably removed from the true value when the instruction to terminate iteration is based on the magnitude of the increment.

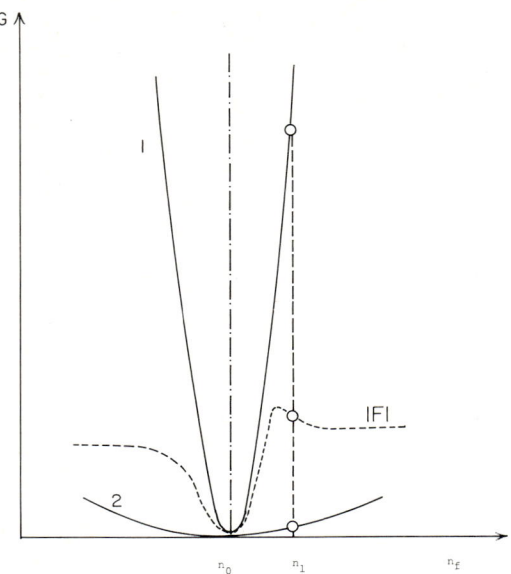

Fig. 4.36. Cross sections of the error function G [eq. (4.116)] for thick (1) and thin (2) films. The curve $|F|$ is discussed in the text. Air–SiO$_2$–Si system, $\lambda = 5461$ Å. After ref. 175.

To illustrate the importance of the choice of error function, fig. 4.36 displays the function F of eq. (4.118) which was used by Schueler[173] and Oldham[172]. Notice that if the function F is used, an initial estimate n_1 will lead to a completely divergent sequence (for $d = 1220$ Å).

4.5.3.3. *The air–SiO$_2$–Si system – an example*

Parameter correlation

From table 4.1 it is evident that, at 5461 Å measuring wavelength, $\partial \Delta/\partial d$, $\partial \Delta/\partial n_F$, and $\partial \Delta/\partial k_S$ have almost identical angle-of-incidence variation. This implies correlation among n_F, d, and k_S. Although at first glance it appears that the ψ dependence on n_F and d might be useful, because their angle-of-incidence dependences are different, the magnitudes of $\partial \psi/\partial n_F$ and $\partial \psi/\partial d$ (table 4.2) are too small to be useful. Because of the correlation among n_F, d, and k_S, an infinity of acceptable solutions are available that lead to the same least-squares residual. By use of values of the relative derivatives in table 4.1 in eqs. (4.131), the following uncertainties are found at a wavelength of 5461 Å (where d is in angstroms)

$$\delta n_F + 0.053 \, \delta d = 0,$$
$$n_F = 1.47 - 0.053 \, (d - 20), \tag{4.142}$$

Table 4.1. The variation with angle-of-incidence of the normalized derivatives of Δ and ψ for the air–Si–SiO$_2$ system at a wavelength of 5461 Å and film thickness of 20 Å. After ref. 175.

ϕ	$\partial\Delta/\partial k_S$ ($\times 34$)	$\partial\Delta/\partial n_F$ ($\times 6$)	$\partial\Delta/\partial d$ ($\times 0.3$)	$\partial\psi/\partial n_S$ ($\times 6.8$)	$\partial\psi/\partial k_F$ ($\times 1.2$)	$\partial\Delta/\partial n_S$ ($\times 3.3$)	$\partial\Delta/\partial k_F$ ($\times 0.25$)
55°	0.280	0.281	0.282	0.586	0.587	0.066	−0.15
60	0.386	0.386	0.387	0.720	0.721	0.124	−0.13
65	0.568	0.569	0.569	0.864	0.860	0.285	0.0
70	1.000	1.000	1.000	1.000	1.000	1.000	1.000
75	3.760	3.767	3.689	0.945	0.959	22.35	41.22
80	−1.186	−1.169	−1.176	−1.015	0.991	2.56	6.20

Table 4.2. The variation with angle-of-incidence of the derivatives of ψ and the relative derivatives of Δ for the air–SiO$_2$–Si system at a film thickness of 20 Å. After ref. 175.

Wavelength	ϕ	$\dfrac{\partial\psi}{\partial n_F}$	$\dfrac{\partial\psi}{\partial d}$	$\dfrac{(\partial\Delta/\partial d)}{(\partial\Delta/\partial n_F)}$	$\dfrac{(\partial\Delta/\partial d)}{(\partial\Delta/\partial k_S)}$	$\dfrac{(\partial\Delta/\partial n_S)}{(\partial\Delta/\partial n_F)}$	$\dfrac{(\partial\Delta/\partial k_F)}{(\partial\Delta/\partial n_F)}$
5461 Å	60°	−0.04 × 0.08	0.16 × 0.005	0.0526	0.0092	−5.65	0.016
	70	1.0 × 0.08	1.0 × 0.005	0.0525	0.0091	−1.81	−0.049
	80	3.8 × 0.08	2.6 × 0.005	0.0528	0.0090	0.82	0.26
2967 Å	60	−0.09 × 0.35	0.20 × 0.025	0.0423	−0.75	−1.16	−0.021
	70	1.0 × 0.35	1.0 × 0.025	0.0420	−0.26	−1.17	0.10
	80	6.9 × 0.35	4.6 × 0.025	0.0403	−0.064	−1.22	0.64

and

$$\delta k_S + 0.009\, \delta d = 0,$$
$$k_S = 0.03 - 0.009\,(d - 20). \tag{4.143}$$

From eq. (4.143), one of the possible solutions that satisfies the MAI measurements is $d = 0$, $k_S = 0.21$, which corresponds to the apparent (pseudo) index of the film-substrate combination.[22] This result led Burge and Bennett [156] to conclude that MAI measurements alone cannot be used to detect the presence of a very thin film on a silicon substrate at a wavelength of 5461 Å. Although n_F, k_S, and d are correlated at 5461 Å, n_S and k_F are uncorrelated with any of the other parameters because $\partial\Delta/\partial n_S$ and $\partial\Delta/\partial k_F$ vary differently with angle of incidence (see table 4.1). This lack of correlation between n_S and k_F is also evident in table 4.2, where the relative

---

[22] The apparent or pseudo refractive index of a film-substrate system is obtained when the values of ψ and Δ for such a system, at a given angle of incidence, are substituted into eq. (4.20b) to give the refractive index of an equivalent film-free substrate.

derivatives

$$\frac{\partial\Delta/\partial n_S}{\partial\Delta/\partial n_F} \quad \text{and} \quad \frac{\partial\Delta/\partial k_F}{\partial\Delta/\partial n_F}$$

vary with angle. Therefore, MAI measurements can be used to determine accurately n_S, k_F and one of the correlated parameters when good estimates are available for the other two correlated parameters.

From table 4.2, data at 2967 Å indicate that there are three correlated (n_F, n_S, and d) and two uncorrelated (k_F and k_S) parameters. Also, at this wavelength, the correlation between n_F and d is not as great as at 5461 Å because the derivatives $\partial\psi/\partial n_F$ and $\partial\psi/\partial d$ (table 4.2) are larger.

To summarize, independent estimates of two or three correlated parameters are needed at wavelengths 5461 Å and 2967 Å, although the requirement is relaxed slightly at 2967 Å because of the larger magnitudes of the derivatives of ψ with respect to n_F and d. At 5461 Å, independent estimates are needed for n_F, d, or k_S, and for d, n_F and n_S at 2967 Å. For a natural (20 Å) oxide film on silicon, good approximations can be obtained for n_F and d.

Although an infinite number of solutions may satisfy the system of equations for MAI measurements on very thin films, the range of solutions is restricted on physical grounds. For example, the constraint that the extinction coefficient k_S must be positive leads to an upper limit on the uncertainty of film thickness. From eq. (4.143), such uncertainty of film thickness is $\delta d \leq (0.030/0.009) = 3$ Å, for a thin SiO_2 film on Si at 5461 Å.

MAI measurements at different values of the SiO_2 film thickness
Table 4.3 shows the sensitivity of the DEHM to film thickness, a factor that influences directly the uncertainty of estimation of the optical parameters. One example of this is the relative error of estimating the refractive index of the substrate as compared to that of the film.

From eq. (4.139), note that the error δb_j in estimating the optical parameter b_j is inversely proportional to the square root of the corresponding DEHM. Hence the relative error of estimating the refractive index of the substrate, n_S, compared to that of the film n_F will be

$$R = \frac{\delta n_S}{\delta n_F} = \left[\frac{\partial^2 G/\partial n_F^2}{\partial^2 G/\partial n_S^2}\right]^{\frac{1}{2}}. \tag{4.144}$$

Values of this ratio determined from the DEHM (such as those in table 4.3) are plotted in fig. 4.37 for thickness values up to 900 Å. From fig. 4.37, it can be seen that for the air-SiO_2-Si system the relative error of determining the refractive index of the substrate compared to that of the film is smaller for thinner films, *i.e.*, MAI will give better estimates of refractive index for the

Table 4.3. DEHM values for the air–SiO$_2$–Si system at a wavelength of 4358 Å*

d	ϕ	$\partial^2 G/\partial n_F^2$	$\partial^2 G/\partial n_S^2$	$\partial^2 G/\partial k_S^2$	$\partial^2 G/\partial d^2$	$\delta\Delta$ °†
20 Å	60°	0.005	0.008	0.044	0.06×10^{-4}	−0.003
	70	0.026	0.020	0.209	0.30	−0.013
	74	0.081	0.063	0.637	0.94	−0.018
	76	0.196	0.367	1.52	2.29	−0.10
	80	0.183	0.765	1.62	2.05	−0.16
100 Å	60	0.135	0.008	0.042	0.06	−0.01
	70	0.523	0.033	0.015	0.23	−0.03
	74	0.880	0.147	0.245	0.37	−0.07
	76	0.787	0.350	0.199	0.31	−0.11
	80	0.074	0.403	0.036	0.04	−0.12
1000 Å	60	0.13×10^3	0.051	0.041	0.35	0.007
	70	1.12×10^3	0.581	0.370	2.24	0.042
	74	0.86×10^3	4.60	0.163	0.30	0.249
	76	2.45×10^3	0.973	2.290	0.59	0.172
	80	372×10^3	0.034	0.468	1.90	0.023

*$n_S - jk_S = 4.85 - j0.14$, $n_F = 1.47$. DEHM are the diagonal elements of the Hessian matrix.
†The error in Δ caused by an angle-of-incidence error of 0.01°. After ref. 175.

substrate than for the film when thin surface films are present. Conversely, the relative uncertainty of refractive index of the film decreases for thicker films. This can also be seen from table 4.4.

Effect of experimental errors
The subsequent procedure was followed to investigate the effects of systematic and random errors on the determination of the optical properties of the SiO$_2$–Si system. First, ellipsometric measurements were simulated by calculating values of Δ and ψ from eq. (4.60) assuming $n_S = 4.85$, $k_S = 0.14$, $n_F = 1.47$, and $k_F = 0$[23] and then modifying these values by neglecting digits beyond the first two decimal places. This corresponds to the introduction of a systematic error of 0.005° and a random error of ±0.005°. These values are typical of the accuracy of ellipsometric measurements using high-quality instruments. Secondly, systematic errors of +0.1 and random errors of ±0.2° were added to Δ and ψ, to simulate the case of large systematic and random experimental errors, respectively.

The modified values of ψ and Δ were used as input data in Marquardt's least-squares algorithm [178]. Solutions were obtained for all of the parameters of the system by use of initial guesses that were within ±15%

[23]These values correspond to a wavelength $\lambda = 4358$ Å [175].

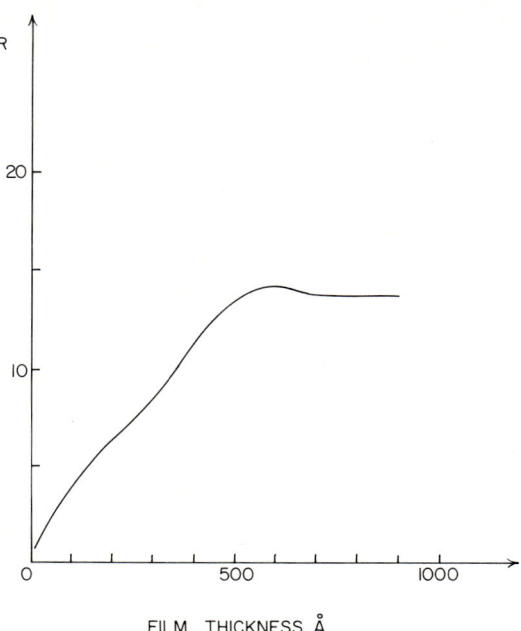

Fig. 4.37. The relative error $\delta n_S/\delta n_F$ of estimation of the substrate refractive index (n_S) compared to the film refractive index (n_F) as a function of film thickness for the air–SiO$_2$–Si system at $\lambda = 4358$ Å. After ref. 175.

of the above-cited values for the simulated model at oxide films 20 Å, 100 Å, and 1000 Å thick. The results in table 4.4 show clearly that (1) even in the presence of experimental errors as large as $-0.1°$ to $0.3°$, it is still possible to obtain solutions for the parameters almost identical to the solution obtained when the error range is 0 to $0.01°$, which is the lower limit in practice. (2) In the case of very thin films (20 Å and 100 Å), it is possible to obtain convergent solutions over a wide range of assumed initial guesses. (3) For a thick film (1000 Å), it is difficult to fit the experimental data even with very accurate measurements unless very accurate initial guesses are used. (4) Even when experimental errors are small, acceptable solutions cannot be obtained without suitable initial estimates.

Choice of angles of incidence
For a low-loss substrate, there is an appreciable dependence of the DEHM sensitivity factors on the angle of incidence. At angles of incidence near the principal angle, the factors are large. Care must be taken as to choice of angle so as to realize a large value of the sensitivity factor without incurring large errors because of an inaccuracy of determining the angle of incidence. An

Table 4.4. Effect of initial guesses and experimental errors on the final solutions for the air-SiO$_2$-Si system at 4358 Å*

Film thickness	G_0	0.005° Systematic Error ±0.005° Random Error						0.1° Systematic Error ±0.2° Random Error					
		G_F ($\times 10^{-4}$)	n_S	k_S	n_F	k_F	d(Å)	G_F ($\times 10^{-1}$)	n_S	k_S	n_F	k_F	d(Å)
20 Å	0.2	2.9	4.85	0.142	1.457	0.002	20.16	2.0	4.853	0.139	1.448	0.021	20
	4.5	2.7	4.852	0.128	1.551	0.004	19.29	2.0	4.851	0.125	1.534	0.038	19
	15	3.0	4.85	0.162	1.558	0.004	17.35	2.0	4.846	0.159	1.541	0.055	17
	27	2.6	4.853	0.109	1.394	0.006	23.98	2.0	4.860	0.107	1.387	0.002	24
100 Å	0.8	1.38	4.852	0.152	1.457	0.003	100.74	2.3	4.858	0.152	1.454	−0.005	101
	1.5	1.14	4.86	0.128	1.488	0.004	98.87	2.3	4.658	0.038	1.438	0.097	107
	24	0.98	4.865	0.076	1.554	0.001	96.23	2.3	4.862	0.032	1.561	0.004	98
	67	1.0	4.855	0.252	1.554	0.00	86.59	2.3	4.856	0.250	1.550	0.000	87
1000 Å	14	3.3	4.853	0.138	1.47	0.00	999.87	1.7	5.024	0.060	1.469	−0.005	1006
	94	8.7	4.79	0.13	1.47	0.003	995	1.7	5.017	0.042	1.471	−0.004	1003
	340	18.2	4.76	0.127	1.475	0.004	992	1.7	5.015	0.036	1.472	−0.003	1003
	≥5×10^3	⟶ Difficult to obtain a convergent solution ⟶											

*G_0 is the initial residual sum of squares, G_F is the residual when iteration was terminated. After ref. 175.

example of this is shown in table 4.3 where the uncertainty of Δ becomes larger at a large angle of incidence. The optimum choice of angles for measurement is a compromise between the need for increased sensitivity of the DEHM values, the absence of cross correlation, and the uncertainty of Δ caused by inaccuracy of measurement of the angle of incidence.

The above computational considerations of MAI measurements have been successfully applied [175] to resolve numerous questions concerning this technique in earlier investigations particularly those of McCrackin and Colson [167], Burge and Bennet [156], Oldham [172], and Johnson and Bashara [179].

4.6. Reflection and transmission by isotropic stratified planar structures

The method of addition of multiple reflections becomes impractical when considering the reflection and transmission of polarized light at oblique incidence by a multilayer film between semi-infinite ambient and substrate media. A more elegant approach that employs 2×2 matrices will now be discussed, and is based on the fact that the equations that govern the propagation of light are linear and that the continuity of the tangential fields across an interface between two isotropic media can be regarded as a 2×2-linear-matrix transformation.[24]

Consider a stratified structure that consists of a stack of $1, 2, 3, \ldots, j, \ldots m$ parallel layers (strata) sandwiched between two semi-infinite ambient (0) and substrate $(m + 1)$ media, fig. 4.38. Let all media be linear homogeneous and isotropic, and let the complex index of refraction of the jth layer be N_j and its thickness d_j. N_0 and N_{m+1} represent the complex indices of refraction of the ambient and substrate media, respectively. An incident monochromatic plane wave in medium 0 (the ambient) generates a resultant reflected plane wave in the same medium and a resultant transmitted plane wave in medium $m + 1$ (the substrate). The total field inside the jth layer, which is excited by the incident plane wave, consists of two plane waves: a forward-travelling plane wave denoted by (+), and a backward-travelling plane wave denoted by (−). The wave-vectors of all plane waves lie in the same plane (the plane of incidence), and the wave-vectors of the two plane waves in the jth layer make equal angles with the z-axis which is perpendicular to the plane boundaries directed toward the substrate. When the incident wave in the ambient is linearly polarized with its electric vector

---

[24]Work on the use of 2×2-matrix methods to study the reflection and transmission of light by multilayered structures was pioneered by Abelès. (For an account of Abelès work see, for example, ref. 23, pp. 51ff.) The present development is similar to that used by Hayfield and White [146].

§4.6]   STRATIFIED ISOTROPIC PLANAR STRUCTURES   333

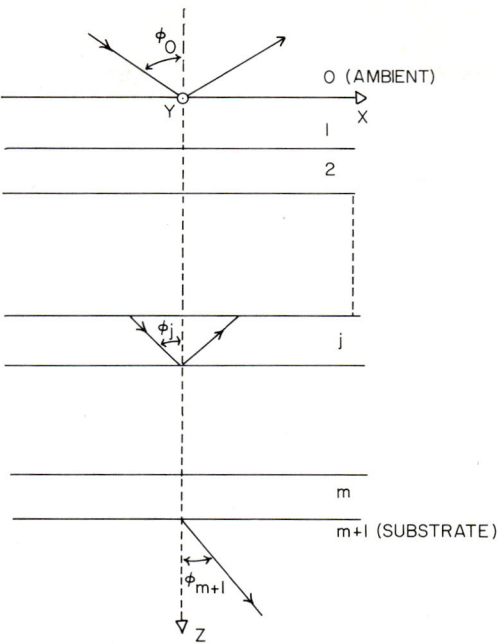

Fig. 4.38. Reflection and transmission of a plane wave by a multi-film structure (films 1, 2, ... m) sandwiched between semi-infinite ambient (0) and substrate ($m+1$) media. ϕ_0 is the angle of incidence; ϕ_j and ϕ_{m+1} are the angles of refraction in the jth film and substrate, respectively.

vibrating parallel (p) or perpendicular (s) to the plane of incidence, all plane waves excited by that incident wave in the various layers of the stratified structure will be similarly polarized, parallel or perpendicular to the plane of incidence, respectively. In the following, it will be assumed that all waves are either p- or s-polarized.

Let $E^+(z)$ and $E^-(z)$ denote the complex amplitudes of the forward- and backward-travelling plane waves at an arbitrary plane z. The total field at z can be described by a 2×1-column vector

$$\mathbf{E}(z) = \begin{bmatrix} E^+(z) \\ E^-(z) \end{bmatrix}. \tag{4.145}$$

If we consider the fields at two different planes z' and z'' parallel to the layer boundaries then, by virtue of system linearity, $\mathbf{E}(z'')$ and $\mathbf{E}(z')$ must be related by a 2×2-matrix transformation

$$\begin{bmatrix} E^+(z') \\ E^-(z') \end{bmatrix} = \begin{bmatrix} S_{11} & S_{12} \\ S_{21} & S_{22} \end{bmatrix} \begin{bmatrix} E^+(z'') \\ E^-(z'') \end{bmatrix}. \tag{4.146}$$

More concisely, eq. (4.146) can be written as

$$\boldsymbol{E}(z') = \boldsymbol{S}\boldsymbol{E}(z''), \tag{4.147}$$

where

$$\boldsymbol{S} = \begin{bmatrix} S_{11} & S_{12} \\ S_{21} & S_{22} \end{bmatrix}. \tag{4.148}$$

Note that $\boldsymbol{S}$ must characterize that part of the stratified structure confined between the two parallel planes at z' and z''.

By choosing z' and z'' to lie immediately on opposite sides of the $(j-1)j$ interface, located at z_j between layers $j-1$ and j, eq. (4.147) becomes

$$\boldsymbol{E}(z_j - 0) = \boldsymbol{I}_{(j-1)j}\boldsymbol{E}(z_j + 0), \tag{4.149}$$

where $\boldsymbol{I}_{(j-1)j}$ is a 2×2 matrix characteristic of the $(j-1)j$ *interface* alone. On the other hand, if z' and z'' are chosen inside the jth layer at its boundaries, eq. (4.147) becomes

$$\boldsymbol{E}(z_j + 0) = \boldsymbol{L}_j \boldsymbol{E}(z_j + d_j - 0), \tag{4.150}$$

where $\boldsymbol{L}_j$ is a 2×2 matrix characteristic of the jth *layer* alone whose thickness is d_j. Only the reflected wave in the ambient medium and the transmitted wave in the substrate are accessible for measurement, so that it is necessary to relate their fields to those of the incident wave. By taking the planes z' and z'' to lie in the ambient and substrate media, immediately adjacent to the 01 and $m(m+1)$ interfaces respectively, eq. (4.147) will read

$$\boldsymbol{E}(z_1 - 0) = \boldsymbol{S}\boldsymbol{E}(z_{m+1} + 0). \tag{4.151}$$

Equation (4.151) defines a *scattering matrix* $\boldsymbol{S}$ which represents the overall reflection and transmission properties of the stratified structure. $\boldsymbol{S}$ can be expressed as a product of the interface and layer matrices $\boldsymbol{I}$ and $\boldsymbol{L}$ that describe the effects of the individual interfaces and layers of the entire stratified structure, taken in proper order, as follows:

$$\boldsymbol{S} = \boldsymbol{I}_{01}\boldsymbol{L}_1\boldsymbol{I}_{12}\boldsymbol{L}_2 \ldots \boldsymbol{I}_{(j-1)j}\boldsymbol{L}_j \ldots \boldsymbol{L}_m\boldsymbol{I}_{m(m+1)} \tag{4.152}$$

Equation (4.152) may be proved readily by repeated application of eq. (4.147) to the successive interfaces and layers of the stratified structure, starting with the ambient-first film (01) interface and ending by the last film-substrate interface $[m(m+1)]$.

From eq. (4.152) it is evident that to determine the stratified structure scattering matrix S, the individual interface and layer matrices I and L have to be calculated.

The matrix I of an interface between two media a and b relates the fields on both its sides as

$$\begin{bmatrix} E_a^+ \\ E_a^- \end{bmatrix} = \begin{bmatrix} I_{11} & I_{12} \\ I_{21} & I_{22} \end{bmatrix} \begin{bmatrix} E_b^+ \\ E_b^- \end{bmatrix}. \tag{4.153}$$

Consider the special cases when one plane wave is incident on the ab interface. In terms of the complex amplitude E_a^+ of an incident plane wave in medium a, the complex amplitudes of the transmitted and reflected plane waves in media b and a, respectively (fig. 4.39–*left*), are given by

$$E_b^+ = t_{ab} E_a^+, \tag{4.154a}$$
$$E_a^- = r_{ab} E_a^+, \tag{4.154b}$$

and $E_b^- = 0$, where r_{ab} and t_{ab} are the Fresnel reflection and transmission coefficients of the ab interface. However, in accordance with eq. (4.153) we have

$$\begin{bmatrix} E_a^+ \\ E_a^- \end{bmatrix} = \begin{bmatrix} I_{11} & I_{12} \\ I_{21} & I_{22} \end{bmatrix} \begin{bmatrix} E_b^+ \\ 0 \end{bmatrix},$$

whose expansion gives

$$E_a^+ = I_{11} E_b^+, \tag{4.155a}$$
$$E_a^- = I_{21} E_b^+. \tag{4.155b}$$

Comparison between eqs. (4.154) and (4.155) leads to

$$I_{11} = 1/t_{ab}, \tag{4.156}$$
$$I_{21} = r_{ab}/t_{ab}. \tag{4.157}$$

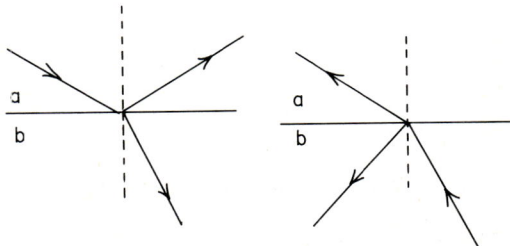

Fig. 4.39. *Left*: reflection and transmission (or refraction) of a plane wave at a two-media (ab) interface. *Right*: the direction of propagation of the transmitted (refracted) wave is assumed to have been reversed.

Now consider a plane wave incident on the ba interface from medium b, fig. 4.39–*right*, at an angle of incidence equal to the angle of refraction in medium b in the above case described in fig. 4.39–*left*. In this case, the fields immediately adjacent to the ba interface are related by

$$E_b^+ = r_{ba} E_b^-, \qquad (4.158a)$$

$$E_a^- = t_{ba} E_b^-, \qquad (4.158b)$$

and $E_a^+ = 0$, where r_{ba} and t_{ba} are the Fresnel reflection coefficients of the ba interface, respectively. On the other hand, eq. (4.153) tells us that, in this case,

$$\begin{bmatrix} 0 \\ E_a^- \end{bmatrix} = \begin{bmatrix} I_{11} & I_{12} \\ I_{21} & I_{22} \end{bmatrix} \begin{bmatrix} E_b^+ \\ E_b^- \end{bmatrix},$$

$$0 = I_{11} E_b^+ + I_{12} E_b^-, \qquad (4.159a)$$

$$E_a^- = I_{21} E_b^+ + I_{22} E_b^-. \qquad (4.159b)$$

Substitution of I_{11} and I_{21} from eqs. (4.156) and (4.157) into eqs. (4.159) transforms the latter equations into

$$E_b^+ = -I_{12} t_{ab} E_b^-, \qquad (4.160a)$$

$$E_a^- = \left[\frac{r_{ab} r_{ba}}{t_{ab}} + I_{22} \right] E_b^-. \qquad (4.160b)$$

Identification of the results shown by eqs. (4.158) and (4.160) leads to

$$I_{12} = -r_{ba}/t_{ab} \qquad (4.161)$$

$$I_{22} = (t_{ab} t_{ba} - r_{ab} r_{ba})/t_{ab}. \qquad (4.162)$$

Finally, use of the relationships between the interface Fresnel coefficients for both directions of propagation, namely $r_{ba} = -r_{ab}$ and $t_{ba} = (1 - r_{ab}^2)/t_{ab}$ [eqs. (4.29) and (4.30)], produces a final interface matrix of the form

$$\mathbf{I}_{ab} = \begin{bmatrix} 1/t_{ab} & r_{ab}/t_{ab} \\ r_{ab}/t_{ab} & 1/t_{ab} \end{bmatrix}$$

$$= (1/t_{ab}) \begin{bmatrix} 1 & r_{ab} \\ r_{ab} & 1 \end{bmatrix}. \qquad (4.163)$$

The interface Fresnel reflection and transmission coefficients that appear in eq. (4.163) must be evaluated using the complex indices of refraction of the two media that define the interface and the *local* angle of incidence. The latter can be found by the repeated application of Snell's law

$$N_0 \sin \phi_0 = N_1 \sin \phi_1 = \ldots = N_j \sin \phi_j = \ldots = N_{m+1} \sin \phi_{m+1}. \qquad (4.164)$$

Now that we have determined the interface matrix I, we turn our attention to the effect of propagation through a homogenous layer of index of refraction N and thickness d on the relationship between the fields inside that layer at both ends. This can be simply described by

$$\begin{bmatrix} E_0^+ \\ E_0^- \end{bmatrix} = \begin{bmatrix} e^{j\beta} & 0 \\ 0 & e^{-j\beta} \end{bmatrix} \begin{bmatrix} E_d^+ \\ E_d^- \end{bmatrix}, \tag{4.165}$$

where the subscripts 0 and d identify the beginning and end of the layer (along the direction of the forward-travelling wave) and the phase shift (layer phase thickness) β is given by

$$\beta = \frac{2\pi dN}{\lambda} \cos \phi, \tag{4.166}$$

with ϕ representing the angle between the direction of propagation in the layer and the perpendicular to its boundaries (the z-axis). The layer matrix L can therefore be written as

$$L = \begin{bmatrix} e^{j\beta} & 0 \\ 0 & e^{-j\beta} \end{bmatrix}. \tag{4.167}$$

From the interface and layer matrices I and L of eqs. (4.163) and (4.167) the overall scattering matrix S of the stratified structure can be found by straightforward matrix multiplication as indicated by eq. (4.152). Equation (4.151) can be expanded as

$$\begin{bmatrix} E_a^+ \\ E_a^- \end{bmatrix} = \begin{bmatrix} S_{11} & S_{12} \\ S_{21} & S_{22} \end{bmatrix} \begin{bmatrix} E_s^+ \\ 0 \end{bmatrix}, \tag{4.168}$$

where, for simplicity, the subscripts a and s refer to the ambient and substrate media respectively, and $E_s^- = 0$. Further expansion of eq. (4.168) yields the overall reflection and transmission coefficients of the stratified structures as

$$R = \frac{E_a^-}{E_a^+} = \frac{S_{21}}{S_{11}}, \tag{4.169}$$

$$T = \frac{E_s^+}{E_a^+} = \frac{1}{S_{11}}, \tag{4.170}$$

respectively. From eq. (4.169) and (4.170) it is clear that only the elements of the first column of scattering matrix S determine the overall reflection and transmission coefficients of the stratified structure.

For the purpose of ellipsometry, the stratified-structure scattering matrix S (or its elements S_{11} and S_{21}) have to be calculated for both linear polarizations parallel (p) and perpendicular (s) to the plane of incidence. Such calculations are readily programmable for execution on a digital computer. Let S_p and S_s represent the scattering matrices for the p and s polarizations. The p and s reflection and transmission coefficients become

$$R_p = \frac{S_{21p}}{S_{11p}}, \tag{4.171}$$

$$R_s = \frac{S_{21s}}{S_{11s}}, \tag{4.172}$$

$$T_p = \frac{1}{S_{11p}}, \tag{4.173}$$

$$T_s = \frac{1}{S_{11s}}. \tag{4.174}$$

The quantities that are measurable by the ellipsometer are the ratio of the complex reflection coefficients for the p and s polarizations

$$\rho_r = R_p/R_s = \frac{S_{21p}}{S_{11p}} \times \frac{S_{11s}}{S_{21s}}, \tag{4.175}$$

and the ratio of the complex transmission coefficients for the same polarizations

$$\rho_t = T_p/T_s = \frac{S_{11s}}{S_{11p}}. \tag{4.176}$$

Note that S_p and S_s are different because each of the interface matrices I that appear in eq. (4.152) is different for the p and s polarizations. The layer matrices L appearing in eq. (4.152) are the same for these two polarizations.

As an example of the application of the above procedure, consider the case of a single film (1) sandwiched between semi-infinite ambient (0) and substrate (2) media, fig. 4.14. From eq. (4.152) the scattering matrix S in this case is given by

$$S = I_{01} L_1 I_{12}, \tag{4.177}$$

which upon substitution from eqs. (4.163) and (4.167) becomes

$$S = \left(\frac{1}{t_{01} t_{12}}\right) \begin{bmatrix} 1 & r_{01} \\ r_{01} & 1 \end{bmatrix} \begin{bmatrix} e^{j\beta} & 0 \\ 0 & e^{-j\beta} \end{bmatrix} \begin{bmatrix} 1 & r_{12} \\ r_{12} & 1 \end{bmatrix}, \tag{4.178}$$

or

$$S = \left(\frac{e^{j\beta}}{t_{01} t_{12}}\right) \begin{bmatrix} (1 + r_{01} r_{12} e^{-j2\beta}) & (r_{12} + r_{01} e^{-j2\beta}) \\ (r_{01} + r_{12} e^{-j2\beta}) & (r_{01} r_{02} + e^{-j2\beta}) \end{bmatrix}, \tag{4.179}$$

when matrix multiplication is carried out. From eq. (4.179) we have

$$S_{11} = \left(\frac{e^{j\beta}}{t_{01}t_{12}}\right)(1 + r_{01}r_{12}e^{-j2\beta}), \tag{4.180a}$$

$$S_{21} = \left(\frac{e^{j\beta}}{t_{01}t_{12}}\right)(r_{01} + r_{12}e^{-j2\beta}). \tag{4.180b}$$

Substitution of these values of S_{11} and S_{21} into eqs. (4.169) and (4.170) yields

$$R = \frac{r_{01} + r_{12}e^{-j2\beta}}{1 + r_{01}r_{12}e^{-j2\beta}},$$

$$T = \frac{t_{01}t_{12}e^{-j\beta}}{1 + r_{01}r_{12}e^{-j2\beta}},$$

which are the same expressions as previously found in §4.3 [eqs. (4.34) and (4.36)] by use of the method of adding multiply-reflected and transmitted partial waves.

For another example, consider the case of two films (1 and 2) between semi-infinite ambient (0) and substrate (3) media. The method of adding multiply-reflected and transmitted waves becomes quite awkward in this case, although such calculation has been made [180]. In contrast, by use of eq. (4.152) the scattering matrix for this system can be easily obtained as

$$S = I_{01}L_1I_{12}L_2I_{23}. \tag{4.181}$$

Upon substitution from eqs. (4.163) and (4.167) eq. (4.181) becomes

$$S = \left(\frac{1}{t_{01}t_{12}t_{23}}\right)\begin{bmatrix} 1 & r_{01} \\ r_{01} & 1 \end{bmatrix}\begin{bmatrix} e^{j\beta_1} & 0 \\ 0 & e^{-j\beta_1} \end{bmatrix}\begin{bmatrix} 1 & r_{12} \\ r_{12} & 1 \end{bmatrix}\begin{bmatrix} e^{j\beta_2} & 0 \\ 0 & e^{-j\beta_2} \end{bmatrix}\begin{bmatrix} 1 & r_{23} \\ r_{23} & 1 \end{bmatrix}, \tag{4.182}$$

or

$$S = \left(\frac{e^{j(\beta_1+\beta_2)}}{t_{01}t_{12}t_{23}}\right)\begin{bmatrix} [(1+r_{01}r_{12}e^{-j2\beta_1}) & [(1+r_{01}r_{12}e^{-j2\beta_1})r_{23} \\ +(r_{12}+r_{01}e^{-j2\beta_1})r_{23}e^{-j2\beta_2}] & +(r_{12}+r_{01}e^{-j2\beta_1})e^{-j2\beta_2}] \\ [(r_{01}+r_{12}e^{-j2\beta_1}) & [(r_{01}+r_{12}e^{-j2\beta_1})r_{23} \\ +(r_{01}r_{12}+e^{-j2\beta_1})r_{23}e^{-j2\beta_2}] & +(r_{01}r_{12}+e^{-j2\beta_1})e^{-j2\beta_2}] \end{bmatrix} \tag{4.183}$$

after matrix multiplications have been carried out explicitly. From eq. (4.183) we have

$$S_{11} = \left(\frac{e^{j(\beta_1+\beta_2)}}{t_{01}t_{12}t_{23}}\right)[(1+r_{01}r_{12}e^{-j2\beta_1})+(r_{12}+r_{01}e^{-j2\beta_1})r_{23}e^{-j2\beta_2}], \tag{4.184a}$$

$$S_{21} = \left(\frac{e^{j(\beta_1+\beta_2)}}{t_{01}t_{12}t_{23}}\right)[(r_{01} + r_{12}\,e^{-j2\beta_1}) + (r_{01}r_{12} + e^{-j2\beta_1})r_{23}\,e^{-j2\beta_2}]. \tag{4.184b}$$

Substitution of eqs. (4.184a) and (4.184b) into eqs. (4.169) and (4.170) yields

$$R = \frac{(r_{01} + r_{12}\,e^{-j2\beta_1}) + (r_{01}r_{12} + e^{-j2\beta_1})r_{23}\,e^{-j2\beta_2}}{(1 + r_{01}r_{12}\,e^{-j2\beta_1}) + (r_{12} + r_{01}\,e^{-j2\beta_1})r_{23}\,e^{-j2\beta_2}}, \tag{4.185}$$

and

$$T = \frac{t_{01}t_{12}t_{23}\,e^{-j(\beta_1+\beta_2)}}{(1 + r_{01}r_{12}\,e^{-j2\beta_1}) + (r_{12} + r_{01}\,e^{-j2\beta_1})r_{23}\,e^{-j2\beta_2}}. \tag{4.186}$$

Equations (4.185) and (4.186) apply to the p and s polarizations by simple attachment of a subscript p or s to R, T and the individual interface Fresnel reflection (r_{01}, r_{12}, r_{23}) and transmission (t_{01}, t_{12}, t_{23}) coefficients. The ratios of overall reflection and transmission coefficients that can be measured ellipsometrically are given by eqs. (4.175) and (4.176).

Although the 2×2-matrix method can be easily extended to derive explicit expressions for R and T in cases that involve three or more films, such explicit expressions are of little practical value because all calculations can be handled by a computer without them.

Throughout this section, we assumed that the optical properties are uniform within each layer of the stratified structure, and that they change abruptly at the sharp interfaces between layers. When the optical properties change continuously (and not in a discontinuous step-like fashion) within an inhomogeneous layer, the method of this section can still be applied. This requires, however, that we divide the inhomogeneous layer into an adequately large number of sublayers, each of which is approximately homogeneous. In fact, this is the most general approach for a problem of this kind, because closed-form solutions are only possible for a few simple cases as, for example, when the refractive index of the layer changes linearly or exponentially along the direction of stratification [181].

4.7. Reflection and transmission by anisotropic stratified planar structures

A 4×4-matrix method has recently been developed by Billard [182], Teitler and Henvis [183], and Berreman and Scheffer [184] to study the reflection and transmission of obliquely incident polarized light by stratified anisotropic planar structures. The method is a generalization of the Abelès 2×2-matrix method [185] applicable to stratified isotropic media and previously discussed in §4.6. The following presentation of the 4×4-matrix method is based on the general exposition given by Berreman [186]. Notice, however, that we use Maxwell's equations in rationalized units, that we

§4.7] STRATIFIED ANISOTROPIC PLANAR STRUCTURES

adhere to the $e^{j\omega t}$ time dependence, and that we use different notation in some places.

With the $e^{j\omega t}$ time dependence assumed, the two Maxwell's "curl" equations take the form

$$-\mathbf{curl}\,\mathbf{E} = j\omega \mathbf{B}, \qquad \mathbf{curl}\,\mathbf{H} = j\omega \mathbf{D} \tag{4.187}$$

where $\mathbf{E}$, $\mathbf{H}$, $\mathbf{D}$, and $\mathbf{B}$ are the electromagnetic-field vectors, as usual. In Cartesian coordinates, eqs. (4.187) can be combined in matrix form as

$$\begin{bmatrix} 0 & 0 & 0 & 0 & -\partial/\partial z & \partial/\partial y \\ 0 & 0 & 0 & \partial/\partial z & 0 & -\partial/\partial x \\ 0 & 0 & 0 & -\partial/\partial y & \partial/\partial x & 0 \\ 0 & \partial/\partial z & -\partial/\partial y & 0 & 0 & 0 \\ -\partial/\partial z & 0 & \partial/\partial x & 0 & 0 & 0 \\ \partial/\partial y & -\partial/\partial x & 0 & 0 & 0 & 0 \end{bmatrix} \begin{bmatrix} E_x \\ E_y \\ E_z \\ H_x \\ H_y \\ H_z \end{bmatrix} = j\omega \begin{bmatrix} D_x \\ D_y \\ D_z \\ B_x \\ B_y \\ B_z \end{bmatrix}, \tag{4.188}$$

or shortly,

$$\mathbf{OG} = j\omega \mathbf{C}. \tag{4.189}$$

In eq. (4.189), $\mathbf{O}$ is a 6×6 symmetric matrix operator which can be partitioned into four 3×3 submatrices to take the form

$$\mathbf{O} = \begin{bmatrix} \mathbf{0} & \mathbf{curl} \\ -\mathbf{curl} & \mathbf{0} \end{bmatrix}. \tag{4.190}$$

$\mathbf{0}$ denotes the 3×3 zero matrix, and $\mathbf{curl}$ is the curl operator

$$\mathbf{curl} = \begin{bmatrix} 0 & -\partial/\partial z & \partial/\partial y \\ \partial/\partial z & 0 & -\partial/\partial x \\ -\partial/\partial y & \partial/\partial x & 0 \end{bmatrix}. \tag{4.191}$$

$\mathbf{G}$ is a 6×1 column vector whose elements are the Cartesian components of $\mathbf{E}$ followed by those of $\mathbf{H}$, and $\mathbf{C}$ is a 6×1 column vector whose elements are the Cartesian components of $\mathbf{D}$ followed by those of $\mathbf{B}$.

In the absence of nonlinear optical effects and spatial dispersion, the constitutive relation between $\mathbf{C}$ and $\mathbf{G}$ can be generally put as

$$\mathbf{C} = \mathbf{MG}, \tag{4.192}$$

where the 6×6 matrix $\mathbf{M}$ carries all the information about the anisotropic optical properties of the medium that supports the electromagnetic fields. $\mathbf{M}$, to be called the optical matrix, can be conveniently partitioned as follows

$$M = \begin{bmatrix} \epsilon & \rho \\ \rho' & \mu \end{bmatrix}, \tag{4.193}$$

where $\epsilon = (M_{ij})$ and $\mu = (M_{i+3,j+3})$, $i, j = 1, 2, 3$ are the (dielectric) permittivity and (magnetic) permiability tensors, respectively, and $\rho = (M_{i,j+3})$ and $\rho'(M_{i+3,j})$, $i, j = 1, 2, 3$ are optical-rotation tensors.[25]

Substitution of eq. (4.192) into eq. (4.189) yields

$$OG = j\omega MG.$$

Next, if in the above equation we replace G by

$$G = e^{j\omega t}\Gamma, \tag{4.194}$$

where Γ is the spatial part of G, it becomes

$$O\Gamma = j\omega M\Gamma, \tag{4.195}$$

which is the spatial wave equation for frequency ω.

The particular problem under consideration involves the reflection and transmission of a monochromatic plane wave obliquely incident from an isotropic ambient medium ($z < 0$) onto an anisotropic planar structure ($z > 0$) stratified along the z-axis. The x-axis of the reference xyz Cartesian coordinate system is assumed to coincide with the line of intersection of the plane of incidence (the plane of the incident wave-vector and the z-axis) and the $z = 0$ interface (see fig. 4.38). From the symmetry of the problem, there is no variation in the y direction of any field component, so that

$$\partial/\partial y = 0. \tag{4.196a}$$

For the tangential fields to match across the boundary $z = 0$, at all of its points at all time, all the waves that are excited by the incident plane wave must have the same spatial dependence in the x-direction as the incident wave. Therefore, if ξ denotes the x component of the wave-vector of the incident wave, all fields should vary in the x-direction as $e^{-j\xi x}$, hence

$$\partial/\partial x = -j\xi. \tag{4.196b}$$

In terms of the refractive index N_0 of the ambient medium and the angle of incidence ϕ_0, ξ is given by

$$\xi = \frac{\omega}{c} N_0 \sin \phi_0, \tag{4.197}$$

---

[25]The bold face letter ρ which denotes the optical-rotation tensor should not be confused with ρ which denotes the ellipsometric ratio of reflection or transmission coefficients.

where c is the free-space wave velocity. Use of eqs. (4.196) simplifies the curl operator of eq. (4.191)

$$\mathbf{curl} = \begin{bmatrix} 0 & -\partial/\partial z & 0 \\ \partial/\partial z & 0 & j\xi \\ 0 & -j\xi & 0 \end{bmatrix}. \tag{4.198}$$

The possibility of using a 4×4-matrix method to study the reflection and transmission of polarized light by anisotropic planar structures is a consequence of the special form assumed by the curl operator in eq. (4.198). In particular, if eq. (4.198) is substituted into eq. (4.190), the resulting operator O generates two linear homogeneous algebraic equations and four linear homogeneous first-order differential equations in the six components of Γ when eq. (4.195) is expanded. The two linear homogeneous algebraic equations can be solved for the field components $E_z(\Gamma_3)$ and $H_z(\Gamma_6)$ along the z-axis in terms of the other four field components $E_x(\Gamma_1)$, $E_y(\Gamma_2)$, $H_x(\Gamma_4)$, $H_y(\Gamma_5)$, along the x- and y-axes. The values E_z and H_z thus obtained are subsequently substituted into the remaining four differential equations to produce four linear homogeneous first-order differential equations in the four field variables E_x, E_y, H_x, and H_y. These can be cast in 4×4 matrix form as follows

$$\frac{\partial}{\partial z} \begin{bmatrix} E_x \\ H_y \\ E_y \\ -H_x \end{bmatrix} = -j\omega \begin{bmatrix} \Delta_{11} & \Delta_{12} & \Delta_{13} & \Delta_{14} \\ \Delta_{21} & \Delta_{22} & \Delta_{23} & \Delta_{24} \\ \Delta_{31} & \Delta_{32} & \Delta_{33} & \Delta_{34} \\ \Delta_{41} & \Delta_{42} & \Delta_{43} & \Delta_{44} \end{bmatrix} \begin{bmatrix} E_x \\ H_y \\ E_y \\ -H_x \end{bmatrix}, \tag{4.199}$$

or

$$\frac{\partial}{\partial z} \psi = -j\omega \Delta \psi. \tag{4.200}$$

Equation (4.200) is the wave equation for the 4×1 generalized field vector

$$\psi = \begin{bmatrix} E_x \\ H_y \\ E_y \\ -H_x \end{bmatrix}, \tag{4.201}$$

with $\boldsymbol{\Delta}$ defining a 4×4 differential propagation matrix of the medium.[26] The elements of $\boldsymbol{\Delta}$ are functions of the elements of the 6×6 optical matrix $\boldsymbol{M}$ [eq. (4.193)] obtained by carrying out the operations indicated above. The relations between the elements of $\boldsymbol{\Delta}$ and the elements of $\boldsymbol{M}$ are

$$\begin{aligned}
\Delta_{11} &= M_{51} + (M_{53} + \eta)A_1 + M_{56}A_5, \\
\Delta_{12} &= M_{55} + (M_{53} + \eta)A_4 + M_{56}A_8, \\
\Delta_{13} &= M_{52} + (M_{53} + \eta)A_2 + M_{56}A_6, \\
-\Delta_{14} &= M_{54} + (M_{53} + \eta)A_3 + M_{56}A_7, \\
\Delta_{21} &= M_{11} + M_{13}A_1 + M_{16}A_5, \\
\Delta_{22} &= M_{15} + M_{13}A_4 + M_{16}A_8, \\
\Delta_{23} &= M_{12} + M_{13}A_2 + M_{16}A_6, \\
-\Delta_{24} &= M_{14} + M_{13}A_3 + M_{16}A_7, \\
-\Delta_{31} &= M_{41} + M_{43}A_1 + M_{46}A_5, \\
-\Delta_{32} &= M_{45} + M_{43}A_4 + M_{46}A_8, \\
-\Delta_{33} &= M_{42} + M_{43}A_2 + M_{46}A_6, \\
\Delta_{34} &= M_{44} + M_{43}A_3 + M_{46}A_7, \\
\Delta_{41} &= M_{21} + M_{23}A_1 + (M_{26} - \eta)A_5, \\
\Delta_{42} &= M_{25} + M_{23}A_4 + (M_{26} - \eta)A_8, \\
\Delta_{43} &= M_{22} + M_{23}A_2 + (M_{26} - \eta)A_6, \\
-\Delta_{44} &= M_{24} + M_{23}A_3 + (M_{26} - \eta)A_7,
\end{aligned} \quad (4.202)$$

where

$$\begin{aligned}
A_1 &= (M_{61}M_{36} - M_{31}M_{66})/D, \\
A_2 &= [(M_{62} - \eta)M_{36} - M_{32}M_{66}]/D, \\
A_3 &= (M_{64}M_{36} - M_{34}M_{66})/D, \\
A_4 &= [M_{65}M_{36} - (M_{35} + \eta)M_{66}]/D, \\
A_5 &= (M_{63}M_{31} - M_{33}M_{61})/D, \\
A_6 &= [M_{63}M_{32} - (M_{62} - \eta)M_{33}]/D, \\
A_7 &= (M_{63}M_{34} - M_{33}M_{64})/D, \\
A_8 &= [(M_{35} + \eta)M_{63} - M_{33}M_{65}]/D,
\end{aligned} \quad (4.203)$$

---

[26] The differential wave equation for the Jones vector $\boldsymbol{E}$ discussed in §2.10 [eq. (2.164)] may be considered as a special case of the more general wave equation for the generalized field vector $\boldsymbol{\psi}$. Instead of the differential propagation Jones matrix N in the 2×2-matrix formalism, we now have the differential propagation matrix $-j\omega\boldsymbol{\Delta}$ in the 4×4-matrix formalism. Obviously, the bold face symbols $\boldsymbol{\psi}$ and $\boldsymbol{\Delta}$ which denote the wave vector and differential propagation matrix respectively, should not be confused with the ordinary ellipsometric parameters ψ and Δ.

$$D = M_{33}M_{66} - M_{36}M_{63}, \tag{4.204a}$$

$$\eta = \xi/\omega = N_0 \sin \phi_0/c. \tag{4.204b}$$

Recall that the 6×6 optical matrix M is structured from the permittivity ϵ, permiability μ and optical-rotation ρ, ρ' tensors, according to eq. (4.193). From ϵ, μ, ρ and ρ', as well as η, the differential propagation matrix Δ can be calculated using eqs. (4.202)–(4.204). With Δ known, the law of propagation (wave equation) for the generalized field vector ψ (or, equivalently, its elements, the components of E and H parallel to the x- and y-axes) is specified by eq. (4.200). Berreman calculated Δ for a number of special cases: (i) an orthorhombic crystal with its principal axes parallel to the x, y, and z coordinates axes, (ii) an isotropic optically active medium [187], (iii) an isotropic medium subjected to a magnetic field along the z-axis and exhibiting Faraday effect [188], and (iv) single-domain, cholesteric or twisted-nematic, liquid-crystal cell with parallel surfaces [186]. These results are summarized in table 4.5.

In the general case of a stratified anisotropic structure, M is some arbitrary function of z and the wave equation (4.200) does not, in general, have an analytical solution. In the special case when M is constant independent of z (over some continuous interval of z), eq. (4.200) is directly integrable to yield

$$\psi(z+h) = L(h)\psi(z), \tag{4.205}$$

where

$$L(h) = e^{-j\omega h \Delta}$$
$$= \left[I - j\omega h \Delta - \frac{(\omega h)^2}{2!} \Delta^2 + j \frac{(\omega h)^3}{3!} \Delta^3 + \ldots \right]. \tag{4.206}$$

Equation (4.205) represents a linear matrix relationship between the generalized field vectors ψ [eq. (4.201)] at two different parallel planes, separated by a distance h, in a homogeneous anisotropic medium whose fields are excited by an incident plane wave. The layer matrix $L(h)$ is determined by the differential propagation matrix Δ in accordance with eq. (4.206). In the latter equation, I represents the 4×4 identity matrix, and the summation of the exponential series can be carried out analytically in some simple cases (when Δ^n has a closed-form expression), or numerically.

An alternative expression for the layer matrix $L(h)$ can be determined from the fact that when Δ is constant independent of z, eq. (4.200) has four particular plane-wave solutions of the form

$$\psi(z) = \psi_\ell(0) e^{-jq_\ell z}, \quad \ell = 1, 2, 3, 4. \tag{4.207}$$

In eq. (4.207), $\psi(0)$ is the value of the generalized field vector of the plane

Table 4.5. The differential propagation matrix $\boldsymbol{\Delta}$ for certain anisotropic media*

Orthorombic Crystal (Principal axes parallel to the xyz coordinate axes)	Isotropic Optical Activity	Faraday Rotation (Magnetic field parallel to z-axis)	Cholesteric Liquid Crystal (Helical axis along the z-axis)
$\boldsymbol{\Delta} = \begin{bmatrix} 0 & a^2 & 0 & 0 \\ b^2 & 0 & 0 & 0 \\ 0 & 0 & 0 & u^2 \\ 0 & 0 & v^2 & 0 \end{bmatrix}$	$\begin{bmatrix} 0 & 1-(\eta^2/\epsilon) & -jg\eta^2/\epsilon & 0 \\ \epsilon & 0 & 0 & -jg \\ 0 & 0 & 0 & 1 \\ 0 & +jg & \epsilon-\eta^2 & 0 \end{bmatrix}$	$\begin{bmatrix} 0 & 1-(\eta^2/\epsilon) & 0 & 0 \\ \epsilon & 0 & 0 & j\gamma \\ 0 & 0 & 0 & 1 \\ 0 & -j\gamma & \epsilon-\eta^2 & 0 \end{bmatrix}$	$\begin{bmatrix} 0 & 1-(\eta^2/\epsilon') & 0 & 0 \\ (\epsilon+\delta\cos 2\beta z) & 0 & 0 & \delta\sin 2\beta z \\ 0 & 0 & 0 & 1 \\ \delta\sin 2\beta z & 0 & (\epsilon-\eta^2-\delta\cos 2\beta z) & 0 \end{bmatrix}$

$a^2 = \mu_{22} - (\eta^2/\epsilon_{33})$
$b^2 = \epsilon_{11}$
$u^2 = \mu_{11}$
$v^2 = \epsilon_{22} - (\eta^2/\mu_{33})$
$\boldsymbol{\epsilon}$ and $\boldsymbol{\mu}$, the dielectric and permeability tensors, are diagonal

ϵ = isotropic dielectric constant
g = isotropic optical-activity parameter

γ = Faraday-rotation parameter proportional to the magnetic field

$\epsilon + \delta = \epsilon_{11}$, $\epsilon - \delta = \epsilon_{22}$, $\epsilon' = \epsilon_{33}$ are the principal values of the dielectric tensor of the molecular planes. $\beta = 2\pi/p$, where p is the pitch of the helical structure

*Compiled from ref. 158. η is given by eq. (4.204b).

wave at $z = 0$, and q_ℓ equals the component of the propagation vector of the plane wave parallel to the z-axis. Substitution of eq. (4.207) into eq. (4.200), gives the matrix-eigenvalue equation

$$[\omega\Delta - qI]\psi(0) = 0, \qquad (4.208)$$

whose eigenvalues q_ℓ are the roots of the quartic polynomial equation

$$\det[\omega\Delta - qI] = 0, \qquad (4.209)$$

where det stands for the "determinant of". Insertion of each eigenvalue q_ℓ ($\ell = 1, 2, 3, 4$) into eq. (4.208), leads to four homogeneous linear equations that can be solved for the elements of the corresponding eigenvector $\psi_{k\ell}(0)$ (where $k = 1, 2, 3, 4$). In terms of the 4×4 matrix $\Psi = [\psi_{k\ell}(0)]$, which is constructed from the four eigenvectors of eq. (4.208) as columns, it can be readily shown that the layer matrix $L(h)$ [relating the fields inside an anisotropic slab of thickness h at its two boundaries, eq. (4.205)] is given by

$$L(h) = \Psi K(h)\Psi^{-1}, \qquad (4.210)$$

where K is a diagonal matrix with elements determined by the eigenvalues q_ℓ

$$K_{\ell\ell} = e^{-jq_\ell h}, \qquad \ell = 1, 2, 3, 4. \qquad (4.211)$$

Examples of the application of this procedure are given in ref. 186.

In an inhomogeneous anisotropic medium, where M is a continuous function of z, eq. (4.205) can be applied if we divide the medium into layers that are sufficiently thin to make M independent of z within each layer. By recursive application of eq. (4.205) to the successive layers, the fields at two planes distance d apart are related by

$$\psi(z + d) = \mathcal{L}(z, d)\psi(z), \qquad (4.212)$$

where

$$\mathcal{L}(z, d) = L(z + d - h_m, h_m) \ldots L(z + h_1 + h_2, h_3)L(z + h_1, h_2)L(z, h_1),$$

$$d = \sum_{i=1}^{m} h_i. \qquad (4.213)$$

For generality, the layer thicknesses $h_1, h_2, \ldots h_m$, are assumed different. [Notice that $L(z, h)$ characterizes a thin homogeneous layer located between z and $z + h$.] Alternative expressions from which the inhomogeneous layer matrix can be numerically computed with increased accuracy and more rapid convergence than that provided by eq. (4.213) have been discussed by Berreman in a more recent publication [189].

4.7.1. Reflection and transmission by a finite anisotropic layer between semi-infinite isotropic ambient and substrate media

Consider the case shown in fig. 4.40 of a stratified layer or slab of an optically anisotropic medium sandwiched between two isotropic ambient and substrate media of refractive indices N_0 and N_2, respectively [183, 186]. Let $z = 0$ and $z = d$ coincide with the interfaces between the layer and the ambient and substrate media, respectively. A plane wave incident from the ambient onto the layer at an angle of incidence ϕ_0 generates a resultant reflected wave in the same medium, at angle of reflection from the z-axis ϕ_0, and a transmitted wave in the substrate medium at an angle of refraction from the z-axis ϕ_2. The relationship between ϕ_0 and ϕ_2 is given by Snell's law

$$N_2 \sin \phi_2 = N_0 \sin \phi_0. \tag{4.214}$$

According to eq. (4.212) a 4×4 matrix $\mathscr{L}$

$$\mathscr{L} = (\ell_{ij}), \quad i, j = 1, 2, 3, 4, \tag{4.215}$$

relates the generalized field vectors [defined by eq. (4.201)] $\psi(d)$ and $\psi(0)$ inside the layer at its two boundary surfaces according to

$$\psi(d) = \mathscr{L}\psi(0). \tag{4.216}$$

Starting from the permittivity ϵ, permiability μ and optical-rotation ρ, ρ' tensors of the medium, the 6×6 optical matrix M is constructed according to eq. (4.193). Next, the 4×4 differential propagation matrix Δ is calculated

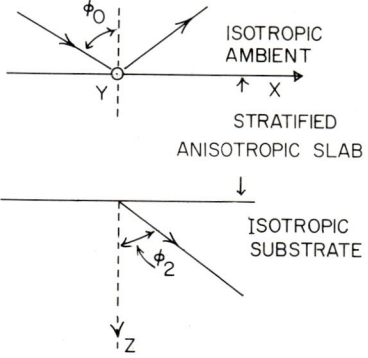

Fig. 4.40. Reflection and transmission of a plane wave by a stratified anisotropic slab sandwiched between semi-infinite isotropic ambient and substrate media. ϕ_0 is the angle of incidence in the ambient and ϕ_2 is the angle of refraction in the substrate. The mutually orthogonal axes x and y are in the ambient/slab interface parallel and perpendicular to the plane of incidence, respectively; the z axis is in the direction of stratification.

§4.7] STRATIFIED ANISOTROPIC PLANAR STRUCTURES 349

from the 6×6 optical matrix M using eqs. (4.202)–(4.204). Finally, the 4×4 layer matrix $\mathscr{L}$ is computed from Δ using eq. (4.210) [or (4.206)] and eq. (4.213) (if the layer is inhomogeneous).

The total field in the ambient medium at $z = 0$ is made up of incident and reflected components, so that the condition of matching the generalized field vector ψ across the $z = 0$ interface can be put in the form

$$\psi(0^+) = \psi_i(0^-) + \psi_r(0^-), \tag{4.217}$$

where the $-$ and $+$ superscripts distinguish the ambient and layer sides, respectively, of the $z = 0$ interface. The total field in the substrate is due to a single transmitted plane wave, and matching the generalized field vector ψ across the $z = d$ interface leads to

$$\psi(d - 0) = \psi_t(d + 0), \tag{4.218}$$

where $d - 0$ and $d + 0$ identify the layer and substrate sides, respectively, of the $z = d$ interface.

Let (E_{ip}, E_{is}), (E_{rp}, E_{rs}), and (E_{tp}, E_{ts}) represent the components of the electric field vectors of the incident, reflected, and transmitted waves, respectively, parallel (p) and perpendicular (s) to the plane of incidence. In a nonmagnetic ($\mu = 1$) optically isotropic medium, the magnetic field components are simply related to their associated orthogonal electric field components through the index of refraction N

$$H_p/E_s = H_s/E_p = N. \tag{4.219}$$

Thus, using eqs. (4.201) and (4.219), the three generalized field vectors ψ_i, ψ_r, and ψ_t of the incident, reflected, and transmitted waves, respectively, can be calculated from the p and s electric field components alone, without explicit reference to the magnetic field components

$$\psi_i = \begin{bmatrix} E_{ip} \cos \phi_0 \\ N_0 E_{ip} \\ E_{is} \\ N_0 E_{is} \cos \phi_0 \end{bmatrix}, \quad \psi_r = \begin{bmatrix} -E_{rp} \cos \phi_0 \\ N_0 E_{rp} \\ E_{rs} \\ -N_0 E_{rs} \cos \phi_0 \end{bmatrix}, \tag{4.220}$$

and

$$\psi_t = \begin{bmatrix} E_{tp} \cos \phi_2 \\ N_2 E_{tp} \\ E_{ts} \\ N_2 E_{ts} \cos \phi_2 \end{bmatrix}. \tag{4.221}$$

Notice that eqs. (4.220) and (4.221) are based on the convention that the p and s directions are as shown in fig. 4.1 and that $\boldsymbol{E} \times \boldsymbol{H}$ is in the direction of propagation.

Substitution of eqs. (4.220) and (4.221) into the boundary condition eqs. (4.217) and (4.218), respectively, gives the internal generalized field vectors $\boldsymbol{\psi}(0)$ and $\boldsymbol{\psi}(d)$ inside the anisotropic layer at its $z = 0$ and $z = d$ boundaries, in terms of the external fields $\boldsymbol{\psi}_i$, $\boldsymbol{\psi}_r$, and $\boldsymbol{\psi}_t$. The internal fields inside the layer at $z = 0$ and $z = d$ are subsequently interrelated by the layer matrix $\mathscr{L}$ according to eq. (4.216). The resulting equation reads

$$\begin{bmatrix} E_{tp} \cos \phi_2 \\ N_2 E_{tp} \\ E_{ts} \\ N_2 E_{ts} \cos \phi_2 \end{bmatrix} = \begin{bmatrix} \ell_{11} & \ell_{12} & \ell_{13} & \ell_{14} \\ \ell_{21} & \ell_{22} & \ell_{23} & \ell_{24} \\ \ell_{31} & \ell_{32} & \ell_{33} & \ell_{34} \\ \ell_{41} & \ell_{42} & \ell_{43} & \ell_{44} \end{bmatrix} \begin{bmatrix} (E_{ip} - E_{rp}) \cos \phi_0 \\ N_0(E_{ip} + E_{rp}) \\ (E_{is} + E_{rs}) \\ N_0(E_{is} - E_{rs}) \cos \phi_0 \end{bmatrix}. \quad (4.222)$$

Equation (4.222) can further be expanded into four separate linear algebraic equations in the six field components (E_{ip}, E_{is}), (E_{rp}, E_{rs}), and (E_{tp}, E_{ts}). E_{tp} can be readily eliminated from the first and second of these four equations, while E_{ts} can be eliminated from the third and fourth equations. This yields two linear algebraic equations connecting the incident and reflected fields alone

$$a_{ip}E_{ip} + a_{is}E_{is} + a_{rp}E_{rp} + a_{rs}E_{rs} = 0,$$
$$b_{ip}E_{ip} + b_{is}E_{is} + b_{rp}E_{rp} + b_{rs}E_{rs} = 0, \quad (4.223)$$

where

$$\frac{a_{ip}}{a_{rp}} = \pm \cos \phi_0 (\ell_{11} N_2 - \ell_{21} \cos \phi_2) + N_0 (\ell_{12} N_2 - \ell_{22} \cos \phi_2),$$

$$\frac{a_{is}}{a_{rs}} = \pm N_0 \cos \phi_0 (\ell_{14} N_2 - \ell_{24} \cos \phi_2) + (\ell_{13} N_2 - \ell_{23} \cos \phi_2)$$

$$\frac{b_{ip}}{b_{rp}} = \pm \cos \phi_0 (\ell_{31} N_2 \cos \phi_2 - \ell_{41}) + N_0 (\ell_{32} N_2 \cos \phi_2 - \ell_{42}),$$

$$\frac{b_{is}}{b_{rs}} = \pm N_0 \cos \phi_0 (\ell_{34} N_2 \cos \phi_2 - \ell_{44}) + (\ell_{33} N_2 \cos \phi_2 - \ell_{43}).$$

(4.224)

In the above equations, the upper and lower symbols on the left correspond, respectively, to the upper (+) and lower (−) signs on the right.

Equations (4.223) can be recast in the form

$$\begin{bmatrix} E_{rp} \\ E_{rs} \end{bmatrix} = \begin{bmatrix} R_{pp} & R_{ps} \\ R_{sp} & R_{ss} \end{bmatrix} \begin{bmatrix} E_{ip} \\ E_{is} \end{bmatrix},$$ (4.225)

or

$$\mathbf{E}_r = \mathbf{R} \mathbf{E}_i.$$ (4.226)

$\mathbf{R}$ is the 2×2 complex-amplitude reflection matrix

$$\mathbf{R} = \begin{bmatrix} R_{pp} & R_{ps} \\ R_{sp} & R_{ss} \end{bmatrix},$$ (4.227)

and is given by

$$\mathbf{R} = (a_{rs}b_{rp} - a_{rp}b_{rs})^{-1} \begin{bmatrix} (a_{ip}b_{rs} - a_{rs}b_{ip}) & (a_{is}b_{rs} - a_{rs}b_{is}) \\ (a_{rp}b_{ip} - a_{ip}b_{rp}) & (a_{rp}b_{is} - b_{rp}a_{is}) \end{bmatrix}.$$ (4.228)

By substituting, $E_{rp} = R_{pp}E_{ip} + R_{ps}E_{is}$ and $E_{rs} = R_{sp}E_{ip} + R_{ss}E_{is}$ [eq. (4.225)], into the second and third equations that are obtained from the expansion of eq. (4.222), the reflected field components are eliminated giving two linear equations that relate the transmitted fields (E_{tp}, E_{ts}) to the incident fields (E_{ip}, E_{is}) only. The two equations thus obtained can be readily put in the form

$$\begin{bmatrix} E_{tp} \\ E_{ts} \end{bmatrix} = \begin{bmatrix} T_{pp} & T_{ps} \\ T_{sp} & T_{ss} \end{bmatrix} \begin{bmatrix} E_{ip} \\ E_{is} \end{bmatrix},$$ (4.229)

or

$$\mathbf{E}_t = \mathbf{T} \mathbf{E}_i.$$ (4.230)

where

$$\mathbf{T} = \begin{bmatrix} T_{pp} & T_{ps} \\ T_{sp} & T_{ss} \end{bmatrix},$$ (4.231)

is the 2×2 complex-amplitude transmission matrix with elements

$$T_{pp} = [(\ell_{21} \cos \phi_0 + \ell_{22} N_0) \\ + R_{pp}(-\ell_{21} \cos \phi_0 + \ell_{22} N_0) + R_{sp}(\ell_{23} - \ell_{24} N_0 \cos \phi_0)]/N_2,$$

$$T_{ps} = [(\ell_{23} + \ell_{24} N_0 \cos \phi_0) \\ + R_{ps}(-\ell_{21} \cos \phi_0 + \ell_{22} N_0) + R_{ss}(\ell_{23} - \ell_{24} N_0 \cos \phi_0)]/N_2,$$ (4.232)

(4.232 cont.)

$$T_{sp} = (\ell_{31} \cos \phi_0 + \ell_{32} N_0)$$
$$+ R_{pp}(-\ell_{31} \cos \phi_0 + \ell_{32} N_0) + R_{sp}(\ell_{33} - \ell_{34} N_0 \cos \phi_0),$$
$$T_{ss} = (\ell_{33} + \ell_{34} N_0 \cos \phi_0)$$
$$+ R_{ps}(-\ell_{31} \cos \phi_0 + \ell_{32} N_0) + R_{ss}(\ell_{33} - \ell_{34} N_0 \cos \phi_0).$$

The determination of the reflection R and transmission T matrices completes the solution of the problem. This is because R and T represent the external measurable entities that are accessible to the ellipsometer. A summary of the steps that are required to calculate R and T from given values of the permittivity ϵ, permeability μ and optical rotation ρ, ρ' tensors (and given geometry) is schematically shown in fig. 4.41.

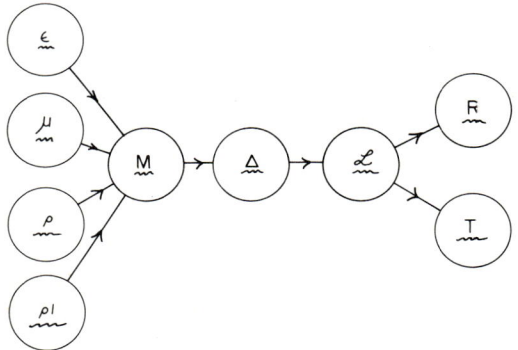

Fig. 4.41. Schematic flow chart of the steps required for the calculation of the reflection and transmission matrices R and T from the dielectric ϵ, magnetic permeability μ and optical activity (rotation) ρ, ρ' tensors.

4.7.2. Reflection and transmission by a semi-infinite anisotropic substrate in an isotropic ambient

We restrict our attention to the special, but important, case when the semi-infinite anisotropic medium is homogeneous, so that the optical matrix M is constant independent of z. No general solution exists when M is an arbitrary function of z [186].

When the optical matrix M of the substrate is constant, the differential propagation matrix Δ is also constant. Under this condition, we have seen that the wave equation for the generalized field vector ψ [eq. (4.200)] has four plane-wave solutions [eq. (4.207)]. Of these four plane-wave solutions, only two could be excited in the semi-infinite substrate by the incident plane wave. These are the two plane waves that propagate (and have wave-vector components) in the positive z-direction. Let q_1 and q_2 denote the only two eigenvalues that have a positive real part of all the four eigenvalues (roots) of the quartic polynomial eq. (4.209). Also, let $C_1\psi_1(0)$ and $C_2\psi_2(0)$ be their associated eigenvectors [obtained from eq. (4.208)], which are known up to the constant amplitude factors C_1 and C_2. C_1 and C_2 will be determined from matching the tangential electric and magnetic field components in the ambient and in the substrate at their common interface $z = 0$. In terms of the generalized field vectors, the boundary conditions assume the form

$$\psi_i(0^-) + \psi_r(0^-) = C_1\psi_1(0^+) + C_2\psi_2(0^+), \tag{4.233}$$

where the $-$ and $+$ superscripts indicate the ambient and substrate sides of the $z = 0$ interface and the subscripts i and r indicate the incident- and reflected-wave components of the total field in the ambient, as before.

If we substitute into eq. (4.233) the values of $\psi_i(0^-)$ and $\psi_r(0^+)$ given by eq. (4.220), and subsequently expand both sides, we obtain

$$\begin{aligned}
(E_{ip} - E_{rp}) \cos \phi_0 &= C_1\psi_{11} + C_2\psi_{12}, \\
N_0(E_{ip} + E_{rp}) &= C_1\psi_{21} + C_2\psi_{22}, \\
(E_{is} + E_{rs}) &= C_1\psi_{31} + C_2\psi_{32}, \\
N_0(E_{is} - E_{rs}) \cos \phi_0 &= C_1\psi_{41} + C_2\psi_{42}.
\end{aligned} \tag{4.234}$$

In eqs. (4.234), ψ_{k1} and ψ_{k2} ($k = 1, 2, 3, 4$) are the components of the two column eigenvectors $\psi_1(0^+)$ and $\psi_2(0^+)$, respectively. Let us introduce a vector C

$$C = \begin{bmatrix} C_1 \\ C_2 \end{bmatrix}, \tag{4.235}$$

whose components determine the amplitudes of the two refracted (transmitted) waves in the substrate. In terms of C, E_i, and E_r (where E_i and E_r are the 2×1 column vectors defined by eqs. (4.225), (4.226)), the first and fourth, and the second and third of eqs. (4.234) can be combined in two 2×2 matrix equations that can be solved for C as follows

$$\begin{aligned}
C &= S_1(E_i - E_r), \\
C &= S_2(E_i + E_r),
\end{aligned} \tag{4.236}$$

where

$$S_1 = \cos \phi_0 \begin{bmatrix} \psi_{11} & \psi_{12} \\ \psi_{41}/N_0 & \psi_{42}/N_0 \end{bmatrix}^{-1}, \quad S_2 = \begin{bmatrix} \psi_{21}/N_0 & \psi_{22}/N_0 \\ \psi_{31} & \psi_{32} \end{bmatrix}^{-1}. \tag{4.237}$$

Elimination of C between eqs. (4.236) leads to

$$E_r = RE_i,$$

where

$$R = (S_1 + S_2)^{-1}(S_2 - S_1), \quad (4.238)$$

is the required 2×2 reflection matrix.

A transmission or refraction matrix T can be defined which relates the amplitudes of the two refracted waves in the substrate to the amplitude of the incident wave as follows

$$C = TE_i, \quad (4.239)$$

From eqs. (4.236), it follows immediately that the transmission matrix T is given by

$$T = S_1(I - R) = S_2(I + R), \quad (4.240)$$

where I is the 2×2 identity matrix. Determination of the reflection (R) and transmission (T) matrices completes the solution of the problem.

4.7.3. Explicit expressions for the reflection matrix in simple special cases

The 4×4-matrix method given above represents a general unified approach to the computation of the reflection R and transmission T matrices of any anisotropic substrate or film-substrate system of interest in ellipsometry. There are a number of frequently encountered simple special cases of anisotropic systems for which explicit expressions of the reflection matrix have already been derived. A summary of these special cases is given below.

4.7.3.1. Isotropic ambient and uniaxially anisotropic substrate

We assume that light is obliquely incident at angle ϕ_0 from an isotropic ambient (medium 0) of refractive index N_0 onto a uniaxial anisotropic substrate (medium 1) with ordinary and extraordinary (complex) refractive indices N_{1o} and N_{1e}, respectively. First, consider the case when the optic axis is *perpendicular* to the ambient-substrate interface. In this orientation, the interface reflection matrix r is diagonal, so that

$$r_{ps} = r_{sp} = 0, \quad (4.241)$$

while the diagonal reflection coefficients r_{pp} and r_{ss} are given by [71]

$$r_{pp} = \frac{N_{1o}N_{1e} \cos \phi_0 - N_0(N_{1e}^2 - N_0^2 \sin^2 \phi_0)^{\frac{1}{2}}}{N_{1o}N_{1e} \cos \phi_0 + N_0(N_{1e}^2 - N_0^2 \sin^2 \phi_0)^{\frac{1}{2}}}, \quad (4.242)$$

$$r_{ss} = \frac{N_0 \cos \phi_0 - (N_{1o}^2 - N_0^2 \sin^2 \phi_0)^{\frac{1}{2}}}{N_0 \cos \phi_0 + (N_{1o}^2 - N_0^2 \sin^2 \phi_0)^{\frac{1}{2}}}. \quad (4.243)$$

In the limiting case when $N_{1o} = N_{1e} = N_1$, eqs. (4.242) and (4.243) reduce to the expressions for the reflection coefficients at the interface between two isotropic media [eqs. (4.4) and (4.5), after applying Snell's law, eq. (4.3)].

When the optic axis is *parallel* to the interface, and is oriented at an angle α from the plane of incidence, the elements of the reflection matrix (which is nondiagonal) are given by [190]

$$\begin{aligned}r_{pp} &= (F_1 G_4 + F_2 G_3)/(F_1 + F_2),\\ r_{ps} &= F_1 F_2 (G_2 - G_1)/(F_1 + F_2),\\ r_{sp} &= (G_4 - G_3)/(F_1 + F_2),\\ r_{ss} &= (F_1 G_1 + F_2 G_2)/(F_1 + F_2);\end{aligned} \quad (4.244)$$

where

$$\begin{aligned}F_1 &= J/(N_0 \sin^2 \phi_0 + J \cos \phi_0) \tan \alpha,\\ F_2 &= N_{1o} \tan \alpha (I + N_0 N_{1o} \cos \phi_0)/(I N_{1o} \cos \phi_0 + N_0 J^2),\\ G_1 &= (N_0 \cos \phi_0 - J)/(N_0 \cos \phi_0 + J),\\ G_2 &= (N_0 N_{1o} \cos \phi_0 - I)/(N_0 N_{1o} \cos \phi_0 + I),\\ G_3 &= (N_{1o}^2 \cos \phi_0 - N_0 J)/(N_{1o}^2 \cos \phi_0 + N_0 J),\\ G_4 &= (I N_{1o} \cos \phi_0 - N_0 J^2)/(I N_{1o} \cos \phi_0 + N_0 J^2);\end{aligned} \quad (4.245)$$

and

$$\begin{aligned}I^2 &= N_{1o}^2 N_{1e}^2 - N_0^2 \sin^2 \phi_0 (N_{1o}^2 \sin^2 \alpha + N_{1e}^2 \cos^2 \alpha),\\ J^2 &= N_{1o}^2 - N_0^2 \sin^2 \phi_0.\end{aligned} \quad (4.246)$$

In eqs. (4.245) and (4.246), ϕ_0 is the angle of incidence, N_0 is the refractive index of the isotropic ambient, and N_{1o}, N_{1e} are the ordinary and extraordinary (complex) refractive indices of the uniaxially anisotropic substrate, as before. When the optic axis is parallel to the interface, *in* or *normal* to the plane of incidence, the values $\alpha = 0$ or $\alpha = \frac{1}{2}\pi$, respectively, can be substituted into eqs. (4.245) and (4.246), leading to a diagonal reflection matrix [eqs. (4.244)] for these two symmetrical orientations.

4.7.3.2. *Isotropic ambient and biaxially anisotropic substrate*

We restrict ourselves to the case when the biaxially anisotropic substrate belongs to the orthorhombic system with its principal axes of absorption and refraction coincident. Furthermore, two of the three principal axes (x and y) are assumed to lie parallel to the ambient-substrate interface, while the third (z) is perpendicular to it. Light is obliquely incident from the isotropic ambient onto the biaxial substrate at an angle of incidence ϕ_0, with its direction of propagation in the xz plane.[27] Let N_0 be the index of refraction of the ambient, and N_{1x}, N_{1y}, and N_{1z} be the principal indices of

---

[27] This coordinate system is similar to that shown in fig. 4.40.

refraction of the substrate along its principal axes x, y, and z. In this symmetrical orientation, the reflection matrix r is diagonal ($r_{ps} = r_{sp} = 0$), and the diagonal r_{pp} and r_{ss} reflection coefficients are given by [191]

$$r_{pp} = \frac{N_{1x}N_{1z}\cos\phi_0 - N_0(N_{1z}^2 - N_0^2\sin^2\phi_0)^{\frac{1}{2}}}{N_{1x}N_{1z}\cos\phi_0 + N_0(N_{1z}^2 - N_0^2\sin^2\phi_0)^{\frac{1}{2}}}, \qquad (4.247)$$

$$r_{ss} = \frac{N_0\cos\phi_0 - (N_{1y}^2 - N_0^2\sin^2\phi_0)^{\frac{1}{2}}}{N_0\cos\phi_0 + (N_{1y}^2 - N_0^2\sin^2\phi_0)^{\frac{1}{2}}}. \qquad (4.248)$$

Upon substituting $N_{1x} = N_{1y} = N_{1o}$, $N_{1z} = N_{1e}$, eqs. (4.247) and (4.248) become identical with eqs. (4.242) and (4.243), respectively, appropriate to the case of a uniaxial substrate whose optic axis is perpendicular to its interface with the ambient.

Alternatively, if the substitutions $N_{1x} = N_{1e}$, $N_{1y} = N_{1z} = N_{1o}$, or $N_{1x} = N_{1z} = N_{1o}$, $N_{1y} = N_{1e}$ are made in eqs. (4.247), (4.248), we obtain the reflection coefficients for a uniaxial substrate whose optic axis lies parallel to the interface, in or perpendicular to the plane of incidence, respectively.

4.7.3.3. Uniaxially anisotropic film on an isotropic substrate in an isotropic ambient [192]

A special case of interest is that in which the optic axis of the uniaxial film is perpendicular to its boundaries with the ambient and substrate. This condition is encountered, for example, when Langmuir-Blodgett layers [193, 194] (which are monomolecular layers of well-oriented molecules) are deposited on isotropic substrates, or when thin vacuum-deposited films have island-like structure.

Because of symmetry, when the incident wave in the ambient is either p or s polarized, the excited waves in the uniaxial film and in the isotropic substrate will possess the same polarization, i.e., p or s, respectively. (Hence the reflection matrix is diagonal.) The method of §4.3, in which multiply reflected and refracted waves in an isotropic film are superimposed to determine the resultant waves in the ambient and substrate, continues to hold in the present case of a uniaxial anisotropic film. However, the refracted wave in the film is affected by a different index of refraction, dependent on its polarization being p or s. Therefore, we may conclude that eqs. (4.37)–(4.40) apply to the present case, provided that the "phase thickness" β of the uniaxial film is assigned the proper value for each of the p and s polarizations. Rewriting eqs. (4.37) and (4.38), we obtain the diagonal reflection coefficients for the three-phase (ambient-film-substrate) system as

$$R_{pp} = \frac{r_{01pp} + r_{12pp} e^{-j2\beta_p}}{1 + r_{01pp} r_{12pp} e^{-j2\beta_p}}, \qquad (4.249)$$

$$R_{ss} = \frac{r_{01ss} + r_{12ss}\,e^{-j2\beta_s}}{1 + r_{01ss}r_{12ss}\,e^{-j2\beta_s}}. \qquad (4.250)$$

In eqs. (4.249) and (4.250), r_{01pp}, r_{12pp} and r_{01ss}, r_{12ss} are the reflection coefficients at the 0–1 (ambient-film) and 1–2 (film-substrate) interfaces for the p- and s-polarization, respectively. They can be obtained directly from eqs. (4.242) and (4.243), noting that $r_{12} = -r_{21}$, as follows

$$r_{01pp} = \frac{N_{1o}N_{1e}\cos\phi_0 - N_0(N_{1e}^2 - N_0^2\sin^2\phi_0)^{\frac{1}{2}}}{N_{1o}N_{1e}\cos\phi_0 + N_0(N_{1e}^2 - N_0^2\sin^2\phi_0)^{\frac{1}{2}}}, \qquad (4.251)$$

$$r_{12pp} = \frac{-N_{1o}N_{1e}\cos\phi_2 + N_2(N_{1e}^2 - N_2^2\sin^2\phi_2)^{\frac{1}{2}}}{N_{1o}N_{1e}\cos\phi_2 + N_2(N_{1e}^2 - N_2^2\sin^2\phi_2)^{\frac{1}{2}}}, \qquad (4.252)$$

$$r_{01ss} = \frac{N_0\cos\phi_0 - (N_{1o}^2 - N_0^2\sin^2\phi_0)^{\frac{1}{2}}}{N_0\cos\phi_0 + (N_{1o}^2 - N_0^2\sin^2\phi_0)^{\frac{1}{2}}}, \qquad (4.253)$$

$$r_{12ss} = \frac{-N_2\cos\phi_2 + (N_{1o}^2 - N_2^2\sin^2\phi_2)^{\frac{1}{2}}}{N_2\cos\phi_2 + (N_{1o}^2 - N_2^2\sin^2\phi_2)^{\frac{1}{2}}}, \qquad (4.254)$$

where N_0, N_2 are the refractive indices of the isotropic ambient and substrate, respectively, and N_{1o}, N_{1e} are the ordinary and extraordinary refractive indices of the uniaxial film. ϕ_0 is the angle of incidence in the ambient, ϕ_2 is the angle of refraction in the substrate and these two angles are interrelated by Snell's law

$$N_0 \sin\phi_0 = N_2 \sin\phi_2. \qquad (4.255)$$

The phase thicknesses β_p and β_s for the p- and s-polarizations that appear in eqs. (4.249) and (4.250) are given by

$$\beta_p = 2\pi\left(\frac{d_1}{\lambda}\right)\left(\frac{N_{1o}}{N_{1e}}\right)(N_{1e}^2 - N_0^2\sin^2\phi_0)^{\frac{1}{2}}, \qquad (4.256)$$

$$\beta_s = 2\pi\left(\frac{d_1}{\lambda}\right)(N_{1o}^2 - N_0^2\sin^2\phi_0)^{\frac{1}{2}}, \qquad (4.257)$$

where d_1 is the film thickness and λ is the free-space wavelength of light.

Although we have considered above the case when the optic axis of the uniaxial film is perpendicular to its boundaries, analogous expressions can be similarly derived for the two other symmetrical orientations, when the optic axis is parallel to the film boundaries in or normal to the plane of incidence.

4.7.3.4. Biaxially anisotropic film on an isotropic substrate in an isotropic ambient [195]

Let two of the principal axes (x and y) of the biaxial film lie parallel to the film boundaries while the third (z) be perpendicular to them. Furthermore, let the plane of incidence be taken to coincide with the xz plane. In this symmetrical orientation, both interface and overall (three-phase) reflection matrices are diagonal. The three-phase diagonal reflection coefficients R_{pp} and R_{ss} are given by eqs. (4.249) and (4.250), but with the interface reflection coefficients determined from eqs. (4.247) and (4.248). Since the substitutions involved are quite straightforward, the results for this case will be left out. Note, however, that β_p and β_s are now given by

$$\beta_p = 2\pi \left(\frac{d_1}{\lambda}\right)\left(\frac{N_{1x}}{N_{1z}}\right)(N_{1z}^2 - N_0^2 \sin^2 \phi_0)^{\frac{1}{2}}, \tag{4.258}$$

$$\beta_s = 2\pi \left(\frac{d_1}{\lambda}\right)(N_{1y}^2 - N_0^2 \sin^2 \phi_0)^{\frac{1}{2}}, \tag{4.259}$$

where N_{1x}, N_{1y}, and N_{1z} are the principal indices of refraction associated with the principal axes x, y, and z of the biaxial film.

4.7.3.5. Uniaxial film on a uniaxial substrate in an isotropic ambient

When the optic axes of the uniaxial film and the uniaxial substrate are aligned in parallel, expressions for the reflection coefficients can be readily derived, provided that the common direction of the optic axes is either normal to the film boundaries, or parallel to them in or perpendicular to the plane of incidence [71]. Consider, for example, the case when the optic axes of the film and substrate are oriented normal to the film boundaries. Again, because of symmetry, both the interface and overall (three-phase) reflection matrices are diagonal. The three-phase diagonal reflection coefficients are given by eqs. (4.249) and (4.250). In the latter equations, notice that the 0–1 interface reflection coefficients (r_{01pp}, r_{01ss}) are the same as in eqs. (4.251) and (4.253), the phase thicknesses of the film β_p and β_s are the same as in eqs. (4.256) and (4.257), while the 1–2 interface reflection coefficients (r_{12pp}, r_{12ss}) are given by

$$r_{12pp} = \frac{N_{1o}N_{2e} \sin^2 \phi_{1e}(N_{1e}^2 - N_0^2 \sin^2 \phi_0)^{\frac{1}{2}} - N_{2o}N_{1e} \sin^2 \phi_{2e}(N_{2e}^2 - N_0^2 \sin^2 \phi_0)^{\frac{1}{2}}}{N_{1o}N_{2e} \sin^2 \phi_{1e}(N_{1e}^2 - N_0^2 \sin^2 \phi_0)^{\frac{1}{2}} + N_{2o}N_{1e} \sin^2 \phi_{2e}(N_{2e}^2 - N_0^2 \sin^2 \phi_0)^{\frac{1}{2}}}, \tag{4.260}$$

$$r_{12ss} = \frac{(N_{1o}^2 - N_0^2 \sin^2 \phi_0)^{\frac{1}{2}} - (N_{2o}^2 - N_0^2 \sin^2 \phi_0)^{\frac{1}{2}}}{(N_{1o}^2 - N_0^2 \sin^2 \phi_0)^{\frac{1}{2}} + (N_{2o}^2 - N_0^2 \sin^2 \phi_0)^{\frac{1}{2}}}. \tag{4.261}$$

In the above equations, N_0, (N_{1o}, N_{1e}), and (N_{2o}, N_{2e}) are the indices of refraction of the ambient (medium 0), film (medium 1), and substrate (medium 2), respectively. ϕ_0 is the angle of incidence in the ambient, ϕ_{1e} and ϕ_{2e} are the angles of refraction of the extraordinary (p-polarized) wave in the film and substrate respectively, and are given by [71]

$$\tan \phi_{1e} = (N_0 N_{1e}/N_{1o})(N_{1e}^2 - N_0^2 \sin^2 \phi_0)^{-\frac{1}{2}}, \tag{4.262}$$

$$\tan \phi_{2e} = (N_0 N_{2e}/N_{2o})(N_{2e}^2 - N_0^2 \sin^2 \phi_0)^{-\frac{1}{2}}. \tag{4.263}$$

4.8. Ellipsometry on surfaces covered with discontinuous films and on surfaces with rough boundaries

In the previous sections of this chapter we considered the reflection and transmission of light by multi-layer structures assuming that each layer is continuous and that interfaces between layers are perfectly smooth (plane) and parallel. In this section we briefly review studies of the effect of deviations from these idealizations on the interpretation of reflection ellipsometric data.

Discontinuous films often arise when they are prepared by the vacuum evaporation of film material onto a substrate. Especially in the early stages of evaporation, the film assumes the form of disconnected islands of film material separated by clear areas of the substrate. A simplified approach to deal with such films is to use the theory of Maxwell Garnett [196] which is based on the representation of the discontinuous film by a random distribution of small-diameter (compared to the wavelength of light) spherical particles of film material embedded in a dielectric ambient. Such an inhomogeneous system can then be proved to be equivalent to a homogeneous one with an *effective* complex refractive index N_e given by

$$\frac{N_e^2 - N_a^2}{N_e^2 + 2N_a^2} = q \frac{N_f^2 - N_a^2}{N_f^2 + 2N_a^2}, \tag{4.264}$$

where N_f and N_a are the refractive indices of the film and ambient materials, respectively, and q is the volume fraction occupied by the spherical particles in the discontinuous film. Often the ambient is vacuum or air so that $N_a = 1$ and eq. (4.264) becomes

$$\frac{N_e^2 - 1}{N_e^2 + 2} = q \frac{N_f^2 - 1}{N_f^2 + 2} \tag{4.265}$$

from which

$$N_e^2 = \frac{1 + 2fq}{1 - fq}, \qquad f = \frac{N_f^2 - 1}{N_f^2 + 2}. \tag{4.266}$$

A more exact approach to the reflection of light from substrates covered by discontinuous or sub-monolayer films is that of Strachan [197] and Sivukhin [198]. In this case, an island-like or sub-monolayer film is replaced by a two-dimensional distribution of Hertzian oscillators whose strength S per unit area of the substrate is

$$S = \sigma_x E_x + \sigma_y E_y + \sigma_z E_z, \tag{4.267}$$

where E_x, E_y, E_z are the components of the electric field of the light wave driving the oscillators along directions x, y (in the surface) and z (normal to the surface). σ_x, σ_y, σ_z are characteristic scattering parameters that are functions of the shape of the island particles of the discontinuous film or of the molecules of the sub-monolayer film. Assuming a plane wave incident at an angle ϕ on an isotropic substrate with refractive index N_2 covered by a distribution of Hertzian dipoles, Strachan derived first order equations for the complex-amplitude reflection coefficients, whose ratio ρ is given by

$$\rho = \bar{\rho} + \delta\rho, \tag{4.268}$$

$$\delta\rho/\bar{\rho} = -j(8\pi/\lambda)(N_2^2 - 1)^{-1}(N_2^2 \cos^2 \phi - \sin^2 \phi)^{-1}$$
$$\times \left[\sigma_x \left(\frac{1}{N_2^2} - \csc^2 \phi \right) + \sigma_y \left(\cot^2 \phi - \frac{1}{N_2^2} \right) + \sigma_z N_2^2 \right], \tag{4.269}$$

where $\bar{\rho}$ is the bare-substrate value of ρ. From eqs. (4.269) and (4.100), we can write

$$\psi - \bar{\psi} = \sum \alpha_i \sigma_i, \quad \Delta - \bar{\Delta} = \sum \beta_i \sigma_i, \quad i = x, y, z, \tag{4.270}$$

where α_i, β_i are coefficients whose explicit form can be determined.

In the case of a molecular film at sub-monolayer coverage on a crystal face, the oscillator strength per unit area σ can be assumed to be proportional to the coverage θ [199] (number of occupied substrate sites/number of available substrate sites),

$$\sigma = \theta \sigma', \tag{4.271}$$

so that

$$\psi - \bar{\psi} = \theta \sum \alpha'_i \sigma'_i, \quad \Delta - \bar{\Delta} = \theta \sum \beta'_i \sigma'_i. \tag{4.272}$$

Equations (4.272) predict changes of the ellipsometric parameters ψ and Δ that are proportional to the fractional coverage θ of a sub-monolayer film on a substrate, a conclusion borne out by experiment (see §6.3).

Berreman [200] developed a numerical method for computing the reflectance and ellipsometric parameters of surfaces covered by a sparse distribution of submicroscopic bumps, pits, or foreign particles each of which is assumed to approximate a figure of revolution about an axis normal to the surface. It was found that the effect of such particles is equivalent, in optical properties, to a uniform uniaxial thin film with the optic axis perpendicular to the surface. The principal values of refractive index of the uniaxial film were determined by particle shape and the refractive indices of particles and substrate. The film thickness was chosen equal to the volume of all particles, bumps, or pits per unit area of the substrate.

All of the above works neglect inter-particle interaction, an effect recently studied by Yamaguchi [201], Yoshida et al. [202] and Yamaguchi et al. [203].

Surface roughness

A simplified approach to study the effect of surface roughness is to replace the roughened "surface layer" by an equivalent film with plane-parallel boundaries, fig. 4.42, whose thickness is equal to a characteristic roughness height parameter (*e.g.*, the rms value), and whose optical properties are determined from those of the substrate and ambient according to the theory of Maxwell Garnett, eq. (4.264). Two film models can be used: (1) a *homogeneous-film* model in which a single value of the volume fraction q of substrate material is assumed, and (2) an *inhomogeneous-film model* in which q varies from 1 to 0 across the equivalent film. The latter model results from dividing the roughened layer into sublayers each of which is assigned a different value of q dependent on the roughness profile.

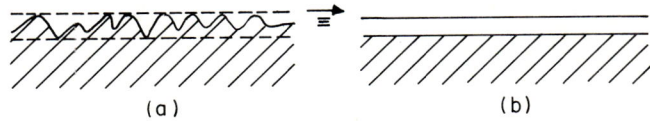

Fig. 4.42. Replacement of the roughened surface layer (a) by an equivalent film with plane parallel boundaries (b).

Fenstermaker and McCrackin [204] used the inhomogeneous-film model to calculate the ellipsometric parameters ψ, Δ of deterministically rough surfaces with square-ridge, triangular ridge, and pyramid topology, for a range of roughness (0–500 Å) and substrate materials (glass, Si, Au, ...). An apparent complex refractive index $n - jk$ of an equivalent perfectly smooth substrate was determined from the calculated ψ and Δ. Figures 4.43 and 4.44 show curves computed by Fenstermaker and McCrackin for the apparent n and apparent κ ($= k/n$) for glass and Si using three model topologies of idealized surface roughness (triangular ridge, square ridge, and pyramid) over a range of roughness heights from 0–500 Å. Inasmuch as values at 0 roughness can be used as reference, these figures show that the effect of surface roughness on the ellipsometrically determined optical properties of substrates is significant.

Measurements on diffraction gratings [205], which are deterministically rough surfaces, have lent support to Fenstermaker and McCrackin's model of the optical effect of surface roughness.

More elaborate studies of the effect of surface roughness on ellipsometry are based on the use of diffraction integrals. Azzam and Bashara [205] used the Stratton-Silver-Chu integral [206] to calculate the ellipsometric parameters of the periodically rough surface of a diffraction grating in the

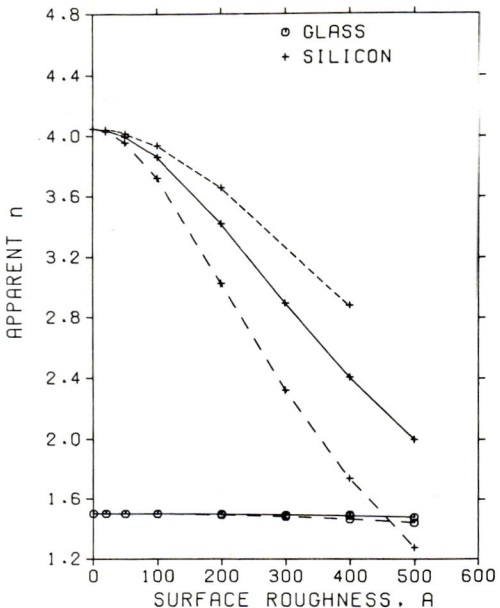

Fig. 4.43. Real part (n) of the apparent complex refractive index ($n - jk$) of a perfectly smooth substrate that is equivalent to rough glass and silicon surfaces. The surface roughness, which ranges from 0–500 Å, is assumed to have three topologies: (1) triangular ridge (———), (2) square ridge (– – –), and (3) pyramid (-------) models. After Fenstermaker and McCrackin, ref. 204.

physical-optics approximation. Ohlídal et al. [207–209] used the scalar Kirchhoff diffraction theory to calculate the ellipsometric parameters of randomly rough bare and filmed surfaces with small values of the rms roughness (σ) and rms surface slope ($\tan \beta$). They developed expressions for the ellipsometric ratio of reflection coefficients ρ in the form

$$\rho = \frac{\bar{\rho} + \tfrac{1}{2} F_\text{p} \tan^2 \beta}{1 + \tfrac{1}{2} F_\text{s} \tan^2 \beta}, \qquad (4.273)$$

where $\bar{\rho}$ is the smooth-surface value of ρ and F_p and F_s are complicated functions of the Fresnel reflection coefficients for p- and s-polarized light, respectively. Ohlídal et al. [209] suggest that ellipsometry be used to determine the rms of surface slopes $\tan \beta$. Furthermore, by examining the dependence of the ellipsometric parameters ψ and Δ ($\rho = \tan \psi \, e^{j\Delta}$) on angle of incidence, they propose to distinguish between roughening and film-growth processes on smooth surfaces. Calculation of the ellipsometric

§4.8] SURFACES COVERED WITH DISCONTINUOUS FILMS 363

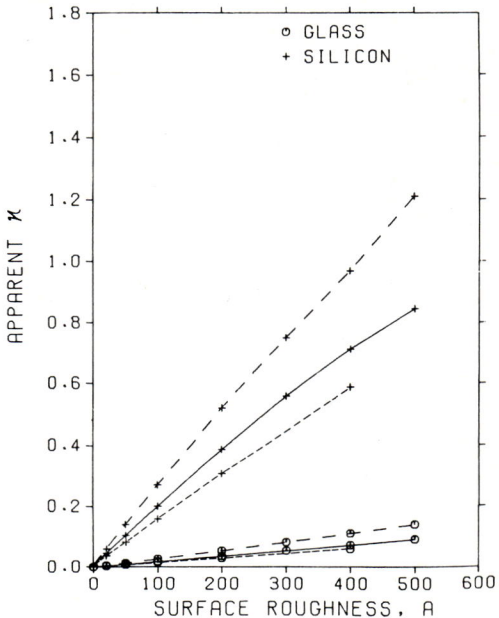

Fig. 4.44. Ratio ($\kappa = k/n$) of the imaginary and real parts of the apparent complex refractive index ($n - jk$) of a perfectly smooth substrate that is equivalent to rough glass and silicon surfaces. The surface roughness, which ranges from 0–500 Å, is assumed to have three topologies: (1) triangular ridge (———), (2) square ridge (- - - -), and pyramid (-------) models. After Fenstermaker and McCrackin, ref. 204.

parameters of randomly rough surfaces has also been made using the Stratton-Silver-Chu integral [210].

Church and Zavada [211] calculated the ellipsometric effect of surface roughness by use of a vector-perturbation theory of scattering of electromagnetic waves from statistically rough surfaces. From their results, an expression for ρ,

$$\rho = \bar{\rho} \frac{1 - \frac{1}{2}(4\pi\sigma \cos \phi)^2 F_p(\phi, \tau)}{1 - \frac{1}{2}(4\pi\sigma \cos \phi)^2 F_s(\phi, \tau)}, \tag{4.274}$$

is obtained, where $\bar{\rho}$ is the smooth-surface value of ρ, σ is the rms roughness, τ is the autocorrelation length, ϕ is the angle of incidence, and the functions F_p and F_s are complicated functions that differ for the p and s polarizations and involve multiple integrations.

The subject of the optical effect of surface roughness is under continued development and more recent papers [212–214] may prove significant for ellipsometry.

CHAPTER 5

Instrumentation and Techniques of Ellipsometry

5.1. Introduction

In this chapter we discuss the instrumental and operational techniques of ellipsometry. We begin with the classic null ellipsometer. Subsequently, we examine the various modifications that are introduced on this basic instrument in order to increase its precision and speed. Finally, we consider the fundamentally different photometric ellipsometers. The basic theory of all of these instruments has already been covered in ch. 3. It should be recognized that the subject of instrumentation in ellipsometry is undergoing rapid development at the present time.

5.2. The basic instrument – The null ellipsometer

The ellipsometer is basically an optical instrument that consists of two arms, or "telescopes", whose axes lie in one plane, fig. 5.1. One telescope is usually stationary and the other rotates around a central axis that passes through the point of intersection of the two telescope axes perpendicular to their plane. The angle between the two telescopes is read on a 360° graduated circle concentric with the central axis by diametrically opposite verniers that are viewed through magnifying lenses. The two diametrically opposite verniers represent a feature common to all other graduated circles of the ellipsometer and are intended to cancel the effect of eccentricity by averaging the two readings.

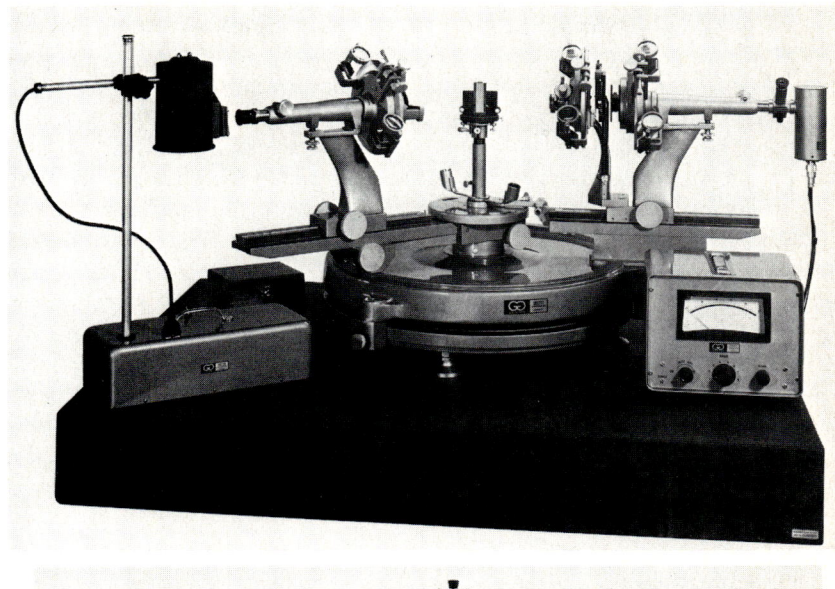

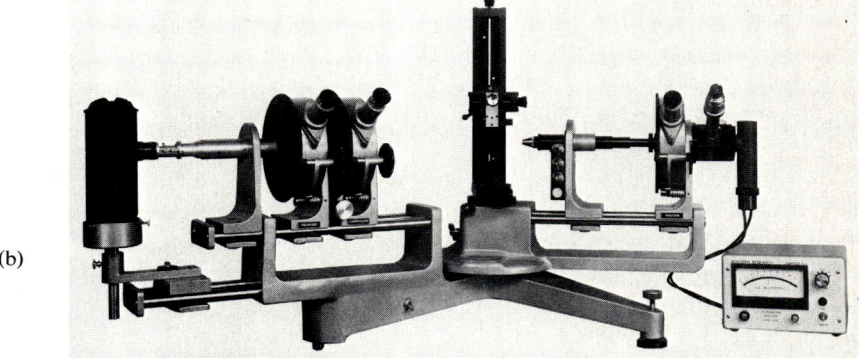

Fig. 5.1. Photographs of ellipsometers with light source and detector attachments: (a) Gaertner Scientific Corp., 1201 Wrightwood Ave., Chicago, Ill. 60614; (b) Rudolph Research, Pier Lane, Fairfield, N.J. 07006.

At the central axis of the instrument there is a cylindrical shaft on which the specimen table[1] would be mounted. The specimen itself is a specularly reflecting surface that is positioned to contain the axis of rotation.

[1] When studying surface reactions in a controlled ambient (liquid, gas, or vacuum), the specimen needs to be enclosed in a test cell (chamber) with entrance and exit optical windows. The chamber has to be located at the central axis of the ellipsometer and it is desirable to be able to manipulate the specimen position and orientation from outside the cell.

Collimated unpolarized or circularly polarized monochromatic light from a suitable source is passed along the center axis (as defined by two suitably placed well-centered pin-holes) of the stationary telescope in which are mounted optical devices that polarize the light incident on the surface under examination in a definite state. Either a linear polarizer alone, or a combination of a linear polarizer and a retarder (compensator), can be used. The polarizer and retarder are both mounted on bearings that permit the rotation of these elements around the telescope axis. The rotational azimuth angles of these elements are determined by 360° graduated circles provided with opposite verniers and magnifying viewing lenses. The state of polarization of the light beam after its reflection from the surface is analyzed by analyzing optical devices placed in the rotatable, reflected-beam, telescope. Either a linear analyzer alone is employed (when a polarizer and a retarder are used in the incident-beam telescope) or a combination of a retarder (compensator) and a linear analyzer (when a polarizer alone is used in the incident-beam telescope). The elements can be rotated around their telescope axis and their azimuths can be read on graduated circles also equipped with verniers and viewing lenses. To minimize eye strain, all scales are usually illuminated.

The operation of the instrument involves rotation of the polarizing and analyzing optical components around their telescope axes until the light leaving the analyzer telescope is totally extinguished. In the earlier days of ellipsometry, the extinction or null condition was visually judged, either directly or, for increased precision, by the use of the so-called half-shade devices [65]. The latter utilize the fact that the human eye is more sensitive to slight differences in brightness of two juxtaposed half-fields than to small changes of brightness of one field. However, when photoelectric detectors became widely available, the visual null-detection methods were generally abandoned. The output current or voltage of the photoelectric detector is measured by a suitable meter, usually after amplification. It is this output which is nulled (minimized) when operating the ellipsometer.

In the following, we discuss in more detail the individual component parts of the ellipsometer.

5.2.1. *The optical devices*

The optical devices needed for an ellipsometer include linear polarizers (analyzers) and phase retarders. A comprehensive recent review of the different types of polarizers, retarders and other polarizing optical devices is that of Bennett and Bennett [10]. Shurcliff [4] and Clarke and Grainger [8] also discuss this subject. It will suffice here to give a brief description of the optical devices used in an ellipsometer.

5.2.1.1. Linear polarizers

An ideal linear polarizer is a device that transforms any state of polarization of light at its input to a linear state at its output. Using the terminology of eigenpolarizations and eigenvalues (§2.6), a linear polarizer can also be defined as a device whose eigenpolarizations are linear with one eigenvalue equal to zero. Because light incident on a polarizer can be resolved into two components "parallel" to these eigenpolarizations, one of these components is always totally rejected while the other is passed, usually with a finite attenuation.

Real linear polarizers have orthogonal linear eigenpolarizations, say x and y, with associated eigenvalues, V_{ex} and V_{ey}, neither one of which is zero. If the y eigenpolarization is almost extinguished, *i.e.*, $|V_{ey}| \ll 1$, the ratio

$$e_r = |V_{ex}|^2/|V_{ey}|^2, \tag{5.1}$$

is called the extinction ratio of the polarizer. This is one of the important parameters that describe the operation of a polarizer [4, 10]. In general, to describe the effect of a polarizer on the state of polarization completely, the ratio of complex eigenvalues

$$\alpha_P = V_{ey}/V_{ex}, \tag{5.2}$$

needs to be specified.

The different physical mechanisms by which one of the two orthogonal resolved components of light can be rejected by a linear polarizer include (1) birefringence, (2) dichroism, and (3) reflection. This provides a suitable basis for the classification of available linear polarizers.

Double-refraction (birefringent) polarizers

When a light beam is incident from air on a planar face of a transparent uniaxially or biaxially anisotropic crystal, it is, in general, refracted into two beams in the bulk of the crystal. These two beams are spatially separated from one another and are orthogonally linearly polarized. If only one of these two beams is utilized, the double refraction mechanism can be considered as perfectly polarizing. A large variety of polarizers have been built on this principle of double refraction or a variation of it. The reader may wish to consult refs. 4, 8–10. One commonly used birefringent polarizer is the Glan-Thompson prism. As shown in fig. 5.2, a Glan-Thompson Prism consists of two sections (of calcite) that are either cemented by an optically isotropic transparent cement (*e.g.*, Canada Balsam) or are spaced by a narrow air gap. The latter type of prism (usually called Glan-Foucault) is suited for operation in the near ultraviolet region of the electromagnetic spectrum. The optic axes of the two halves of the prism are parallel to each

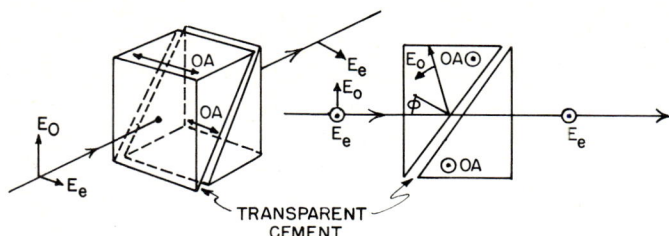

Fig. 5.2. A Glan-Thompson prism consists of two cemented sections of calcite with parallel optics axes (OA). The prism acts as a polarizer because total internal reflection of the ordinary-component vibration E_o leaves only the extraordinary component E_e to be transmitted.

other and to the plane-parallel entrance and exit end faces of the prism. The principle of operation in this case is as follows. Light perpendicularly incident on one of the end faces of the prism propagates without refraction to the interface between the two prisms. This light can be resolved into two components with electric vectors vibrating parallel and perpendicular to the optic axis. These orthogonally polarized component waves experience two different refractive-index discontinuities at the calcite/gap interface. These discontinuities are $n_e - n_g$ and $n_o - n_g$, for the two linear polarizations parallel (extraordinary) and perpendicular (ordinary) to the optic axis, respectively. If both n_o and n_e are greater than n_g (this is certainly true for an air gap, $n_g = 1$, and can be generally satisfied by suitable choice of cement material), and the angle of incidence ϕ on the interface (as determined primarily by the prism dimensions) is such that, $n_g/n_o < \sin \phi < n_g/n_e$, total internal reflection takes place for the ordinary-component beam only while the extraordinary-component beam is transmitted. The totally-internally-reflected beam can be absorbed by blackening the side of the prism on which it is incident.

The above condition of total internal reflection for the ordinary component continues to hold if light is incident on one of the end faces of the prism within a certain solid "acceptance angle". In this case, the incident and emergent beams are not colinear; there will be a lateral displacement which is proportional to the total thickness of the prism. An angular deviation between the incident and emergent beams results if the end faces of the prism are not parallel; the deviation is proportional to the "wedge angle" between these faces. Besides the extinction ratio, which is typically as high as 10^5 for Glan-Thompson prisms, the acceptance and wedge angles need to be specified if such prisms are to be used in an ellipsometer.

Dichroic polarizers

In certain absorbing media, the attenuation experienced by linearly polarized light is dependent on the direction of linear polarization. Such media are said to be linearly dichroic. The absorption is maximum and minimum when the direction of the electric-field vibration is along two orthogonal directions, called the principal axes of dichroism. When an arbitrarily polarized light beam propagates through such a medium, it becomes increasingly linearly polarized, because of the differential absorption of one component polarization. Near-total absorption of one component can be achieved after a sufficient distance of propagation.

Naturally occurring linear dichroism is usually associated with linear birefringence, although one or the other is more pronounced within a given spectral range (a Kramers-Kronig type integral relationship [138] connects these two properties as a function of wavelength). Tourmaline is one example of a crystal that exhibits strong dichroism in the visible.

Dichroism can be induced synthetically by incorporating an oriented array of certain needle-shaped particles (or macromolecules) in an isotropic transparent matrix. In this case, light which is linearly polarized with its electric vector vibrating parallel to the long direction of the needles is more strongly absorbed than light which is polarized orthogonal to the needles. A wide range of different types of dichroic polarizers is available from Polaroid [4]. They take the form of sheets of ~ 1-mm thickness and are sometimes referred to as sheet polarizers. In general, the extinction ratio of these polarizers ($\sim 10^3 - 10^4$) is one-to-two orders of magnitude lower than those of crystal (birefringent) polarizers. Also, the spectral windows (regions of transparency) of Polaroid sheet polarizers are more limited than those of crystal polarizers [4].

A distinct type of polarizer which operates on very much the same principle as the dichroic polarizer is the wire-grid polarizer. In this case, an array of parallel stripes (*i.e.*, a grid) of a suitable metal is deposited on a transparent blank. The interline spacing is chosen to be much less than the wavelength of light, and usually the metal "wires" are deposited on the peaks of a grating structure impressed on the blank. Each "wire" of the wire-grid polarizer may be thought of as an "extended needle" of a dichroic polarizer. The linear electric-field vibration parallel to the wires is the one which is strongly absorbed. The wire-grid polarizers represent an important class of polarizers for ellipsometry in the infrared, although their extinction ratios are modest, $\sim 10 - 10^3$.

Reflection polarizers

It is well known that when light is incident on a perfectly smooth dielectric surface, the reflected wave whose electric vector vibrates parallel to the plane of incidence is totally extinguished at the Brewster angle (§4.2). Although polarizers based on this principle are not commonly used in the visible, they are serious candidates for the infrared [10]. To obtain a reflection polarizer where the incident and emergent beams are colinear (as would be desirable in ellipsometry) a combination of three reflections can be used. Such a polarizer [142] can also be rotated to vary the azimuth of the outgoing linear polarization.

Polarizers that can suppress either the parallel (p) or the perpendicular (s) polarization can be designed using a film-substrate system [109, 215]. This type of polarizer does not seem to be commercially available at the present time and no mention of its use in ellipsometry has been made. However, it holds good promise, especially in wavelength ranges where more classic type polarizers are not available.

5.2.1.2. Retarders
A retarder is an optical device that introduces a relative phase shift between two specific orthogonally polarized components into which light incident on the device can be resolved without affecting their relative amplitude. The two orthogonal components which are characteristic of the retarder are the retarder's eigenpolarizations and, in general, can be linear, circular, or elliptic (§2.6). Retarders that are used in ellipsometers are linear retarders. Associated with a linear retarder are two directions or axes, a fast axis and a slow axis. The component of incident polarized light whose electric vibration is parallel to the slow axis is retarded in phase relative to the component vibration parallel to the fast axis as the light passes through the retarder. When the relative retardation is $\frac{1}{2}\pi$, π, ... the retarder is called a quarter-wave, half-wave, ... retarder, respectively. Linear quarter-wave retarders are the most commonly used retarders in ellipsometers.

There are two important mechanisms by which a certain relative retardation can be introduced between the two resolved orthogonal linear components of polarized light: (1) propagation through a linearly birefringent medium and (2) total internal reflection[2].

Birefringent retarders
The basic principle of operation of such retarder has been discussed in §2.2. Consider a plate of a uniaxially anisotropic crystal whose optic axis is parallel to the end faces of the plate. Light perpendicularly incident on the plate is resolved into two linearly polarized components one parallel and the other perpendicular to the optic axis. If n_o and n_e denote the ordinary and extraordinary refractive indices of the crystal, respectively, the component

---

[2] A third mechanism is reflection by a film-coated substrate (§4.4 and ref. 109).

waves that are linearly polarized parallel and perpendicular to the optic axis propagate through the crystal at speeds of c/n_e and c/n_o, respectively, where c is the free-space wave velocity. If the thickness of the plate is d, the difference in speed between the ordinary and extraordinary components leads to a cumulative phase shift equal to $(2\pi d/\lambda)(n_o - n_e)$ [eq. (2.17)], where λ is the free-space wavelength. For a given material with known birefringence $(n_o - n_e)$, the thickness d that is required to produce any given retardation may be calculated from the above simplified expression. This expression is accurate, however, only if multiple reflections between the parallel-plane end faces of the retarder are either avoided or do not add coherently. A more exact expression that accounts for multiple reflections has been derived by Holmes [31]. The theory on which this derivation is based also predicts that the transmittances along the fast and slow axes are different. This apparent interference-caused "dichroism" is of significance when interpreting ellipsometric measurements [73, 74].

The most common materials that are used for the construction of birefringent retarders are quartz and mica. Quartz is optically active although this optical activity is not evident when light propagates perpendicular to the optic axis, as is encountered in the usual cut of a retarder. Mica is actually biaxially anisotropic. However, in a mica sheet, two of the principle axes of the index ellipsoid are parallel to the sheet so that, for perpendicularly incident light, it behaves as a uniaxial crystal. Calcite is potentially useful for making retarders. However, it has relatively large birefringence which requires impractically thin plates to achieve the retardations that are usually needed, or demands considerable control over the parallelism and roughness of the retarder's end faces.

In spectroscopic ellipsometry, a retarder is desirable whose relative retardation can be tuned so that it continues to operate as a quarter-wave retarder as the wavelength is varied. In this case, the term compensator is often used to describe the retarder. One popular tunable retarder is the Babinet-Soleil compensator. The Babinet-Soleil compensator shown in fig. 5.3 consists of two sections of quartz: a variable-thickness section and a fixed-thickness section. The variable-thickness section uses two wedges with parallel optic axes and the wedges move with respect to one another to vary the total thickness of this section. The fixed-thickness section is made of a single plate of quartz whose optic axis (as with the other section) is parallel to its end faces. The two sections are cemented together with their optic axes orthogonal to one another. When the thicknesses of both sections are equal, the net retardation is zero. When the thickness difference is d, the approximate retardation is $(2\pi d/\lambda)(n_o - n_e)$.

Tuning this compensator is achieved by sliding one of the two wedges of the variable-thickness section with respect to the other, which is usually

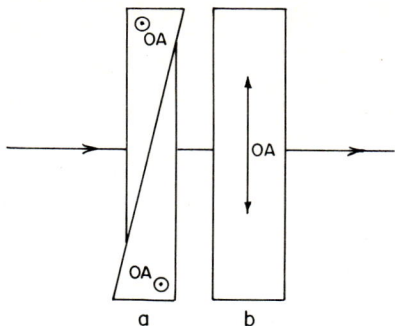

Fig. 5.3. A Babinet-Soleil (BS) compensator consists of two sections of quartz *a* and *b*. Section *a* is composed of two wedges that slide with respect to one another and have parallel optics axes (OA). Section *b* is a single slab with its optic axis (OA) orthogonal to that of *b*.

done by a fine mechanical micrometer calibrated by the manufacturer to indicate the amount of retardation.

Total-internal-reflection retarders: The Fresnel rhomb
This device exploits the difference between the phase shifts that are experienced by the p and s components of light upon total internal reflection, §4.2. The basic design equation is eq. (4.28). One such device is shown in fig. 5.4. It employs three reflections to maintain colinearity of the incident and emergent beams. By suitably coating this rhomb, it is possible to make it essentially achromatic, *i.e.*, its retardation stays close to quarter-wave as the wavelength is varied in the visible [216]. The achromatic nature of the Fresnel rhomb makes it a particularly attractive retarder for spectroscopic ellipsometry.

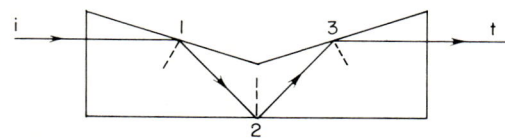

Fig. 5.4. Shows a Fresnel-rhomb retarder employing three internal reflections 1, 2, and 3 to maintain the colinearity of the incident (i) and transmitted (t) beams.

5.2.1.3. *Other polarizing optical devices*
Linear polarizers and linear retarders constitute the essential polarizing optical devices of the ellipsometer. Other "polarizing" optical devices that find occasional use include half-shade devices and depolarizers. When employed, half-shade devices are not permanent elements in the optical train

of the ellipsometer; they are temporarily inserted to increase the precision of setting the polarizer and analyzer for extinction, as the null is visually observed. As we mentioned earlier, the half-shade method of visual null detection has been superceded by photoelectric techniques and will not be discussed here. The interested reader may read more about half-shade devices in refs. 9, 10, and 159.

Depolarizers [8, 10] are devices whose function is to randomize the state of polarization of a light beam. Ideally, they should transform any polarized or partially-polarized light into unpolarized (natural) light. This is practically impossible, although satisfactory depolarizing performance can be realized, for example, by using a thick plate of a highly birefringent material whose thickness fluctuates over the beam cross section. In this case, the light beam is polarized at any point over its cross section but this polarization changes randomly from point to point such that the integrated effect over the entire beam section leads to effective depolarization. A crystalline calcite wedge is one example of this type of "pseudo-depolarizer."

Depolarizers can be helpful when placed before and after the main optical train of the ellipsometer. Thus, a depolarizer between the source and the ellipsometer serves the purpose of cancelling the effect of source polarization, whereas a depolarizer placed after the ellipsometer, in front of the photodetector, removes the dependence of the detector sensitivity on the polarization of the impinging radiation.

5.2.2. Optical components of an ellipsometer that perform no polarizing function

Each one of the optical devices discussed above effects an intended prescribed change in the state of polarization of the ellipsometer light beam. There are a number of other optical elements that are used in an ellipsometer for some purpose other than to change the state of polarization of the light beam. These components may be permanently or temporarily incorporated in the ellipsometer design. The most important of these are: (1) collimating lenses, (2) spatial filters, (3) test-cell windows, and (4) alignment devices.

As their name indicates, collimating lenses are used to collimate the light beam before it passes through the essential optics of the ellipsometer. They are used when the light source itself is uncollimated, *e.g.*, a gas-discharge lamp. A collimating lens is usually mounted in the polarizer telescope a focal distance away from the entrance pinhole to that telescope. Light from the source must be externally focused on the latter pinhole, fig. 5.5. When laser sources are used, collimating lenses are not needed. Because they are placed outside the essential polarizing optical train of the ellipsometer, residual birefringence in collimating lenses is not troublesome. However, collimating

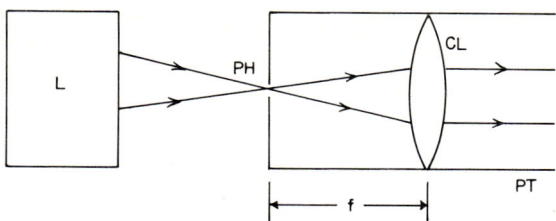

Fig. 5.5. Light from the source L is focused on the entrance pinhole PH of the polarizer telescope PT. A convergent lens CL at focal distance f from PH collimates the ellipsometer light beam.

lenses can introduce additional parasitic multiple reflections between the optical elements of the ellipsometer; they also limit its spectral range. It is therefore advisable to remove them when they are not needed.

To increase the localization of ellipsometric measurements, *i.e.*, to examine surface areas much less than the usual 1 mm^2 (which is limited by the light beam diameter), it has been recently proposed [217] that convergent lenses be used before and after reflection from the surface. The lens placed before reflection focuses the light onto a small spot on the surface. The size of this spot is generally dependent on the quality of the lens, but its linear dimensions are of the order of a few wavelengths across. In this application, the lenses should exhibit little or no birefringence. For further detail on this interesting idea, see ref. 217.

Spatial filters include any device that may be inserted to better define the beam profile or to exclude any spurious light from reaching the photodetector. The simplest type of spatial filter used in an ellipsometer is the pin-hole. This may be of a fixed diameter (~ 1 mm) or of a controllable aperture. They must be mounted concentric with the telescope axes because they serve to align the light beam along the telescope axes and to prevent oblique parasitic beams from reaching the photodetector. For the latter purpose, a special spatial filter may also be used in front of the photodetector to mask all but the center spot [218].

Most of the interesting applications of ellipsometry involve reactions on surfaces that are exposed to environments other than the room atmosphere. In this case, the specimen is mounted in a test cell with entrance and exit optical windows. These windows must obviously be transparent and, by virtue of their location, must introduce the least possible birefringence. The presence of a small wedge angle between the window faces should also be avoided because this introduces, by itself, a change in the state of polarization of light even at normal or near-normal incidence. If off-normal incidence on the windows is desired, to eliminate multiple reflections, this obliquity must be accounted for when the angle of incidence at the sample is calculated [219].

For the purpose of alignment of the ellipsometer telescopes alone, some additional alignment optical devices may be needed. The Gauss eyepiece is one such device, fig. 5.6.

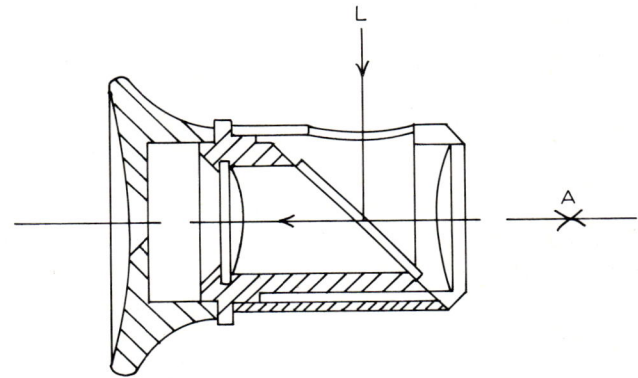

Fig. 5.6. A Gauss eyepiece collimates light from an auxiliary source L parallel to the axis A of an ellipsometer telescope.

5.2.3. Light sources and detectors

In principle, the choice of a light source and a detector will depend on the particular wavelength range of the ellipsometric measurements. Although most applications of ellipsometry have been in the visible, there has been increasing interest in applying it in the infrared, millimetric, and microwave regions of the spectrum, as well as in the near ultraviolet. The primary limiting factor in using ellipsometry in a given spectral interval is the availability of polarizing optical devices that function satisfactorily in that particular interval. Each spectral range will require its own polarizing optical elements that constitute the essential optical train of the ellipsometer, its own sources and its own detectors.

The monochromaticity and collimation of laser beams are ideally suited for ellipsometry. It should be noted that the stability of the polarization state of the laser output is important; there should be no polarization flipping or polarization fluctuations. Unpolarized output is preferred although linearly polarized beams can be easily circularly polarized by a suitable quarter-wave plate. One visible-wavelength laser that has gained wide-spread popularity is the 1 mW-output 6328 Å He-Ne laser which is now commercially available for less than $100.

Classic light sources that continue to be useful in ellipsometry include gas-discharge lamps with a number of strong spectral lines that can be suitably filtered or selected by a monochromator. The mercury-arc lamp is one such source that has been used extensively, especially its strong 5461 Å mercury green line. Gas-discharge sources provide a continuum of radiation between their strong resonant lines. This continuum, when used in conjunction with a suitable monochromator, can be very useful in spectroscopic ellipsometry. For the latter application, a large selection of discretely- or continuously-tunable lasers has also become available.

As with light sources, there is a wide range of photodetectors to choose from, dependent on the wavelength range of interest. Photomultiplier tubes have been most useful in visible ellipsometry (*e.g.*, RCA 1P21, 1P28). To stabilize these detectors, some cooling arrangement may have to be worked out. In addition to photomultiplier tubes, other photodetectors such as Si photodiodes have also been employed. Besides its spectral-response curve, the noise characteristic of a photodetector is important. This noise characteristic directly affects the precision of the ellipsometric measurements (§5.5). Description of photodetectors can be found in ref. 220.

Although not mentioned explicitly, it is understood that suitable power and voltage sources will be required to operate the light sources and the detectors. In addition, the electrical output of the photodetector is usually further amplified and metered by appropriate instruments.

5.3. Operation of the null ellipsometer

5.3.1. Alignment of the telescopes

Alignment of the polarizer (incident-beam) and analyzer (reflected-beam) telescopes of an ellipsometer is a purely mechanical, optically assisted, process by which the geometric axes of these two telescopes are adjusted to become perpendicular to and intersect one another at the central axis of the ellipsometer, fig. 5.7. (The latter is the axis of rotation of the analyzer

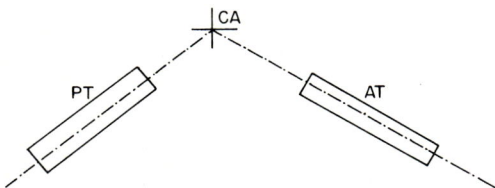

Fig. 5.7. In an aligned ellipsometer, the axes of the polarizer and analyzer telescopes PT and AT intersect at and are normal to the central axis CA of the ellipsometer.

telescope and represents the main reference axis of the ellipsometer.) In a ready-made instrument, this critical mechanical condition should be insured by the manufacturer. However, in the case of a "home-made" ellipsometer or a ready-made one whose alignment has been perturbed, a process of (re)alignment has to be carried out. For this purpose, each of the two telescopes should be equipped with both lateral and tilt screws to permit small movements of their axes in two perpendicular planes.

A satisfactory alignment procedure is summarized below.

(1) To avoid beam-deviation errors, multiple-reflections and parasitic images, remove the polarizing optical devices from the two telescopes.

(2) A plane-parallel plate (PPP) of glass is mounted on the specimen table and a Gauss eyepiece (GEP) is fitted to the outer (exterior) end of the polarizer telescope. By adjusting the position of a light source, the GEP directs a beam of light parallel to the axis of the polarizer telescope. The latter condition can be checked by use of a variable-size aperature (iris) coaxially mounted on the inner (interior) end of the polarizer telescope. As the aperature is made smaller, it should not eclipse the light beam if the beam is exactly transmitted along the polarizer-telescope axis. The reflected beam from one of the two faces of the PPP is autocollimated with the incident beam by coinciding the cross-hairs of the GEP with their reflected image. The position of the reflected image is controlled, in two directions, by rotation and tilt of the sample table. Once autocollimation from one face of the PPP has been achieved, the sample table is rotated exactly by 180°, as determined by the angle-of-incidence graduated circle. If autocollimation from the other face of the PPP is automatically obtained, the polarizer telescope axis is perpendicular to the axis of rotation. In a misaligned telescope, however, this is not the case, fig. 5.8. The tilt screw has to be adjusted and a check made on the preservation of autocollimation after 180° rotation of the sample table. The direction and magnitude of further adjustment of the tilt screw is judged by the change in the degree of

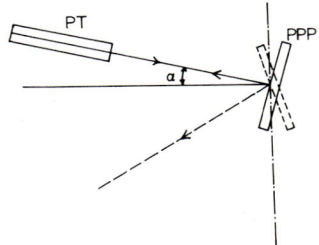

Fig. 5.8. Autocollimation of light along the polarizer telescope PT with the aid of a plane-parallel plate PPP. Shown is the case when the polarizer telescope is tilted by an angle α from its proper position.

coincidence between the GEP cross-hairs and their images achieved in two successive settings of the tilt screw. The whole procedure is subsequently repeated with the analyzer telescope at any angle of incidence. The process is usually rapidly convergent and is not particularly tedious. Its accuracy is limited by the accuracy of the parallelism between the faces of the PPP and its precision is limited by the extent to which the light beam can be directed by the GEP to be truly parallel to the telescope axis.

(3) The autocollimation method described above (step 2) insures the orthogonality of the telescope axes with the central axis of the instrument only. Thus the telescope axes may not intersect the axis of rotation and they may even lie in two different planes perpendicular to the central axes, fig. 5.9. To verify any alignment condition beyond that of the orthogonality of the telescope axes with the central axis, a set of small-aperture optical stops (irises) and a laser with small beam divergence are needed. Two irises, preferably of variable size, are coaxially fitted at both ends of each telescope to define its axis. The laser beam is sent along the axis of the polarizer telescope through the irises mounted at its ends after the latter have been shut down to about 1 mm diameter. It is assumed that either the laser be placed on a table that can be leveled and whose height can be adjusted, or that a beam deflector be used with enough controls to laterally and angularly move the beam until it passes along the axis of the stationary polarizer telescope. As a check on this condition, one should be able to observe (on a piece of paper) that the iris can be symmetrically closed around the beam without eclipsing it.

(4) Rotate the analyzer telescope to the straight-through position until the laser beam passes through the irises at its ends as best as possible. If the analyzer telescope is not grossly misaligned, it will not be necessary to open up the apertures to large diameters. Reduce the size of the apertures to 1 mm each and effect *lateral* movements of the analyzer telescope using the appropriate set screws together with small rotations of the telescope around the central axis until the laser beam travels along the axis of this telescope. This may be either visually observed or photoelectrically detected (the transmitted intensity is maximized). Recheck the normality condition of the analyzer telescope axis with the central axis by autocollimation (step 2).

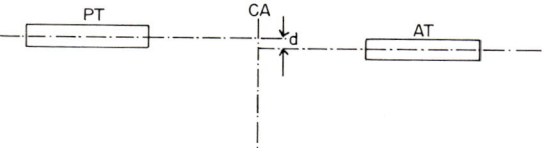

Fig. 5.9. Shows the axes of the polarizer and analyzer telescopes PT and AT orthogonal to the central axis CA of the ellipsometer but vertically displaced with respect to one another by a distance *d*.

(5) After the above steps have been completed, we are sure that the axes of the two telescopes are in one plane perpendicular to the axis of rotation. This assumes, however, that the two pivot points around which the two telescopes can be tilted and laterally adjusted lie in one plane perpendicular to the central axis. If this is not true, an additional control has to be provided to level the pivot point of one telescope with that of the other. It is necessary, however, to make sure that the telescope axes intersect the central axis of the ellipsometer, fig. 5.7. To verify this, a reflector is mounted on the specimen table and adjusted in both position (a moving carriage is necessary) and tilt until the reflected beam, at some angle of incidence, passes through the analyzer telescope which has two small irises at its ends. The analyzer telescope is subsequently rotated to another angle of incidence. If a pure rotation of the specimen table brings the reflected beam along the analyzer telescope axis, the two telescope axes must be intersecting at the central axis. This condition can be further confirmed by observing that the illuminated spot does not walk on the specimen surface as the sample table is rotated. If the above condition is not met, repeated trials of lateral movements of both telescopes followed by a check of their normality with the central axis by autocollimation would be needed until the above test is successful.

(6) The polarizing optical devices (polarizer, compensator, and analyzer) are reinserted in their mounts.

(7) In the above alignment procedure, the autocollimation test for the normality of the telescope axes with the central axis of the ellipsometer (step 2) can be carried out with the help of a small light-weight laser, instead of a Gauss eyepiece. The laser is mounted solidly to the telescope after it is aligned to send its beam along the axis of the telescope as defined by two small apertures. When GEP is used, the autocollimation condition was detected by coincidence of its cross-hairs with their reflected image. When a laser is used, the retroreflected beam from the plane-parallel glass plate can be easily observed for coincidence with the incident laser beam by use of a paper screen with a small hole that passes the laser beam. Autocollimation is completed when the reflected spot on the screen is made to coincide with the small hole.

Another ellipsometer alignment procedure that uses a laser as its basic tool has been recently described by Zeidler *et al.* [221]. Aspnes and Studna [91] proposed a distinctly different alignment procedure based on null measurements at two angles of incidence using a polarizer-dielectric surface-analyzer ellipsometer arrangement. Here, misalignment information (specifically, telescope tilt angles), can be extracted from the nulling azimuth angles of the polarizer and analyzer at the two angles of incidence.

5.3.2. Calibration of the ellipsometer divided circles

The angle-of-incidence divided circle is readily calibrated once the ellipsometer telescopes have been aligned as explained above. By bringing the analyzer telescope to the straight-through position, and transmitting a laser beam through a sequence of four pin-holes at the telescope ends, the angle-of-incidence circle should read 180°. Any difference from 180° represents the calibration correction.

Calibration of the azimuth scales (divided circles) of the polarizer, compensator, and analyzer necessitates a separate procedure. Calibration of the polarizer and analyzer divided circles involves the determination of the circle readings when the elements are positioned to pass an electric vector that lies exactly in the plane of incidence. Calibration of the compensator divided circle involves the determination of the true azimuth (which can be different from the azimuth as read on the circle) of the fast axis of the compensator from the plane of incidence.

It is important to realize that calibration of the azimuth scales of the polarizing optical devices of an ellipsometer is critically dependent on the uniqueness of the plane of incidence, which is the plane formed by the incident and reflected beams. Unless this plane is maintained as different measurements are made on the same reflector or as different specimens are tested, the azimuth calibration parameters determined by any method would not be constant. This requires that alignment of the telescope axes precedes calibration of the azimuthal scales. It means also that the telescope axes have to be well defined by use of small irises in the course of any experiment on the ellipsometer, that beam deviation by the optical elements is minimal, and that the specimen is accurately positioned.

Three methods have been used to calibrate the azimuth scales of the ellipsometer divided circles.

Method 1: This method [68] is based on the simple fact that when light incident on an optically isotropic reflector is linearly polarized parallel (p) or perpendicular (s) to the plane of incidence, the reflected light is, likewise, p- and s-polarized, respectively. With the compensator removed and an optically isotropic specimen accurately positioned for reflection, we search for the polarizer setting that makes the reflected beam linearly polarized, as can be detected by total extinction of that beam by the analyzer. Under this condition, the transmission axes of the polarizer and analyzer are oriented so that one is parallel and the other is perpendicular to the plane of incidence and we have the lowest-level minimum of all the minima that are obtained by setting the polarizer (analyzer) at a fixed azimuth and adjusting the analyzer (polarizer) for minimum intensity. Accurate execution of this calibration

§5.3] OPERATION OF THE NULL ELLIPSOMETER

method for the polarizer and analyzer azimuth scales is based on the following analysis.

Consider the polarizer-specimen-analyzer (PSA) ellipsometer arrangement. The polarizer and analyzer are assumed to be ideal in the sense that they transmit exactly linearly polarized light. The specimen surface is assumed to be optically isotropic so that the p ↔ s cross-reflection coefficients are zero. With these assumptions, the detected signal $\mathcal{I}_D$ is given by [see eq. (3.283)]

$$\mathcal{I}_D(P, A) = F[\sin^2 P \sin^2 A + \tan^2 \psi \cos^2 P \cos^2 A \\ + \tan \psi \cos \Delta \sin 2P \sin 2A] \tag{5.3}$$

where F is independent of P and A. Let P_c and A_c be the readings of the polarizer and analyzer graduated scales when these elements are positioned to transmit an electric vector vibrating parallel to the plane of incidence. In practice, P_c and A_c are two small angles which define the calibration corrections to be determined.

When the polarizer azimuth is near zero and the analyzer azimuth is near $\frac{1}{2}\pi$, we may write

$$P = P_s - P_c = \beta, \\ A = A_s - A_c = \tfrac{1}{2}\pi + \alpha, \tag{5.4}$$

where P_s and A_s are the azimuths as read on the scales and β and α are small angles

$$|\alpha|, |\beta| \lesssim 10°. \tag{5.5}$$

In terms of α and β, eq. (5.3) becomes

$$\mathcal{I}_D(\beta, \alpha) = F[\beta^2 + \alpha^2 \tan^2 \psi - 4\alpha\beta \tan \psi \cos \Delta], \tag{5.6}$$

to second order in the small angles β and α. If the polarizer is set (β = constant) and the analyzer is adjusted for minimum signal, we have

$$\left.\frac{\partial \mathcal{I}_D}{\partial \alpha}\right|_{\beta=\text{const}} = 0 = 2\alpha \tan^2 \psi - 4\beta \tan \psi \cos \Delta. \tag{5.7a}$$

On the other hand, if the analyzer is set (α = constant) and the polarizer is adjusted for minimum signal, we have

$$\left.\frac{\partial \mathcal{I}_D}{\partial \beta}\right|_{\alpha=\text{const}} = 0 = 2\beta - 4\alpha \tan \psi \cos \Delta. \tag{5.7b}$$

Expressed in terms of the scale readings P_s, A_s and the calibration angles P_c and A_c, eqs. (5.7) become

$$\overline{(A_s - 90° - A_c)} \tan \psi - 2(P_s - P_c) \cos \Delta = 0 \tag{5.8a}$$

$$-2\overline{(A_s - 90° - A_c)} \tan \psi \cos \Delta + (P_s - P_c) = 0. \tag{5.8b}$$

From eq. (5.8a), it is evident that if repeated measurements are taken with P_s set at several values near P_c and at each polarizer setting the analyzer angle A_s for minimum signal is recorded, a plot of $(A_s - 90°)$ vs P_s yields a straight line. Similarly, from eq. (5.8b), if measurements are taken with A_s set at several values near $A_c + 90°$ and at each analyzer setting, the polarizer angle P_s for minimum signal is recorded, a plot of $(A_s - 90°)$ vs P_s yields another straight line. The two lines intersect at a point where

$$\begin{aligned} P_s &= P_c, \\ (A_s - 90°) &= A_c, \end{aligned} \tag{5.9}$$

thus the calibration angles P_c and A_c can be determined. At the point of intersection, the signal level should be the lowest of all minima. Figure 5.10 shows an example of the results obtained by McCrackin et al. [68] using this method.

It should be noticed that the above procedure can be carried out in a somewhat different manner if settings for the polarizer are chosen near 90°

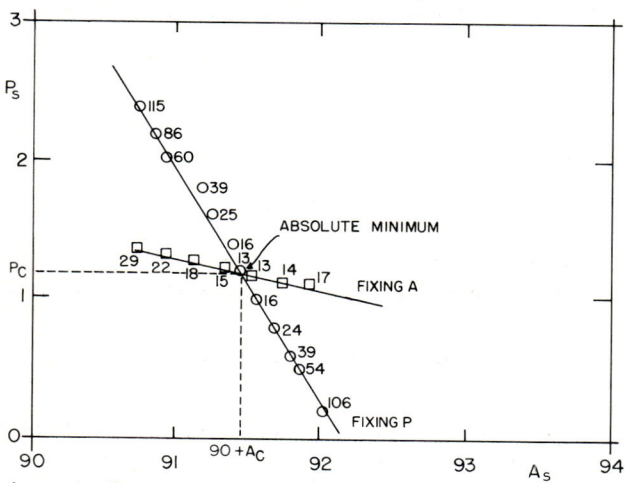

Fig. 5.10. Experimental calibration curves for the polarizer and analyzer graduated circles, after McCrackin et al. [68]. The numbers next to the points indicate the level of the intensity minima in relative units.

and those for the analyzer near zero. Plots of A_s vs $(P_s - 90°)$ in this case, corresponding to the two different ways by which the measurements can be taken (set P adjust A for minimum or set A and adjust P for minimum), yield two straight lines whose intersection yields the calibration angles P_c and A_c.

If the polarizer and analyzer are imperfect in that they transmit elliptically polarized light (instead of linearly polarized light), or if the surface is anisotropic [86], the angles P_c and A_c as determined by the above method do not represent the true calibration angles. In this case, P_c and A_c lump together both the true calibration errors and imperfection parameters that characterize the polarizer, analyzer, and the surface. Variations of Method 1 that account for one source of imperfection or the other have been published [80, 85, 87, 89, 90, 101].

Method 2: This method [159] is based upon the fact that when parallel (p) polarized light is incident on an isotropic film-free smooth dielectric surface it is totally extinguished at the Brewster angle $\phi_B = \tan^{-1} n$ where n is the refractive index of the dielectric sample. In this method, the analyzer telescope is set for reflection at the Brewster angle and the specimen is accurately positioned. With the compensator and analyzer removed, the polarizer is rotated until the reflected beam is extinguished. The reading of the polarizer graduated scale at this extinction setting is the calibration angle P_c. The calibration angle A_c for the analyzer (which has been reinserted) can then be determined from a null measurement in the straight-through (PA) arrangement, in which P is set at P_c and A is adjusted for null.

In the above two methods, the compensator graduated circle can be easily calibrated once the polarizer and analyzer calibration angles P_c and A_c have been determined. By arranging the ellipsometer for the straight-through polarizer-compensator-analyzer (PCA) operation, the polarizer is set at P_c, the analyzer at $A_c + 90°$ and the compensator is rotated until its fast axis is aligned with the transmission axis of the polarizer at which point the transmitted light is extinguished. The reading of the compensator divided circle C_c gives the required calibration angle. If extinction is obtained with the fast axis of the compensator aligned parallel to the transmission axis of the analyzer (instead of the polarizer), the reading of the compensator azimuth graduated scale will be $C_c + 90°$.

Method 3: In contrast with Methods 1 and 2, this third method [97] does not require any special arrangement of the ellipsometer or any special measurements. The ellipsometer is operated in the normal mode in the PCSA arrangements. The (nominally) quarter-wave compensator C is set with its azimuth near either $+45°$ or $-45°$ before reflection from the sample S which is mounted in air (not in a test cell, so that window birefringence is avoided). The two distinguishable nulls (P', A'), (P'', A'') are determined by adjustment of the polarizer and analyzer. If θ_C denotes the deviation of the relative retardation of the compensator from 90°, the two-zone residual Res A defined by

$$\text{Res } A = A' + A'' - 180°, \tag{5.10}$$

will be given by

$$\text{Res } A^\pm = \mp \theta_C \sin 2\psi \sin \Delta + 2C_c \sin 2\psi \cos \Delta - 2A_c \tag{5.11}$$

where the $+$ and $-$ signs correspond to the $+45°$ and $-45°$ azimuths of the compensator. Equation (5.11) is based on the error analysis of §3.7 [see also the following §5.4]. It is also assumed that the surface S is isotropic and that it is accurately positioned on the sample table.

If the specimen is chosen to be an isotropic, film-free and smooth dielectric, we have $\sin \Delta = 0$ and the first term on the right-hand side of eq. (5.11) disappears,

$$\text{Res } A \Big|_{\substack{\text{dielectric} \\ \text{surface}}} = 2C_c \sin 2\psi \cos \Delta - 2A_c. \tag{5.12}$$

By repeating the ellipsometric measurements at different angles of incidence and plotting the measured residual Res A at each angle versus the quantity $2 \sin 2\psi \cos \Delta$, a straight line is obtained, eq. (5.12), whose slope and intercept determine the compensator and analyzer calibration angles C_c and A_c, respectively. An example of this plot is shown in fig. 5.11. The polarizer

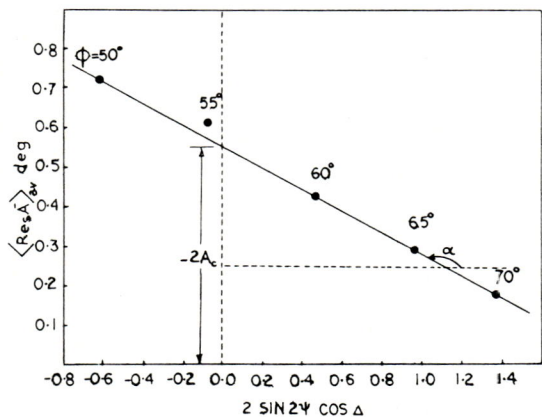

Fig. 5.11. Average of the two-zone residual Res A vs ($2 \sin 2\psi \cos \Delta$) for a clean supersmooth fused-quartz specimen. The slope ($\tan \alpha$) and intercept with the ordinate ($-2A_c$) give C_c and A_c, the compensator- and analyzer-circle calibration errors, respectively.

calibration angle P_c is subsequently determined from the four-zone residual Res $1P$ [see §5.4]

$$\text{Res } 1P = (P' + P'')_+ + (P' + P'')_- \tag{5.13}$$

which is determined from null measurements at both the $+$ and $-$ azimuths of the compensator. By neglect of a small polarizer imperfection parameter, it can be shown that (see §3.7 and §5.4)

$$\text{Res } 1P = -4(P_c - C_c), \tag{5.14}$$

from which P_c is calculated. Alternatively, P_c can be determined from a straight-through PA null measurement in which A is set at A_c and P is adjusted for null.

5.3.3. Nulling schemes

In the polarizer-compensator-surface-analyzer (PCSA) or the PSCA ellipsometer arrangement, extinction of the light leaving the analyzer can be generally obtained by alternative adjustment of two of the three azimuth angles P, C, and A, or of one of these azimuth angles and the compensator's relative retardation δ_C. If a nulling scheme is identified by the two parameters that are adjusted in reaching the null, we would have six such schemes: the (P, A), (P, C), (C, A), (δ_C, P), (δ_C, C) and (δ_C, A) nulling schemes. During a nulling process employing any two parameters, the other two parameters are held constant.

In presently available ellipsometer designs, the azimuth angles of the polarizing components P, C, or A can be determined with a resolution of $0.005° - 0.01°$, whereas the compensator (*e.g.*, Babinet-Soleil) relative retardation δ_C can be controlled with a resolution of $0.1° - 1°$. Because the resolution of δ_C is one-to-two orders of magnitude poorer than that of any of the azimuth angles P, C, or A, use of δ_C as a nulling parameter is not commonly practiced. We only compare the first three of the above-listed six nulling schemes.

Two considerations are important in discussing a nulling scheme (1) the availability and number of physically distinguishable nulls, and (2) the mode of convergence and speed of nulling. The availability and number of nulls for the (P, A), (P, C) and (C, A) schemes have been studied in detail in §3.9. In general, two, one, and four nulls, respectively, are available for the (P, A), (P, C), and (C, A) schemes.

In the PCSA ellipsometer, the compensator C is usually set at a fixed azimuth of $+\frac{1}{4}\pi$ or $-\frac{1}{4}\pi$ and its retardation is adjusted to quarter-wave. Under such conditions, the null is most rapidly reached by rotating the polarizer and the analyzer assuming that measurements are made on optically isotropic surfaces. In fact, the null can be reached in two steps, first by adjusting the polarizer to minimum intensity, then adjusting the analyzer to the null. This can be seen from the expression of the detected signal[3]

$$\mathscr{I}_D = \text{const} [\sin^2 (A - A_n) - \sin 2A_n \sin 2A \sin^2 (P - P_n)], \quad (5.15)$$

where P_n, A_n are the null settings. $\mathscr{I}_D$ can be brought to zero first by setting $P = P_n$, then $A = A_n$. To attain higher precision, the ellipsometer is usually nulled by the following sequence of steps [159]. (1) Rotate the polarizer P to minimize the detected signal, (2) with the polarizer left at the above setting, reduce the detected signal further to a still deeper minimum by rotating the analyzer A, (3) repeat steps 1 and 2 until no further reduction in signal is observable, (4) adjust the polarizer by use of the fine tangent screw to obtain readings P^+ and P^- of the polarizer-azimuth scale at two equal signal levels on both sides of the minimum; the true polarizer null setting will be $P_n = \frac{1}{2}(P^+ + P^-)$, (5) with the polarizer set at P_n, adjust the analyzer A to two settings A^+ and A^- at equal signal levels on both sides of the minimum; the true analyzer null setting will be $A_n = \frac{1}{2}(A^+ + A^-)$. This completes the nulling procedure. The validity of this "method of swings" is based on the fact that the detected signal $\mathscr{I}_D$ [eq. (5.15)] is a symmetrical parabolic function of the polarizer off-null azimuth deviation $(P - P_n)$, for small values of $P - P_n$; and that only when $P = P_n$, is $\mathscr{I}_D$ likewise a symmetrical parabolic function of the analyzer off-null azimuth deviation $(A - A_n)$, for small values of $(A - A_n)$.

When the compensator is set at an azimuth other than $\pm \frac{1}{4}\pi$, its retardation is different from quarter-wave, or if the surface under measurement is anisotropic, null convergence becomes possible through a prolonged sequence of alternate adjustments of the polarizer and analyzer. In fact, this sequence is theoretically infinite, although it is practically stopped when azimuth changes between successive steps fall below the resolution limit of the graduated circles, or cause intensity changes below the detection limit. The same observations apply to the (P, C) and (C, A) nulling schemes. Figures 5.12–5.14 show examples of the mode of convergence to the null in the (P, A), (P, C), and (C, A) schemes.

5.4. Sources of error and their correction

The ratio of reflection coefficients (eigenvalues) $\rho = \tan \psi \, e^{j\Delta}$ measured by an ellipsometer is susceptible to systematic errors from various sources. Most important among these are (1) azimuth-angle errors, (2) component and

---

[3] Equation (5.15) is an alternate form of eq. (3.48) [or (3.49)] obtained by use of the null condition of eq. (3.45) [or (3.46)].

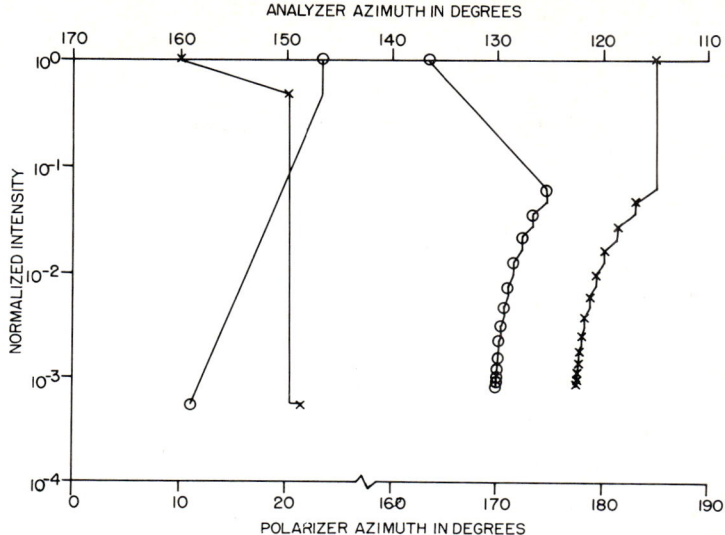

Fig. 5.12. Ellipsometer null convergence when the compensator is held at a fixed azimuth [$C = 45°$ (left) and $C = 10°$ (right)] and the polarizer and analyzer are adjusted. A gold-coated glass slide was tested at $\lambda = 6328$ Å and an angle of incidence of 70°. O and X represent the measured polarizer and analyzer azimuths, respectively, at the successive minima leading to the null.

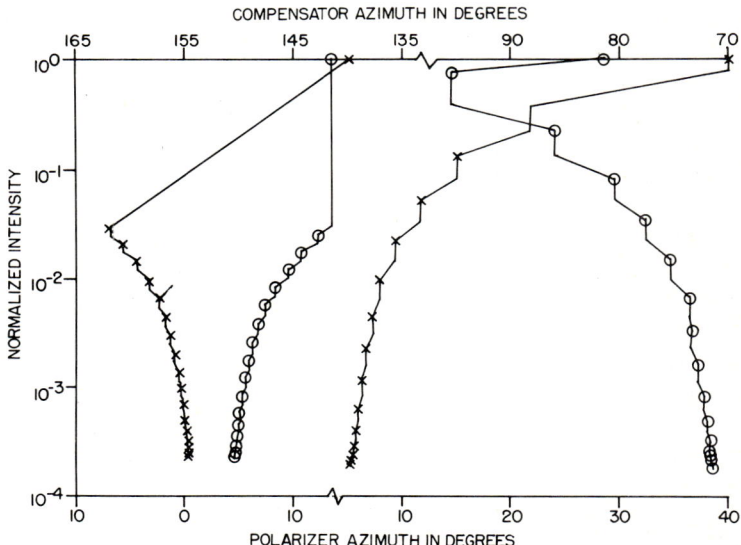

Fig. 5.13. Ellipsometer null convergence when the analyzer is held at a fixed azimuth [$A = 40°$ (left) and $A = 130°$ (right)] and the polarizer and compensator are adjusted. A gold-coated slide was tested at $\lambda = 6328$ Å and angle of incidence of 70°. O and X represent the measured polarizer and compensator azimuths, respectively, at the successive minima leading to the null.

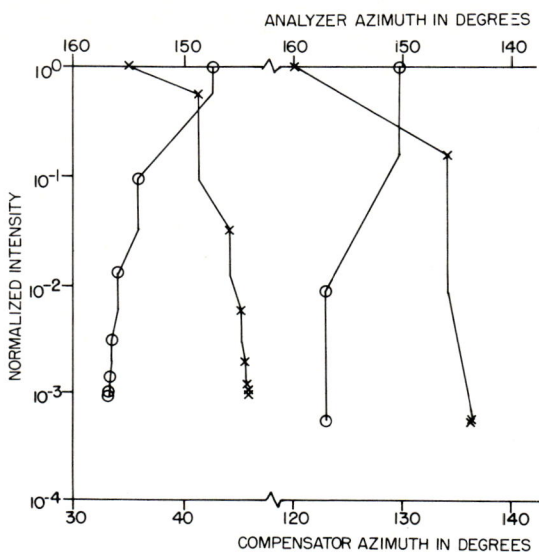

Fig. 5.14. Ellipsometer null convergence when the polarizer is held at a fixed azimuth [$P = 1°$ (left) and $P = 64.9°$ (right)] and the compensator and analyzer are adjusted. A gold-coated slide was tested at $\lambda = 6328$ Å and angle of incidence of 70°. O and X represent the measured compensator and analyzer azimuths respectively, at the successive minima leading to the null.

cell-window optical imperfections, (3) beam deviation errors, (4) parasitic beams, (5) source-beam polarization and collimation errors, (6) polarization-dependent detector sensitivity and (7) residual mechanical imperfections. In the following, we examine the effect of the above sources of error on the calculated ψ and Δ and present methods for correcting such errors.

5.4.1. Azimuth-angle errors, component and cell-window imperfections

The effect of azimuth-angle errors, of component and cell-window imperfections on the measured value of ρ has been comprehensively studied in §3.7 and §3.8. In §3.8 we have also examined the effect of source-beam polarization, component depolarization and of surface imperfections (anisotropy and roughness).

Azimuth-angle errors may arise from (1) residual calibration errors of the graduated circles which are constant, and (2) specimen mispositioning errors that vary as the specimen is reset or changed. Component imperfections (finite extinction ratios for the polarizing prisms, stress birefringence in windows, optical activity and defect scattering in compensators, etc.) are unavoidable although, for ellipsometric applications, quality optical elements should always be used.

When the imperfection parameters characteristic of a given ellipsometer are known or measured, correction of their effect on one-zone (single-null) measurement is possible by use of the analysis of §3.7 and §3.8. However, it is more common practice to average data from two- or four-null measurements in such a manner as to cancel the effect of many of the imperfections. The usual scheme is to set the compensator azimuth at (or near) $C = +\frac{1}{4}\pi$ and its retardation at (or near) $\frac{1}{2}\pi$ then obtain two different nulls (P_2, A_2) and (P_4, A_4) by adjustment of the polarizer and analyzer. By rotating the compensator 90° to the new azimuth $C = -\frac{1}{4}\pi$, another pair of nulls (P_1, A_1) and (P_3, A_3) are recorded. The ellipsometric parameters ψ and Δ are obtained from the nulling azimuth angles (P_i, A_i), $i = 1, 2, 3, 4$ by two- or four-zone averages using eqs. (3.191), (3.192) and (3.198), (3.199). The outcome of *two-zone averaging* can be summarized as follows.

(a) ψ is free of the effect of (1) all azimuth-angle errors, (2) the deviation from $-j$ of the slow-to-fast complex relative transmittance of the compensator, (3) cell-window birefringence, (4) p ↔ s cross scattering due to spurious sample-surface anisotropy or roughness, and (5) imperfections of the analyzer. The following imperfections, however, continue to affect the two-zone value of ψ: (1) polarizer ellipticity (*i.e.*, deviation from an exact linear state of the polarizer light output), (2) cell-window dichroism, (3) imperfections of the compensator leading to off-diagonal elements in its Jones matrix, and (4) depolarization of the polarizer light output.

(b) Δ is free of the effect of (1) analyzer azimuth-angle error, (2) polarizer and analyzer ellipticities, (3) the deviation from $-j$ of the compensator relative transmittance, (4) cell-window dichroism, (5) p ↔ s cross scattering due to surface anisotropy or roughness, and (6) depolarization of the polarizer light output. The following imperfections, however, continue to affect the two-zone value of Δ: (1) polarizer and compensator azimuth-angle errors, (2) cell-window birefringence, (3) off-diagonal imperfection elements in the fast axis-slow axis Jones matrix of the compensator.

The outcome of *four-zone averaging* is as follows.

(a) ψ is free of all of the above-cited imperfections with the exception of the depolarization of the polarizer light output.

(b) Δ is free of all of the above-cited imperfections with the exception of cell-window birefringence.

5.4.2. *Measurement of ellipsometer imperfection parameters*

In an ellipsometer with imperfections, the nulling angles (P_i, A_i) $i = 1, 2, 3, 4$ in the four zones do not satisfy the ideal zone relations namely the orthogonality of the polarizer azimuths and the complementarity of the analyzer azimuths in two conjugate zones (2 and 4 or 1 and 3). It is therefore possible to define residuals, equal to zero in an ideal ellipsometer, that can be used to determine the various imperfection parameters of an ellipsometer. Without an integral multiple of $\pm\frac{1}{2}\pi$, the *two-zone residuals* are

$$\text{Res } A^+ = A_2 + A_4, \qquad \text{Res } A^- = A_1 + A_3,$$
$$\text{Res } P^+ = P_4 - P_2, \qquad \text{Res } P^- = P_3 - P_1, \tag{5.16}$$

where the + and − superscripts identify the $C = +\frac{1}{4}\pi$ and $C = -\frac{1}{4}\pi$ compensator azimuths, respectively.

When the off-diagonal elements in the principal-frame Jones matrices of all optical components of the ellipsometer can be neglected, the nulling angles in the four zones are given by table 3.8 of ch. 3. From this table, the residuals of eqs. (5.16) are given by

$$\text{Res } A^\pm = \mp t^\pm_{1C} \sin 2\psi \sin \Delta$$
$$+ 2 \sin 2\psi \cos \Delta \delta C - 2\delta A - t_{2w} \sin 2W \sin 2\psi \sin \Delta \tag{5.17a}$$

and

$$\text{Res } P^\pm = - t^\pm_{2C} \sin \Delta + t_{2w'} \sin 2W' \cot 2\psi, \tag{5.17b}$$

where δP, δC, and δA are the polarizer, compensator and analyzer azimuth-angle errors, $t_{1C} + j t_{2C}$ represents the deviation from $-j$ of the compensator relative transmittance, and (t_{2w}, W), $(t_{2w'}, W')$ represent the birefringence (t) and the fast-axis azimuth (W) of the entrance (W) and exit (W′) windows, respectively. ψ and Δ are the ellipsometric angles of the surface, as usual.

No window terms appear in eq. (5.17) when measurements are taken with no windows, so that the deviation from unity of the magnitude of the compensator relative transmittance can be found from

$$t^\pm_{2C} = - \text{Res } P^\pm / \sin \Delta. \tag{5.18}$$

If the compensator- and analyzer-azimuth circles are accurately calibrated and the specimen is exactly positioned, the azimuth angle errors δC and δA can be set equal to zero in eq. (5.17a). The deviation from 90° of the compensator relative retardation can then be determined by

$$t^\pm_{1C} = \mp \text{Res } A^\pm / \sin 2\psi \sin \Delta, \tag{5.19}$$

where, again, measurements are assumed to be taken with no windows present. Any error of the azimuth angle of the compensator or the analyzer will cause an error of the value of t_{1C} calculated from eq. (5.19). Notice that, ideally, the properties of the compensator do not change as it is rotated so that $t^+_{1C} = t^-_{1C}$ and $t^+_{2C} = t^-_{2C}$. However, Oldham [75] has shown that this may

§5.4] SOURCES OF ERROR AND THEIR CORRECTION

not be the case dependent on how the compensator is constructed or mounted.

Equation (5.17a) has been used in §5.3.2 as the basis of a calibration procedure for the compensator- and analyzer-azimuth scales.

If measurements are made first with the windows removed then with the windows in place, we get from table 3.8 and eq. (5.17b), respectively,

$$t_{2W} \sin 2W = -2(A_{iW} - A_{iNW})/\sin \Delta \sin 2\psi,$$

$$\pm t_{2W'} \sin 2W' = [(\text{Res } P^{\pm})_W - (\text{Res } P^{\pm})_{NW}] \tan 2\psi,$$

(5.20)

where $i = 1, 2, 3,$ or 4 and W and NW mean with window and no window, respectively. Equations (5.20) show how the important cell-window imperfection parameters can be measured.

The *four-zone residuals* can be defined as

$$\text{Res } 1A = A_1 - A_2 - A_3 + A_4,$$
$$\text{Res } 2A = A_1 - A_2 + A_3 - A_4,$$
$$\text{Res } 3A = A_1 + A_2 + A_3 + A_4,$$

(5.21)

and

$$\text{Res } 1P = P_1 + P_2 + P_3 + P_4,$$
$$\text{Res } 2P = P_1 + P_2 - P_3 - P_4,$$
$$\text{Res } 3P = P_1 - P_2 - P_3 + P_4.$$

(5.22)

From table 3.8, we obtain

$$\text{Res } 1A = 4t_{2P} \sin 2\psi = 4\xi,$$
$$\text{Res } 2A = 2t_{1C}^{av} \sin 2\psi \sin \Delta,$$
$$\text{Res } 3A = (t_{1C}^{-} - t_{1C}^{+}) \sin 2\psi \sin \Delta + 4 \sin 2\psi \cos \Delta \delta C$$
$$\quad - 4\delta A - 2t_{2W} \sin 2W \sin 2\psi \sin \Delta;$$

(5.23)

and

$$\text{Res } 1P = -4t_{1P} - 4\delta P + 4\delta C = -2\eta,$$
$$\text{Res } 2P = 2t_{2C}^{av} \sin \Delta,$$
$$\text{Res } 3P = (t_{2C}^{-} - t_{2C}^{+}) \sin \Delta + 2t_{2W'} \sin 2W' \cot 2\psi,$$

(5.24)

where $t_{1,2C}^{av} = \frac{1}{2}(t_{1,2C}^{+} + t_{1,2C}^{-})$ and $t_{1P} + jt_{2P}$ is the complex imperfection parameter of the polarizer[4] (t_{1P} is equivalent to a polarizer-azimuth error and t_{2P}

---

[4] See eqs. (3.171) and (3.183).

defines the polarizer ellipticity). ξ and η are the two-zone ψ and Δ correction factors, respectively [see eqs. (3.193)]. From the first of eqs. (5.23), we see that the polarizer ellipticity t_{2P} can be calculated from the four-zone residual Res $1A$,

$$t_{2P} = \text{Res } 1A/4 \sin 2\psi. \tag{5.25}$$

Whereas from the second of eqs. (5.23), we see that the average value t_{1C}^{av} of the deviation of the compensator retardation from 90° can be determined from the four-zone residual Res $2A$ by

$$t_{1C}^{av} = \text{Res } 2A/2 \sin 2\psi \sin \Delta, \tag{5.26}$$

independent of compensator and analyzer azimuth-angle errors. Finally, the two-zone Δ-correction factor η is given by

$$\eta = 2t_{1P} + 2\delta P - 2\delta C = -\tfrac{1}{2}\text{Res } 1P, \tag{5.27}$$

as can be seen from the first of eqs. (5.24).

Table 5.1 shows the measured imperfection parameters of two ellipsometers using eqs. (5.16)–(5.27).

5.4.3. Beam deviation errors

When the end faces of an optical element are not exactly parallel, the ellipsometer light beam is angularly deviated upon transmission through the element. This angular beam deviation, to be denoted by δ,[5] is proportional to the small "wedge angle" between the element's faces. As the component is rotated, the transmitted beam describes a right-circular cone of apex angle δ and whose axis is along the incident beam, fig. 5.15. Beam deviation, especially that caused by the polarizer, has been recognized by several investigators [71, 77, 79] as a source of error in ellipsometry. A recent analysis of beam deviation errors has been published by Zeidler et al. [222].

Perhaps the most important way by which beam deviation causes an error in the measured values of ψ and Δ is through its effect on the angle of incidence [79]. In one-zone (single-null) measurements, the calculated values of ψ and Δ correspond to an angle of incidence different from that read on the scale because of beam deviation. Furthermore, even with the ellipsometer otherwise ideal, beam deviation causes (ψ, Δ) to be different in the different zones because the associated azimuthal settings of the optical elements are different. This creates apparent inter-zone disagreements and the effect on zone averaging should be assessed. Additional errors can arise

---

[5]Not to be confused with the compensator relative retardation δ_C (which is always subscripted by C).

Table 5.1. Imperfection properties of two ellipsometers from measurements in four zones. $\eta_{P,C} = 2t_{1P} + 2\delta P_c - 2\delta C_c$, $t_{1P} + jt_{2P} = \delta\rho_P$, $t_{1C} + jt_{2C} = \delta\rho_C$, where δP_c and δC_c are polarizer and compensator calibration errors, $\delta\rho_P$ and $\delta\rho_C$ are the deviations of the complex relative transmittances of the polarizer and compensator from zero and $-j$, respectively

	$\eta_{P,C}$ (deg)	t_{2P} (deg)	t_{1C}^* (deg)	t_{2C}^+ (rad)	t_{2C}^- (rad)
Ellipsometer I	-0.628 ± 0.032	-0.011 ± 0.002	0.307 ± 0.016 $(0.175 \pm 0.046)^a$	0.0019 ± 0.0006	0.0018 ± 0.0005
Ellipsometer II	-0.352 ± 0.017	0.011 ± 0.004	0.453 ± 0.101 $\left\{\begin{array}{l} t_{1C}^+ = 2.089 \pm 1.300 \\ t_{1C}^- = 1.180 \pm 1.327 \end{array}\right\}^b$	0.0018 ± 0.0005	0.0009 ± 0.0002

[a] The retardation setting of the compensator slightly changed, accidentally, about halfway during taking the data.
[b] These values are determined from the data corresponding to each compensator azimuth separately, using eq. (5.19), neglecting azimuth-angle errors.

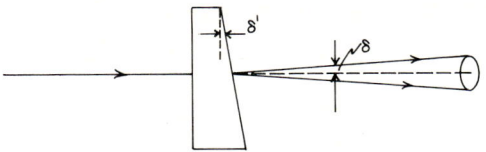

Fig. 5.15. Beam deviation when light passes through an optical element with non-parallel surfaces. As the element is rotated around the incident beam, the transmitted beam describes a cone with apex angle $\delta = \delta'(n-1)/n$, where δ' and n are the wedge angle and refractive index of the element, respectively.

because of the finite precession of the beam over areas of the optical elements, the sample, and the detector that can have nonuniform properties. Whereas the angle-of-incidence errors caused by beam deviation can be analyzed, those due to beam precession and component nonuniformity are difficult to account for.

Consider the effect of beam deviation by the polarizer on the angle of incidence on the surface S of the sample. For simplicity, we assume that no other elements are interposed between the polarizer P and the surface S. (The discussion therefore applies directly to the polarizer-surface-compensator-analyzer (PSCA) ellipsometer arrangement.) Figure 5.16 shows the ellipse of intersection of the cone of the deviated beam with the surface. If the angle of incidence of the undeviated beam is ϕ_0, it can be easily seen that

$$\phi_C = \phi_0 + \delta, \qquad \phi_D = \phi_0 - \delta, \qquad \phi_A = \phi_B = \phi_0$$

where A, B, C, and D are points on the surface at the ends of the minor and major axes of the ellipse. From geometrical considerations, the angle of

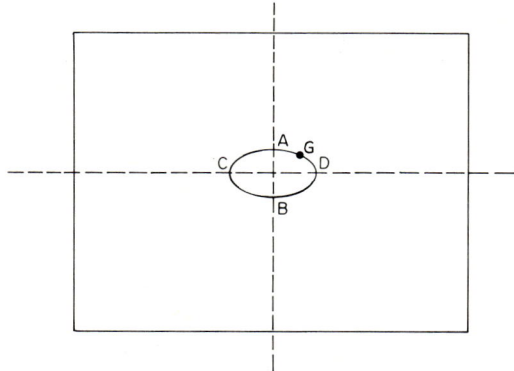

Fig. 5.16. Ellipse of intersection of the cone of the deviated beam transmitted by the polarizer with the surface of the specimen.

incidence at a general point G on the ellipse is given by [222]

$$\phi = \phi_0 \pm [\cos \phi_0/(\cos^2 \phi_0 + \tan^2 \theta)^{\frac{1}{2}}]\delta, \tag{5.28}$$

where

$$\theta = P_G - P_D$$

is the angle by which the polarizer should be rotated to move the spot on the sample from D, where the polarizer reading is P_D, to the general point in question G, where the polarizer reading is P_G, fig. 5.16. The $-$ and $+$ signs in eq. (5.28) correspond to the following ranges of θ;

$$0 \leq |\theta| \leq \tfrac{1}{2}\pi \quad \text{and} \quad \tfrac{1}{2}\pi < |\theta| \leq \pi,$$

respectively.

Besides the variation of the angle of incidence ϕ which is described by eq. (5.28), the plane of incidence (which is formed by the incident beam and the surface normal) precesses as the polarizer is rotated. Because azimuth angles of the linear or elliptic electric-field vibrations of the light beam are all measured from the plane of incidence, beam deviation causes azimuth-angle errors that are also functions of the polarizer setting.

In order to be able to use eq. (5.28), the polarizer-deviation angle δ and the polarizer-azimuth setting P_D (at which the planes of incidence of the deviated and undeviated beams are coincident) should be measured. To determine δ, we measure the radii r_1 and r_2 of the intersection circles of the cone of the deviated beam with a screen placed perpendicular to the undeviated beam at two large distances d_1 and d_2 from the polarizer. From such measurements, δ is given by

$$\delta = (r_2 - r_1)/(d_2 - d_1). \tag{5.29}$$

To determine P_D, the polarizer is removed and a laser beam is sent along the polarizer-telescope axis as defined by two small apertures coaxially mounted at its ends. A reflector is accurately positioned on the sample table (to contain the central axis of the ellipsometer) and the reflected beam is observed on a distant screen. The sample table is rotated and the points of intersection of the reflected beam with the screen are marked at different angles of incidence. These points are denoted by X in fig. 5.17. The straight line L through such points represents the line of intersection of the plane of incidence of the undeviated beam with the screen. When the polarizer is subsequently reinserted and rotated, the trajectory of the illuminated spot (where the reflected beam hits the screen) becomes a circle C, fig. 5.17, because of beam deviation. The azimuthal setting of the polarizer at which the reflected spot lies on the straight line L is P_D.

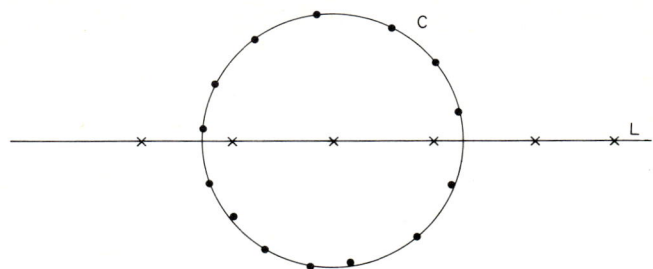

Fig. 5.17. Construction to determine the scale reading when the transmitted beam from a deviating polarizer is in the unperturbed plane of incidence. L represents the locus of the reflected spot when the polarizer is removed and a well aligned sample is rotated. C is the corresponding locus when the polarizer is inserted.

If (P_n, A_n) represent the polarizer- and analyzer-azimuth angles from one-zone (single-null) measurement, a sure way to cancel the effect of the change of angle of incidence caused by beam deviation is to search for another null in the neighbourhood of $(P_n + \pi, A_n)$ and subsequently average the calculated values of ψ and Δ from these two nulls. That the averaged values of ψ and Δ truly correspond to values at the undeviated angle of incidence ϕ_0 can be seen from eq. (5.28). In eq. (5.28), changing θ by π changes the sign of the angle of incidence error only. Hence, at the polarizer azimuths P and $P + \pi$, the angles of incidence are equally above and below ϕ_0 and averaging ψ and Δ at these angles will yield the values at ϕ_0, to first order.

In the PCWSW'A ellipsometer arrangement where the compensator C and entrance window W are interposed between the polarizer P and the surface S, the possible additional beam deviation caused by C and W should also be considered. However, because these elements are fixed, their effect will be to cause a constant shift in ϕ_0 superimposed on the variable deviation by the rotating polarizer. If the wedge angles of C and W as well as the orientations of the unit normals to their surfaces are known, the constant shift in ϕ_0 can be calculated.

Beam deviation by the analyzer causes the light beam incident on the photodetector to precess as the analyzer is rotated. Due to possible nonuniform sensitivity of the photodetector, this precession causes spurious analyzer-azimuth dependent variations in the detected signal. However, the effect is similar to that due to a polarization-sensitive photoelectric response which will be shown below to have little or no effect on the null azimuth angles.

It is needless to emphasize the importance of specifying close tolerances ($\sim 0.01°$–$0.05°$) on the wedge angles of the components of the ellipsometer especially for the polarizer. It is also to be remembered that the angle-of-incidence errors caused by beam deviation are particularly important at angles of incidence where the derivatives $\partial\Delta/\partial\phi$ and $\partial\psi/\partial\phi$ are high (*e.g.*, near the pseudo-Brewster angle of low-absorption substrates $\partial\Delta/\partial\phi \to \infty$ as the extinction coefficient $k \to 0$).

5.4.4. *Effect of auto-collimated parasitic beams from multiple reflections*

In an ordinary (non-interferometric) ellipsometer, the light beam is supposed to make a single path through the optical train from the source to the detector. With the exception of the oblique reflection at the specimen surface, the light beam travels (nearly) perpendicular to the surfaces of all other optical elements in its path. The possibility thus arises that several autocollimated beams (that are folded on the primary forward-travelling beam), originating from inter-component multiple reflections, will find their way to the photodetector. The intensities of these overlapping parasitic beams at the photodetector will depend on the number of multiple reflections that they have experienced, and their polarization states before the analyzer will depend on their detailed folded path. Because the normal-incidence reflectances from the surfaces of the transparent materials (glass, calcite, quartz, etc.) that make the ellipsometer components is small ($\sim 4\%$), the intensity of the strongest parasitic beam is only a small fraction of that of the primary beam. Also, the intensity of the parasitic beams decreases rapidly with the number of multiple reflections that they have experienced.

The effect of these overlapping parasitic beams can be exactly analyzed by use of the Mueller calculus (§2.12). Each parasitic beam is followed along its folded path with each reflection at a surface or transmission through an element accounted for by an appropriate Mueller matrix. By direct addition of the first elements [S_0, see eq. (1.119a)] of the Stokes vectors of all the parasitic beams after the analyzer, a complete expression for the transmitted intensity as a function of the azimuth angles and component optical properties is obtained. By manipulation of such an expression, the shift in the azimuth location of the intensity minima due to the parasitic beams can be calculated. Winterbottom [71] used a simpler (approximate) approach for the calculation of the effect of the parasitic beams by calculating their intensities and polarization states before the analyzer. The states of polarization of the primary and parasitic beams are subsequently represented by points on the Poincaré sphere and each point is assigned a "weight" equal to the corresponding beam intensity. An "average" state of polarization of the total light incident on the analyzer is then found from the "center of gravity" of such a set of points on the sphere. The associated intensity of this average state is taken to be the sum total of the intensities of all beams.

Instead of trying to correct for the effect of the parasitic beams as explained above (which is not an easy task), it is advisable to avoid such beams altogether. This can be effectively done by applying anti-reflection coatings to all surfaces normal to the beam. Alternatively, the individual devices can be slightly tilted so that the parasitic beams travel obliquely and do not overlap with the primary beam. The oblique beams are then easily rejected by use of small apertures at the telescope ends or by a spatial filter in front of the detector. Component tilts, however, should be small to avoid significant lateral displacements of the primary beam.

5.4.5. *Errors that arise from certain assumptions concerning the ellipsometer light beam*

Subtle errors can arise because the ellipsometer light beam has properties different from those usually assigned to it. It is generally assumed that the light beam is (1) monochromatic with zero or negligible bandwidth, (2) unpolarized or circularly polarized before it passes through the ellipsometer, (3) totally polarized after it passes through the polarizer and remains so throughout, (4) perfectly collimated, and (5) that it is uniformly polarized over its cross section. Each of these assumptions represents an idealization. The effect of a departure from such an idealization has to be considered.

5.4.5.1. *Bandwidth*

All light sources used in ellipsometry are quasi-monochromatic with finite (nonzero), but very small, bandwidths. The optical properties of the individual components of the ellipsometer, as well as those of the sample under measurement, that determine their effect on the state of polarization of light should be understood to be average values over the spectral width of the beam. It is to be remembered that all materials from which the optical devices or the sample are made are dispersive, *i.e.*, they have wavelength-dependent optical properties. An analytical study of the effect of bandwidth on the ellipsometrically measured ratio of reflection coefficients (eigenvalues) $\rho = \tan\psi e^{j\Delta}$ is feasible but difficult. Smith [77] made an experimental study of the effect of bandwidth on the measured values of ψ and Δ and some of his results are shown in fig. 5.18.

5.4.5.2. *Source polarization and component depolarization*

In §3.8, we have proved that small residual linear polarization of the source beam before it enters the ellipsometer does not cause errors in ψ and Δ to first order, hence can be tolerated. In general, the azimuth settings of the optical components of the ellipsometer at zero or minimum overall transmission of the optical train are independent of the state of polarization of the source beam provided that such a state is not linear. In other words, the exact azimuth location of the null should be entirely independent of the

§5.4] SOURCES OF ERROR AND THEIR CORRECTION 399

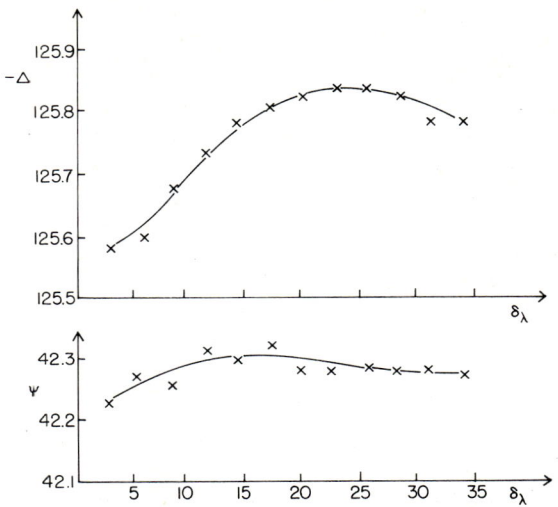

Fig. 5.18. The two-zone averaged ellipsometric angles Δ and ψ for an aluminum film sample at $\lambda = 5461$ Å as functions of the spectral bandwidth δ_λ (Å) of the incident light (as controlled by slit width), after Smith [77].

source polarization. Mathematically, we can prove the above statement as follows. Let $\tau(P, C, A)$ represent the transmittance of the optical train as a function of the polarizer (P), compensator (C), and analyzer (A) azimuths, and let $S(P)$ represent the variation with the polarizer azimuth of the light intensity after the polarizer caused by source polarization. The intensity of the light falling on the photodetector will vary as

$$I_D = \tau(P, C, A) \times S(P). \tag{5.30}$$

When

$$S(P) \neq 0, \tag{5.31}$$

for all values of P, we have $I_D = 0$ only if

$$\tau(P, C, A) = 0, \tag{5.32}$$

which is determined by the optical train and is independent of $S(P)$.

The source polarization (other than linear) will not affect the null points (P_n, C_n, A_n) as long as the transmittance function τ has exact zeros, i.e., $\tau(P_n, C_n, A_n) = 0$. However, τ may possess non-zero minima instead of exact zeros (nulls), in which case the azimuth settings at the minimum will be affected. Because of the multiplicative nature of τ and S in eq. (5.30), only the setting of P at the minimum is shifted due to $S(P)$. The settings of

C and A which are determined by

$$\frac{\partial I_\mathrm{D}}{\partial C, A} = 0 = \frac{\partial \tau}{\partial C, A}, \qquad (5.33)$$

are independent of $S(P)$.

Because it is common practice to determine the azimuth angles at the minimum by averaging two settings giving equal intensity levels above the minimum (the so-called method of swings), an additional error can arise from the distortion of the I_D curve by the function $S(P)$ in the neighbourhood of the minimum. This distortion will only affect the polarizer azimuth but not the compensator or analyzer azimuths.

The effect of incomplete but near total polarization of the light output of the polarizer has been examined in §3.8. There we found that small depolarization ($\sim 1\%$) can lead to an error in ψ (but not in Δ) ($\sim 0.15°$) which does not cancel upon two- or four-zone averaging. Also we have identified the elements in the Mueller imperfection matrices of the components of the ellipsometer which lead to further depolarization as light passes through the optical train (table 3.14). Some of these depolarizing imperfection-matrix elements cause errors in ψ and Δ (tables 3.11 to 3.13).

5.4.5.3. Collimation and non-uniform polarization

In practice, light beams have finite cross section and can only be approximately modeled as a single infinite plane wave. One consequence of truncating the lateral extent of an infinite plane wave is that its wavefronts do not remain plane. The rays are no longer parallel but diverge by an amount dependent on the lateral dimensions of the beam in units of wavelength. Thus perfect collimation (parallel rays) is not a property of light beams with finite cross section. This effect is based on the fundamental phenomenon of diffraction. For a quasimonochromatic beam of visible light of 1 mm diameter (typically used in ellipsometry), the diffraction-limited beam divergence is small (~ 0.4 mr) but leads to an angle-of-incidence error. Another consequence of the finite lateral extent of the beam is that the wave cannot be uniform over its wavefronts. A wavefront, which is by definition an equiphase surface, is not simultaneously an equi-amplitude surface. The wave amplitude must drop to zero outside the space occupied by the beam. A common distribution of wave amplitude is the Gaussian distribution as in the TEM_{00} output mode of a laser source. A nonuniform irradiance distribution is of no direct consequence in ellipsometry. On the other hand, a nonuniform polarization distribution over the beam wavefront is critical. This may occur if any of the elements placed in the path of the light beam possess nonuniform polarizing properties over the area covered by the beam. A nonuniform distribution of polarization over the beam cross section is a source of error in ellipsometry. It will lead to broad minima whose position will be a function of the distribution.

Attempts have been made recently to increase the area-resolution capability of the ellipsometer by using light beams of greatly reduced sectional areas (few microns diameter). This is brought about by focusing a well collimated beam by a positive lens[217] or, alternatively, by using an aperture-limited beam of high-power density. Here beam-size effects become important.

A rigorous mathematical approach to analyze beam-size effects and the associated lack of collimation is to replace the beam by the equivalent superposition of plane waves into which its spatial structure can be resolved. These component plane waves have different directions, amplitudes and probably different polarizations. In principle, any light beam, for which such an expansion is known, can be used in ellipsometry. However, the interpretation of the measurements is necessarily more difficult.

5.4.6 Polarization-dependent photodetector sensitivity

The response of a photoelectric detector can, in general, be a function of the polarization of the impinging light. The light leaving the ellipsometer is linearly polarized by the analyzer which is the last element in the optical train. In the nulling process, the analyzer is rotated so that the azimuth A of the linear electric-field vibration incident on the photodetector is variable. If $D(A)$ denotes the polarization-dependent sensitivity of the photodetector and $\tau(P, C, A)$ is the transmittance of the optical train of the ellipsometer, the detected signal (apart from a constant multiplier) can be written as

$$\mathcal{I}_D = \tau(P, C, A)D(A). \tag{5.34}$$

If we compare eq. (5.34) with eq. (5.30), we immediately realize that the effect of polarization-dependent photodetector sensitivity is analogous to that of residual linear polarization of the source beam. From a development similar to that of eqs. (5.30)–(5.33), we can readily conclude that the azimuth location of the null is unaffected by $D(A)$, provided that the ellipsometer transmittance $\tau(P, C, A)$ has exact zeros. If τ has non-zero minima, the azimuth setting of the analyzer alone at the minimum will be shifted; the polarizer and compensator azimuths at the minimum are unchanged. Additional error of the analyzer azimuth arises from the asymmetry of $\mathcal{I}_D(A)$ around the minimum caused by $D(A)$ when the method of swings is used.

A cure for this problem of polarization-dependent detector sensitivity is to depolarize the light before it falls on the photodetector. This is easily done by inserting a depolarizer between the analyzer and the detector. Alternatively, the fixed-analyzer nulling scheme may be used.

5.4.7. Residual mechanical imperfections

The mechanical construction of an ellipsometer should be as precise as possible. Although alignment of the telescopes should be the primary responsibility of the manufacturer, enough controls should be provided to effect an exact alignment if necessary. The polarizing optical elements should be rotatable around the exact telescope axes with no wobble. If this is hard to achieve, two diametrically opposed readouts of the graduated circles should be provided whose readings could be averaged to cancel the effect of eccentricity. Residual telescope misalignment causes errors in both the angle of incidence and the azimuth angles. Eccentricity of the rotating elements causes systematic azimuth-angle errors.

5.4.8. Other errors

Although we have considered the most common sources of error in a null ellipsometer (many of which are shared by all other types of ellipsometers), additional error may be encountered in particular experiments. For example, if measurements are made on a surface immersed in a liquid ambient the effect of refraction at the window/liquid interface on the angle of incidence at the sample should be considered [219]. Also, if the liquid possesses residual optical activity, the optical rotation as the light traverses the liquid medium should be corrected for [223]. In very high-accuracy ellipsometric work, the effect of stray magnetic fields (including the earth's magnetic field) may have to be considered because of the finite (nonzero) Verdet constant of air. Spurious polarization effects caused by slits or apertures in the ellipsometer may also need to be assessed [115].

5.4.9. Model errors

In the above, we have been concerned with obtaining the most accurate values of ψ, Δ (and ϕ) from a given ellipsometer set-up by analyzing the effect of all the possible sources of error and showing how they can be eliminated. The subsequent use of ψ, Δ and ϕ to calculate other optical characteristics of the surface is a distinct separate step in which we are liable to incur additional errors. These errors arise when we assign to the surface ideal properties that it does not have and are called *model errors*. For example, it is common practice when we measure the optical constants of "bare" substrates to overlook the inevitable presence of a surface film, anisotropy, or surface roughness. Errors in calculating the optical constants n and k from the accurately measured ψ, Δ and ϕ that are caused by the neglect of one or the other of the above-mentioned characteristics of a real surface represent an entirely different problem which is not instrumental in nature. It may be said that such errors arise from the use of the wrong equations (which, in turn, arise from the use of the wrong model) to relate n and k to ψ, Δ and ϕ.

5.5. Precision of null ellipsometers

The ultimate precision of null ellipsometers is mechanically limited by the smallest division of the graduated circles that determines the least resolvable azimuth change. Interdivision read-out is possible by the use of verniers. Recently, Moiré grating goniometers have been introduced which allow accurate interdivision interpolation and permit automatic photometric read-out of the azimuth scales by stepping motors and fringe counting [224].

To attain the ultimate precision of a null ellipsometer ($\delta\psi \approx \delta\Delta \approx 0.005 - 0.01°$) in real measurements, azimuth changes around the null position of the order of the least resolvable azimuth should produce photometric signals distinguishable from the background noise. The photometric signal $\Delta\mathcal{I}_D$ near null is a parabolic function of the azimuth changes $\Delta\theta$

$$\Delta\mathcal{I}_D = S(\Delta\theta)^2, \tag{5.35}$$

where S is a coupling or sensitivity coefficient. In general, S is directly proportional to the source intensity, detector sensitivity, and surface reflectivity, the latter being a function of the incident polarization. S also depends on the particular optical component whose azimuth is being changed. When $\Delta\theta$ is equal to the least resolvable azimuth, S should be large enough to make $\Delta\mathcal{I}_D$ [(eq. (5.35)] detectable in the presence of noise $\mathcal{I}_N$. Because of the very low light level falling on the photodetector at null, shot noise is relatively unimportant and the noise component $\mathcal{I}_N$ is primarily produced by the photodetector itself and its associated electronics.

In addition to the brute-force approaches of using intense light sources and high-sensitivity photodetectors, the precision of a null ellipsometer can be improved by other techniques. The method of swings, whereby azimuth settings leading to equal signals on both sides of the null are averaged, has been said to improve the precision of null ellipsometers ten times [159].

Another useful procedure is to use a mechanical chopper to interrupt the ellipsometer light beam at a suitable frequency. The ac (alternating current) component of the photodetector output is fed to the signal channel of a phase-sensitive detector which is simultaneously supplied by a reference signal internally generated by the chopper. This results in precise detection of $\Delta\mathcal{I}_D$ with considerable discrimination against noise. Combined with the method of swings, this procedure is capable of achieving precision comparable to that of the azimuth scales.

Faraday rotators (modulators) fed with alternating currents have been used to automate the method of swings. In the common PCSA ellipsometer (where the compensator C is set and the polarizer P and analyzer A are adjusted for null), two Faraday modulators are used, one after the polarizer (M_P) and the other (M_A) before the analyzer, fig. 5.19. The Faraday modulators may be rigidly attached to, and rotate as one unit with, their associated polarizing prisms. The effect of the small sinusoidal optical rotations induced by the Faraday modulators M_P and M_A is equivalent to small mechanical oscillation of the polarizer and analyzer azimuths, respectively. Therefore, we may write

$$P - P_n = \beta_0 + \beta_1 \sin(\omega_P t + \delta_P),$$
$$A - A_n = \alpha_0 + \alpha_1 \sin(\omega_A t + \delta_A), \tag{5.36}$$

where (P_n, A_n) are the nulling azimuth angles (for zero dc light flux falling on the photodetector) and (β_0, α_0), (β_1, α_1) are the constant and sinusoidal components of off-null azimuth offsets of the polarizer and analyzer, respectively. (ω_P, δ_P) and (ω_A, δ_A) define the angular frequency (ω) and temporal phase (δ) of the polarizer and analyzer optical rotations. If eq. (5.36) is substituted into eq. (5.15), the detected signal $\mathscr{I}_D$ is seen to have dc, fundamental, second- and higher-harmonic components of frequencies 0, (ω_P, ω_A), ($2\omega_P$, $2\omega_A$), ... However, when

$$\beta_0 = \alpha_0 = 0, \tag{5.37}$$

(*i.e.*, when the polarizer and analyzer are set at their required null positions), the dc and first-harmonic components disappear; the modulation generates only second- and higher-harmonic components of the detected signal. The procedure to reach the null in the PM_PCSM_AA ellipsometer (which incorporates the Faraday modulators M_P and M_A) is to adjust the azimuthal settings of the polarizer and analyzer until the fundamental components of the

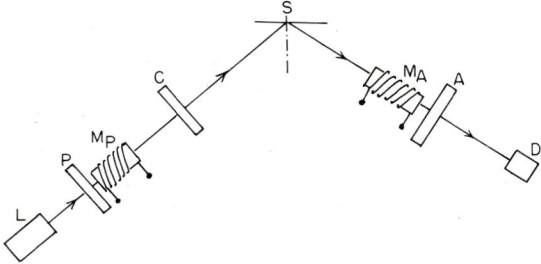

Fig. 5.19. Increased precision polarizer(P)-compensator(C)-surface(S)-analyzer(A) null ellipsometer with Faraday modulators M_P and M_A incorporated after the polarizer and before the analyzer, respectively. L and D are the light source and detector, respectively.

detector signal at the corresponding modulator frequencies disappear. This condition can be precisely reached by use of a phase-sensitive detector whose signal and reference channels are fed from the photodetector and modulators, respectively. For generality, we have assumed above that the modulators M_P and M_A are excited at two different frequencies ω_P and ω_A and that their associated temporal phases δ_P and δ_A are arbitrary. In practice, it is sufficient and convenient to choose

$$\omega_P = \omega_A \quad \text{and} \quad \delta_P - \delta_A = \pm \tfrac{1}{2}\pi. \tag{5.38}$$

In the above PM_PCSM_AA ellipsometer, the polarizer P and analyzer A azimuths are electronically oscillated around their null settings, so that the method is essentially an automation of the "method of swings" in the PCSA static ellipsometer. The polarization modulators also represent the temporal analogs of the old spatial half-shade devices.

Although Faraday modulators improve the precision of ellipsometric measurements, care must be taken to avoid any adverse effect on accuracy due to their presence. In particular, the cylindrical glass rods (cores) commonly used in Faraday rotators should exhibit the least possible residual birefringence. Also, the planar end faces of the rods should be parallel to minimize beam deviation errors and should be anti-reflection coated to minimize the effect of additional auto-collimated parasitic beams from inter-component multiple reflections.

5.6. Automation of null ellipsometers

The manual nulling of an ellipsometer (by adjustment of its optical components) and the subsequent visual reading of the scales requires several minutes for completion. For increased accuracy, additional null measurements are often required to cancel the effect of various sources of error (for example, by two- or four-zone averaging; see §3.7, §3.8, and §5.4). The time and effort involved in manual nulling and the visual reading of the scales are of no major concern when no (or very slow) changes occur at the surface under measurement, or when the measurements need not be repeated several times. Frequently, however, a rapid surface reaction is to be followed or a large set of data is to be acquired as, for example, in multiple-null, multiple-angle-of-incidence, multiple-wavelength (spectroscopic) or multiple-sample ellipsometry. Under such circumstances, the time and effort of measurement become important and automation of the ellipsometer becomes essential.

The design of automatic ellipsometers has progressed along two distinct lines: (1) to adhere to the concept of null measurement and to use a servo-system to achieve self-nulling (compensation), (2) to give up the null method altogether and to use photometric or interferometric methods. Figure 5.20 shows a schematic diagram of the different types of automatic ellipsometers. In this section we consider automatic null ellipsometers and in §5.7 we discuss automatic photometric ellipsometers.

Automatic null ellipsometers are of two different kinds: (1) those that employ motors to drive the polarizing elements to null, (2) those that are entirely electro-optic with no moving parts.

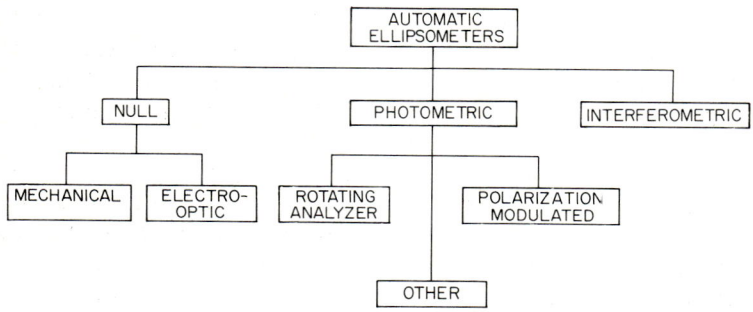

Fig. 5.20. Schematic classification of automatic ellipsometers.

5.6.1. Motor-driven self-nulled ellipsometers

Examples of motor-driven self-nulled ellipsometers are those designed by Takasaki [226] and by Ord and Wills [227]. Another version of the same ellipsometer has been recently described by Roberts and Meadows [224] who incorporated grating goniometers for azimuth read-out.

A schematic diagram of Takasaki's ellipsometer is shown in fig. 5.21. This arrangement is obtained from the classic PSCA ellipsometer (polarizer-sample-compensator-analyzer) by the addition of two ADP[6] cells M_P and M_A as modulators. Each ADP cell is mounted to rotate with the polarizer or analyzer as one unit, with the electrically induced fast axis at 45° with the transmission axis. If ϵ represents the electrooptically-induced birefringence in the modulator, then

$$\epsilon = KV \sin(\omega t + \delta) \tag{5.39}$$

where K is the linear electrooptic coefficient and $V \sin(\omega t + \delta)$ is the applied sinusoidal voltage. Each cell may be driven by its own voltage of proper amplitude V, frequency ω, and phase δ. With the compensator

---

[6] ADP is ammonium dihydrogen phosphate.

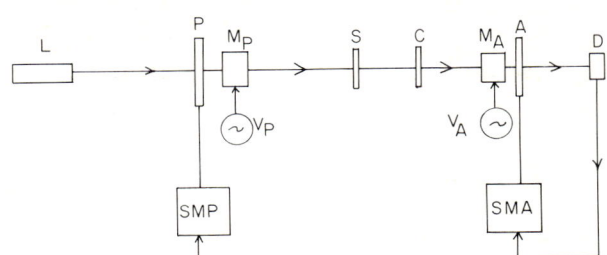

Fig. 5.21. Automatic polarizer (P)-sample(S)-compensator(C)-analyzer(A) null ellipsometer. The polarizer and analyzer are controlled by servomotors SMP and SMA, respectively, which are actuated by feedback signals from the photo-electric detector D. The feedback signals are ac signals generated by small modulation of the light beam (from the source L) by modulators M_P and M_A which rotate with P and A, respectively. V_P and V_A represent the ac voltage sources driving the corresponding modulators, after Takasaki[226].

azimuth at $\pm 45°$ and its retardation at $90°$, the birefringence signals ϵ_P and ϵ_A of the polarizer and analyzer modulators M_P and M_A lead to detector signals

$$\mathscr{I}_D|_{M_P} = \pm \text{const } \epsilon_P \cos(2A \pm \Delta), \quad \mathscr{I}_D|_{M_A} = \pm \text{const } \epsilon_A (\cos 2P - \cos 2\psi)$$
(5.40)

where ψ and Δ are the ellipsometric parameters of the surface S and the + and − signs correspond to the +45° and −45° compensator azimuths, respectively. From eq. (5.40), note that the detector signal $\mathscr{I}_D|_{M_P}$, due to the polarizer ADP cell, vanishes when $\cos(2A \pm \Delta) = 0$ and switches polarity as the analyzer azimuth A is swept past this angle. On the other hand, the detector signal $\mathscr{I}_D|_{M_A}$, due to the analyzer ADP cell, vanishes when $(\cos 2P - \cos 2\psi) = 0$ and switches polarity as the polarizer azimuth P is swept past this angle. However, the conditions $\cos(2A \pm \Delta) = 0$ and $(\cos 2P - \cos 2\psi) = 0$ also lead to zero[7] dc (steady) detected signal [eqs. (3.63) and (3.64)], hence correspond to the desired null condition that would be sought in the absence of the modulators.

The principle of Takasaki's ellipsometer is to use $\mathscr{I}_D|_{M_P}$ and $\mathscr{I}_D|_{M_A}$ as feedback signals to actuate servomotors which, in turn, rotate the analyzer and polarizer, respectively, to null.

By use of proper switching techniques (exploiting the signs of the derivatives of the control signals) Takasaki adapted his ellipsometer for measurements in four zones. The ac voltages applied to the ADP cells were chosen of the same frequency (60 Hz/sec), of equal amplitude (~200 V) and 90° apart in phase. The short-term reproducibility of the data was ~0.01° and

---

[7] It can be proved that the same conditions also correspond to polarizer and analyzer azimuths that maximize the transmittance of the optical train. These settings can be easily avoided in practice by observing that the dc light level is near zero and not maximum.

the response speed was rather slow (5 sec were needed to balance the system with the polarizer or analyzer off null by $\sim 1°$).

An important limitation of Takasaki's servo-operated null ellipsometer is the need for visual reading of the scales by an operator once the null has been automatically reached. This mode of data acquisition is slow and makes the net speed of the system unsatisfactory in many cases. This limitation can be overcome if the servo-motors are replaced by stepping motors to drive the polarizer and analyzer, as has been achieved by Ord and Wills [227]. By counting the pulses that drive the stepping motors, an automatic positional read-out is obtained. Equally important is the fact that such motors can be connected to and be controlled by a digital computer. This permits on-line data acquisition and on-line data reduction. The process by which the null is reached is simply an automatic version of the manual "method of swings"[8]. Thus the computer steps the polarizer between two positions leading to equal detected (dc) intensity levels above the minimum and drives the motor back to their center point. After the polarizer has been set, the analyzer is subsequently adjusted to null in a similar manner. This computer-operated instrument[227] has 0.01° resolution and one second nulling time. Although used as a following ellipsometer for *in situ* studies of anodic oxide films, its range of applications is wide and is limited only by its precision and response time.

5.6.2. *Electro-optic self-nulled ellipsometers*

If speed and precision higher than those achieved by the above motor-driven self-nulled ellipsometers are required, while at the same time adhering to the principle of the null method, the stationary (with no moving parts) electro-optic design suggested by Winterbottom [225] is one answer. The principle of the method is to use Faraday cells to magneto-optically rotate the azimuth angles of the polarizer and analyzer, instead of doing the same manually or by motors. A block diagram of an ellipsometer of this type built by Mathieu *et al.* [228] is shown in fig. 5.22. The Faraday coils are excited by (1) a variable dc current, (2) an alternating modulation current of small amplitude. The component of the detector signal at the modulation frequency is fed back to control the dc current levels in the polarizer and analyzer Faraday cells, until null is reached. Some of the performance characteristics obtained by Mathieu *et al.* for their ellipsometer are shown in table 5.2. In this table, the slew rate represents the speed by which azimuths can be magneto-optically rotated, the dynamic range is the maximum change of azimuth for a given Faraday-cell design, and the resolution gives the least resolvable

[8] See §5.3.3.

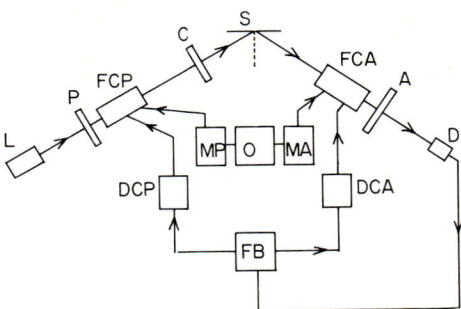

Fig. 5.22. Automatic magneto-optic null ellipsometer using Faraday rotators. P, C, S, and A represent the basic polarizer-compensator-surface-analyzer optical train between the light source L and detector D. FCP and FCA are Faraday cells after the polarizer and before the analyzer, respectively. FCP, FCA are excited by dc currents from the voltage sources DCP, DCA and by small ac currents from modulator sources MP and MA (derived from the same oscillator O), respectively. Automatic nulling is achieved by feedback control (FB) of the dc current levels in FCP and FCA using, as feedback signal, the detector output at the modulator frequency. After Mathieu et al. [228].

Table 5.2. Performance characteristics of automatic ellipsometer*

Slew rate for polarizer and analyzer azimuth	1.6°/msec to 0.33°/sec
Dynamic range for analyzer azimuth	51°
Dynamic range for polarizer azimuth	55°
Resolution	7×10^{-4} to 8×10^{-2} deg

*Table III, ref. 228.

change in the ellipsometric parameters ψ or Δ. The read-out of this type of ellipsometer can be derived directly from the compensating Faraday cell current supplies. These signals give the magneto-optic azimuth rotations that must be added to the fixed azimuths of the polarizer and analyzer. In a given experiment (where a reaction is to be followed as a function of time), the latter azimuths may be chosen to correspond to the null at the initial conditions (*i.e.*, before a reaction evolves on the surface, thereby making full use of the dynamic range of the ellipsometer). The two signals correspond to the dc current levels in the two Faraday cell coils at null and can be automatically recorded, graphed, or digitized and fed to a digital computer for more involved, on-line, data reduction. The most important aspect of the design of an ellipsometer of this type lies with the Faraday cells. To test the performance of this and other automatic self-nulled ellipsometers, Muller and Mathieu [229] proposed the use of a rotating disc (substrate) with a special pattern of deposited film, to simulate changes of the optical properties that can take place when a real surface reaction takes place. By control of the pattern and the speed of rotation, a variety of situations can be reproduced, thus making possible the determination of significant performance parameters.

A different type of fast electro-optic ellipsometer is that developed by Yamaguchi and Hasunuma [230]. A schematic diagram of this ellipsometer system is shown in fig. 5.23. Two KDP Z-cut crystals (KDP_1 and KDP_2) are used as variable phase plates whose retardations are controlled by the dc voltages V_1 and V_2 (half-wave retardation is produced by an applied voltage of about 75 kV at $\lambda = 5461$ Å). The electrically induced fast axes of KDP_1 and KDP_2 are chosen at 45° and 0 (parallel) azimuths from the plane of incidence; their respective retardations are δ_1 and δ_2, respectively. When the light incident on KDP_1 is circularly polarized, it can be proved that the relative amplitude of the p and s components incident on the sample is controlled by δ_1, whereas the relative phase of the same components is controlled by δ_2. Therefore, by controlling the voltages applied to KDP_1 and KDP_2, any state of polarization for the incident light can be generated. In the nulling mode of operation, the voltages V_1 and V_2 are varied until the incident polarization is changed by reflection into a linear state crossed with the fixed analyzer A. Instead of feeding back a dc signal (detected after the analyzer A) to control the voltage supplies V_1, and V_2 to achieve self-nulling, Yamaguchi and Hasunuma [230] used ac photoelectric control signals, resulting from small modulation voltages superimposed on the dc voltages V_1 and V_2, in a manner

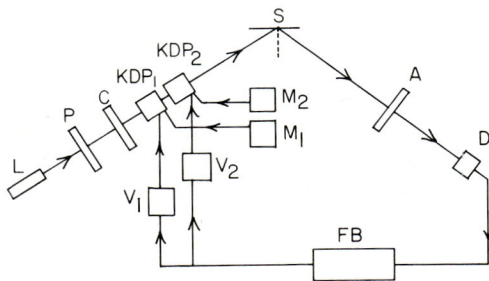

Fig. 5.23. Automatic electro-optic null ellipsometer using KDP crystals. PCSA is the basic polarizer-compensator-surface-analyzer optical train between the light source L and photodetector D. KDP_1 and KDP_2 act as voltage-controlled variable-phase retardation plates that are excited by the voltage sources V_1 and V_2. Automatic nulling is achieved by feedback (FB) control of the dc voltage levels V_1 and V_2 applied to KDP_1 and KDP_2 using, as feedback signal, the detector output at the frequency of the modulators M_1 and M_2 which also excite KDP_1 and KDP_2. After Yamaguchi and Hasunuma [230].

similar to that described earlier in connection with Takasaki's ellipsometer. The voltage read-out V_1 and V_2 of the present ellipsometer directly give ψ and Δ, respectively. These can be automatically recorded, graphically plotted (as a surface reaction evolves), or digitized and fed to a computer for complicated, on-line data reduction. The total response time of this instrument is of the order of one second.

5.7. Automatic photometric ellipsometers

Two such systems have received special attention recently, namely the rotating-analyzer ellipsometer (RAE) and the polarization-modulated ellipsometer (PME). Other photometric designs are to be expected.

5.7.1. *The rotating-analyzer ellipsometer*

A rotating-analyzer ellipsometer (RAE) is one that uses a synchronously rotating analyzer to detect the state of polarization of light after its reflection from a surface under measurement. A conventional instrument intended for operation as a null ellipsometer can be readily modified to function as a RAE. The major change is the addition of a new assembly for the analyzer so that it can be rotated around the beam axis by a synchronous motor. The automatic RAE also requires an electronic detection system capable of deriving the parameters that describe the reflected polarization ellipse from the periodic variations of the light flux transmitted by the rotating analyzer.

Several automatic RAE systems have been described that employ both analog [111–113] and digital [114, 115, 231, 232] signal detection. We describe below one such system that incorporates a number of novel features. Designed by Hauge and Dill [115], primarily for the measurement of dielectric films on silicon wafers in a manufacturing environment, this ellipsometer (called ETA for Ellipsometric Thickness Analyzer) can also be used for the conventional research applications of ellipsometry.

Figure 5.24 is a schematic of the mechanical and optical layout of the ellipsometer. The instrument is constructed to operate at a fixed angle of incidence of 70° and also in the straight-through (or direct-path) configuration. Because a RAE does not require continuous control of the azimuth angles of the polarizer or the compensator, pinned-position mounts for these components can be used. Three settings of 0° (for initial alignment), 12° (for increased sensitivity for the detection of films less than 300 Å in thickness) and 45° (SiO_2 films 300–1300 Å) are used for the polarizer. The compensator is inserted in one setting with its fast axis at 90° azimuth, *i.e.*, perpendicular to the plane of incidence. The analyzer is driven by a synchronous motor at a

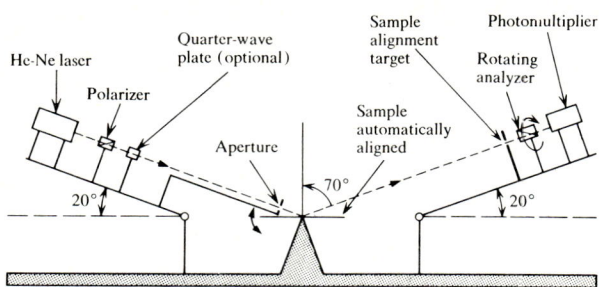

Fig. 5.24. Optical and mechanical components of an automatic rotating-analyzer ellipsometer. After Hauge and Dill [115].

50 rps constant speed. The light beam leaving the rotating analyzer passes through a diffuser and fiber-optic bundle to the photo-multiplier tube (PMT). The diffuser and fiber-optic bundle depolarize the light incident on the cathode of the PMT, thereby avoiding a polarization-dependent photoelectric effect. The sample is automatically aligned by the arrangement shown in fig. 5.25. Tilt control around two orthogonal axes parallel to the surface of the sample is provided by two servomotors that are actuated by imbalance signals derived when the reflected light beam is not centered on a control aperature that has a surrounding four-sector fiber-optic arrangement. Height control for the sample stage is not necessary when samples of sufficiently constant thickness (*e.g.*, silicon wafers) are used.

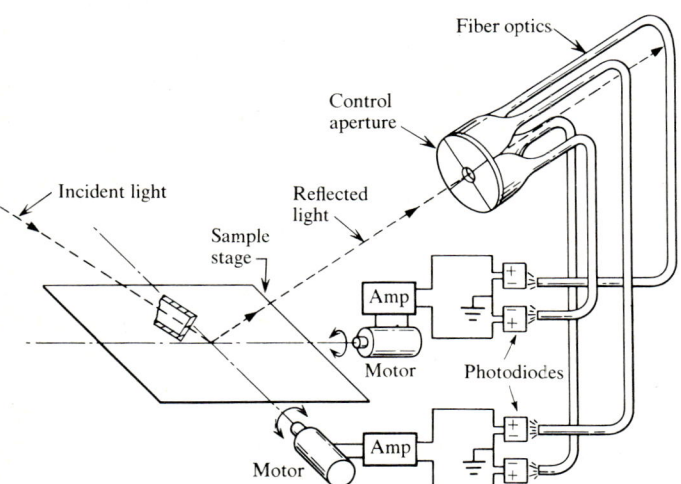

Fig. 5.25. Automatic sample alignment using a central aperature with fiber optics. After Hauge and Dill [115].

The scheme for data acquisition is shown in fig. 5.26. An optical angular encoder mounted on the same shaft with the rotating analyzer senses the instantaneous azimuth of the analyzer. One output of the encoder provides a synchronizing pulse each complete rotation, and the other output gives 256

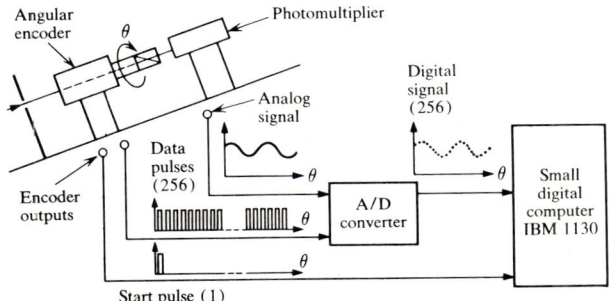

Fig. 5.26. Electrical components of an automatic rotating-analyzer ellipsometer. After Hauge and Dill [115].

equally spaced pulses per revolution. These pulses trigger an analog-to-digital (A/D) converter to sample the analog signal from the PMT, converting it to digital form, to be read by an on-line computer. The digital intensity data at 256 equally spaced angular positions of the rotating analyzer is subsequently Fourier analyzed to determine the reflected polarization by a least-square fit of a sine-wave to such data, fig. 5.27. With the computer supplied by other information concerning the incident polarization and the optical model of the surface (including all known optical parameters), the unknown optical parameters of the surface can also be calculated by suitable programming. The speed of Hauge and Dill's ellipsometer is such that the total time required for sample alignment, data acquisition, analysis, and recording of film thickness is typically 5 seconds, while it has a precision of thickness measurement of about 1 angstrom (for SiO_2 and Si_3N_4 films on Si).

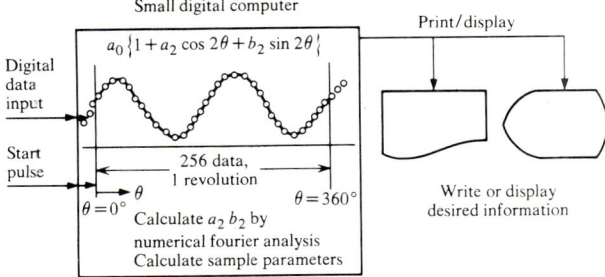

Fig. 5.27. Data analysis in a rotating-analyzer ellipsometer. After Hauge and Dill [115].

A wavelength-scanned RAE for spectroscopic applications has been described by Aspnes [114] and is shown in fig. 5.28. The discrete angular azimuth positions at which the intensity of the light leaving the rotating analyzer is sampled are determined by interrupting an auxiliary laser beam with a chopper mounted to rotate with the analyzer. The first data point in each cycle is synchronized by a second beam (derived from the same laser) transmitted only when that point is being taken. In this RAE system, the monochromator is under computer control and is advanced from one wavelength to the next automatically. Other details concerning the automatic data acquisition and on-line digital computation are similar to those discussed in connection with Hauge and Dill's ellipsometer. This ellipsometer is capable of recording entire spectra of the real and imaginary parts of the dielectric function of a solid surface in the visible and near-visible spectrum with precision in the order of one part per 10^5. The high precision and speed of such a system also makes it of considerable value in the spectroscopy of very thin film reactions on surfaces.[9]

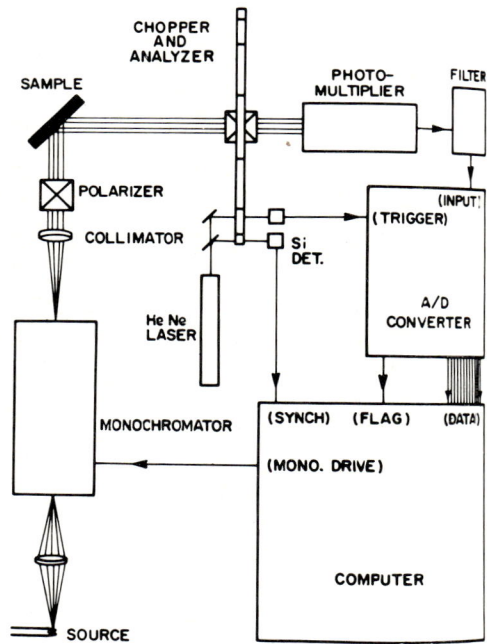

Fig. 5.28. Schematic diagram of an automatic wavelength-scanning rotating-analyzer ellipsometer. After Aspnes [114].

[9] A more complete discussion of this ellipsometer has been recently published by Aspnes and Studna [233].

5.7.2. The polarization-modulated ellipsometer

The principle of operation of the polarization-modulated ellipsometer (PME) has already been discussed in §3.10 and will not be repeated here. At the heart of this ellipsometer is the polarization modulator shown in fig. 5.29 designed by Jasperson and Schnatterly [116]. A uniaxial sinusoidal standing strain wave is established in a rectangular block of fused quartz by means of

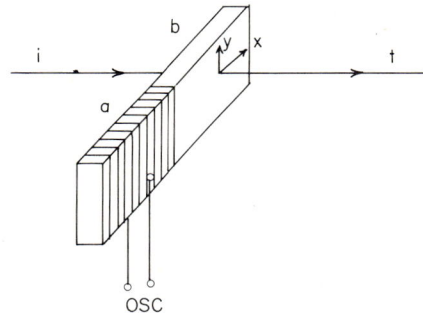

Fig. 5.29. A piezobirefringence polarization modulator is composed of cemented crystal and fused-quartz blocks a and b, respectively. Under the action of an ac voltage from an oscillator (OSC), a stress standing wave is established in block b and the associated oscillating birefringence modulates the state of polarization of a light beam as it passes through the block. x and y indicate the principal strain directions. After Jasperson and Schnatterly [116].

an ac-driven piezoelectric crystal quartz transducer cemented to one end of the block. The oscillating strain is accompanied by an oscillating induced birefringence such that the fused quartz block acts as a linear retarder with an alternating (time-varying) relative retardation. The direction of the strain determines the direction of the fast (slow) axis of this modulated retarder; the amplitude of the sinusoidal retardation is proportional to that of the strain and, consequently, to the voltage applied to the transducer.

An automatic ellipsometer system based on the principle of polarization modulation has been described recently by Jasperson et al. [117] and is shown in fig. 5.30. In this system the fundamental and second-harmonic components of the detector photoelectric current are synchronously detected by two separate lock-in amplifiers (LI) while the dc component is separated by a low-pass filter (LPF). The reference signal to these amplifiers comes from the oscillator that derives the polarization modulator. The outputs of the three (dc, fundamental, and second-harmonic) channels are sequentially sampled and digitized by an analog-to-digital converter C, measured by a digital voltmeter (DV) and stored in a programmable calculator PC. On-line data

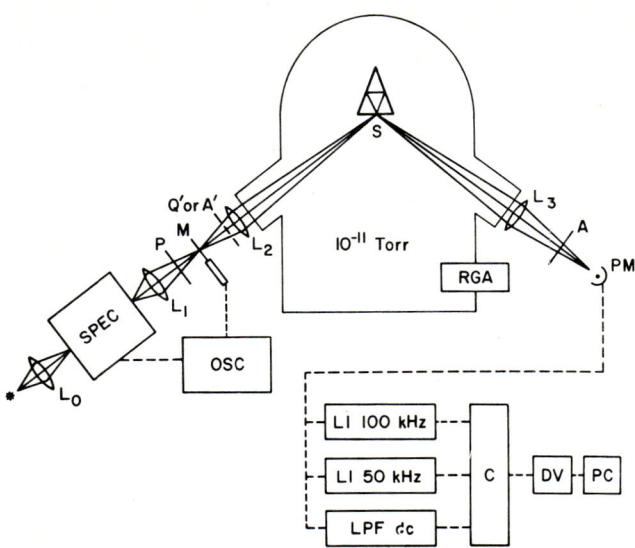

Fig. 5.30. Automatic polarization-modulated ellipsometer used for the study of *in situ* vacuum-deposited films. P, M, S, and A represent the polarizer, modulator, surface, and analyzer, respectively. The L's are collimating lenses. The surface S is mounted in a vacuum chamber whose atmosphere is analyzed by a residual gas analyzer RGA. SPEC is a spectrometer with a wavelength-scanning mechanism controlled by the same oscillator OSC that drives the modulator M. The signal from the photomultiplier tube PM is Fourier-analyzed by lock-in (LI) amplifiers and low-pass filter (LPF). C, DV, and PC represent coupler-converter, digital voltmeter, and programmable calculator, respectively. After Jasperson *et al.*[117].

analysis and display is done by the system. For spectroscopic applications, a monochromator can be used whose drive (for wavelength scan) is coupled to the oscillator that drives the polarization modulator to change its output voltage to keep the retardation amplitude constant as a function of wavelength. The reported precision of such a system is of the order of 0.001° in the ellipsometric angles ψ and Δ and the time required for the determination of the optical constants n and k of a thick substrate at one wavelength is approximately 30 seconds.

CHAPTER 6

Applications of Ellipsometry

6.1. Introduction

In this chapter we give representative examples of the applications of ellipsometry to the characterization of interfaces and films. The applications are divided into a number of major categories. In §6.2 we consider the determination of optical properties of materials by ellipsometry with emphasis on spectroscopy. §6.3 is concerned with adsorption, §6.4 with oxidation, and §6.5 with electrochemistry and corrosion. Applications in biology and medicine are given in §6.6. Other miscellaneous applications of ellipsometry are reviewed in §6.7.

6.2. Optical properties of materials and spectroscopic ellipsometry

The transformation of the state of polarization of light by reflection is the basis of an important method for the determination of the optical properties of materials. In this section we discuss this application of ellipsometry. Distinction will be made between the cases when the sample of a material constitutes a bulk or a thin-film phase and whether it is optically isotropic or anisotropic.

6.2.1. Optical properties of materials in bulk phase

In this case light is reflected at the planar interface between a "substrate" phase of the given material and an "ambient" phase of known optical properties. The ambient must, obviously, be light transmitting (transparent) and it should, ideally, not perturb the substrate. The latter condition is satisfied if the ambient is chosen to be vacuum or an inert gas. But even in these cases we must remember that the mere termination of the bulk phase of the substrate at its interface with the ambient creates a surface layer of the material of different optical properties than in the bulk. In the terminology of solid state surface physics, we have to consider the effect of surface states on the optical properties of the material. For non-inert ambients, additional modification of the substrate at the interface is expected *e.g.*, in the form of oxide or contamination films. Of course, this brief discussion of the effect of the ambient on the substrate brings up the question of how the substrate was prepared in the first place. Here, again, many subtle effects have to be kept in mind. Among them are the possible presence of (1) a contaminant or oxide film, (2) a stressed (Beilby) layer (with high density of dislocations) due to mechanical forces involved in the preparation (*e.g.*, cleaning or polishing) of the sample, and (3) surface roughness. All of these should be considered as potential sources of error in the determination of optical properties by a light-reflection technique such as ellipsometry.

For solid samples prepared by vacuum evaporation of *thick* films, important differences in the *bulk* optical properties usually arise from structural differences in the film (state of aggregation) that, in turn, are caused by different evaporation conditions. These differences are nicely, but only qualitatively, explained by the theory of Maxwell Garnett [196]. A single crystal, rather than a thick evaporated film, is therefore preferable (whenever possible) for the determination of the intrinsic optical properties of a given material. This is particularly important in spectroscopic studies, such as band-structure analysis from spectra of the dielectric function.

In spite of the above-mentioned basic limitations that are common to all reflection methods, ellipsometry is a valuable technique for the determination of the optical properties of materials, especially in wavelength regions where the materials are strongly absorbing, so that transmission measurements are precluded. Ellipsometry is also suited to very small samples (*e.g.*, of new materials) because it requires reflections from a very small area only (1 mm^2 or less). In spectroscopic applications, both the real and imaginary parts of the complex refractive index (or dielectric function) can be determined as a function of wavelength, without recourse to Kramers-Kronig [138] dispersion integrals. Methods for analytic continuation to other wavelengths of the ellipsometrically determined dielectric function over a limited spectral interval have been recently suggested [234].

Several examples will demonstrate the capabilities of ellipsometry as a tool for the determination of the optical properties of materials.

Figure 6.1 shows the optical properties (refractive index n and extinction

§6.2] SPECTROSCOPIC ELLIPSOMETRY 419

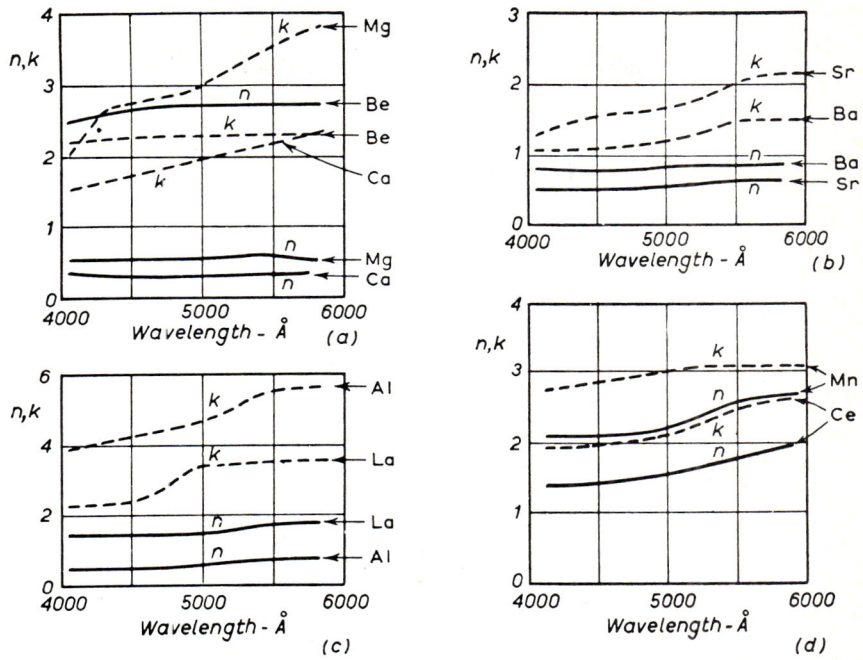

Fig. 6.1. The refractive index n and extinction coefficient k of various metals obtained by a return-path, principal-angle, null ellipsometer. After ref. 107, 129.

coefficient k) of several metals obtained by O'Bryan[107, 235] using a novel return-path, principal-angle, null ellipsometer. The results for calcium (Ca), magnesium (Mg) and beryllium (Be) are shown in a; for barium (Ba) and strontium (Sr) in b; for aluminum (Al) and lanthanum (La) in c; and those for cerium (Ce) and manganese (Mn) are shown in d. All measurements were made *in vacuum* on (presumably thick) freshly evaporated films, without any exposure to the atmosphere. O'Bryan's technique was novel in that the effect of instrumental systematic errors was minimal (*e.g.*, cell-window stress birefringence was entirely avoided), and in its *in-situ* measurements on *as-evaporated* films under vacuum.

The optical properties of two noble metals, silver (Ag) and gold (Au), measured by two different ellipsometric techniques, are shown in figs. 6.2 and 6.3, respectively. The results in fig. 6.2 by Holcomb and Bashara [127] show the real (ϵ_1) and imaginary (ϵ_2) parts of the complex dielectric function of a *thick* vacuum-evaporated Ag film as measured in air, shortly after evaporation, by a conventional null ellipsometer. The results in fig. 6.3 by Aspnes [114] are also for ϵ_1 and ϵ_2 (plotted versus photon energy, instead of

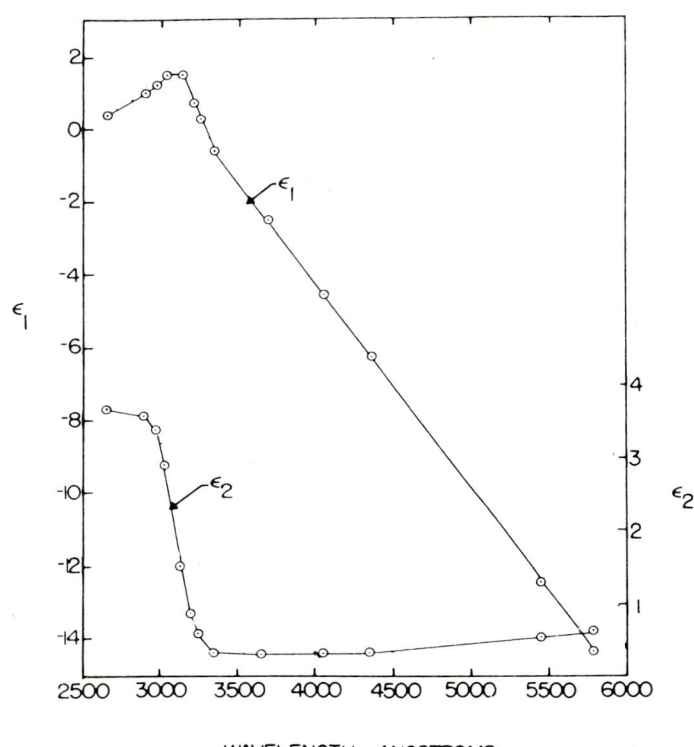

Fig. 6.2. The real (ϵ_1) and imaginary (ϵ_2) parts of the complex dielectric function of a thick vacuum-evaporated Ag film measured in air, shortly after evaporation, by a conventional null ellipsometer. After ref. 127.

wavelength) of a vacuum-evaporated Au film measured by a fully-automated, wavelength-scanned, rotating-analyzer ellipsometer employing digital Fourier detection. By numerically calculating the third derivative (with respect to photon energy) of spectra of the dielectric function like those shown in fig. 6.3, Aspnes was able to directly verify their relationship to low-field electroreflectance (ER) spectra[236], as was predicted earlier by theory[237].

Ellipsometry has also been used to determine the optical properties of liquid metals; e.g., fig. 6.4 shows ϵ_1 and ϵ_2 of mercury in the visible, as obtained by Faber and Smith[238] using a rotating-analyzer–fixed-analyzer ellipsometer. Measurements on freshly distilled free surfaces and on free surfaces cleaned by glow discharge are represented by the open and full circles, respectively. The full squares are for measurements taken with

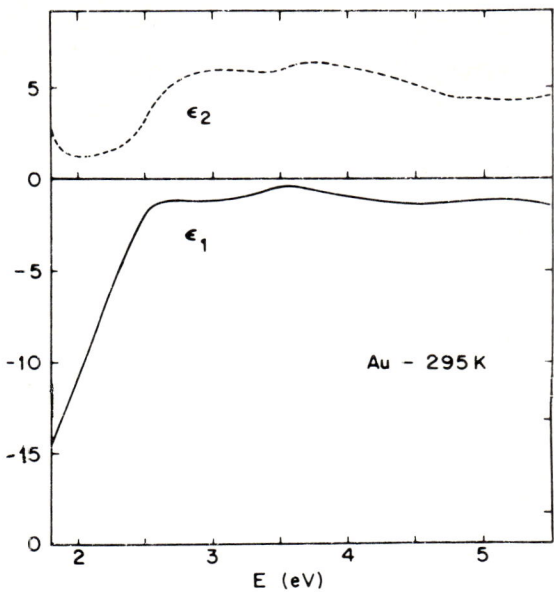

Fig. 6.3. The components of the complex dielectric function, plotted versus photon energy, for a vacuum-evaporated Au film measured by an automated rotating-analyzer ellipsometer. After ref. 114.

internal reflection at a quartz-mercury interface. Superimposed on fig. 6.4, and shown by continuous curves, are the predictions of the free-electron Drude theory [136–138]. Faber and Smith considered the fact that ϵ_1 and ϵ_2 from their careful ellipsometric experiments are consistently above their corresponding values from the free-electron theory as evidence of the inadequacy of such a theory for liquid metals.

Figure 6.5 shows Archer's [239] results for the refractive index n and extinction coefficient k of chemically cleaned (etched) germanium (Ge, a semiconductor) at 300°K plotted as functions of wavelength of light in the visible. Measurements were made by a conventional (null) ellipsometer, with the sample mounted in a test cell with open windows which were used to pass a clean and dry stream of nitrogen to maintain the sample surface cleanliness. Absence of the cell windows eliminated instrumental errors that could otherwise have arisen from stress birefringence. Figure 6.5 shows the "apparent" n and k of Ge when no surface film is assumed (dashed line) and also n and k when account is taken of a suspected oxide film 10 Å thick of 1.9 refractive index (continuous line). The optical properties (n and k) of Ge

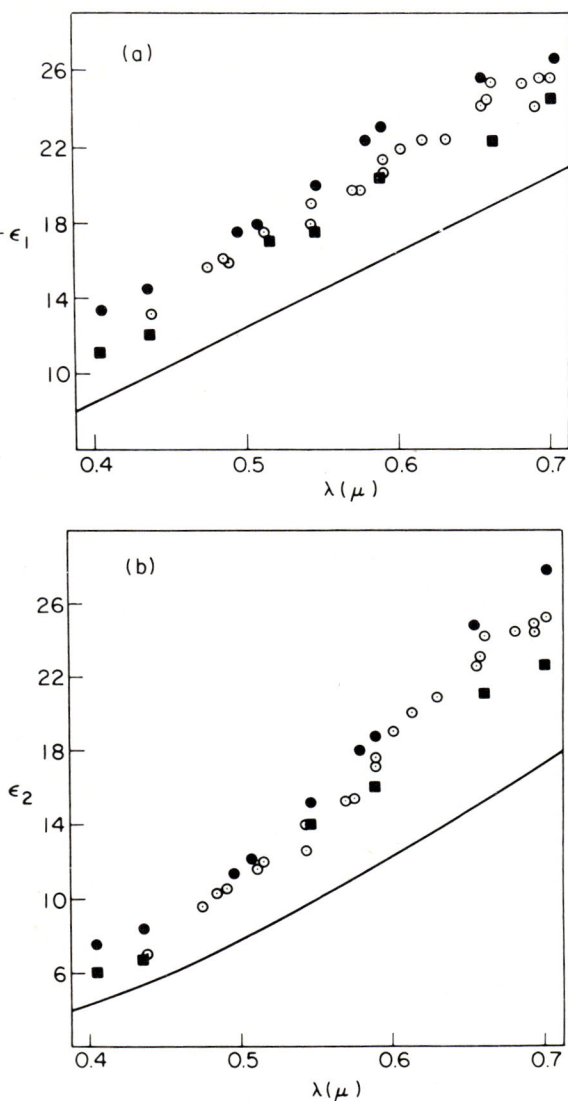

Fig. 6.4. The wavelength variation of the components of the complex dielectric function of mercury (a) for ϵ_1 and (b) for ϵ_2. The open circles are the results of experiments on freshly distilled surfaces and the full circles were obtained after glow-discharge cleaning; the full squares refer to values obtained from internal reflection on a quartz/mercury interface. The continuous curves show the prediction of the simple free-electron Drude theory. After ref. 238.

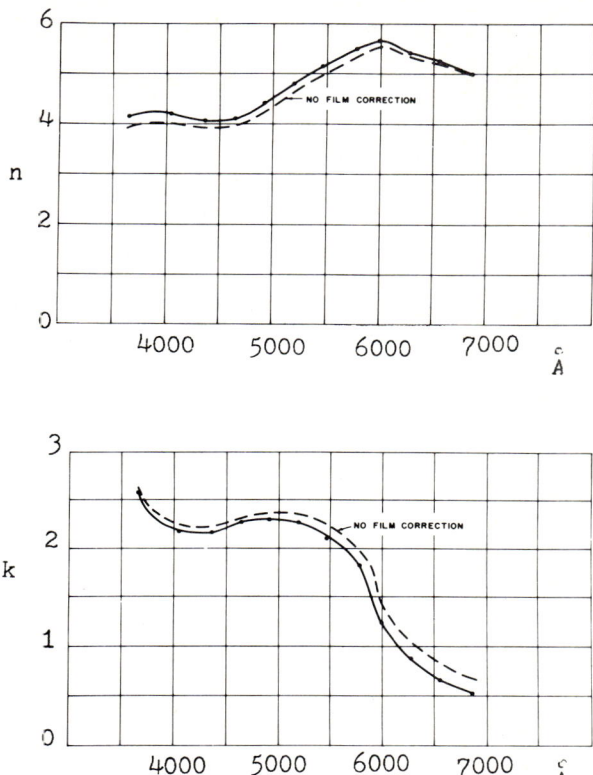

Fig. 6.5. The "apparent" n and k of Ge at room temperature plotted versus wavelength when no surface film is assumed (dashed line) and also when account is taken of a suspected oxide film 10 Å thick of 1.9 refractive index (continuous line). After ref. 159.

at the liquid-nitrogen temperature of 77°K, also measured by Archer [240] using null ellipsometry, are shown in fig. 6.6 in the photon energy range between 1.95 and 2.80 eV. In this range, dispersion of the optical properties is attributed to excitation of electrons from the valence band to the conduction band across a band gap. Spin-orbit interactions cause the double peak in k and the corresponding changes of slope in n.

The temperature dependence of the refractive index n and extinction coefficient k of silicon (Si) was measured by van der Meulen and Hien [232] using an automatic (digital Fourier) rotating-analyzer ellipsometer. Measurements on single-crystal wafers of Si oxidized to different thicknesses [14 Å(+), 38 Å(×) and 169 Å(○)] led to results in good agreement for $n(T)$ and $k(T)$ of Si over the temperature range from 300 to 1350°K, fig. 6.7.

All the above cases are for optically isotropic materials. The use of ellipsometry to determine the optical properties of anisotropic crystals has been rather limited, primarily because methods for the acquisition of appropriate ellipsometric data and the interpretation of such data were not available. However, this picture may soon be changed by the recent development of such techniques as generalized ellipsometry [§3.9] and the 4×4 matrix formalism for the study of light reflection from anisotropic structures [§4.7]. We present two examples of the application of ellipsometry to determine the optical properties of anisotropic crystals. In fig. 6.8, we show the results of Meyer et al. [241] for the ordinary $(n_o - jk_o)$ and extraordinary $(n_e - jk_e)$ complex refractive indices of gallium selenide (GaSe) which is uniaxially anisotropic. Null ellipsometric measurements in the wavelength range 2000–8000 Å were made on two crystal faces one perpendicular and the other parallel to the optic axis, at incidence angles of 60° and 40°, respectively. For the latter crystal face, the optic axis was oriented perpendicular to the plane of incidence.

Figure 6.9 shows the three principal complex refractive indices $(n_1 - jk_1)$,

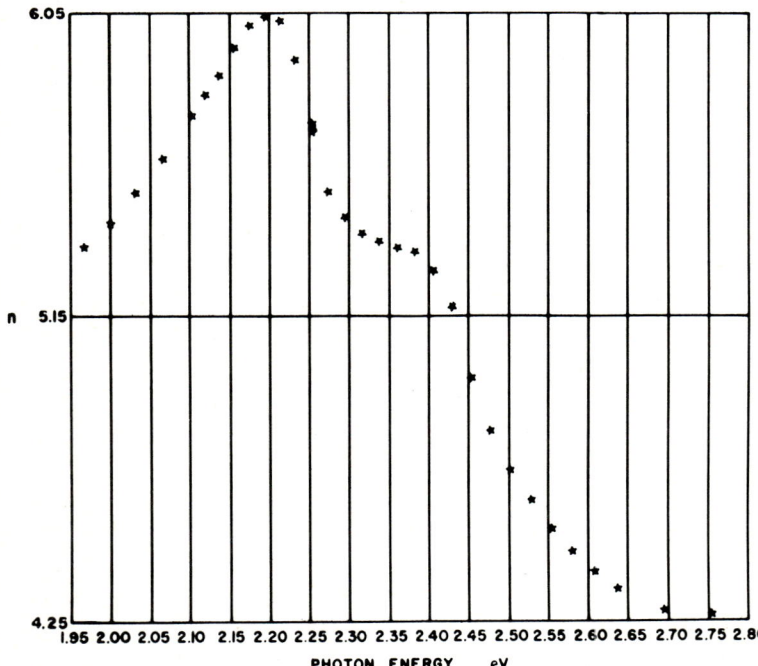

Fig. 6.6a. The refractive index n of germanium as a function of photon energy at 77°K. After ref. 159.

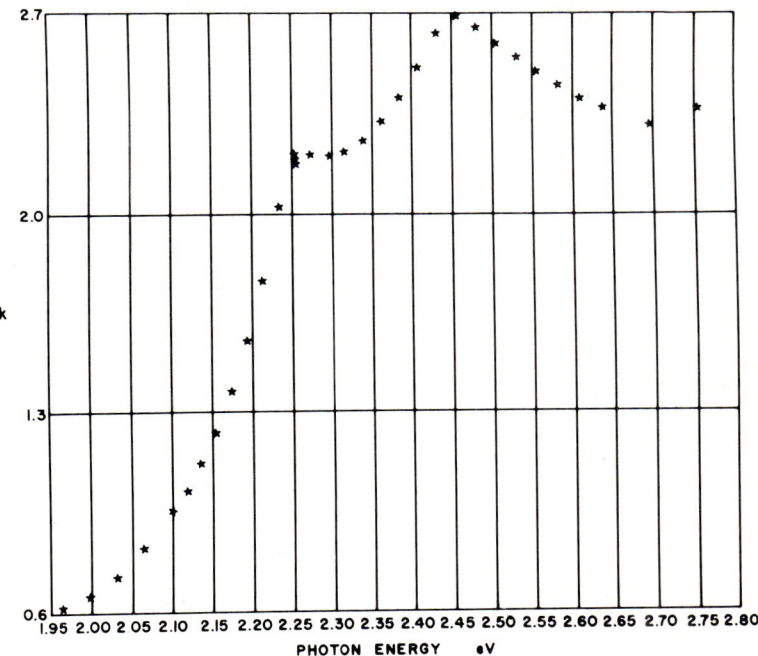

Fig. 6.6b. The extinction coefficient k of germanium as a function of photon energy at 77°K. After ref. 159.

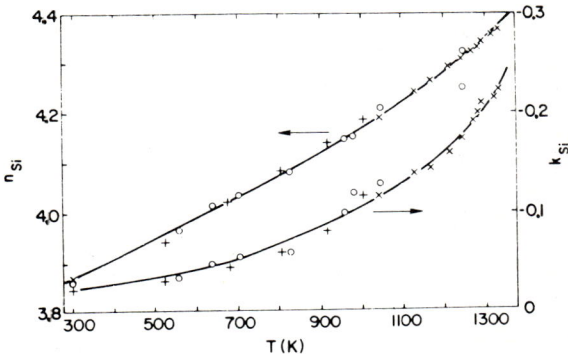

Fig. 6.7. The temperature dependence of the refractive index n and extinction coefficient k of silicon measured by an automated rotating-analyzer ellipsometer on single-crystal wafers of Si oxidized to different thicknesses [14 Å(+), 38 Å(×) and 169 Å(○)]. After ref. 232.

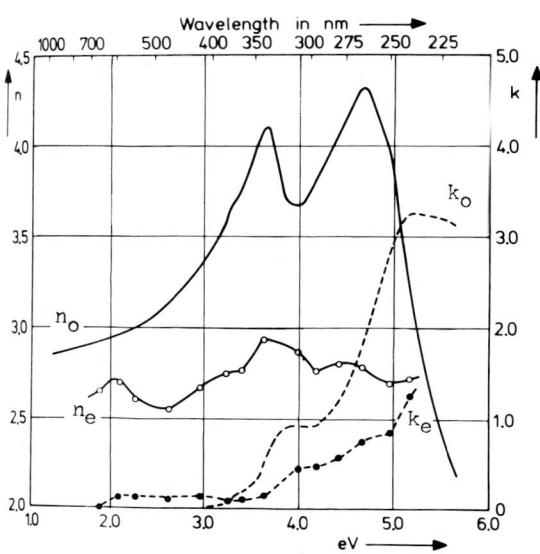

Fig. 6.8. Spectra of the ordinary $(n_o - jk_o)$ and extraordinary $(n_e - jk_e)$ complex refractive indices of gallium selenide measured by null ellipsometry on two crystal faces, one perpendicular and the other parallel to the optic axis. After ref. 241.

$(n_2 - jk_2)$ and $(n_3 - jk_3)$ of a biaxial crystal of vanadium pentoxide measured by ellipsometry at seven spectral lines of a mercury source in the visible (405, 436, 489, 546, 578, 659 and 709 nm). These results, by Jacobsen and Kerker[242], were derived from data obtained on three crystal faces each perpendicular to one of the major crystallographic axes.

By use of the extension of ellipsometry called modulated ellipsometry [§3.11], changes in the optical properties of materials produced by periodic perturbations such as stress, electric field or temperature can be measured. Figures 6.10 and 6.11 show the results obtained by Buckman and Bashara [125] for the electric-field-induced changes $\delta\epsilon_1$ and $\delta\epsilon_2$ of the dielectric function of opaque polycrystalline (optically isotropic) films of Au and Ag measured by electro-modulated ellipsometry (the electric field was applied in an electrolytic cell with the sample as one of its electrodes). These results agree with the usual interpretation of the band structure of Au and Ag; specifically, the $L_{32} \rightarrow L'_2$ transition at 2.1 eV for Au and the hybrid interband transition and plasma resonance at 3.9 eV for Ag are evident in figs. 6.10 and 6.11, respectively. Compared with modulated reflectance techniques, modulated ellipsometry has the advantage of point-by-point determination of both $\delta\epsilon_1$ and $\delta\epsilon_2$ at each photon energy, without recourse to Kramers-Kronig analysis.

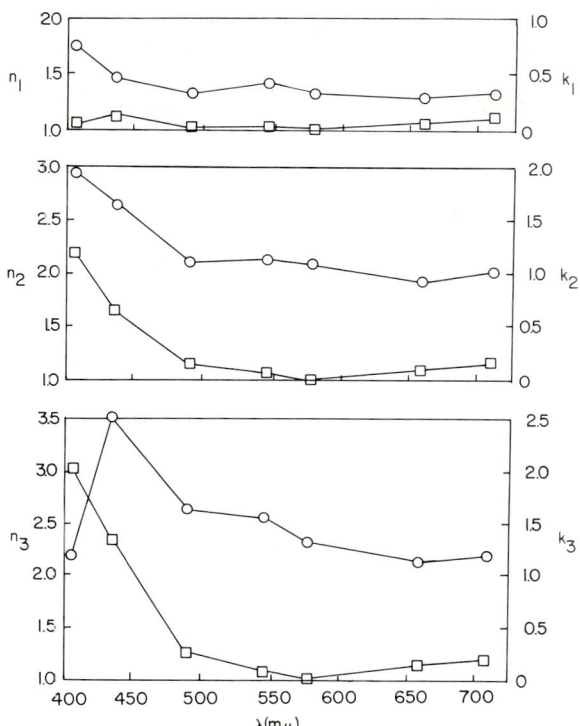

Fig. 6.9. The principal complex refractive indices of vanadium pentoxide measured by ellipsometry at wavelengths of 405, 436, 489, 546, 578, 659, and 709 mμ: n_i—○, k_i—□, $i = 1, 2, 3$. After ref. 242.

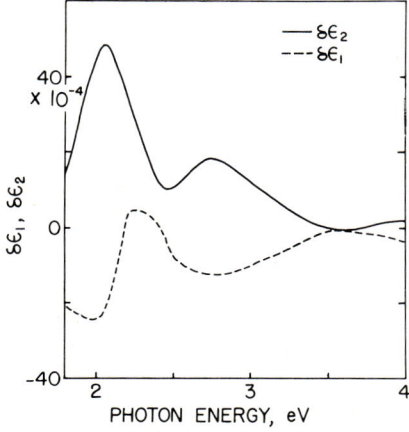

Fig. 6.10. The change in the complex dielectric function for Au vs photon energy as measured by electro-modulated ellipsometry. After ref. 125.

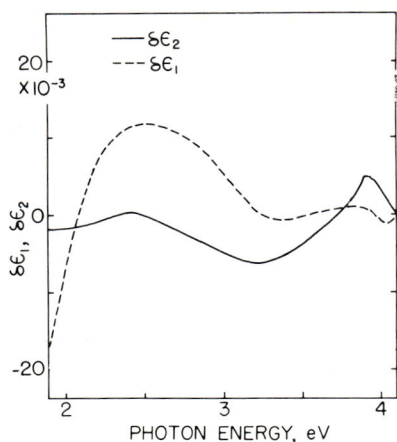

Fig. 6.11. The change in the complex dielectric function for Ag vs photon energy as measured by electro-modulated ellipsometry. After ref. 125.

6.2.2. *Optical properties of materials in thin-film phase*

Ellipsometric measurements on a thin-film phase formed at the interface between a substrate and an ambient are usually conducted to determine both the thickness and optical properties of the film. In this section, we are interested in those applications of ellipsometry with the primary purpose of measuring the optical properties (complex refractive index $n - jk$, dielectric function $\epsilon_1 - j\epsilon_2$, or any other related quantity) of thin films at an ambient/substrate interface. Measurements are often done as a function of wavelength, making ellipsometry a valuable spectroscopic tool of thin films. Applications in this area have been rather limited; the prospects, however, are great.

Kruger and Ambs [243] used ellipsometry to measure the refractive indices of thin films of condensed gases (oxygen, nitrogen, carbon dioxide, water, argon, neon, and krypton) on metal substrates (gold and copper) at the liquid-helium temperature of 4.2°K. Measurements were restricted to the mercury green line of 5461 Å and the results for condensation on a gold mirror are shown in table 6.1. In some experiments, the gases admitted through a capillary to the Dewar system in which the sample was mounted (with pressure $< 5 \times 10^{-5}$ torr) were subjected before entry to a microwave discharge. This led to condensed films of slightly different refractive indices as shown in table 6.1. These differences were explained by suggesting that the discharge dissociates the molecules of O_2, N_2, CO_2 and H_2O and excites the atoms of the other inert gases. Thus, the condensed films consist of both normal and dissociated (atomic) or excited species, even though the

Table 6.1. Refractive indices of thin films of gases condensed on a gold mirror at $\lambda = 5461$ Å [243]

Gas	Treatment	Temp. of deposition	n
argon	undischarged	4.2°K	1.26
	discharged	4.2°K	1.27
carbon dioxide	undischarged	77°K	1.34
	discharged	4.2°K	1.26
	undischarged	4.2°K	1.22
krypton	undischarged	4.2°K	1.28
	discharged	4.2°K	1.28
neon	undischarged	4.2°K	1.23
	discharged	4.2°K	1.24
nitrogen	undischarged	4.2°K	1.22
	discharged	4.2°K	1.22
oxygen	undischarged	4.2°K	1.25
	discharged	4.2°K	1.28
water	undischarged	4.2°K	1.19
	discharged	4.2°K	1.30
	undischarged	77°K	1.20
	discharged	77°K	1.25

percentage of the later is quite small (<0.1). We should mention that the refractive index of the condensed film of a given gas was determined by a computer search for the value of the index that would best fit (ψ, Δ) values obtained at multiple thicknesses of the same film. This is one of the early examples of computer application to ellipsometric data reduction.

Bartell and Churchill [244] were probably first to recognize the potential of ellipsometry as a spectroscopic tool for thin films on substrates. Figure 6.12 shows the ellipsometrically determined visible spectra of refractive index n and extinction coefficient k (which Bartell and Churchill call *absorption coefficient*) of a 1M dispersion of the dye tetraphenylporphine (TPP) in a film matrix of collodion 31 Å-thick on a chromium substrate. (The collodion was introduced to randomize the orientation and prevent the aggregation of the dye molecules which would otherwise be aligned perpendicular to the interface. Thus, the assumption could be made that the film was isotropic and the results could also be compared with spectra of the dye in solution.) Figure 6.13 shows the bulk absorption spectrum of TPP in benzene. The agreement between the spectra of the extinction coefficient k of the 31 Å film of TPP in collodion on Cr, fig. 6.12 (bottom), with the bulk spectrum of TPP in benzene, fig. 6.13, is impressive, considering the differences in solvents and concentrations of TPP. The agreement includes not only the strong principal peak but also the weak subsidiary peaks. In this study,

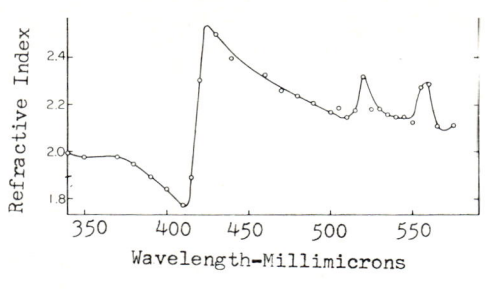

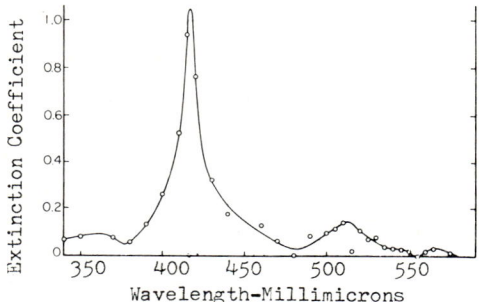

Fig. 6.12. The ellipsometrically determined visible spectra of refractive index n (top) and extinction coefficient k (bottom) of a 1M dispersion of the dye tetraphenylporphine (TPP) in a film matrix of collodion 31 Å thick on a chromium substrate. After ref. 244.

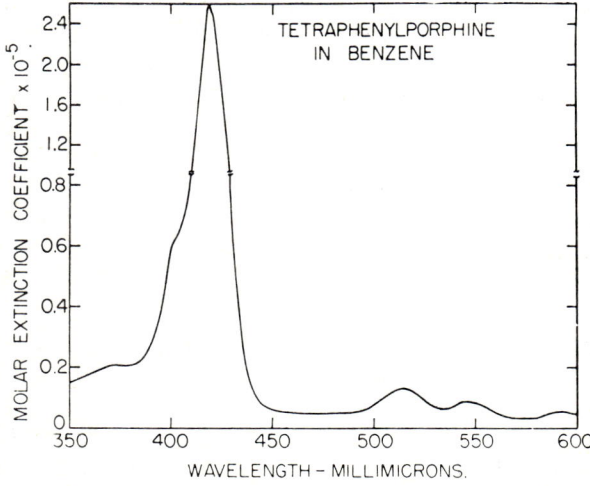

Fig. 6.13. The bulk absorption spectrum of TPP in benzene. After ref. 244.

§6.2] SPECTROSCOPIC ELLIPSOMETRY 431

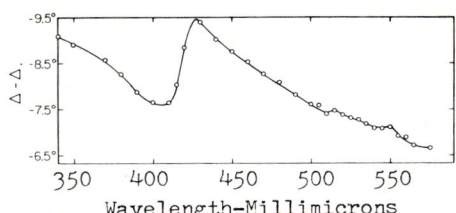

Fig. 6.14. The spectrum of $\Delta - \bar{\Delta}$ reproduces the essential structural features of the spectrum of the refractive index n of fig. 6.12 (top). After ref. 244.

Bartell and Churchill observed that a simple plot of the spectrum of $\Delta - \bar{\Delta}$ (where $\bar{\Delta}$ is the bare-metal value of the ellipsometric angle Δ), fig. 6.14, reproduces the essential structural features of the spectrum of the refractive index n, fig. 6.12 (top).

Bashara and Peterson [245] observed anomalous thickness-dependent dispersion of the refractive index n and extinction coefficient k (which they call *absorption* coefficient) of polybutadiene films (a dielectric) on Au and Cr substrates. Figure 6.15 shows their results for n (a) and k (b) at three different film thicknesses (22, 55, and 136 Å), over a wavelength span between 3500 and 6500 Å.

Dignam et al. [246] extended spectroscopic ellipsometry to the infrared. Figure 6.16 shows their first data for CH_3OH adsorbed on Ag using an automatic wavelength-scanned ellipsometer with sixteen reflections between two parallel silver mirrors at an angle of incidence of 80°. The quantities plotted versus the wavelength in microns in fig. 6.16 are the real (R) and imaginary (I) parts of the complex optical density function D_N, defined by Dignam et al. as $\log_e (\rho/\bar{\rho})$, where ρ $(= \tan \psi \, e^{j\Delta})$ and $\bar{\rho}$

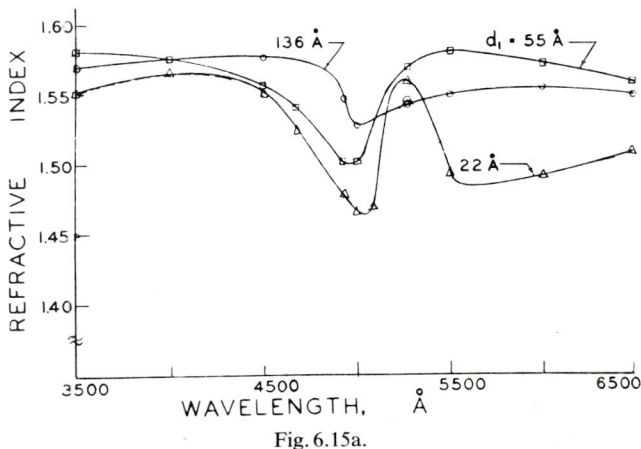

Fig. 6.15a.

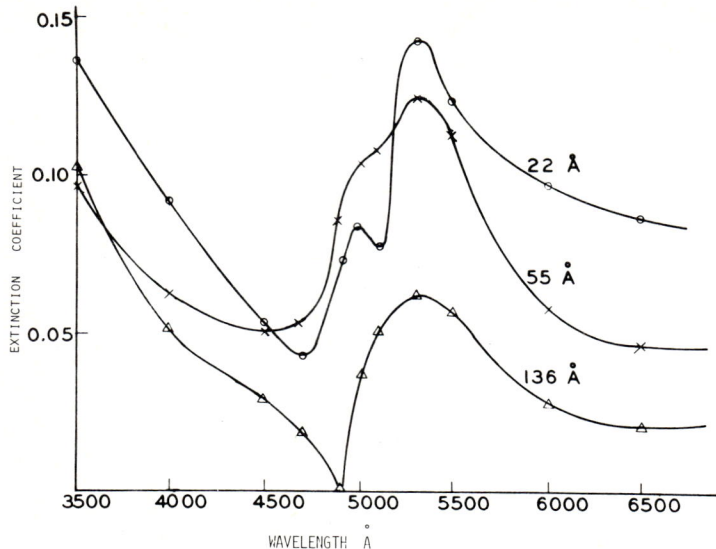

Fig. 6.15. The spectra of the refractive index n (a) and extinction coefficient k (b) of polybutadiene films on opaque Au substrate for film thicknesses of 22, 55 and 136 Å. After ref. 245.

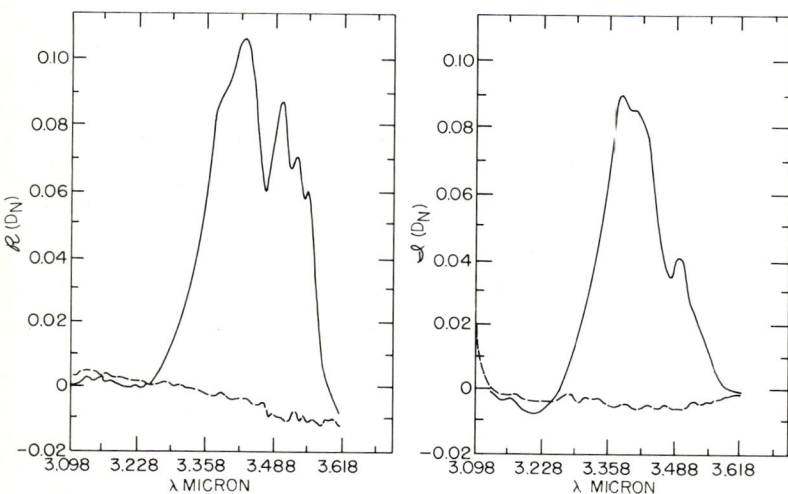

Fig. 6.16. The IR spectra of the real part (left) and imaginary part (right) of the ellipsometric complex optical density function D_N for CH_3OH adsorbed on Ag. The solid curve is for one adsorption scan and the dashed curve follows desorption. After ref. 246.

§6.3]   PHYSICAL AND CHEMICAL ADSORPTION   433

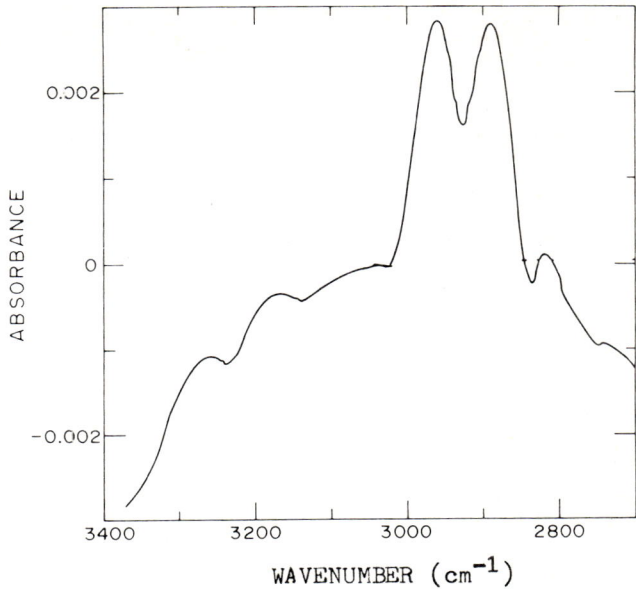

Fig. 6.17. IR absorbance spectrum for butanol adsorbed onto vacuum-deposited silver. After ref. 247.

($= \tan \bar{\psi} \, e^{j\bar{\Delta}}$) are the ellipsometric functions of the film-covered and film-free substrates, respectively. Five peaks occur in the spectrum of $R(D_N)$ at 3.38, 3.43, 3.50, 3.52, and 3.55 μm, all in the vicinity of the C–H stretching vibrational frequency. The adsorption of CH_3OH on Ag was reversible; desorption was observed by the nearly flat scan in fig. 6.16.

A more recent result by Dignam et al. [247] is shown in fig. 6.17 for the absorbance spectrum of butanol adsorbed on Ag. Here, the sensitivity of the automatic ellipsometer has been increased to 10^{-5} absorbance units. The prominent bands around 2890 and 2960 cm^{-1} were identified as the C–H stretching bands of butanol. Other spectral features were explained as interference from preadsorbed impurities on the Ag surface.

6.3. Physical and chemical adsorption

The adsorption of molecular or atomic species on surfaces in contact with gaseous or liquid ambients has been studied non-destructively and *in situ* by ellipsometry. Dependent on the interaction between the adsorbate and the substrate being weak or strong, adsorption is generally classified as physical or chemical, respectively. The term adsorption itself implies reversibility; an adsorbed layer can be *desorbed*, for example, by heating. This is in contrast with the formation of permanent films, such as oxides, whose study by ellipsometry will be discussed in §6.4.

The physical adsorption of water vapour on chemically cleaned Si surfaces at constant temperature (25°C) was studied by Archer [248], using ellipsometry, as a function of the relative humidity (water vapor pressure) of a controlled atmosphere. The resulting isotherms are shown in fig. 6.18,

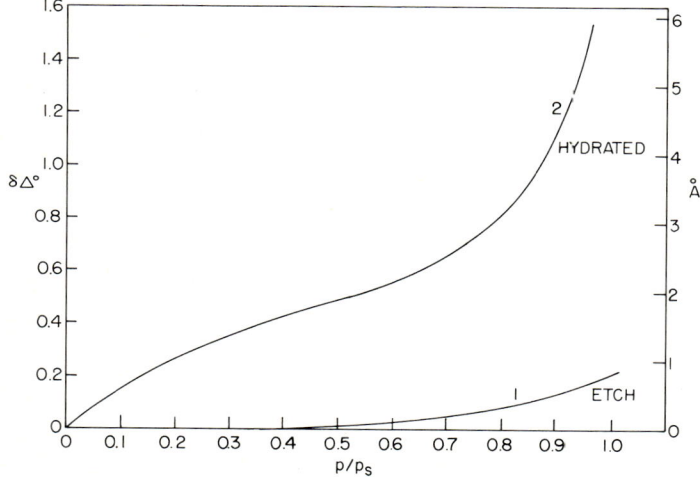

Fig. 6.18. The change in Δ caused by the physical adsorption of water vapor onto etched and hydrated silicon surfaces and the thickness of the adsorbed layer as a function of normalized pressure p/p_s where p_s is the saturation vapor pressure. After ref. 248.

where $\delta\Delta$ and the corresponding adsorbed-film thickness (calculated assuming the refractive index of the film to be that of water) are plotted against the normalized pressure p/p_s; p_s being the saturation vapor pressure. Notice the significant effect that surface-chemical preparation has on the adsorption curve. The very low adsorption on etched surfaces, curve 1, is consistent with their *hydrophopic* nature. With a special hydrating treatment, these surfaces become *hydrophilic* and their increased ability to adsorb water is evident from curve 2. The shape of curve 2 follows the type II isotherm according to the Brunauer, Emmett and Teller (BET) theory [249].

For the same chemical treatment, different methods of substrate preparation lead to different adsorption. Figure 6.19 shows the results for water vapor adsorption on three physically different (1-mechanically polished, 2-cleaved, and 3-fast-etched) surfaces subjected to the same final chemical

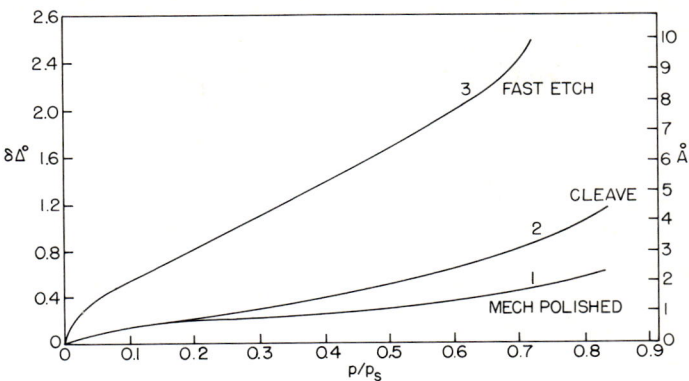

Fig. 6.19. Shows the isotherms ($T = 25°C$) for water vapor adsorption on three physically different silicon surfaces (1-mechanically polished, 2-cleaved, and 3-fast-etched) subjected to the same final chemical hydrating treatment. After ref. 248.

hydrating treatment. By analyzing each of the curves shown in fig. 6.19 (and others) by the BET theory, only curve 3 (for fast-etched Si) yielded a calculated film thickness at monolayer coverage (3.7 Å) that agreed reasonably well with the diameter of a water molecule estimated from the density of liquid water (3.1 Å). Similar results were obtained for the adsorption of carbon tetrachloride and acetone on the same types of Si surfaces. The experimental finding that methods of preparing the substrate surface other than fast etching lead to ellipsometric thickness at monolayer coverage that is less than the diameter of the adsorbed molecule was explained by Archer as a consequence of surface roughness.

The chemisorption (chemical adsorption) of oxygen on cleaved silicon under high vacuum at room temperature (24°C) was also studied by Archer [248]. Figure 6.20 represents a typical chemisorption curve of $\delta\Delta$ versus the time of exposure. After an initial steep rise, $\delta\Delta$ saturates when a complete monolayer has been formed. (The dashed curve superimposed on fig. 6.20 shows the pressure rise as O_2 was suddenly admitted to the vacuum chamber in which the sample was mounted. After about 8 minutes, the pressure stayed constant at 3.1×10^{-7} torr.) If the coverage (defined as the number of adsorbed atoms or molecules per single atom of the substrate surface) at the completion of one monolayer of O_2 on Si is θ_o, then the partial (sub-monolayer) coverage $\theta(t)$ at any time t during adsorption can be calculated from

$$\frac{\theta(t)}{\theta_o} = \frac{\delta\Delta(t)}{\delta\Delta_o} \tag{6.1}$$

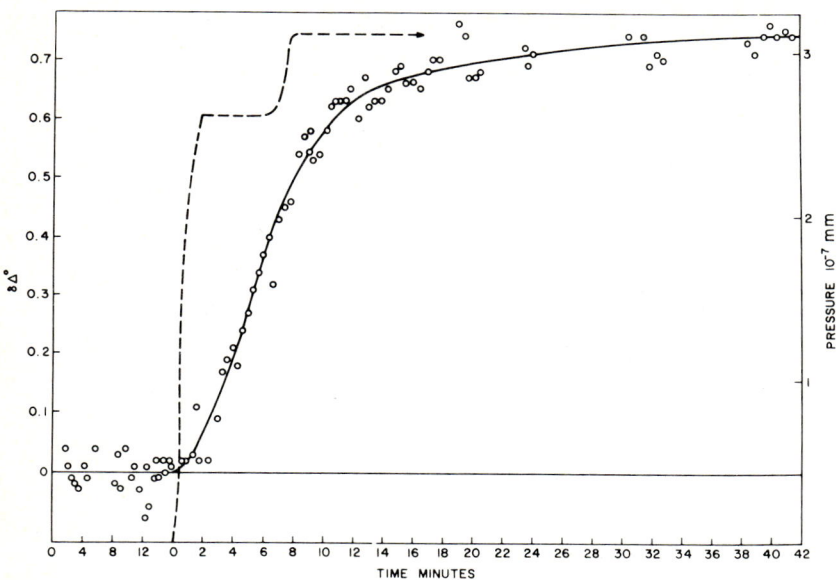

Fig. 6.20. The change in Δ during chemisorption of oxygen on cleaved silicon under high vacuum at room temperature vs exposure time. After an initial steep rise, $\delta\Delta$ saturates when a complete monolayer has been formed. Also shown by the dashed curve is the rise in pressure as O_2 was suddenly admitted. After ref. 248.

where $\delta\Delta(t)$ and $\delta\Delta_o$ are the changes of Δ at partial and complete coverage, respectively. The rate of change of θ is related to the sticking probability S (that a colliding molecule sticks or adsorbs to the surface) by

$$d\theta/dt = (\nu p/N_o)S, \qquad (6.2)$$

where $\nu = nc/4$ (n is the molecular density per unit pressure and c is the average molecular velocity), p is the pressure, and N_o is the total number of molecules required to form a complete monolayer. Figure 6.21 shows the sticking probability as a function of coverage for the data shown in fig. 6.20, calculated by use of the above two equations (assuming $\theta_o = 1$ and $\delta\Delta_o = 0.71°$).

Smith [250] studied the adsorption of several molecules of known dimensions (including caproic, lauric, stearic and behenic acids and pentane) on a clean mercury surface in a Langmuir trough at and below monolayer coverage. This investigation combined surface tension and contact potential measurements with ellipsometry (at $\lambda = 5461$ Å and 70° angle of incidence). Figure 6.22 shows the ellipsometric signal $\bar{\Delta} - \Delta/\alpha$ ($\bar{\Delta}$ is the bare-surface value of the relative phase shift Δ and α is a constant) plotted against the

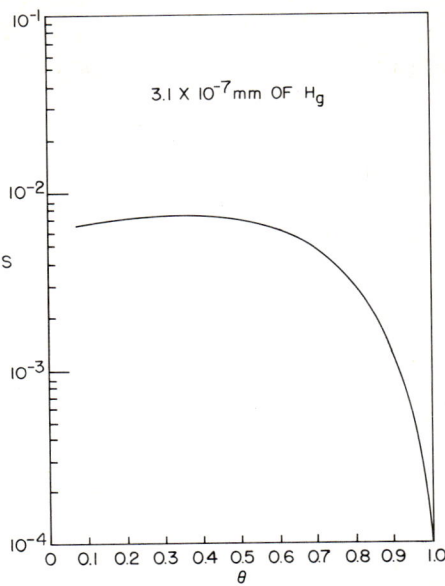

Fig. 6.21. The sticking probability S of oxygen molecules colliding with a Si surface as a function of coverage θ for the indicated pressure at room temperature. After ref. 248.

experimental value of the surface concentration of molecules Γ (molecules/cm^2) on a log-log scale. The measurements were taken during the two-dimensional (2-D) compression of a spread film of known quantity of molecules. The results of fig. 6.22 verified that $\bar{\Delta} - \Delta$ is a *linear* function of the coverage θ (also shown in fig. 6.22) below continuous (close-packed) monolayer formation. The coverage θ is defined in this case as the product of the projected area per molecule A_o (assumed to be known) times the molecular concentration Γ; $\theta = A_o \Gamma$. The average film thickness d in fig. 6.22 is given by $d = \theta d_o$, where d_o is the molecular dimension perpendicular to the surface. The values of the constant α that appear in fig. 6.22 were calculated as $\Delta - \bar{\Delta}/d$. From these values of α, the indices of refraction n of the films at complete monolayer coverage ($\theta = 1$) were determined assuming the continuum 3-D ellipsometric model of Drude. Table 6.2 compares n thus calculated with the handbook refractive indices of the bulk phase of each material n_B; the agreement is excellent lending support to the applicability of the 3-D theory to an essentially 2-D problem.

Meyer and Bootsma [251] studied the chemisorption of several gases on clean silicon (111) and (100) surfaces by ellipsometry at $\lambda = 5461$ Å and 70° angle of incidence. Figure 6.23 shows the saturation part of three chemisorp-

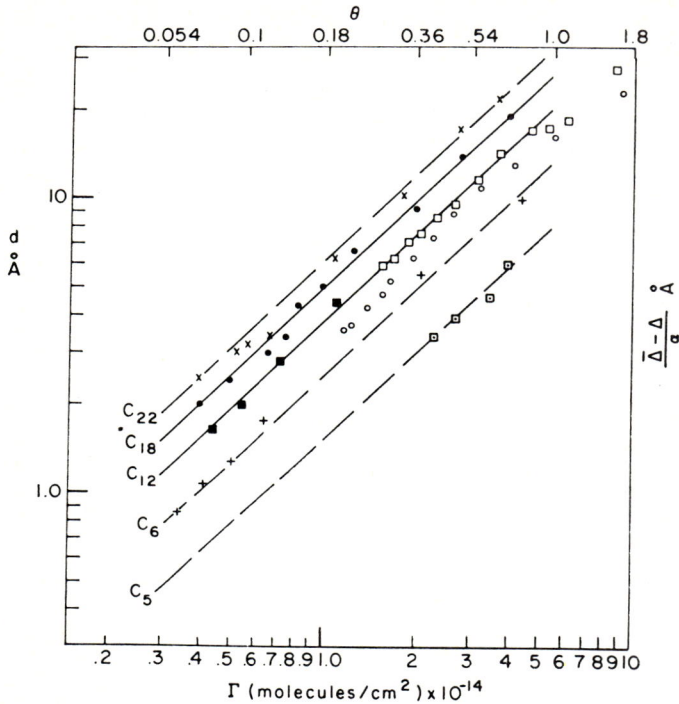

Fig. 6.22. Log-log plot of ellipsometer reading $(\bar{\Delta} - \Delta)/\alpha$ vs Γ for behenic, stearic, lauric, caproic acids, and pentane on Hg. $\bar{\Delta} - \Delta$ is directly proportional to average film thickness or θ between $d \sim 0.2$ and 24 Å or $\theta \sim 0.05$ and 1.

	Symbol	α
Pentane C_5	⊡	0.130 ± 0.005
Caproic acid C_6	+	0.139 ± 0.013
Lauric acid C_{12}	■ ○ □	0.14 ± 0.003
Stearic acid C_{18}	●	0.18 ± 0.010
Behenic acid C_{22}	X	0.16 ± 0.015

After ref. 250.

tion $\delta\Delta$-vs-exposure time curves, following the initial rapid rise which took 1 minute or less for completion at a gas pressure of 0.1 torr. Figure 6.24 shows $\delta\Delta$ vs $\delta\psi$ during adsorption. A significant observation is that there is a change of ψ which is larger than, and is in the opposite direction to, what would be expected in physical adsorption. The latter is shown in fig. 6.24 for the case of Xe, for comparison. Moreover, this change of ψ is virtually the same at monolayer coverage (beyond the breakpoint in the $\delta\Delta$-vs-$\delta\psi$ curve of fig. 6.24) for all gases considered; the change of $\delta\Delta$ is quite different from

Table 6.2. Comparison of calculated values of the film index of refraction n at $\theta = 1$, with bulk values n_B [250]

Compound	No. of C atoms	Shape	α deg/Å	n	n_B at 20°C
Pentane	5	rod	0.130 ± 0.005	1.36	1.36
Caproic acid	6	rod	0.139 ± 0.013	1.42	1.42
Lauric acid	12	rod	0.140 ± 0.003	1.43	1.43
Stearic acid	18	rod	0.180 ± 0.010	1.66	1.68
Behenic acid	22	rod	0.160 ± 0.015	1.5	—

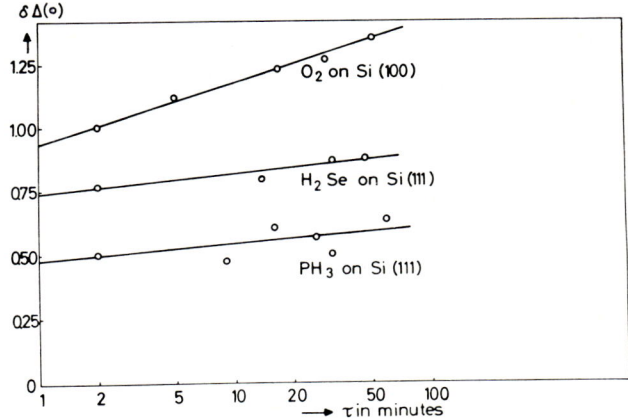

Fig. 6.23. Change in Δ as a function of exposure time in the saturation region for the chemisorption of O_2, H_2Se, PH_3 on silicon surfaces at 0.1 torr. After ref. 251.

gas to gas. Table 6.3 summarizes the observed changes $\delta\psi_{obs}$ and $\delta\Delta_{obs}$ of the ellipsometric parameters ψ and Δ that arise from the chemisorption of monolayer of several gases on the clean Si (111) surface. $\delta\Delta_{obs}$ is compared with a value $\delta\Delta_{calc}$ calculated using the macroscopic theory of Drude. In the latter, literature values of atomic polarizabilities (α) and diameters are used to calculate an effective refractive index (n_{eff}) for the adsorbed layer. Notice the good agreement between the observed values ($\delta\Delta_{obs}$) and those ($\delta\Delta_{calc}$) estimated from the extrapolated macroscopic theory. As in the study by Smith that we quoted earlier, this data by Meyer and Bootsma supports the applicability of the macroscopic theory to film coverages at and below the monolayer level.

In order to explain the observed anomalous change of $\psi(\delta\psi_{obs})$, Meyer

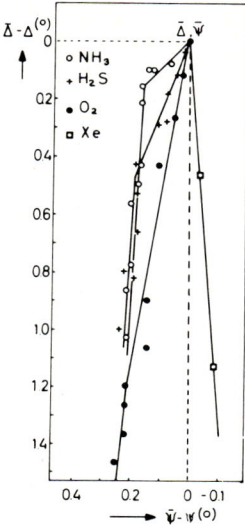

Fig. 6.24. Experimentally determined $\delta\Delta$ vs $\Delta\psi$ for the chemisorption of NH_3, H_2S and O_2 and for the physisorption of Xe on a clean Si(111) surface. After ref. 251.

Table 6.3. Gas adsorption on clean silicon (111) surface [251]

Compound	α^* (Å^3)	diam* (Å)	θ molecules per Si-atom	n_{eff}	$\delta\psi_{\text{obs}}$ (deg) ± 0.03	$\delta\Delta_{\text{obs}}$ (deg) ± 0.06	$\delta\Delta_{\text{calc}}$ (macrosc.) (deg)
HCl	2.65	3.6	0.5	1.19	0.17	0.77	0.64
HBr	3.64	3.9	0.5	1.24	0.17	0.77	0.83
HI	6.63	4.3	0.5	1.42	0.22	1.33	1.27
Cl_2	4.50	3.6	0.5	1.33	0.15	1.00	0.94
H_2S	3.78	3.7	0.25	1.13	0.17	0.43	0.46
H_2Se	5.33	4.0	0.25	1.17	0.20	0.73	0.67
NH_3	2.06	3.0	0.17	1.05	0.17	0.17	0.20
PH_3	4.71	3.8	0.17	1.10	0.17	0.47	0.42
AsH_3	5.60	4.0	0.17	1.12	0.17	0.53	0.50
O_2	6.0†	2.8	0.5	1.62	0.20	1.07	~1.0

*In the H_xA compounds the Van der Waals diameter of the A-atom is given, and the polarizability is taken as the sum of the atomic polarizabilities.
†Twice the polarizability of the O^{2-} ion.

and Bootsma suggested that at the surface of a crystal there is a thin layer (~5–10 Å thick for semiconductors) of different electronic, hence optical, properties due to the presence of uncompensated (dangling) bonds and their associated surface states. Chemisorption of a foreign layer leads to compensation of these bonds and the optical properties of the surface layer shift towards those of the bulk. Using this model, it was possible to explain all ellipsometric observations. In particular, the finding that $\delta\psi_{obs}$ is independent of the adsorbed gas is consistent with the fact that compensation of the dangling bonds is effected by all gases, so that the shift of the optical properties of the surface layer to bulk values is the same for all gases and causes the same $\delta\psi_{obs}$. The direct contribution to $\delta\psi_{obs}$ from the *transparent* chemisorbed gas layer itself is known to be zero, to first order, although such a layer contributes most of the observed change of Δ, $\delta\Delta_{obs}$.

Other related work by Meyer and co-workers includes the use of ellipsometry for the quantitative calibration of Auger electron spectroscopy [252] and the study of the density function of surface states [253] from the ellipsometric spectra of $\delta\psi_{obs}$ (and $\delta\Delta_{obs}$).

The physical adsorption isotherms of krypton, oxygen and acetylene on a single crystal (110) face of silver at very low pressures ($<10^{-6}$ torr) are shown in fig. 6.25. These results by Muller et al. [163] were obtained by null

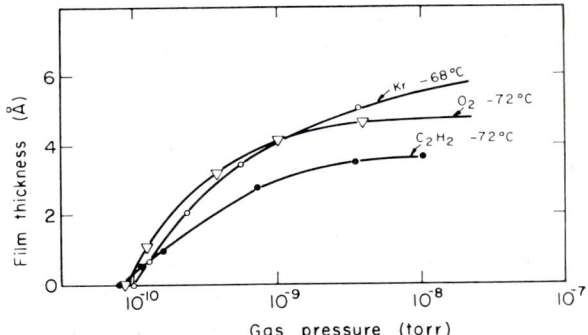

Fig. 6.25. Adsorption isotherms of krypton, oxygen and acetylene on silver (110) in terms of film thickness determined by ellipsometry. After ref. 163.

ellipsometry at $\lambda = 5461$ Å and 45° angle of incidence. The experimental set-up used by these investigators permitted simultaneous complementary observations by low-energy electron diffraction. Each film thickness shown in fig. 6.25 corresponded to an equilibrium value after an initial steep rise at each pressure. From the isotherms of fig. 6.25, the heat of adsorption can be deduced as a function of coverage from the Clausius-Claperyon relation. This is shown in fig. 6.26. This study may be considered as a surface-thermodynamic application of ellipsometry.

Most adsorption studies conducted so far by ellipsometry have been concerned with adsorption at a gas/solid or a gas/liquid interface, as

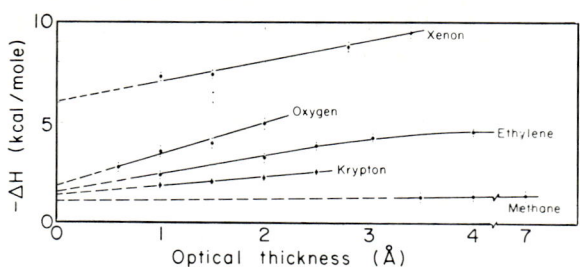

Fig. 6.26. Isosteric heats of adsorption derived from adsorption isotherms with extrapolation to zero coverage. After ref. 163.

represented by the foregoing examples. By comparison, corresponding ellipsometric studies at liquid/solid (or liquid/liquid) interfaces have been rather limited. An illustration of this application is the study of polymer adsorption at a liquid/solid interface by Stromberg et al. [254, 255]. Because a polymer is a long-chain flexible molecule, it can adsorb on a surface in many different possible conformations, and the conformation may vary with time. In general, it is thought that a long polymer macromolecule will attach itself to a solid surface along segments of its length that are separated by loops extending into the solution. Ellipsometry offers a nondestructive probe to measure *in situ* the extensions of the substrate-adsorbed polymer in solution. In interpreting the ellipsometric data, Stromberg et al. considered the segments of polymer attached to the surface and the loops projecting into the solution, together with the permeating solvent, as forming a film over the surface. The effective refractive index of such a film is slightly different from that of the solution, because of the low polymer content of the film. Furthermore, because the polymer content is variable throughout the thickness of the film, an inhomogeneous-film model is more descriptive of this situation. Stromberg et al. chose an exponential distribution based on a theoretical study by DiMarzio and McCrackin [256]. Figure 6.27 shows the thickness as a function of time for the adsorption of 5.37×10^5 molecular-weight fraction of polystyrene from 0.23 mg/ml solutions of cyclohexane on chrome ferrotype surface at 34°C. The indicated thickness is that of an equivalent homogeneous film which has almost the same ellipsometric effect as the exponentially inhomogeneous one. As seen in fig. 6.27, the adsorbed film thickness first increases rapidly with time, then levels off at a plateau. Stromberg et al. explained the results by assuming that subsequent adsorption of polymer molecules induces desorption of attached segments of previously adsorbed polymer molecules, leading to larger loops, extending deeper into the solution, so that the film thickness is increased until a plateau is established under steady state conditions. Other examples of adsorption

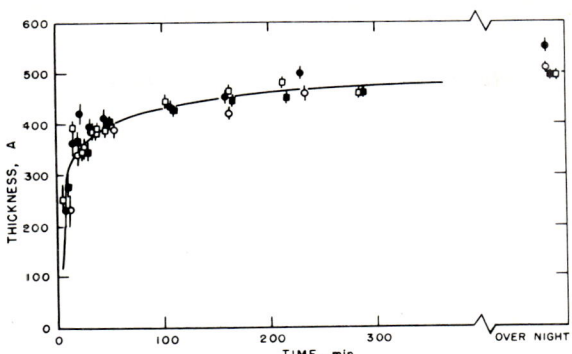

Fig. 6.27. Thickness of adsorbed film on Cr substrate of polystyrene (M.W. = 537,000), vs time, in cyclohexane solution at a concentration of 0.23 mg/ml, at 34°C. After ref. 256.

reactions at a liquid/solid interface will be discussed in §6.6 on biological applications. The related phenomenon of ionic adsorption on electrodes will be dealt with in §6.5 on electrochemical applications.

6.4. Oxidation of semiconductor and metal surfaces

Ellipsometry has found extensive application in the study of the oxidation of semiconductor and metal surfaces in various environments. In this section we present examples of investigations by ellipsometry of oxide-film growth on surfaces of semiconductors and metals in atmospheric, controlled-gaseous, and liquid ambients. We defer discussion of anodic oxidation to §6.5 on electrochemical applications.

Figure 6.28 shows Archer's [257] results for the atmospheric oxidation of Si and Ge as a function of time (on a logarithmic scale) immediately (200 sec) after mechanical polishing and hydrofluoric-acid rinsing (etching). The relative humidity and temperature remained within $54 \pm 4\%$ and $28 \pm 2°C$, respectively. Conventional manual null ellipsometry was used at wavelength $\lambda = 5461$ Å and angle-of-incidence $\phi = 61.26°$. The film thickness was determined from the change of Δ from its "bare-surface" value $\bar{\Delta}[\bar{\Delta} = 179.43°(\text{Si}), \bar{\Delta} = 166.28°(\text{Ge})]$ using a calculated proportionality factor $\alpha = 0.14°$ deg./Å. The linear parts of the curves in fig. 6.28 indicate that film growth eventually follows Elovich equation

$$dL/dt = A \exp(-BL), \tag{6.3}$$

which becomes

$$L = a + b \log(t + t_o), \tag{6.4}$$

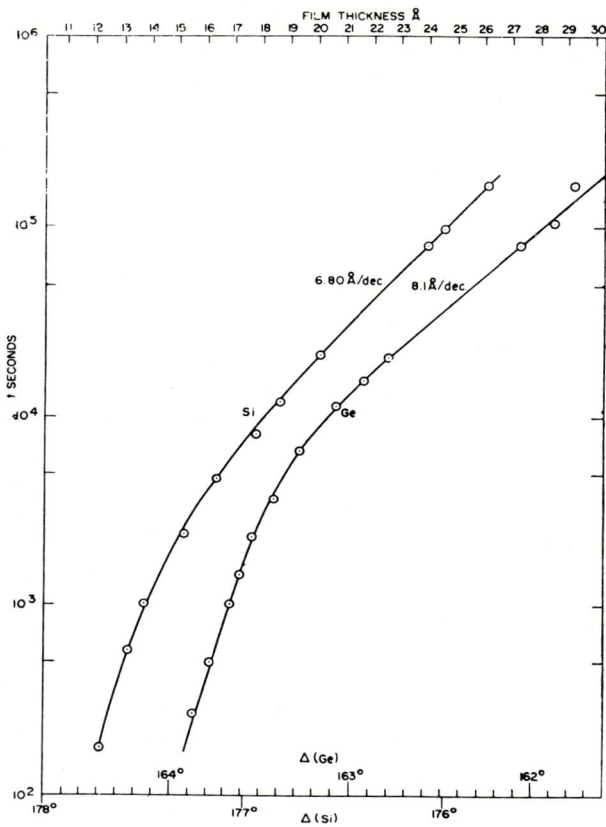

Fig. 6.28. Relative phase difference Δ and film thickness as a function of time during the atmospheric oxidation of Si and Ge after an HF rinse. After ref. 257.

after integration. The values of a, b and t_o were estimated from the linear portions of the curves in fig. 6.28 to be -9.74 Å, 6.86 Å/decade, and 1500 sec, respectively, for Si; and -17.2 Å, 9.05 Å/decade, and 4500 sec, respectively, for Ge.

Figure 6.29 shows the film thickness-versus-time data by Lukeš [258] for the oxidation of the (110) face of a cleaved GaAs crystal in air at room temperature (the humidity and temperature were not regulated in this experiment but the effect of their fluctuations was presumably unimportant). As in Archer's work (fig. 6.28), measurements were made by null ellipsometry at $\lambda = 5461$ Å (at an angle of incidence of 70°). Figure 6.29 illustrates that film growth obeys the logarithmic law [eq. (6.4)] very well for six decades of time (up to a million seconds) with an estimated slope of

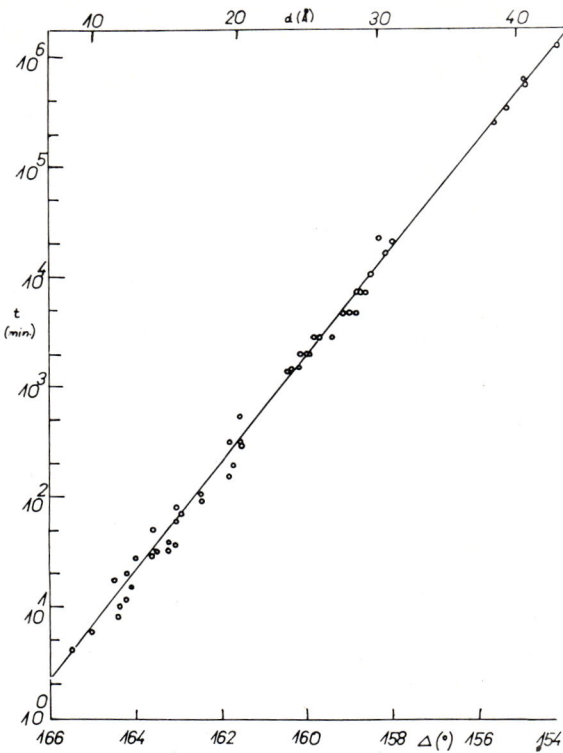

Fig. 6.29. Δ and film thickness versus time for a GaAs cleaved single-crystal with (110) oriented surface oxidized in air at room temperature. After ref. 258.

6.2 ± 0.5 Å/decade. Measurements on crystal faces other than (110) yielded nearly the same results, within experimental error.

We now cite examples of the oxidation of metals in controlled gaseous ambients. Figure 6.30 represents Smith's [259] data on the oxidation of titanium (Ti) at a temperature of 325°C for different pressures of oxygen (admitted to the vacuum chamber in which the sample was mounted). Again, a logarithmic time law [eq. (6.4)] is obeyed at each pressure. The insert in fig. 6.30 shows the oxide (TiO$_2$) film thickness at $t = 10^3$ sec, plotted versus $-\ln P_o$, where P_o is the oxygen pressure. The slope of this line confirmed an expected $P_o^{1/2}$-dependence of the oxide-film thickness on pressure. The effect of temperature on the oxidation of Ti at constant oxygen pressure is shown in fig. 6.31. Here the oxide-film thickness in angstroms is plotted versus time in seconds (on a logarithmic scale, as usual) for different temperatures at the same pressure, $P_o = 3 \times 10^{-6}$ torr. Notice that the rate of oxide growth (slope)

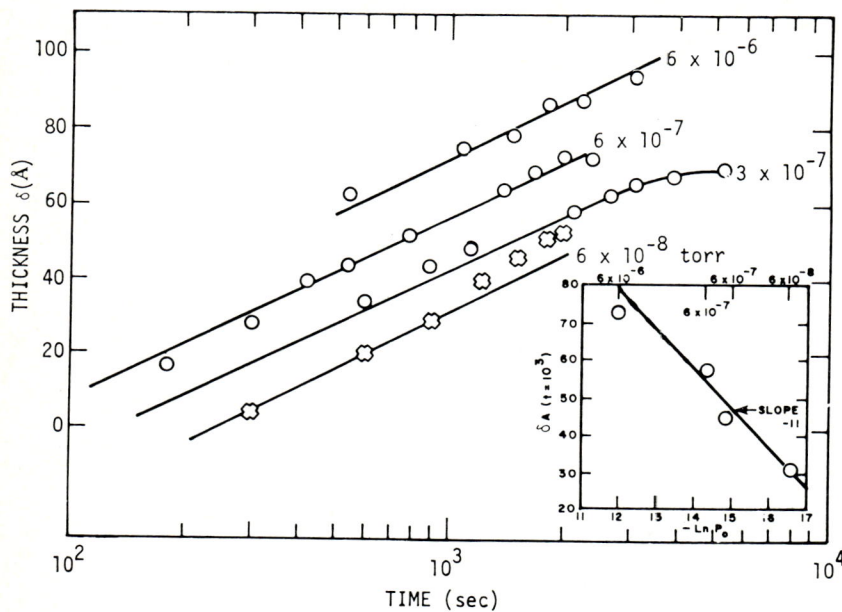

Fig. 6.30. Oxide-film thickness versus log t for non-doped Ti at various oxygen pressures and at 325°C. Insert shows the value of film thickness $\delta(\text{Å})$ at $t = 10^3$ sec vs-ln P_o. The slope of this line confirmed an expected $P_o^{1/2}$ dependence of the oxide-film thickness on pressure. After ref. 259.

first increases with temperature to a maximum (at 391°C for the data shown in fig. 6.31), then decreases to nearly zero (at 510°C). At the latter temperature, it is believed that the oxide film sublimates as fast as it is formed. The exact temperature dependence of slope is not completely understood. It should be mentioned that in addition to ellipsometry, Smith employed three other techniques (Auger spectroscopy, surface potential and ion bombardment) in his study of the oxidation of Ti. This simultaneous integration of different and independent techniques has become increasingly important in gaining detailed information on the complex processes involved in gas/solid-surface interactions.

By employing oxygen pressures below 10^{-6} torr, Chou et al. [260] succeeded in using null ellipsometry to resolve the initial growth of ultra-thin (0–8 Å) lead oxide (PbO) on clean lead (Pb) surfaces. (Auger electron spectroscopy was also used in these studies.) Figure 6.32 shows their results at different temperatures and pressures. The thickness of the growing oxide film is closely proportional to the oxygen exposure (pressure-time product) until a limiting thickness of about 6.3 Å, irrespective of pressure. The ellipsometrically determined refractive index of 2.8 at $\lambda = 5461$ Å for these

OXIDATION OF TITANIUM

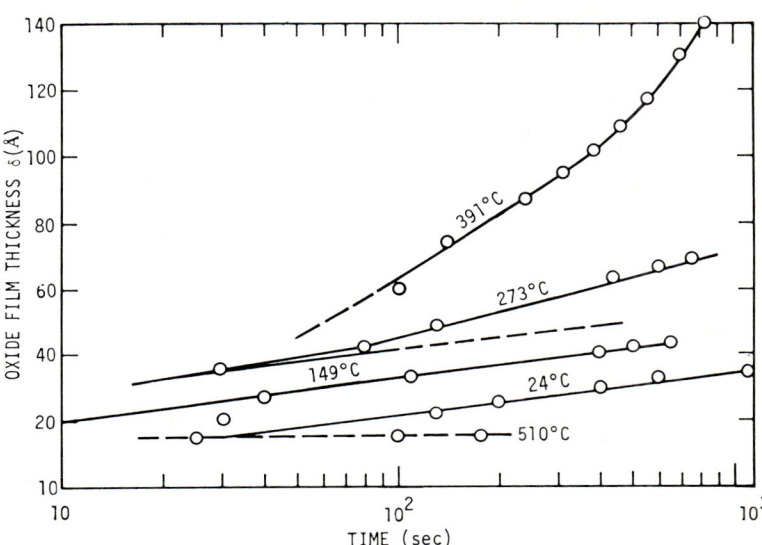

Fig. 6.31. Plot of oxide film thickness vs log t at various temperatures with $P_o = 3 \times 10^{-6}$ torr, for non-doped Ti. After ref. 259.

ultra-thin films agreed remarkably well with the bulk refractive index of orthorhombic PbO crystal structure. (The crystallinity of the film was checked independently by X-ray diffraction.) Chou et al. provided a theoretical explanation of the growth kinetics shown in fig. 6.32.

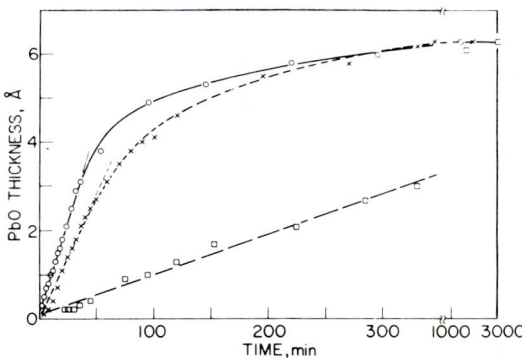

Fig. 6.32. Variation of oxide-film thickness with exposure time during the oxidation of Pb at various temperatures and oxygen pressures: ○—296°K, 1.5×10^{-7} torr; ×—323°K, 1.5×10^{-7} torr; □—296°K, 5×10^{-8} torr. After ref. 260.

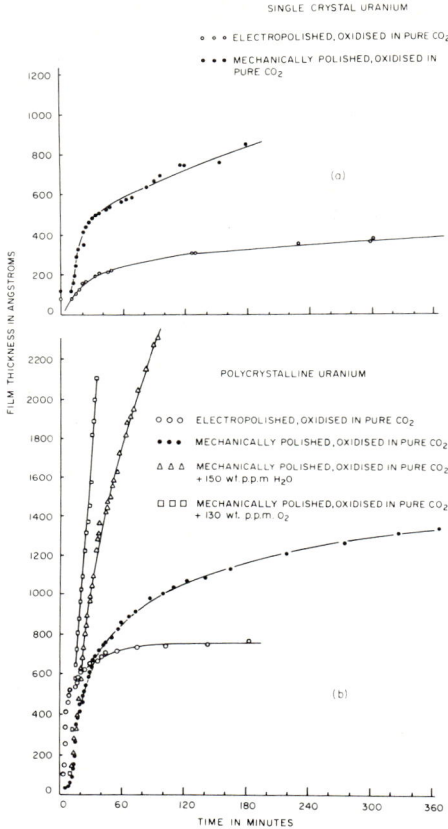

Fig. 6.33. Oxidation of single and polycrystalline uranium at 300°C in carbon dioxide. After ref. 146.

The oxidation of single crystal and polycrystalline uranium at 300°C in carbon dioxide atmospheres has been studied by Hayfield and White [146] using null ellipsometry at $\lambda = 5461$ Å. Their results appear in fig. 6.33 and show that surface preparation and the existence of small amounts of oxygen and water vapour have significant influence on the rate of oxidation of uranium and on the resulting oxide-film thickness. Although the effect of O_2 and H_2O on oxidation kinetics may be expected, that of surface preparation is more difficult to explain.

The above examples were for the gaseous oxidation of *bulk* semiconductors and metals. Ellipsometry has also been used to study the oxidation of *thin films* of metals. Whereas a single-film model is applicable in the former

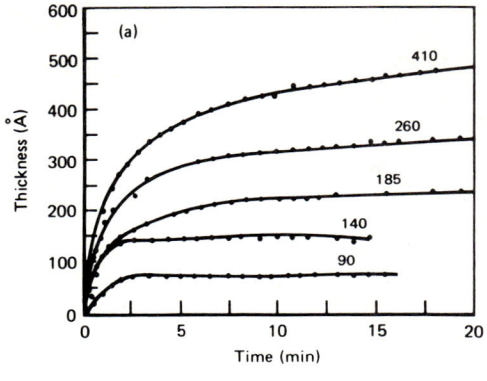

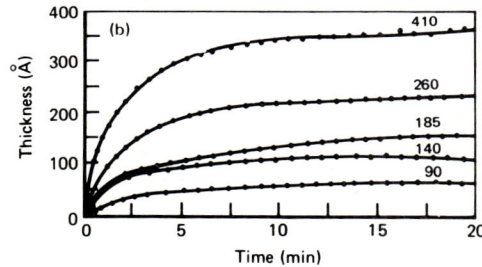

Fig. 6.34. (a) The oxide-growth-rate curves for the oxidation of molybdenum films of five different thicknesses at 2×10^{-4} torr O_2 and 575°C. (b) The decrease of thickness of these same metal films upon oxidation. The number that labels each curve is the initial thickness of the metal film in Å. After ref. 261.

case, a double-film model is required to analyze the data in the latter. Figure 6.34 shows the results by Lederich [261] for the kinetics of oxide-film growth (a) and metal-film loss (b) of 80–400 Å-thick molybdenum films of different initial thicknesses (indicated by each curve) on quartz substrates exposed to 2×10^{-4} torr O_2 at 575°C. Although quantitative comparison was not possible, there is good agreement between the kinetics of oxide-film growth and metal-film loss in fig. 6.34 (parts a and b).

In addition to oxidation in gases, ellipsometry has also been used, to a limited extent, to study oxidation of semiconductors and metals in liquid ambients. An example is Kruger's work [262] on the oxidation of copper single crystal surfaces in water. Figure 6.35 represents the oxidation of several crystallographic planes (all available on one sample) of a Cu single crystal immersed in unstirred (stagnant) water in equilibrium with pure

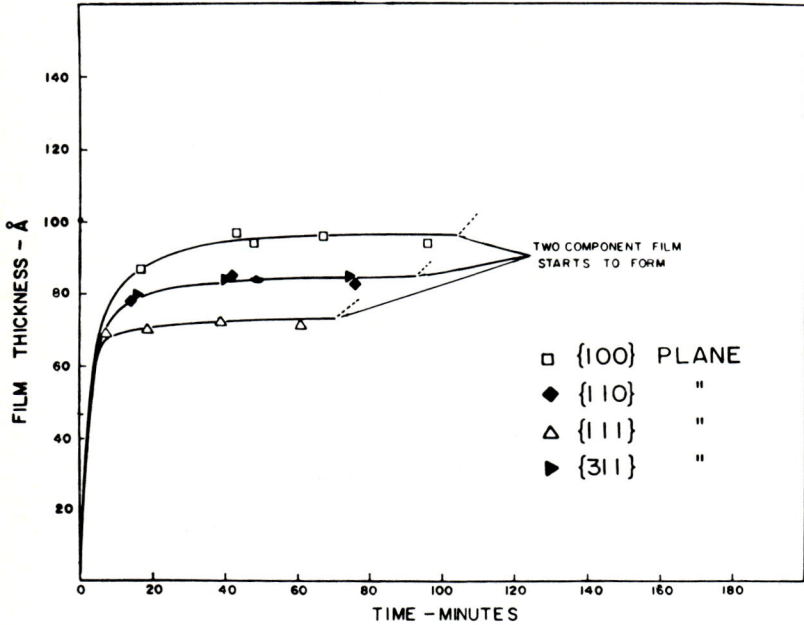

Fig. 6.35. Growth of oxide film on copper in unstirred water in equilibrium with 1 atm of O_2. After ref. 262.

oxygen at 1 atm. The oxidation is initially very rapid (in fact, too rapid to be followed by a manual null ellipsometer) and levels off at a limiting thickness of 65–95 Å after about one hour. The limiting thickness of the Cu_2O film depends on the crystal plane on which it is formed and is least for the (111) surface. Continued exposure of Cu to water beyond 2 hr led to the formation of a second film of CuO. The formation of a *different* film was evidenced by an *increase* in the ellipsometric parameter Δ, instead of a continued decrease as would be expected if the oxide were the same. This is shown in fig. 6.36. The film was most probably CuO, as was indicated by independent X-ray analysis. Kruger observed that stirring only increased the time at which CuO started to form. Similar results were observed when oxygen-helium mixtures with different oxygen concentrations were used instead of oxygen alone.

Ellipsometry is obviously equally applicable to monitoring the growth of films other than oxides (*e.g.*, nitrides or sulfides) on semiconductor and metal surfaces in different ambients. For example, fig. 6.37 shows the growth of silver sulfide tarnish films on silver in room air at atmospheric pressure as measured by Burge et al. [263]. The growth rate of initially 0.6 Å/hr dropped

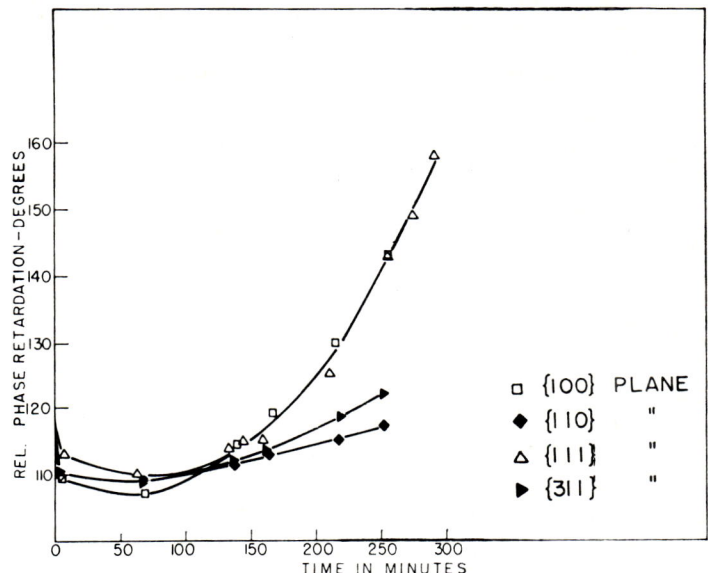

Fig. 6.36. Change in relative phase retardation for film growing on copper in unstirred water in equilibrium with 1 atm of O_2. After ref. 262.

to about 4 Å/day after a 2 Å thick film had formed. The arrows in fig. 6.37 indicate the beginning of a daytime period. The nighttime growth rate was somedays one-fourth the daytime rate, indicating a possible connection to

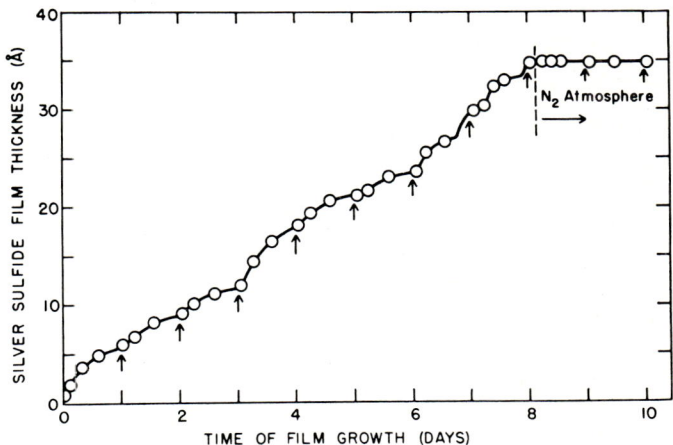

Fig. 6.37. Growth of silver sulfide film on silver in room air at atmospheric pressure. The vertical arrows indicate the beginning of a daytime period. After ref. 263.

increased availability of sulphur-containing compounds during daytime. When the samples were contained in a nitrogen atmosphere, the growth of the silver-sulfide film stopped completely, as shown in fig. 6.37. In this investigation, independent measurement of film thickness using transmission electron microscopy correlated well with values obtained by ellipsometry.

6.5. Electrochemistry

As in the adsorption and oxidation studies discussed in §6.3 and §6.4, ellipsometry has also been a valuable probe of electrochemical electrode processes. The principal advantages of ellipsometry in electrochemistry are: (1) that the electrode/electrolyte interface is examined *in situ*, (2) that it is essentially non-perturbing (in the absence of photochemical reactions), and (3) that it can be readily integrated and carried out simultaneously with other conventional electrochemical measurements (*e.g.*, of voltage, current or capacitance). In this section we provide illustrative examples of a range of electrochemical processes that have been investigated by ellipsometry such as ionic adsorption, anodic oxidation, passivation and corrosion, and electropolishing.

Paik *et al.* [264] studied the adsorption of ions from solutions of sodium salts of Cl^-, Br^-, ClO_4^-, SO_4^-, and OH^- onto electrodes of Au, Ag, Rh and Ni. Figure 6.38 shows the observed change of the relative phase shift, $\delta\Delta_{obs}$,

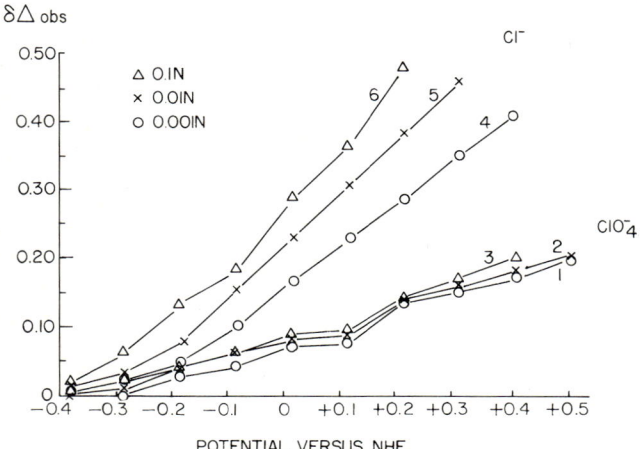

Fig. 6.38. Shows the observed change of the relative phase shift, $\delta\Delta_{obs}$, resulting from the adsorption of ClO_4^- and Cl^- from solutions of different concentration onto an Ag electrode at different potentials. Plotted from table 2 of ref. 264.

that resulted from the adsorption of Cl⁻ and ClO₄⁻ from solutions of different concentration onto an Ag electrode at different potentials. Off-null dynamic photometric ellipsometry was used at $\lambda = 5500$ Å and 64.20° angle of incidence. $\delta\Delta_{obs}$ was attributed to two sources: (1) the adsorbed layer of ions, and (2) the change of the optical properties of the electrode induced by the electric field at the electrode/electrolyte interface. (It is interesting to observe the similarity of this interpretation to that used by Meyer and Bootsma in analyzing the ellipsometric data for gaseous chemisorption on Si surface; see §6.3.) Using a suitable model for the optical properties of metals, Paik et al. found that the change of Δ due to electromodulation, $\delta\Delta_{mod}$, was linearly proportional to the surface charge density on the electrode (in the range from -20 to $100\ \mu\text{C/cm}^2$). By subtraction of a calculated $\delta\Delta_{mod}$ from $\delta\Delta_{obs}$, they determined the change of Δ due to adsorption alone, $\delta\Delta_{ads}$. From $\delta\Delta_{ads}$, and an optical model for the adsorbed layer that used a modified Lorentz–Lorenz equation, the amount of adsorbed ions was calculated in units of charge, q. The results thus obtained are shown in fig. 6.39 for the data given in fig. 6.38.

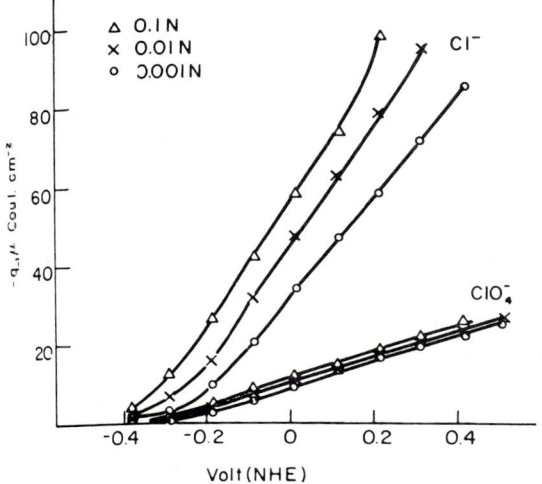

Fig. 6.39. Degree of adsorption from solution (expressed as surface charge density) of Cl⁻ and ClO₄⁻ ions on an Ag electrode of varying potential; calculated from the data shown in fig. 6.38. After ref. 264.

Young and Zobel [265] studied the anodic oxidation of tantalum, niobium and silicon at constant potential by *in situ* ellipsometry. Their specific aim was to examine the law that governs high-field ionic conduction in anodic oxide films under steady state conditions. Figure 6.40 shows their results for

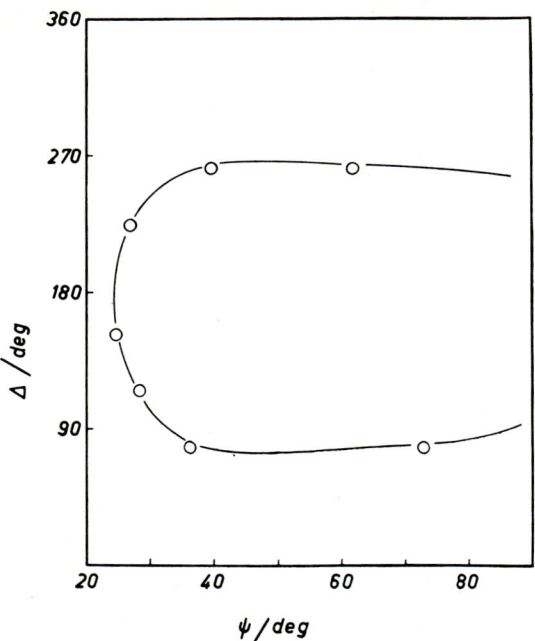

Fig. 6.40. Δ vs ψ during the anodic oxidation of niobium in 0.2N sulfuric acid solution at constant potential measured by *in situ* ellipsometry. After ref. 265.

Δ vs ψ during the course of anodic oxidation of niobium in 0.2N sulfuric acid solution at constant potential (the formation voltage is 60V) and 25°C temperature. Null ellipsometry was used at $\lambda = 5461$ Å and 67.5° angle of incidence. The continuous curve in fig. 6.40 is calculated using 1.334, 2.37, and $3.6 - j3.6$ as the refractive indices of the ambient, film, and substrate, respectively. From the film thickness L (measured by ellipsometry) as a function of time, the derivative $\partial L/\partial t$ was numerically evaluated. The ionic current density I was subsequently obtained from $\partial L/\partial t$ using Faraday's Law. Figure 6.41 plots $\log_{10} I$ for measurements at two temperatures against the electric field strength E, where $E = V/L$, and V is the potential of the niobium electrode less the reversible potential for oxide formation. The experimental points in fig. 6.41 were fitted by continuous curves calculated from a relationship between I and E of the form $I = I_0 \exp[-W(E)/kT]$, where the activation energy $W(E)$ is a *quadratic* function of E, instead of the linear function expected from classical theory [266]. (k is Boltzmann's constant and T is the absolute temperature, as usual; and I_0 is the value of I when $W = 0$.) Young and Zobel advanced models to explain this deviation of

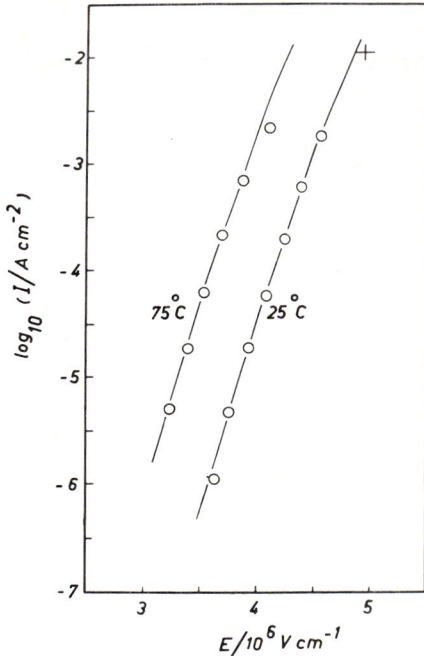

Fig. 6.41. The logarithm of the ionic current density, $\log_{10} I$, vs electric field strength, E, determined from ellipsometric data obtained during the anodic oxidation of niobium in 0.2N sulfuric acid solution at 25°C and 75°C. After ref. 265.

the behavior of high-field ionic conduction from the prediction of classical theory.

The oxidation of platinum anodes in 0.1N sulfuric acid solution was investigated as a function of potential by Reddy *et al.* [267] using *in situ* ellipsometry. By potentiostating the electrode at the desired potential, and allowing time for steady state conditions to be reached, ψ and Δ were measured by null ellipsometry as a function of potential V; the results are shown in fig. 6.42. From fig. 6.42 it is seen that the anodic oxidation is initiated only when the potential exceeds about 0.95 V, as is evidenced by the distinct break of the Δ, ψ-vs-V curves at this potential. The data beyond the break in fig. 6.42 could be fitted to a model of a growing oxide film of constant complex refractive index $2.625 - j1.5$ (at $\lambda = 5461$ Å) on Pt. The film thickness calculated from such a model is shown in fig. 6.43 as a function of potential. Reddy *et al.* suggested a mechanism for the anodic oxidation of Pt to explain the linear relation between film thickness and the applied potential minus the formation potential (0.95 V) which is seen in fig. 6.43.

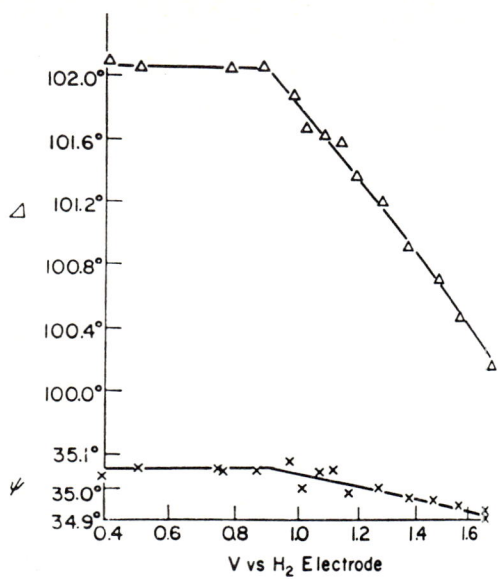

Fig. 6.42. Steady-state values of Δ and ψ as a function of potential. Platinum in 0.1N H_2SO_4. After ref. 267.

Recent experiments by Horkans *et al.* [268] on a Pt-in-1N $HClO_4$ system gave results for film thickness vs potential that are similar to those of fig. 6.43. The latter group used periodic potential scanning and an off-null dynamic photometric ellipsometer to obtain the optical properties (n and k) of the oxide film as functions of wavelength, in addition to film thickness.

An example of a more complex situation that has been successfully examined by ellipsometry is the study by Dell'Oca [269] of the anodic oxidation of silicon nitride films (Si_3N_4). Figure 6.44 shows Dell'Oca's results for the anodization of a Si_3N_4 film 916 Å-thick on a Si substrate in a solution of 0.04 M KNO_3 in ethelene glycol at a constant current density of 20 mA/cm². The wavelength of measurements was 6328 Å and the angle of incidence was 70°. The crosses (×) are experimental points and the dashed curve is computed using a SiO_2–Si_3N_4 two-film model, with parameters as indicated. In particular, it was found that the experimental points are best fitted by the computed curve if the growth of the oxide film at the expense of the nitride film is assumed to take place at a conversion ratio (rate) CR of 1.81 Å SiO_2 for each 1 Å Si_3N_4. Figure 6.45 illustrates the kinetics of oxide-film growth and nitride-film loss during anodization at 20 mA/cm². Figure 6.46 gives the calculated oxide growth-rate dx_1/dt (obtained from data such as that of fig. 6.45) as a function of the electric field E in the oxide

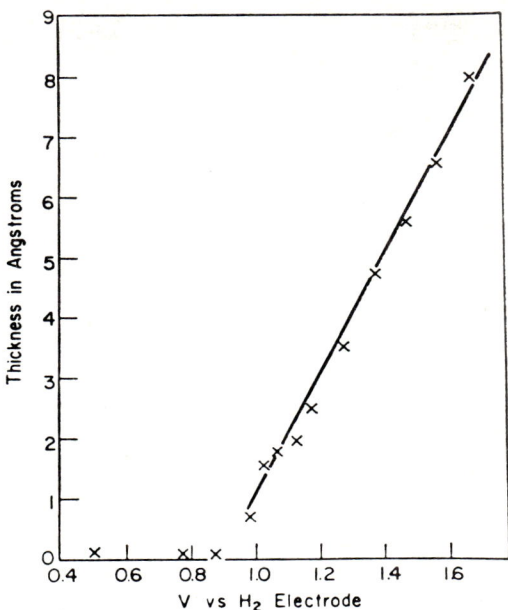

Fig. 6.43. Dependence of oxide film thickness on potential. Platinum in 0.1N H_2SO_4. After ref. 267.

(evaluated from the voltage drop across that film) from measurements taken at several constant current densities. A linear relation between dx_1/dt (on a logarithmic scale) and E provides an adequate fit, so that the classical theory of high-field ionic conductivity [266] is reasonably applicable in this case. Dell'Oca noted that the growth characteristics of anodic SiO_2 on Si_3N_4 films are similar to those of anodic SiO_2 on Si [265].

Passivation (or passivity) generally signifies the formation of a protective film on the surface of a metal that maintains its stability and prevents its corrosion in an aggressive environment [148, 270]. Ellipsometry has contributed significant insight into the mechanism of passivation of several metals in acid solutions by providing new non-electrochemical evidence as to the nature of the film that forms on the metal surface during passivation, and also by resolving the growth kinetics of film thickness. Figure 6.47 shows results by Bockris et al. [271] of their ellipsometric investigation of the mechanism of passivity of nickel in 0.5 M SO_4^{--} acid solution of pH = 3.15. The ellipsometric parameters Δ and ψ, measured under conditions of potentiostatic control, are plotted as functions of potential V during passivation. Using this data, Bockris et al. calculated the thickness L_F,

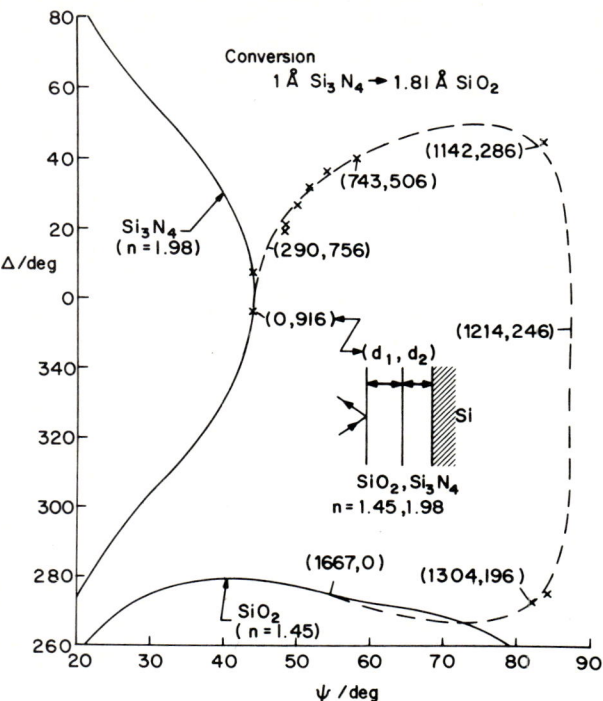

Fig. 6.44. Ellipsometric results for the anodization of a 916 Å Si_3N_4 film at 20 mA/cm². Dashed line is computed from the model and values given in the figure. Curves for Si_3N_4 and SiO_2 are also shown (solid lines). After ref. 269.

refractive index n_F, and absorption coefficient $\kappa_F(= k_F/n_F)$ of the passivating film that forms on the Ni surface; these film parameters together with the steady state current density i appear in fig. 6.48 as functions of potential V. A film starts to form at the first inflection of $i - V$ curve, increases rapidly in thickness as i reaches its maximum value, then stays essentially constant afterwards while i decreases. However, concomitant with the rapid decrease of i after it reaches its maximum is a rapid increase in κ_F, hence in conductivity. Bockris et al. suggested that the mechanism of passivity is associated with this increase of conductivity because it diminishes the field that assists anodic dissolution and transport through the film.

Kruger and Calvert [272] studied the passivation of iron in a sodium borate-boric acid solution (which was chosen to minimize oxide-film dissolution), with the iron electrode held at a constant voltage required for passive film formation. Figure 6.49 gives the kinetics of film growth for two values of

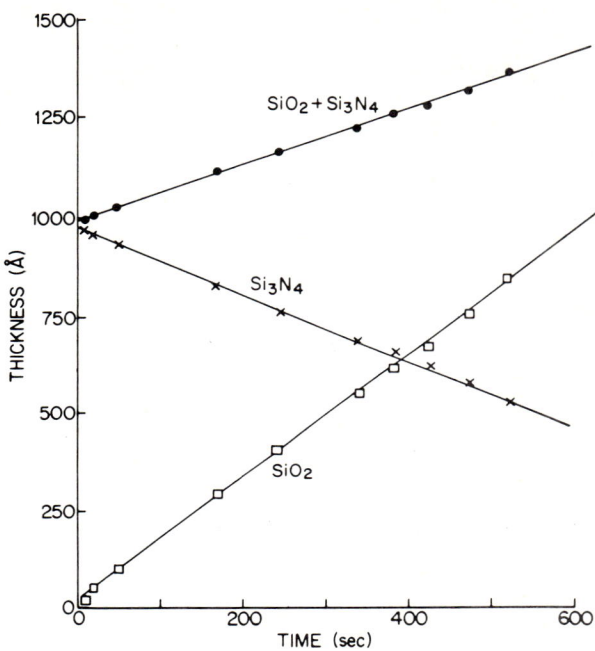

Fig. 6.45. Typical thickness development with time for the anodization of Si_3N_4 at 20 mA/cm². After ref. 269.

pH (7.4 and 9.1) over a period of time (shown on a logarithmic scale) that ranged from 0.01 to 10^4 sec. Three separate stages of film formation can be distinguished in the curves of fig. 6.49. In stage 1, the film thickness is linearly proportional to the square root of time, as is shown on an expanded scale in fig. 6.50. Because the rate of film growth (slope) is dependent on the pH of the solution (and is higher for pH = 9.1 than for pH = 7.4), it was inferred that during this stage film growth was controlled by the solution and not by the film itself. Stage 2 shows a deviation from a linear relation between thickness and the square root of time and the growth rate seems to be controlled by both the solution and film. Stage 3 is described by a steady growth rate that is virtually independent of the solution.

Kruger [270, 273] studied the dissolution of a passive film on iron by changing the potential to a value where passivity breaks down. Figure 6.51 shows the variation of Δ when the potential of an iron surface covered by a passive film is lowered to a value below the passivation potential. Notice that Δ first *increases* while the current is negative, which is indicative of film dissolution. This continues until Δ reaches its maximum value at which point the entire film has presumably dissolved. The subsequent *decrease* in

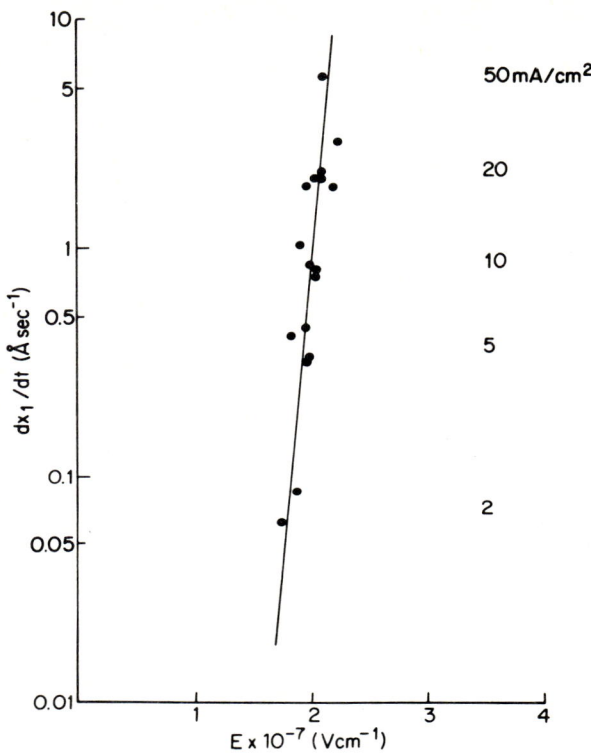

Fig. 6.46. Silicon oxide growth rate dx_1/dt (Å sec^{-1}) as a function of the electric field E (10^{-7} V cm^{-1}) in the film. The formation current density is indicated. After ref. 269.

Δ was attributed by Kruger to surface roughening. Dissolution of passive films usually signals the initiation of pitting and the onset of corrosion.

McBee and Kruger [270, 274] observed subtle effects of chloride ions on passive films on iron prior to breakdown when a measuring wavelength $\lambda = 4100$ Å was used. Figure 6.52 shows their results for the variation of ψ, Δ and current i (at constant potential) as Cl$^-$ ions were added and removed periodically. Continuous changes of ψ and Δ do occur prior to breakdown and, significantly, the films recover their optical properties upon the removal of the Cl$^-$ ions from solution.

Ellipsometry has also been used to study surface films on electrodes during electropolishing. In particular, Novak *et al.* [275] examined the electropolishing of copper electrodes in a 65% aqueous solution of phosphoric acid by *in situ* ellipsometry. Figure 6.53 shows their results for current I (top), the change of Δ (middle), and the change of ψ (bottom) as a

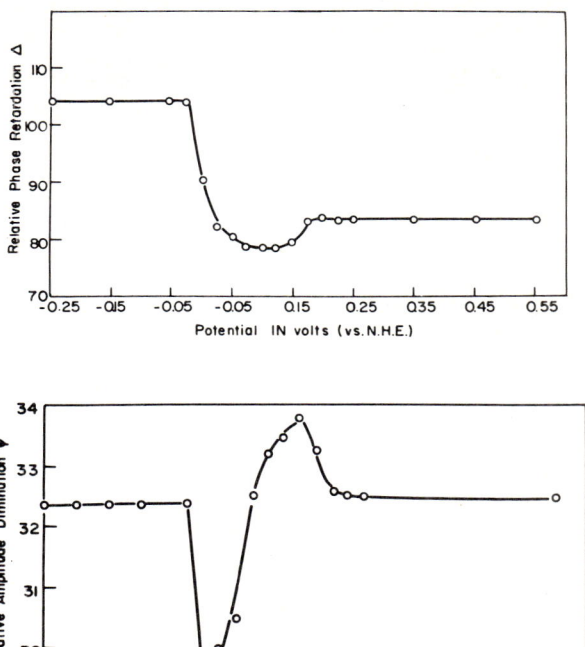

Fig. 6.47. Variation of the relative phase retardation Δ (top) and the arc tangent of the relative amplitude attenuation ψ (bottom) as a function of potential during the steady-state anodic passivation of nickel in 0.5M sulfuric acid solution (pH = 3.15). After ref. 271.

function of V under steady state conditions. Curves c and e were obtained by displacing the acid by glycerol while maintaining the same potential applied. For potentials up to 540 mV (vs NHE), there is a rise of I but Δ and ψ remain constant. This range represents a region of surface etching but not electropolishing. The jumps in Δ and ψ, as the potential is increased by 50 mV, from 540 to 590 mV, indicate the formation of surface films required for electropolishing. The steady behaviour of all parameters afterwards is consistent with steady state conditions of electropolishing. The large drop in Δ (13–17°) as the electrolyte was replaced by glycerol resulted from the displacement of a loose nonadherent viscous layer. The remaining change of Δ (middle, curve c) is due to a solid adherent film. The thickness and refractive index of the adherent film were computed as a function of potential from the glycerol data, the results appear in fig. 6.54. Notice that

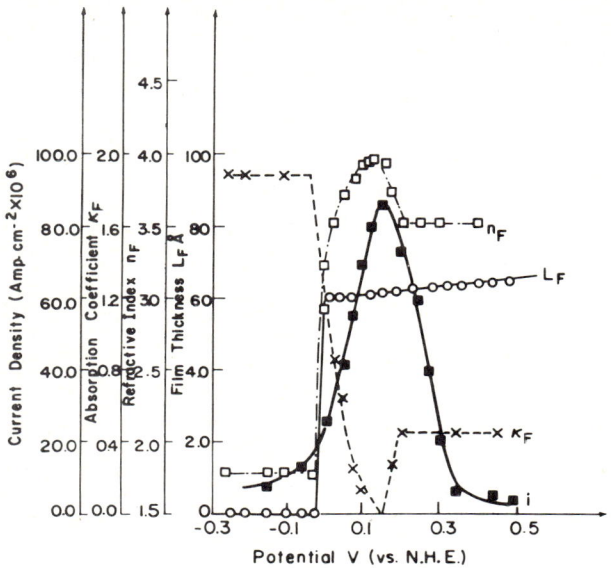

Fig. 6.48. Variation of the film optical properties n_F, κ_F, and film thickness L_F with potential during the passivation of nickel in 0.5M sulfuric acid solution at pH = 3.15. The current density i is also indicated. After ref. 271.

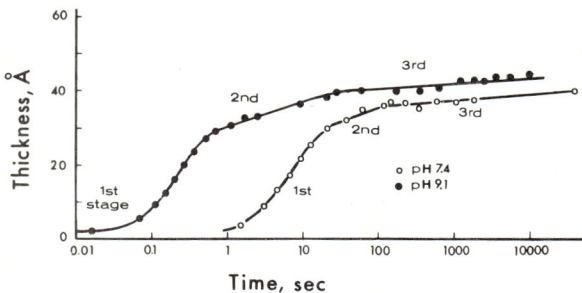

Fig. 6.49. The kinetics of oxide film growth on iron during passivation in a sodium borate-boric acid solution at two values of pH. After ref. 272.

the thickness of the film reaches a stable value of ~60 Å and the refractive index shows some fluctuation between ~2–3. The thickness of the viscous displaceable film, obtained from both the acid and glycerol data, was estimated to be in the range 2000–3500 Å with a refractive index close to that of the bulk electrolyte.

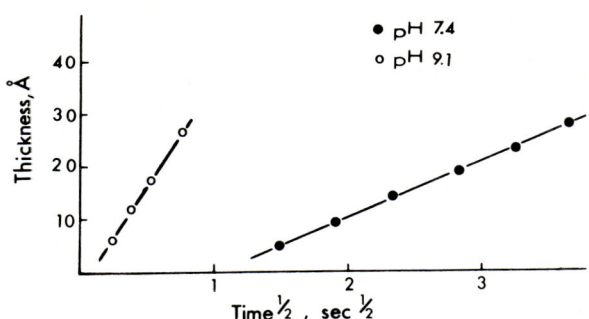

Fig. 6.50. The very initial stage of film growth for the curves of fig. 6.49. After ref. 272.

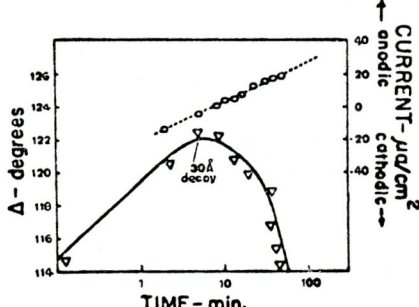

Fig. 6.51. The change in the relative phase retardation Δ when a film formed potentiostatically at +0.8 V (S.C.E.) on iron in a sodium borate-boric acid solution (pH 8.1) is dissolved by changing the potential to a value where passivity breaks down. After ref. 270.

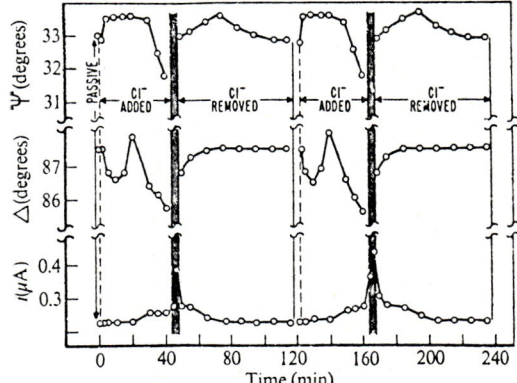

Fig. 6.52. Changes occurring in Δ and ψ for a passive film on iron in the presence of Cl⁻ prior to breakdown. These changes were observed at a wavelength of 4100 Å. As shown, the removal of Cl⁻ from solution causes the parameters to go back to their original values. After ref. 270.

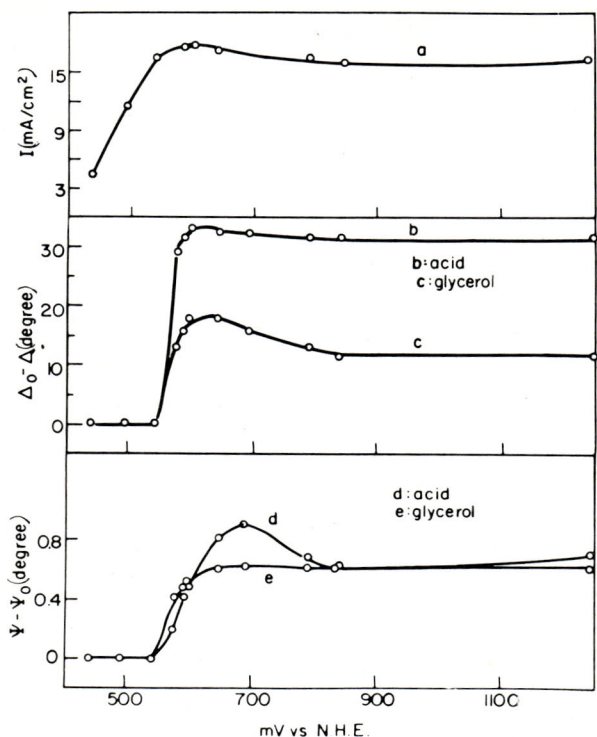

Fig. 6.53. *Top*-steady-state current-potential relation during the electropolishing of copper, a; *middle*-decrease of Δ caused by the total film, b, and by the adherent film, c; *bottom*-increase of ψ caused by the total film, d, and by the adherent film, e. After ref. 275.

6.6. Applications of ellipsometry in biology and medicine

There have been a number of significant applications of ellipsometry to biology and medicine. They include the pioneering studies by Rothen [276] of immunological antigen-antibody reactions in thin films (the word "ellipsometer" was coined by Rothen [65]); by Vroman and associates [277] of the adsorption of blood (plasma) proteins on foreign surfaces (a subject of considerable importance to the understanding of the intrinsic mechanism of blood coagulation); and, more recently, by Poste and associates [278] of monitoring the synthesis of cell-surface materials. New applications of ellipsometry to biology and medicine are anticipated in view of the fact that many *in vivo* biological reactions take place at interfaces. In the following we will summarize some of the results obtained in the above-mentioned applications.

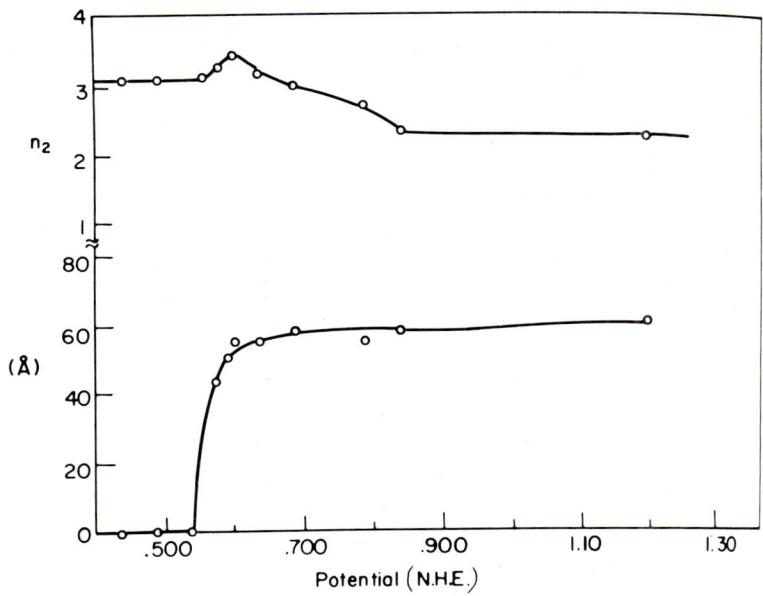

Fig. 6.54. Refractive index (n_2) and thickness (Å) of the adherent film as a function of potential computed from the data of fig. 6.53. After ref. 275.

6.6.1. Interaction of blood with foreign surfaces—Blood coagulation

Whenever blood comes in contact with a foreign surface (*i.e.*, any surface other than the endothelium lining of blood vessels), the complex process of intrinsic blood coagulation [279] is initiated. This coagulation or clotting leads to the transformation of the soluble blood protein fibrinogen into polymerized insoluble filaments of fibrin. This transformation proceeds via a complicated chain reaction that involves many other blood proteins, called the blood-clotting factors. The interaction of blood with a foreign surface is evidently of critical importance in the design of prosthetic implants (such as artificial heart valves), in the realization of artificial-organ systems (*e.g.*, heart or kidney), and in blood banking. Vroman and his associates [277] made extensive use of ellipsometry in an attempt to reveal some of the kinetics of protein adsorption to the foreign surface. Below is a summary of some of their findings.

(1) Rapid adsorption of protein films, readily detectable by ellipsometry, takes place within a few seconds when either hydrophobic or hydrophilic surfaces are exposed to intact plasma or to plasma deficient in one or more of the clotting factors.

(2) On continued exposure, the adsorption from normal intact (whole)

plasma is followed by partial desorption which is sensed as a decrease of film thickness. This partial desorption does not occur when the plasma is deficient in the Hageman factor (factor XII) and reappears when this factor is added.

(3) Fibrinogen is the major protein component of the film adsorbed when a solid surface contacts blood plasma. This is verified by the nearly identical adsorption curves of films formed onto slides immersed in solutions of fibrinogen alone and those exposed to whole plasma. Furthermore, fibrinogen films and films deposited from intact plasma behave similarly when exposed to either intact or factor XII-deficient plasma, showing partial desorption and no desorption in these two cases, respectively. Additional evidence of the abundance of fibrinogen in films adsorbed from whole plasma is found in the ability of such films to specifically bind animal antiserum prepared against human fibrinogen.

Whereas these experiments by Vroman *et al.* [277] were made using (cell-free) plasma, Baier and Dutton [280] confirmed the third observation in the above list by measurements on germanium substrates exposed to fresh-flowing blood from the jugular veins of anaesthesized dogs. By ellipsometry, they found that a 50 Å film coats the surface in 5 seconds. This thickness agrees with values obtained from parallel measurements using electron microscopy. Furthermore, infrared total-internal-reflection spectra of films deposited on germanium prisms by exposure to whole blood (as explained above) were similar to spectra of fibrinogen films deposited on the internal reflection elements from solutions containing fibrinogen alone. This proved that fibrinogen is the major constituent of proteinaceous films deposited on the surface in contact with whole blood.

In addition to the fact that the fibrinogen→fibrin transformation takes place on the surface, the adsorption of fibrinogen and other proteins modifies the surface properties in such a way as to enhance the attachment of platelets, the major cellular component of the blood clot. Generally, the growth of cells on a solid surface (as is common in the practice of tissue culture [281]) seems to be significantly influenced by the presence of pre-adsorbed protein films. This aspect of tissue culture deserves further study in which ellipsometry can play an important role [278].

6.6.2. *Antigen-antibody immunological reactions in thin films*

An antigen is usually defined as any foreign molecular species which when introduced into a living host stimulates the production of corresponding antibodies that specifically interact (complex) with the antigen [282]. *In vivo*, these immunological reactions (as they are called) take place either at (cell) surfaces or in bulk (in blood). Although antigens may or may not be proteins, antibodies generally are. Most laboratory techniques of immunological reactions for serologic diagnosis are based on *in-volume* reactions of antigens and antibodies. However, the possibility of observing these reactions on surfaces (in thin films) had been recognized in the late 1930's. Perhaps the first of such investigations were those due to Langmuir and Schaefer [283] who found a method to condition slides coated by transferred built-up films of barium stearate [193, 194] so that they adsorb different biological materials from solution. Interestingly enough, films deposited (adsorbed) on top of the barium stearate multilayer were detected by an ingenious nonellipsometric optical technique (Blodgett and Langmuir [194]) as a shift of the angle of incidence at which incident s-polarized light experiences minimum reflectance. A less precise version of this optical technique (based only on color comparison) was used by Shaffer and Dingle [284] to investigate antigen-antibody reactions between adsorbed monolayers of egg albumin and anti-egg albumin and also between pneumococcus polysaccharide and anti-polysaccharide specific antibody. The latter system was subsequently studied by Porter and Pappenheimer [285] using the more precise technique. Further study of immunological reactions in thin films by the optical technique of Blodgett and Langmuir were carried out by Rothen and Landsteiner [287] (using egg albumin and various specific antisera) and by Bateman *et al.* [286] (using an antigen prepared from streptococcus pyogenes with homologous antistreptococcal serum).

Although the potential of ellipsometry as a sensitive optical tool for the detection of very thin films on surfaces was recognized by Drude [63] as early as 1889, and was extensively used by Tronstad and Winterbottom [71] for the study of the oxidation of metals in the 1930's, its use for the study of antigen-antibody reactions was initiated by Rothen and Landsteiner [287] only in 1942. In addition to its higher sensitivity, ellipsometry has the important advantage over the earlier optical method of Blodgett and Langmuir [194] of not requiring the pre-deposition of a stack of many layers (up to 50 or more) of barium stearate on the metal slides before they are used for the antigen-antibody reaction. A metalized slide coated by a film of a polysaccharide from type III or type VIII pneumococcus, as thin as 5 Å, was found by Rothen [276] to specifically adsorb relatively thick (up to 500 Å) films from antiserum, as measured by ellipsometry. Bovine-serum albumin (BSA) deposited films were found to specifically react with antisera leading to the growth of films whose thickness was anomalously proportional to the thickness of the BSA film (the BSA film was a transferred multilayer from a Langmuir trough).

More recent work on antigen-antibody reactions by ellipsometry was done by Giaever [288]. A Ni-coated slide was immersed in a tank filled by a saline solution. When BSA-containing solution was dripped into the tank, BSA continued to adsorb onto the slide until a limiting thickness of 30 Å, corresponding to a complete monomolecular layer of BSA. When rabbit antiserum against BSA was subsequently dripped into the tank, another layer grew on top of the BSA layer to 50 Å thickness, fitting the size of the anti-BSA molecule. Additional work by ellipsometry and other optical techniques is being tried by Giaever to assay the amount of carcinoembryonic antigen (CEA) in the blood of patients with gastrointestinal cancer [288].

Two points need to be made in connection with experiments on antigen-antibody reactions in thin films. First, while the forces at a liquid-solid interface may not be strong enough to alter the structure of the adsorbed protein molecules, those at a liquid-air interface usually cause decoiling (denaturing) of the protein's polypeptide chain. Thus, if a protein film is transferred to an ellipsometer slide from the water surface of the Langmuir trough, it could not be assumed that such film of denatured protein has all the biological activity of a film of intact molecules. Of course if the thickness of a transferred film measured by ellipsometry is much smaller (~ 5 Å) than any of the molecular dimensions of the native protein molecule, denaturation should be strongly suspected; the small thickness, in this case, corresponds to the thickness of the polypeptide chain only [276].

The second point concerns the fact that not all of the second adsorption, in the two-step immunological reaction experiment, can be considered specific (when whole antisera are used, as is usually the case). Care must therefore be taken to correct for the nonspecific component of the total adsorption. This can be done by control experiments in which a slide is coated with a heterologous antigen (instead of a homologous one) before dipping it into the same antiserum solution. The thickness of any film adsorbed in this control experiment is an indicator of the extent of nonspecific adsorption and should be subtracted, as a correction, from the thickness obtained in an immunological-reaction experiment by ellipsometry.

6.6.3. The immunoelectroadsorption test

This is a test developed by Rothen and Mathot [289] in which the adsorption of one or two reactants in an immunological reaction is assisted by the application of an electric field. In this case, the metalized ellipsometer slide acts as one electrode and another counter electrode is mounted in the test cell. Because the charge carried by a particular protein depends on the pH of the solution containing that protein as well as the protein's isoelectric point (*i.e.*, the pH value at which the net charge on the protein molecule is zero), the polarity of the ellipsometer slide can be made either positive or negative, dependent on the antigen or antibody under consideration. Because of the directed movement of the desired protein towards the slide under the action of the applied electric field, greater dilutions of these proteins in solution can be used in the immunological reaction. It is reported that the detection sensitivity of the immunoelectroadsorption test can be six orders of magnitude higher than would be realizable with no field applied. The required current is usually a fraction of a milliampere and needs to be applied only for one or two minutes, during which a monolayer of antigen from a dilute solution ($< 10^{-7}$ g/ml) can be adsorbed. By use of nickel slides "activated" by a magnetic field, an immunological reaction between polysaccharide from type III or VIII pneumococcus and the corresponding immune anti-sera has been detected by Rothen at concentration levels as low as 10^{-14} g/ml of polysaccharide.

The immunoelectroadsorption method has been used to identify eight arthropod-borne viruses, to measure levels of growth hormones in human and bovine blood, and for the serodiagnosis of *Schistosoma mansoni* in man [276]. In spite of its reported interesting applications, the immunoelectroadsorption test has been criticized because of several complicating surface reactions that can occur during the electroadsorption process [278].

6.6.4. Measurement of cell-surface (coat) materials

One of the important techniques of modern biological research is that of *in-vitro* cell cultures [281], where a population of cells (usually of the same type) is sustained in a controlled nutrient environment (solution) so that many cellular processes can be observed. It is a common observation that when initially separated (dispersed) cells are placed into the culture vessel, they tend to either attach to each other in suspension, or adhere to the surfaces of the culture dish. Furthermore, normal cells that adhere to the solid surface of the culture dish (as a substrate) tend to grow in the form of a confluent monolayer one-cell thick. Such a monolayer population of cells attached to an inert solid substrate provides a suitable model for many studies.

Rosenberg [290] was first to observe that cells mechanically sheared off from a solid substrate on which they are grown leave behind a thin film coating the substrate. Such a film was not observable by light microscopy, but was readily detectable by ellipsometry. Rosenberg found that when several cells (embryonic chick skin and retina, human epidermoid carcinoma and human conjunctiva) cultured for several hours on metalized slides are removed by a jet of isotonic saline, an adsorbed layer about 40 Å thick was found to coat the slide after careful washing of the slide with distilled water and drying.

The exact nature of the thin molecular carpet remaining after cells are detached from a solid substrate was not identified by Rosenberg who used the term "microexudate" to describe it.

More recently, Poste and his associates [278] published an extensive study of films left by cells on slides and advanced considerable evidence to prove that such microexudate films previously observed by Rosenberg can, in fact, be identified with the glycoprotein coat material at the surface of the cell. In addition to measurements on a wide range of different cells, Poste and Poste and Greenham examined the changes that occur in this cell-coat material left on the ellipsometer slides as a result of viral infection, cell fusion and the release of lysosomal enzymes onto the cell surface.

Although the exact picture of the molecular composition and structure of the membranes of mammalian cells is far from complete, several useful models have been proposed based on available data. Perhaps the most important model at the present time is that of Nicholson and Singer [291]. As shown in fig. 6.55, the cell membrane, according to this model, consists of a

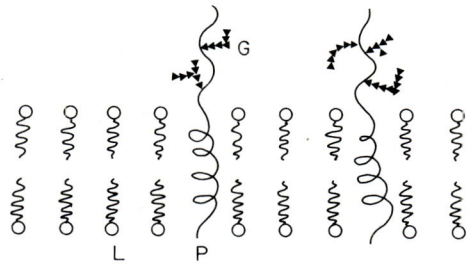

Fig. 6.55. The cell membrane according to the model of Nicholson and Singer [291] consists of a lipid bilayer with protein molecules incorporated in it. L represents a lipid molecule with its polar-head group (represented by a circle) and its hydrocarbon chain (by a curly tail). P is the polypeptide chain of a protein and G glycose residues. The complex of P and G is a glycoprotein.

lipid bilayer with protein molecules incorporated in it. The portion of the polypeptide chain of the protein molecule that is imbedded inside the bilayer is mostly helical in configuration, as suggested by optical-rotatory-power (ORP) and circular-dichroism (CD) spectral data of membrane segments [291]. The portion of the polypeptide chain that lies outside the membrane (*i.e.*, on the outer surface of the cell) has complexed with it several glycose residues. (The entire polypeptide chain together with the glycose residues is often called a "glycoprotein".)

The above model of the cell membrane shows that a carbohydrate-rich layer exists external to the permeability-controlling lipid-protein bilayer.

The word "coat" has often been used to describe this external carbohydrate layer. According to Poste, it is this coat (or most of it) which is left behind when cells grown on a solid substrate are detached by a shearing force. This conclusion was supported by detailed ellipsometric studies on many cells which showed that the properties of the Rosenberg's "microexudate" molecular carpet are indeed identical to those of the cell-coat material measured by other techniques.

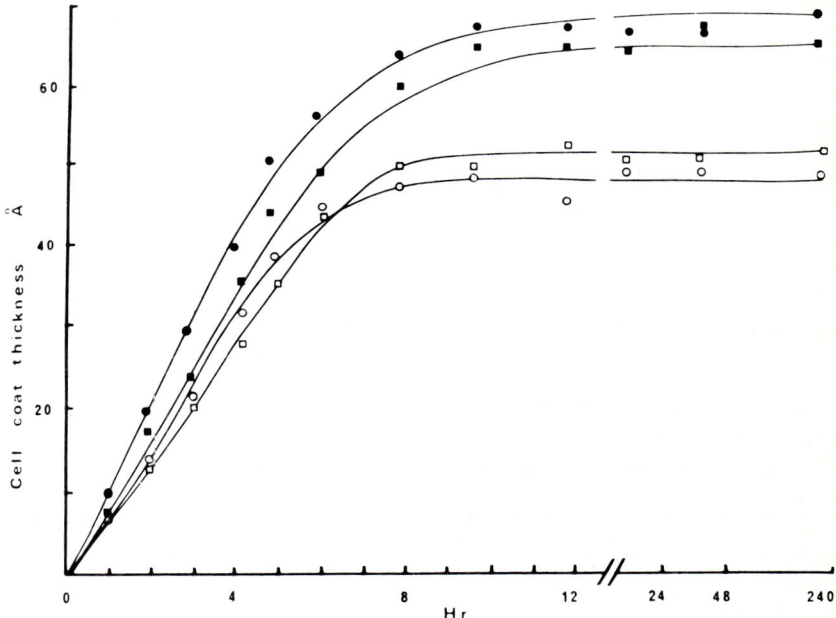

Fig. 6.56. The deposition of cell coat materials on ellipsometer slides *in vitro* by dog (DKC) and ferret (FKC) kidney cells as a function of time. ■—■ = primary DKC; □—□ = secondary DKC; ●—● = primary FKC; ○—○ = secondary FKC. After ref. 292.

Figure 6.56 shows the ellipsometrically measured thickness of cell-coat material "deposited" on slides by different types of cells, plotted as a function of the time that the cells have stayed on the slide *in vitro* [292, 278]. The rate of synthesis of cell-coat material, as measured by ellipsometry, was found to agree with parallel measurements obtained by an isotope-incorporation technique. The fact that the rate of synthesis of cell-coat material is significantly reduced at low temperatures, fig. 6.57, and also when metabolic inhibitors are present in the culture medium, fig. 6.58, prove that the "microexudate" is an active metabolic product of the cell [292, 278].

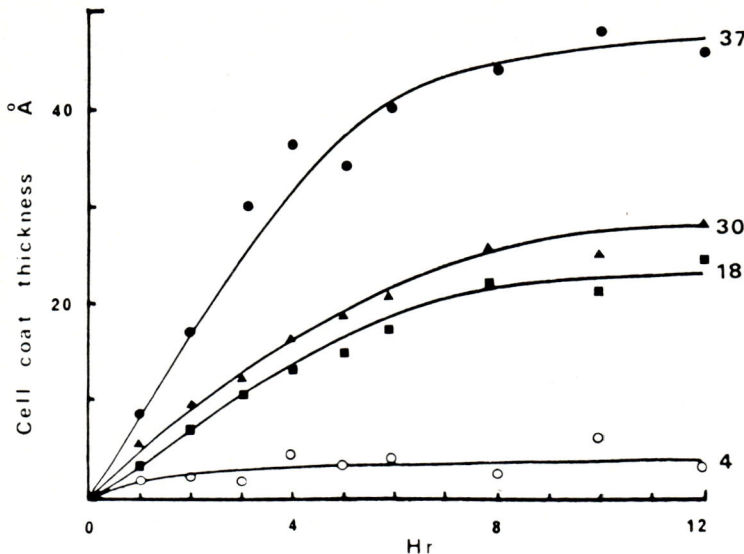

Fig. 6.57. The deposition of cell coat materials on ellipsometer slides *in vitro* by dog kidney cells cultivated at 37°C, 30°C, 18°C and 4°C. After ref. 292.

The ellipsometrically measured thickness was also found to agree reasonably well with values of cell-coat thickness obtained by electron microscopy.

Very interesting results have been published by Poste et al. concerning the changes that occur in the cell-coat thickness (as measured by ellipsometry) upon infection of cells by viruses. Dependent on the type of cell-virus system under consideration, either (a) a decrease or (b) an increase in cell-coat thickness is observed [278].

Figure 6.59 shows the decrease in cell-coat thickness of chick-embryo cells as a function of time, following their infection by Newcastle-disease virus. In this case, a very characteristic "cytopathic effect" takes place, namely that the virally-infected cells fuse together to form multinucleate cells (polykaryocytes). The increase in population of polykaryocytes is shown in fig. 6.59, also as a function of time following viral infection. Figure 6.59 shows strong correlation between the increased production of polykaryocytes and the reduction in cell-coat thickness. Poste related this phenomenon to the proposition that for two cells to fuse, low-radius-of-curvature microvilli (protrusions or bulges) need to be formed on their respective surfaces; which, in turn, require a reduction in cell-coat thickness. Furthermore, Poste found that a critical minimum thickness does exist, approximately equal to 35 Å, below which two cells can fuse, and above which they cannot.

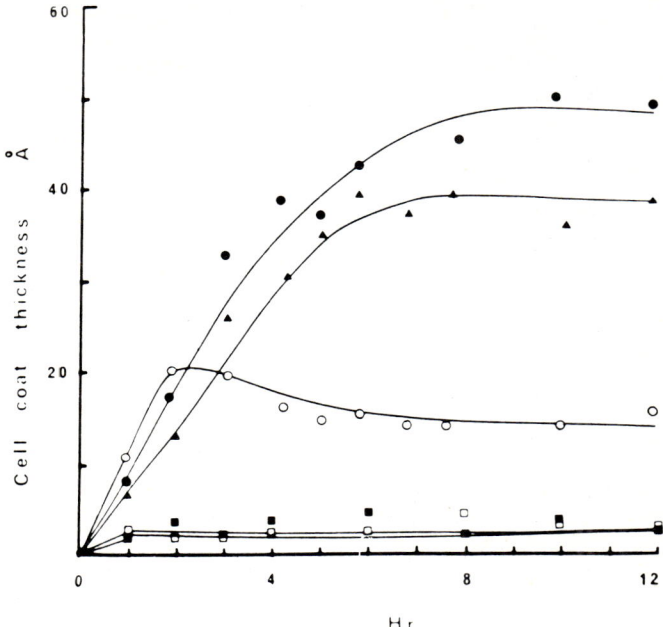

Fig. 6.58. The deposition of cell coat materials on ellipsometer slides by secondary dog kidney cells cultured in the presence of 1 μg/ml actinomycin D, ●—● = control culture without actinomycin D; ■—■ = actinomycin added 2 hr. before subculture; □—□ = actinomycin added at subculture; ○—○ = actinomycin added 2 hr. after subculture; ▲—▲ = actinomycin added 6 hr after subculture. After ref. 292.

Figure 6.60 shows a case when viral infection (by polyoma virus) of cells (mouse embryo fibroblasts) causes an increase of the coat thickness. The cells in this case are said to be "transformed". Similar results were obtained when the cells were chemically (instead of virally) transformed.

6.7. Other applications of ellipsometry

Ellipsometry has been used to study radiation damage in solids. Schroeder and Dieselman [293] showed that the bombardment of vitreous silica by 150 keV argon ions resulted in an increase of the refractive index of silica by 0.02, which was observed by ellipsometry. These authors analyzed their data using an inhomogeneous-film model that accounted for the distribution of implanted argon in silica.

Ibrahim and Bashara [294] studied by ellipsometry the surface damage of silicon caused by low-energy argon ions. Figure 6.61 shows the increase of ψ

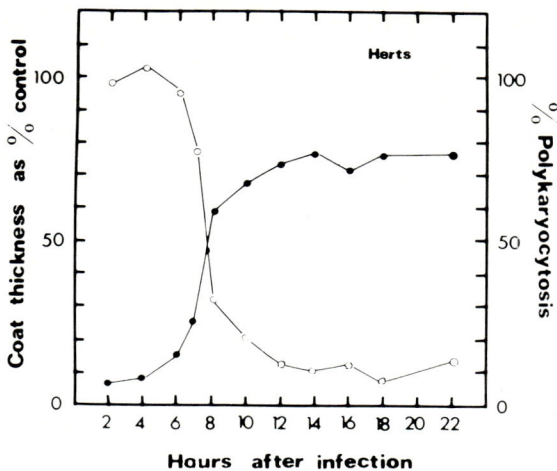

Fig. 6.59. The temporal relationship between changes in coat thickness (O—O) and the development of polykaryocytes (●—●) in chick embryo cells infected with Newcastle disease virus (NDV), strain Herts. Coat thickness in the virus infected cells is expressed as a percentage of the coat thickness in uninfected cells and each point represents the mean from measurements on 50 cells. After ref. 278.

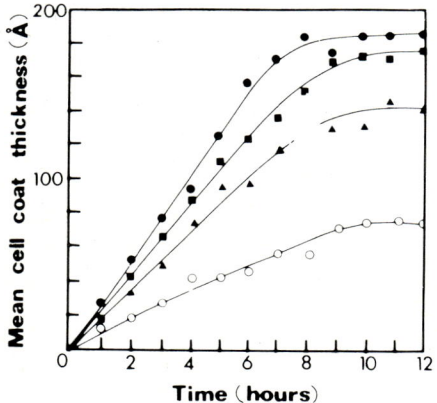

Fig. 6.60. The deposition of cell coat materials on ellipsometer slides by normal mouse embryo fibroblasts (O—O) and cloned derivatives of mouse embryo cells transformed by polyoma virus (●—●; ■—■; ▲—▲). After ref. 278.

(measured at $\phi = 70°$, and $\lambda = 6328$ Å) induced by the bombardment of silicon by argon ions (4×10^{15} ions/cm² or 2μ A/cm²) in the ion-energy range from 150 eV to 400 eV. In fig. 6.61, the increase of ψ is almost proportional

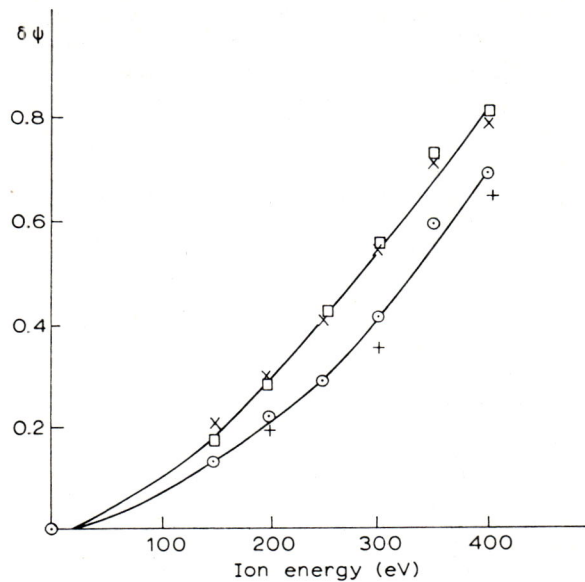

Fig. 6.61. The increase in ψ induced in silicon by ion bombardment at 4×10^{15} ions/cm² ($2\mu A$/cm²). Each point was obtained after 5 min bombardment time at $\lambda = 6328$ Å and 70° angle of incidence. (+)S95(30Ω-cm), (○)S100(8Ω-cm), (X)S118(0.1Ω-cm). (□)S125(0.01Ω-cm). Crystal orientations: (111) for samples S95, S118 and S125; (100) for S100. After ref. 294.

to the ion energy and may be taken as a measure of damage. The change of ψ is higher for the lower-resistivity samples, but is fairly independent of the crystal orientation [(111) or (100)]. Table 6.4 lists the changes of both ψ and Δ resulting from ion damage in two specimens, and also the effect of annealing thereafter. Annealing at 800°C seems to remove the damage, as evidenced by the recovery of ψ and Δ to their initial values before bombardment. Ibrahim and Bashara modeled the damage optically in two ways: in one, the effect of damage was simply assumed to modify the substrate uniformly throughout; in the other, the damage was assumed to be uniformly confined to a damage layer on top of an unperturbed substrate. In both cases, the presence of an oxide film on silicon was accounted for in analyzing the data from multiple-angle-of-incidence (MAI) ellipsometry. Figure 6.62 shows (continuous curves) the refractive index n'_s of the assumed uniformly-damaged substrate and the extinction coefficient k_D, of the damage layer as functions of wavelength between 3000 and 6000 Å. The effect of damage is seen as the difference between the continuous curves and the broken curves that give n and k of Si before bombardment. In addition

Table 6.4. The changes in ψ and Δ caused by ion bombardment at 20°C, a wavelength of 6328 Å and 70° angle-of-incidence; and after temperature annealing [294].

	S100		S125	
	ψ (deg)	Δ (deg)	ψ (deg)	Δ (deg)
Ion energy (eV)				
Zero	10.73	178.17	10.78	178.14
200*	10.93	178.74	11.05	178.94
300*	11.13	179.12	11.34	179.57
400*	11.40	179.28	11.59	179.55
Annealing (°C)				
20†	11.32	178.92	11.43	179.15
500‡	10.93	178.25	10.97	177.80
800‡	10.81	178.18	10.78	177.74

*The specimen was annealed at 800°C for 10 min and allowed to cool before bombardment to determine the ion-induced damage.
†For a period of 15 hr.
‡For 10 min and allowing the specimen to cool.

to these optical parameters, MAI ellipsometry gave an estimate of the thickness of the damage layer; 20 Å at 200 eV and 45 Å at 400 eV.

In the course of the above-mentioned experiments, Ibrahim and Bashara proved, from ellipsometric observations, that simultaneous heating and ion-bombardment is more effective in cleaning silicon surfaces than either heating to elevated temperatures alone or ion-bombardment at room temperature [295]. Figure 6.63 shows their results for the increase of Δ (indicative of oxide-film removal) with cumulative ion-bombardment time (using 400 eV argon ions at $2 \mu A/cm^2$ ion-current density) for silicon samples heated to 800°C. The data was taken at 70° angle of incidence and $\lambda = 6328$ Å. The same workers suggested that ellipsometry can be used to measure the sputtering yield of silicon [296] (subjected to ion bombardment), and also to measure surface temperature [297].

Adams and Bashara [298] applied ellipsometry to the determination of the detailed complex refractive-index profiles of silicon implanted by high-energy (35–70 keV) phosphorous ions (P_{31}^+). Figures 6.64 and 6.65 show the refractive index (n) and extinction coefficient (k) profiles, respectively, for P_{31}^+ ion-implanted Si (dose = 2×10^{14} ions/cm^2), at 35 (left), 52.5 (middle), and 70 keV (right) ion energies. The general shape of the profiles was found to be

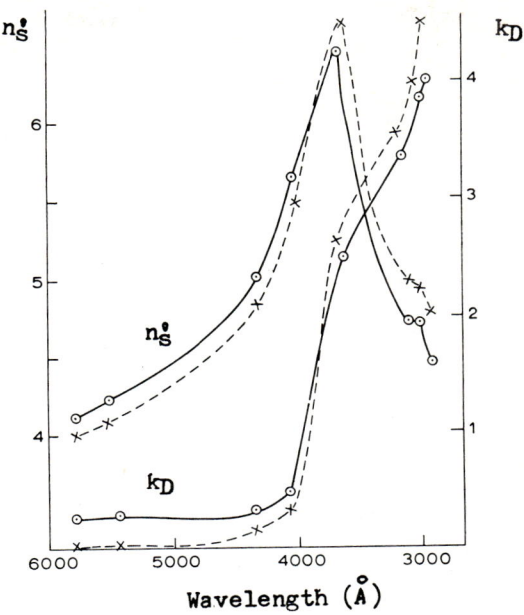

Fig. 6.62. The changes in the optical properties of 0.1 Ω-cm silicon sample due to 350 eV ion bombardment at 5×10^{16} ions/cm^2. Broken and solid lines are for the optical properties before and after bombardment, respectively. After ref. 294.

in good agreement with the damage profiles predicted by Brice [299], and the initial peaks of n and k agree with the position of the damage peaks (indicated by R_{PD}) predicted from the theory of Sigmund and Sanders [300]. These results were obtained by an anodization-stripping technique; they correlated well with profiles computed differently from fitting MAI ellipsometer measurements on the *as-implanted* surface, with the assumption that the change of n and k due to damage by the implanted species obeys a Gaussian distribution.

Besides its application to radiation damage, ellipsometry has also been used to characterize the *mechanical* damage of dielectric and semiconductor surfaces that results from the polishing techniques that are usually employed in their preparation. If we are interested in the determination of the optical properties of materials by light reflection, the presence of such surface-damage layers should be considered (§6.2). Yokota et al. [301] determined the refractive index n_f and thickness d_f of the "polish film" on various glass surfaces that were prepared by standard techniques. Using ellipsometry at the principal angle, Yokota et al. found that the refractive index of the polish film may be either smaller or larger than that of bulk

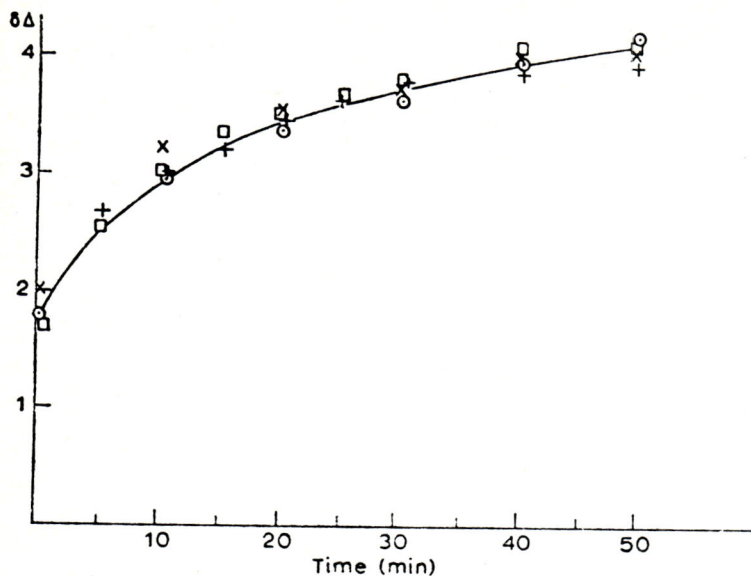

Fig. 6.63. The increase in Δ with cumulative ion-bombardment time measured at 6328 Å wavelength, 70° angle of incidence, 800°C specimen temperature, 400 eV ion voltage and 2μ A/cm² current density. The data was taken on Si samples with different doping concentrations. After ref. 295.

glass by an amount (~0.005–0.1) that is dependent on the type of glass, polish agent, and pressure. The thickness of the polish layer was found to range from about 100–500 Å. Vedam and So [302] observed that the measured value of ψ for mechanically polished silicon differed from sample to sample by as much as 2°, and from the value expected for strain- and film-free surface by about 5°. They attributed such variations to a damaged-silicon layer, about $2\,\mu$m-thick, induced by the final stage of mechanical polishing that was performed using 1 μm diamond paste. Removal of the damage layer by chemical etching led to the same value of ψ for all samples within experimental error.

A recent study that is also related to mechanical damage is that of Smith [303] who used ellipsometry to measure the growth of sub-microscopic cracks during fatigue cycling, interpreting the results in terms of a modification of the model that Fenstermaker and McCrackin used to examine the effect of surface roughness on ellipsometry (§4.8).

Knorr and Leslie [304] used ellipsometry to control and measure the thickness of the very thin insulator film of metal-insulator-metal (MIM) tunnel junctions *during fabrication*, so that a desired value of resistance

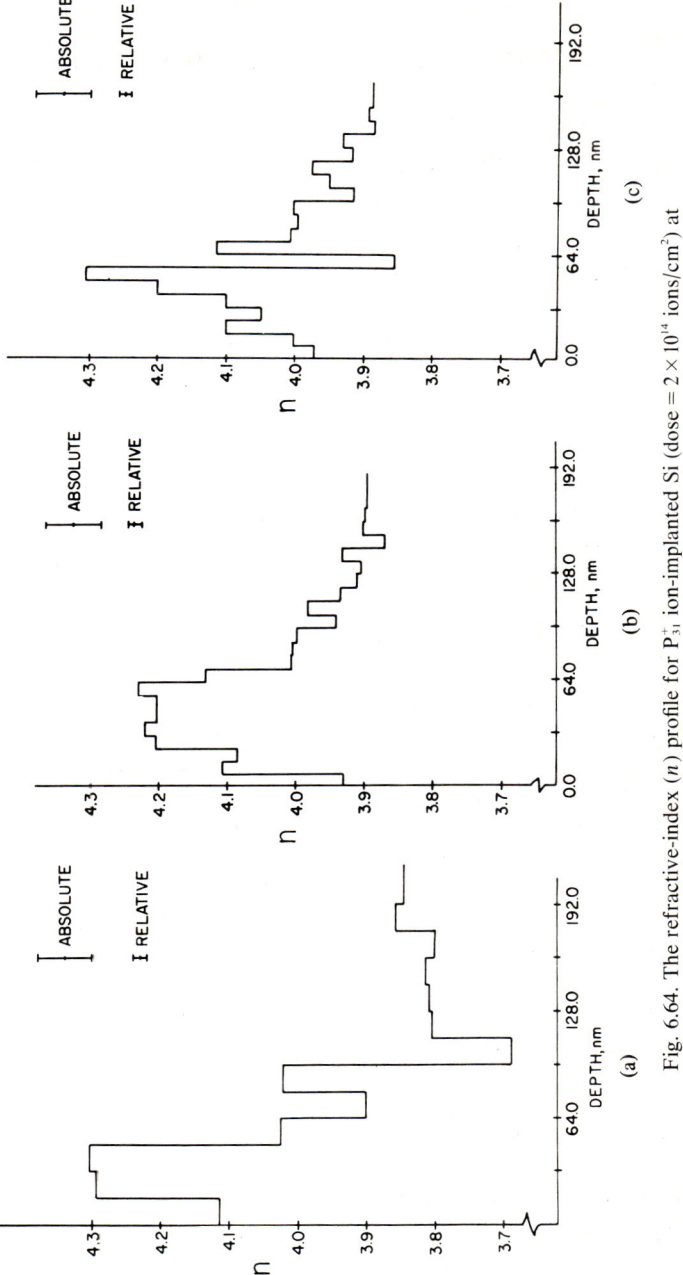

Fig. 6.64. The refractive-index (n) profile for P_{31}^+ ion-implanted Si (dose = 2×10^{14} ions/cm^2) at 35 (left), 52.5 (middle), and 70 keV (right) ion energies. After ref. 298.

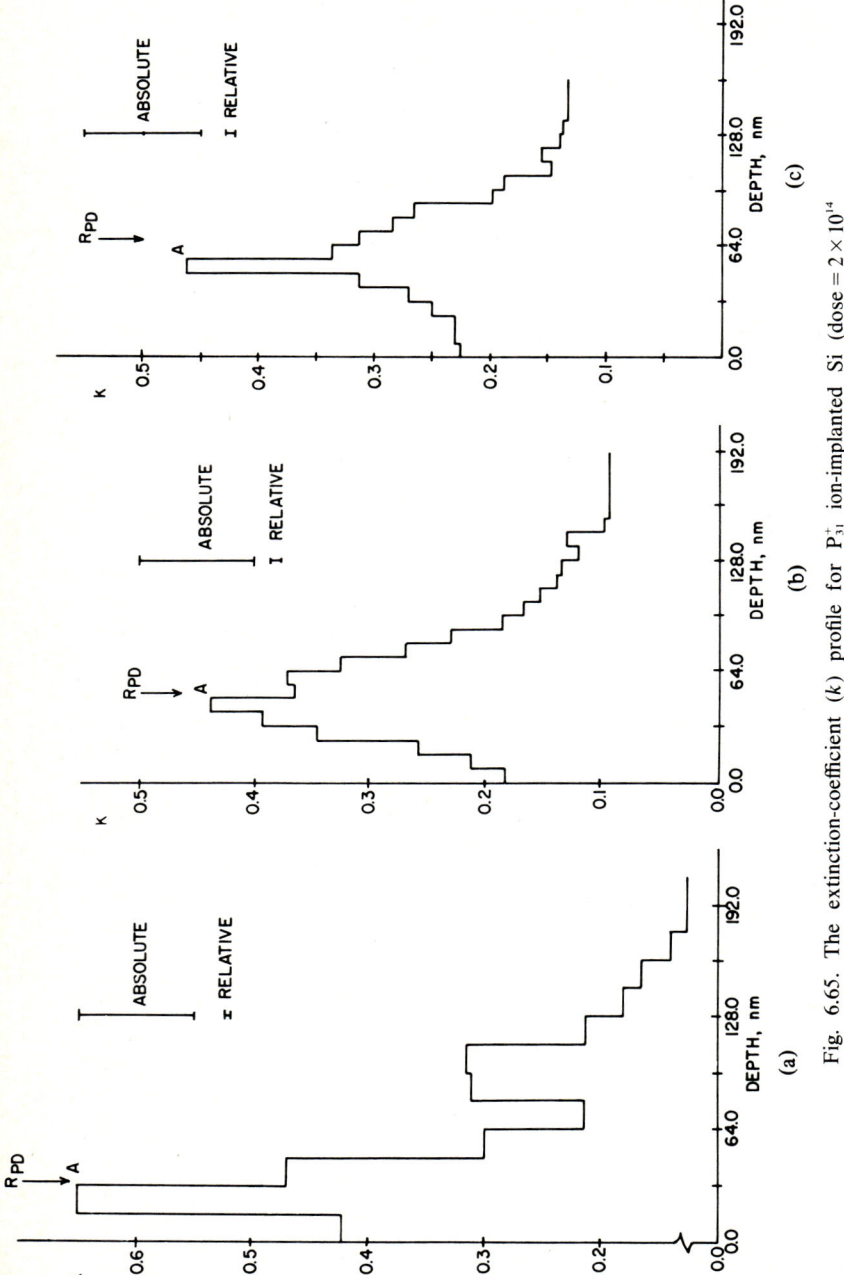

Fig. 6.65. The extinction-coefficient (k) profile for P_{31}^+ ion-implanted Si (dose = 2×10^{14} ions/cm^2), at 35 (left), 52.5 (middle), and 70 keV (right) ion energies. After ref. 298.

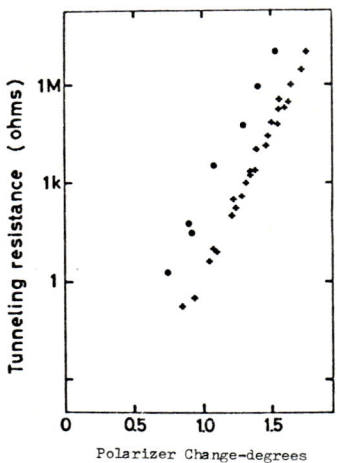

Fig. 6.66. The resistance of Al–Al$_x$O$_y$–Al (crosses) and Al–Al$_x$O$_y$–Pb (circles) tunneling junctions versus the difference of the polarizer null position before and after oxidation (P_{Al}–P_t). (P_{Al}–P_t) can be converted into thickness (0.53° corresponds to 10 Å). After ref. 304.

could be reproducibly obtained. Previously, only estimates of barrier thickness were possible, based on capacitance measurements on *completed* junctions. Figure 6.66 shows their results for the tunnel resistance (on a logarithmic scale) of Al-Al$_x$O$_y$-Al and Al-Al$_x$O$_y$-Pb tunneling junctions versus oxide thickness expressed in terms of the change of the polarizer-null angle in degrees as Al is oxidized to form the dielectric layer of the MIM junction (0.53° correspond to 10 Å). The linearity of the logarithm of the tunneling resistance with barrier thickness (fig. 6.66) is expected from classical tunneling theory [305].

Chou et al. [306] used Auger electron spectroscopy in conjunction with ellipsometry to establish that phosphorus segregates into the oxide layer, when degenerately P-doped silicon is thermally oxidized. The phosphorus pile-up was found to be confined to a thin layer, located at some distance from the ellipsometrically determined SiO$_2$/Si interface. Figure 6.67 shows the AES phosphorus signal (at 120 eV) as a function of the oxide-film thickness measured by ellipsometry. This data was taken at different points on one sample that had a taper-etched oxide with a nearly linear thickness profile (obtained by vertically dipping, at constant rate, the sample with an initial uniform thickness of 60 Å in an etchant solution).

Cornish and Young [307] investigated the electrooptic and electro-strictive effects in anodic oxide films on tantalum by ellipsometry. Figure 6.68 shows the lower part of three Δ-vs-ψ cycles, as the thickness of growing a Ta$_2$O$_5$ film increased. The fact that the Δ-versus ψ curve for the growing transpar-

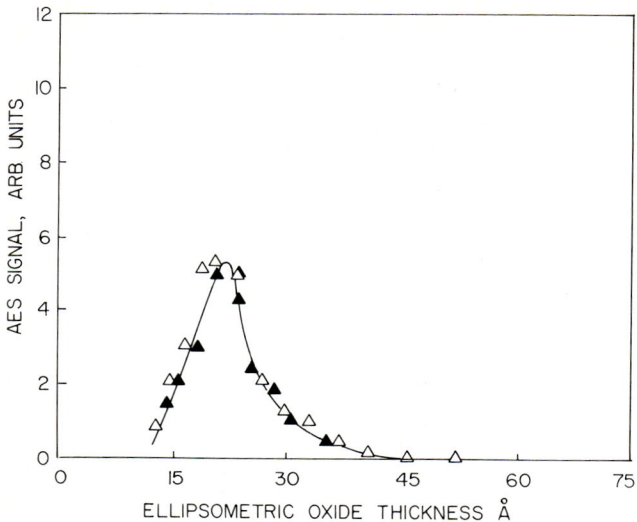

Fig. 6.67. The Auger electron spectroscopic phosphorous signal at 120 eV as a function of the oxide-film thickness measured by ellipsometry at different points on an Si substrate covered by a taper-etched SiO_2 film. After ref. 306.

ent oxide film did not follow the same closed loop, but rather followed open loops that shifted upwards along the Δ axis, was indicative of the uniaxial anisotropy of the oxide along the direction of the applied electric field. By proper analysis of data such as that shown in fig. 6.68, for different values of the applied electric field, Cornish and Young obtained the results shown in

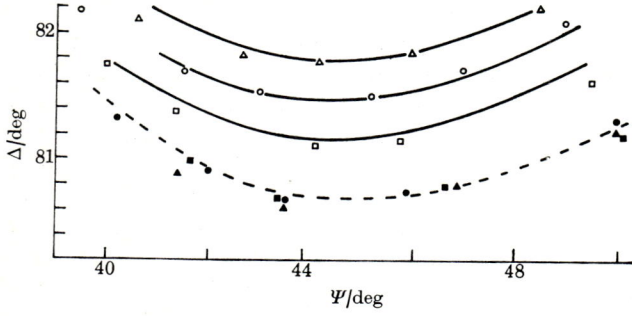

Fig. 6.68. Lower part of Ψ, Δ domain for increasing thickness of films on tantalum up to three cycles. Lower curve (dashed) is computed for an isotropic oxide and the experimental points are for zero field: ■, first cycle; ●, second cycle; ▲, third cycle. Upper three curves (solid lines) represent an anisotropic film with experimental points for field applied: □, first cycle; ○, second cycle; △, third cycle. After ref. 307.

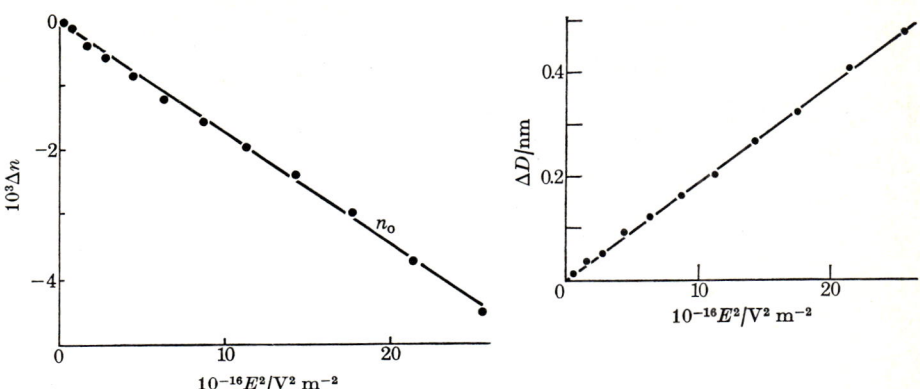

Fig. 6.69 *left*. The steady-state change Δn_o in the ordinary refractive index n_o of an anodic oxide film on Ta plotted against the square of the electric field in the film, E^2. The line is $\Delta n_o = -bE^2$ where $b = 1.74 \times 10^{-20}$ m^{2}V^{-2}. *Right*. Steady-state change ΔD in the thickness of an anodic oxide film on Ta against E^2. The line is $\Delta D = cE^2 D$ with $c = 7.82 \times 10^{-21}$ m^{2}V^{-2}, $D = 236.8$ nm. After ref. 307.

fig. 6.69. Figure 6.69 *left* indicates that the change Δn_o of the ordinary refractive index n_o is a quadratic function of the electric field; fig. 6.69 *right* indicates that the change ΔD of the thickness D of a 2368 Å-thick oxide film is also a quadratic function of the electric field. Such quadratic field dependence of both the electrooptic and electro-strictive effects as shown in fig. 6.69 is expected for a material that is isotropic in the absence of the field.

Buckman et al. [308] observed, by ellipsometry, large refractive-index changes of lead-iodide films induced by photolysis. Figure 6.70 gives their results for the dependence of the ellipsometrically determined n and k on exposure time t, when a 1.45 μm-thick PbI$_2$ film was irradiated by 4 mW/cm^2 light of $\lambda = 4880$ Å, at a temperature of 165°C. The data was taken at different measuring wavelengths indicated by each curve.

Reber and Steiger [309] examined the photolysis of lead-chloride crystal surfaces during their exposure to UV actinic light ($\lambda = 2540$ Å) by ellipsometry, and investigated the photolytic behaviour both in the presence and absence of oxygen. Figure 6.71 shows their data for change of Δ, $\delta\Delta$, as a function of exposure time in air (a), with an estimated incident flux $I = 3.1 \times 10^{13}$ quanta/cm^2 sec when measurements are made at 70° angle of incidence; and in argon (b), with $I = 9.5 \times 10^{12}$ quanta/cm^2 sec at 60° angle of incidence. The data in fig. 6.71 (a) could be explained as indicating the growth of a lead-oxide (or lead-oxychloride) film on PbCl$_2$ in the presence of oxygen, whereas that in fig. 6.71 (b) was found consistent with the formation of a thin

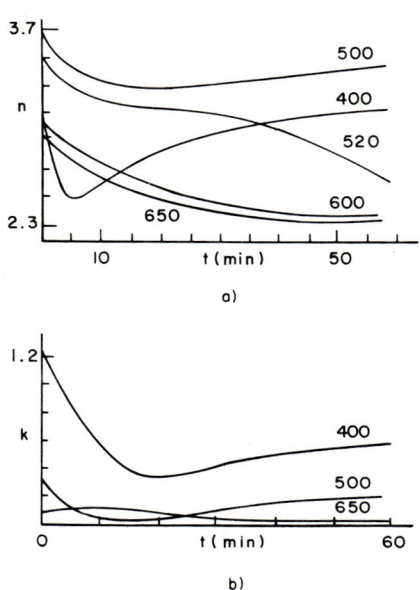

Fig. 6.70. Dependence of n and k of PbI_2 on time of exposure to $\sim 4\,mW/cm^2$ of 488 nm light, at a film temperature of 165°C, with measuring wavelength in nm as a parameter. After ref. 308.

film of lead nuclei on the $PbCl_2$ crystal face. The progress of photolysis was observed to be linear with time in the presence of oxygen (fig. 6.71(a)) and to be linear with the logarithm of time in the absence of oxygen (fig. 6.71(b)).

Nakamura and Kondo [310] studied [by principal-angle (58°) ellipsometry at $\lambda = 5461$ Å] the diffusion to the surface of an antistatic agent incorporated in plastic during its moulding. It was found that the antistatic agent diffused to form a layer 685 Å-thick ($n_f = 1.575$) on the surface of a polyethylene disk ($N_s = 1.601 - j.009$) in thirteen days after moulding. This thickness corresponds to several tens of molecules of the antistatic agent [supposed to be $CH_3(CH_2)_{13}N(C_2H_4OH)_2$]. Washing the disc with ethanol and water removed the layer of antistatic agent; however, almost the same layer (621 Å) recovered in two days after washing. The recovery kinetics are shown in fig. 6.72. The refractive index data during recovery (associated with the film-thickness changes that are shown in fig. 6.72) were used by Nakamura and Kondo to estimate the surface concentration of the antistatic agent in polyethylene by use of the theory of Maxwell Garnett (§4.8), knowing the bulk refractive indices of the agent (1.466) and that of polyethylene (1.601).

In addition to the many research applications of ellipsometry discussed

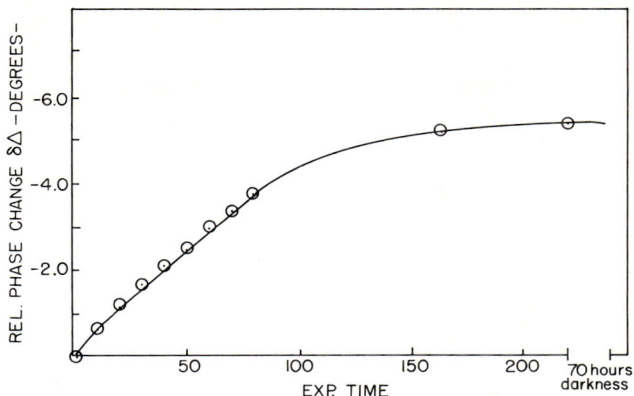

Fig. 6.71. (a) Relative phase change, $\delta\Delta$, as a function of exposure time in minutes for the photolysis of a lead chloride crystal in air for exposure to $\lambda = 254$ nm, $I = 3.1 \times 10^{13}$ quanta cm^{-2} sec^{-1}. Angle of incidence $\phi = 70°$. After ref. 309.

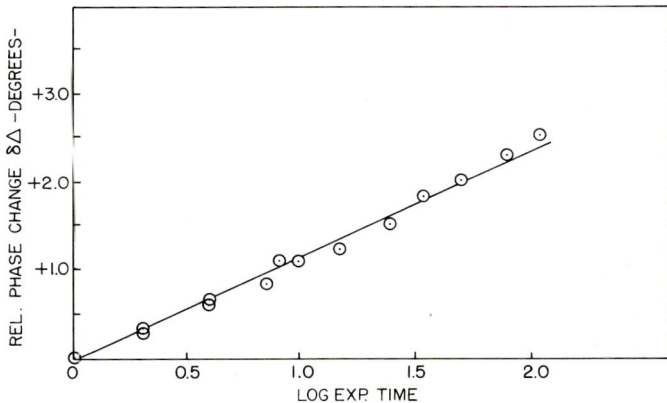

Fig. 6.71. (b) Relative phase change, $\delta\Delta$, as a function of exposure time in minutes for the photolysis of lead chloride in argon for exposure to $\lambda = 254$ nm, $I = 9.5 \times 10^{12}$ quanta cm^{-2} sec^{-1}. $\phi = 60°$. After ref. 309.

above and in §6.2–§6.6, ellipsometry has also proven to be an important tool for film-thickness measurement in industry. In the field of semiconductor technology, the ellipsometer has met the need for measuring the thickness of various films on silicon, such as oxides and nitrides, particularly in the thickness range below about 500 Å where conventional interferometric techniques [311] such as VAMFO (Variable Angle Monochromatic Fringe Observation) and CARIS (Constant Angle Reflection Interference Spectroscopy) become either inaccurate or inapplicable (for the wavelengths typi-

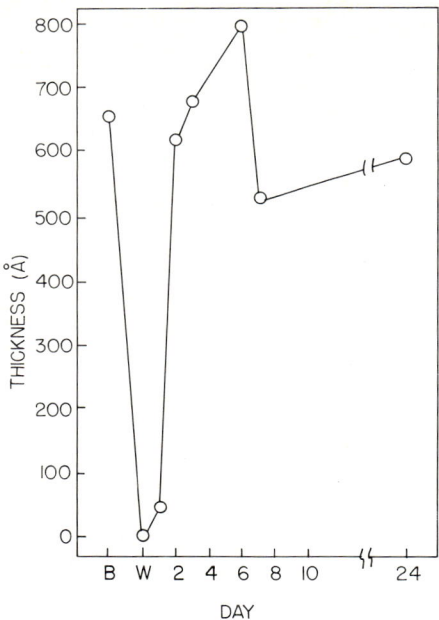

Fig. 6.72. Recovery process of the thickness of an antistatic film on plastic monitored by ellipsometry. B is before washing and W after washing. After ref. 310.

cally employed). The automatic ellipsometer of Hauge and Dill [115] (described in §5.7) was designed so that it could operate in a manufacturing environment, for *on-line* measurement of films (SiO_2 and Si_3N_4) on every product wafer of silicon. This permits both the rejection of "out-of-spec" wafers, and the detection of parameter drifts so that corrective feed-back control can be used to maintain the process within specification.

Kontsevoi *et al.* [312] reported on an "ellipsometric laser microscope", with lateral resolution of 10 μm, for the investigation of the steps involved in making microelectronic circuits. The thickness of films of phosphoro-silicate glass formed on Si surfaces by phosphorous diffusion, the kinetics of epitaxial growth of SiO_2 on Si, and the traces of SiO_2 in windows after photolithography could all be measured.

Hilton and Jones [313] designed an infrared null ellipsometer at $\lambda = 54.6$ μm (from a globar source), with quartz quarter-wave plate and wire-grid polarizers, for measurement of thickness of epitaxial films on Si (n on n^+) and Ge (p on p^+) in the thickness range from 1–10 μm. Using similar optics, DeNicola *et al.* [314] developed an IR ellipsometer for measurement of n-on-n^+ silicon epitaxial layers using a 0.5 mW water-vapor laser at $\lambda = 118.6$ μm. These investigators had the further objective of obtaining a measure of the grading of the n/n^+ junction.

Appendix

This appendix summarizes the Jones and Mueller matrices of the optical devices that constitute the elements of any polarizing optical system. The Jones matrix T of any optical device (or system) is most conveniently expressed in terms of its eigenpolarizations and associated eigenvalues as

$$T = (\chi_{e1} - \chi_{e2})^{-1} \begin{bmatrix} V_{e2}\chi_{e1} - V_{e1}\chi_{e2} & V_{e1} - V_{e2} \\ -\chi_{e1}\chi_{e2}(V_{e1} - V_{e2}) & V_{e1}\chi_{e1} - V_{e2}\chi_{e2} \end{bmatrix}. \tag{A.1}$$

In eq. (A.1) χ_{e1} and χ_{e2} are the complex numbers that represent the eigenpolarizations with respect to any *arbitrary basis states*, and V_{e1} and V_{e2} are the associated complex eigenvalues. When two orthogonal linear polarizations (*e.g.*, x and y) are chosen as basis states, we have

$$\chi = (\tan\theta + j\tan\epsilon)/(1 - j\tan\theta\tan\epsilon) \tag{A.2}$$

where θ and ϵ are the azimuth and ellipticity angle of the polarization ellipse represented by χ. On the other hand, if the right- and left-handed circular polarizations (r and ℓ) are used as basis states, we have

$$\chi = \tan(\epsilon + \tfrac{1}{4}\pi)\, e^{-j2\theta}. \tag{A.3}$$

The eigenvalues V_{e1} and V_{e2} could be generally written as

$$V_{e1} = e^{-\gamma_0} e^{\gamma/2}, \quad V_{e2} = e^{-\gamma_0} e^{-\gamma/2}, \tag{A.4}$$

$$\gamma_0 = \alpha_0 + j\delta_0, \quad \gamma = \alpha + j\delta. \tag{A.5}$$

Table A.1. Normalized† Jones matrices of retarders

Type of Retarder	Cartesian Jones Matrix T_{car}	Circular Jones Matrix T_{cir}	Remarks
Elliptic	$(1+\chi_{\text{ef}}\chi_{\text{ef}}^*)^{-1}\begin{bmatrix}(e^{j\delta/2}+\chi_{\text{ef}}\chi_{\text{ef}}^*e^{-j\delta/2}) & 2j\chi_{\text{ef}}^*\sin\tfrac{1}{2}\delta \\ 2j\chi_{\text{ef}}\sin\tfrac{1}{2}\delta & (e^{-j\delta/2}+\chi_{\text{ef}}\chi_{\text{ef}}^*e^{j\delta/2})\end{bmatrix}$		χ_{ef} is the fast eigenpolarization of the elliptic retarder and is given by eqs. (A.2) and (A.3) for the Cartesian and circular representations, respectively, where θ and ϵ are the azimuth and ellipticity angle of χ_{ef}. δ represents the relative retardation of the retarder throughout this table.
Linear	$\begin{bmatrix}\cos\tfrac{1}{2}\delta+j\cos 2\theta\sin\tfrac{1}{2}\delta & j\sin 2\theta\sin\tfrac{1}{2}\delta \\ j\sin 2\theta\sin\tfrac{1}{2}\delta & \cos\tfrac{1}{2}\delta-j\cos 2\theta\sin\tfrac{1}{2}\delta\end{bmatrix}$	$\begin{bmatrix}\cos\tfrac{1}{2}\delta & je^{j2\theta}\sin\tfrac{1}{2}\delta \\ je^{-j2\theta}\sin\tfrac{1}{2}\delta & \cos\tfrac{1}{2}\delta\end{bmatrix}$	θ is the azimuth of the fast axis of the linear retarder
Circular	$\begin{bmatrix}\cos\tfrac{1}{2}\delta & \mp\sin\tfrac{1}{2}\delta \\ \pm\sin\tfrac{1}{2}\delta & \cos\tfrac{1}{2}\delta\end{bmatrix}$	$\begin{bmatrix}e^{\mp j\delta/2} & 0 \\ 0 & e^{\pm j\delta/2}\end{bmatrix}$	The upper and lower signs correspond to the fast eigenpolarization coinciding with the right- and left-circular states, respectively.

†The term normalized signifies that any overall isotropic (polarization-independent) retardation or attenuation is neglected.

Table A.2. Normalized† Jones matrices of partial polarizers

Type of Partial Polarizer	Cartesian Jones Matrix $\boldsymbol{T}_{\text{car}}$	Circular Jones Matrix $\boldsymbol{T}_{\text{cir}}$	Remarks
Elliptic	$(1+\chi_{e\ell}\chi_{e\ell}^{*})^{-1}\begin{bmatrix}(e^{\alpha/2}+\chi_{e\ell}\chi_{e\ell}^{*}e^{-\alpha/2}) & 2\chi_{e\ell}^{*}\sinh\frac{1}{2}\alpha \\ 2\chi_{e\ell}\sinh\frac{1}{2}\alpha & (e^{-\alpha/2}+\chi_{e\ell}\chi_{e\ell}^{*}e^{\alpha/2})\end{bmatrix}$		$\chi_{e\ell}$ is the low-absorption eigenpolarization of the elliptic partial polarizer and is given by eqs. (A.2) and (A.3) for the Cartesian and circular representations, respectively, where θ and ϵ are the azimuth and ellipticity angle of $\chi_{e\ell}$. α represents the relative attenuation of the partial polarizer throughout this table.
Linear	$\begin{bmatrix}[\cosh\frac{1}{2}\alpha+\cos 2\theta\sinh\frac{1}{2}\alpha] & \sin 2\theta\sinh\frac{1}{2}\alpha \\ \sin 2\theta\sinh\frac{1}{2}\alpha & [\cosh\frac{1}{2}\alpha-\cos 2\theta\sinh\frac{1}{2}\alpha]\end{bmatrix}$	$\begin{bmatrix}\cosh\frac{1}{2}\alpha & e^{j2\theta}\sinh\frac{1}{2}\alpha \\ e^{-j2\theta}\sinh\frac{1}{2}\alpha & \cosh\frac{1}{2}\alpha\end{bmatrix}$	θ is the azimuth of the low-absorption axis of the linear partial polarizer.
Circular	$\begin{bmatrix}\cosh\frac{1}{2}\alpha & \mp j\sinh\frac{1}{2}\alpha \\ \pm j\sinh\frac{1}{2}\alpha & \cosh\frac{1}{2}\alpha\end{bmatrix}$	$\begin{bmatrix}e^{\mp\alpha/2} & 0 \\ 0 & e^{\pm\alpha/2}\end{bmatrix}$	The upper and lower signs correspond to the low-absorption eigenpolarization coinciding with the right- and left-circular states, respectively.

†The term normalized signifies that any overall isotropic (polarization-independent) retardation or attenuation is neglected.

Table A.3. Normalized† Jones matrices of ideal polarizers

Type of Ideal Polarizer	Cartesian Jones Matrix T_{car}	Circular Jones Matrix T_{cir}	Remarks
Elliptic	$T = (1 + \chi_{et}\chi_{et}^*)^{-1} \begin{bmatrix} 1 & \chi_{et}^* \\ \chi_{et} & \chi_{et}\chi_{et}^* \end{bmatrix}$		χ_{et} is the transmitted eigenpolarization of the ideal elliptic polarizer and is given by eqs. (A.2) and (A.3) for the Cartesian and circular representations, respectively, where θ and ϵ are the azimuth and ellipticity angle of χ_{et}.
Linear	$\begin{bmatrix} \cos^2\theta & \sin\theta\cos\theta \\ \sin\theta\cos\theta & \sin^2\theta \end{bmatrix}$	$\begin{bmatrix} \frac{1}{2} & \frac{1}{2}e^{j2\theta} \\ \frac{1}{2}e^{-j2\theta} & \frac{1}{2} \end{bmatrix}$	θ is the azimuth of the transmission axis of the ideal linear polarizer.
Circular	$\begin{bmatrix} \frac{1}{2} & -\frac{1}{2}j \\ +\frac{1}{2}j & \frac{1}{2} \end{bmatrix}$ $\begin{bmatrix} \frac{1}{2} & +\frac{1}{2}j \\ -\frac{1}{2}j & \frac{1}{2} \end{bmatrix}$	$\begin{bmatrix} 0 & 0 \\ 0 & 1 \end{bmatrix}$ $\begin{bmatrix} 1 & 0 \\ 0 & 0 \end{bmatrix}$	The upper and lower matrices correspond to the transmitted eigenpolarization coinciding with the right- and left-circular states, respectively.

†The term normalized signifies that any overall isotropic (polarization independent) retardation or attenuation is neglected.

In eqs. (A.4) and (A.5) α_0 and δ_0 represent the isotropic (polarization independent) attenuation and phase retardation, respectively. α and δ represent the relative attenuation and relative retardation of one eigenpolarization with respect to the other, respectively. (When δ is positive, χ_{e1} represents the fast eigenpolarization.) For simplicity, the factor $e^{-\gamma_0}$ may be dropped from both eigenvalues when the Jones matrix of a device is constructed. The resulting matrix can be called the *normalized Jones matrix* of the device. Such a matrix does not take account of the overall isotropic (polarization-independent) attenuation and phase-shifting properties of the device. In tables A.1–A.3, we list the normalized Jones matrices of retarders, partial polarizers and ideal polarizers in both the Cartesian and circular representations. These matrices are derived by use of eqs. (A.1)–(A.4).

Mueller Matrices

The Mueller matrix of a non-depolarizing optical device or system can be obtained by substituting its Jones matrix from tables A.1–A.3 into eqs. (2.243). The explicit expression of such matrix for a general (elliptic) retarder or polarizer is complicated. It will suffice to tabulate below, table A.4, the Mueller matrices of linear retarders and linear polarizers. The input and output coordinate systems are assumed to be both parallel to the principal directions of the device (*i.e.*, the transmission and extinction axes in case of polarizers, and the fast and slow axes in case of retarders).

To find the Mueller matrix of a device in a general orientation, the law of transformation of the Mueller matrix [eq. (2.246)]

$$M' = R(-\alpha)MR(\alpha), \qquad (A.6)$$

is applied where the rotation matrix $R(\alpha)$ is also given in table A.4. For example, the Mueller matrix of an ideal polarizer at an azimuth θ with respect to parallel input-output coordinate systems is given by substituting the Mueller matrix of an ideal polarizer M in the orientation $\theta = 0$ as well as the Mueller rotation $R(\theta)$ and counter-rotation $R(-\theta)$ matrices, from table A.4 into eq. (A.6). The resulting matrix is given by

$$\begin{matrix} M = \tfrac{1}{2} \\ \text{(ideal polarizer} \\ \text{orientation } \theta) \end{matrix} \begin{bmatrix} 1 & \cos 2\theta & \sin 2\theta & 0 \\ \cos 2\theta & \cos^2 2\theta & \sin 2\theta \cos 2\theta & 0 \\ \sin 2\theta & \sin 2\theta \cos 2\theta & \sin^2 2\theta & 0 \\ 0 & 0 & 0 & 0 \end{bmatrix}. \qquad (A.7)$$

Likewise, the Mueller matrix of a device that exhibits a composite of properties found in table A.4 can be obtained from the expression of the Mueller matrix of a train of devices [eqs. (2.247)]

$$M = M_N M_{N-1} \ldots M_2 M_1. \qquad (A.8)$$

For example, the Mueller matrix of an isotropic surface from which light is obliquely reflected can be obtained by multiplying the Mueller matrices of a linear retarder of retardation Δ and that of a linear partial polarizer of relative amplitude attenuation $(\tau_{min}/\tau_{max})^{1/2} = \tan \psi$, both to be found in table A.4. Normalized to unit reflectivity for unpolarized incident light, the resulting matrix is

$$\begin{matrix} M = \\ \text{(surface with} \\ \text{ellipsometric} \\ \text{parameters } \psi, \Delta) \end{matrix} \begin{bmatrix} 1 & -\cos 2\psi & 0 & 0 \\ -\cos 2\psi & 1 & 0 & 0 \\ 0 & 0 & \sin 2\psi \cos \Delta & \sin 2\psi \sin \Delta \\ 0 & 0 & -\sin 2\psi \sin \Delta & \sin 2\psi \cos \Delta \end{bmatrix}.$$

(A.9)

In §3.8, the derivative of the rotation matrix is needed. Straightforward differentiation of such matrix in table A.4 gives

$$\frac{d\mathbf{R}(\alpha)}{d\alpha} = \begin{bmatrix} 0 & 0 & 0 & 0 \\ 0 & -2\sin 2\alpha & 2\cos 2\alpha & 0 \\ 0 & -2\cos 2\alpha & -2\sin 2\alpha & 0 \\ 0 & 0 & 0 & 0 \end{bmatrix}. \tag{A.10}$$

Table A.4. The Mueller matrices of basic optical devices

Device	Mueller Matrix
Linear Retarder (retardation δ)	$\begin{bmatrix} 1 & 0 & 0 & 0 \\ 0 & 1 & 0 & 0 \\ 0 & 0 & \cos\delta & \sin\delta \\ 0 & 0 & -\sin\delta & \cos\delta \end{bmatrix}$
Linear Partial Polarizer (maximum and minimum transmittances $\tau_{\max}$ and $\tau_{\min}$)	$\frac{1}{2}\begin{bmatrix} (\tau_{\max}+\tau_{\min}) & (\tau_{\max}-\tau_{\min}) & 0 & 0 \\ (\tau_{\max}-\tau_{\min}) & (\tau_{\max}+\tau_{\min}) & 0 & 0 \\ 0 & 0 & 2(\tau_{\max}\tau_{\min})^{\frac{1}{2}} & 0 \\ 0 & 0 & 0 & 2(\tau_{\max}\tau_{\min})^{\frac{1}{2}} \end{bmatrix}$
Linear Ideal Polarizer (maximum transmittance $\tau_{\max}$)	$\frac{1}{2}\tau_{\max}\begin{bmatrix} 1 & 1 & 0 & 0 \\ 1 & 1 & 0 & 0 \\ 0 & 0 & 0 & 0 \\ 0 & 0 & 0 & 0 \end{bmatrix}$
Rotator (angle of rotation α)	$\begin{bmatrix} 1 & 0 & 0 & 0 \\ 0 & \cos 2\alpha & \sin 2\alpha & 0 \\ 0 & -\sin 2\alpha & \cos 2\alpha & 0 \\ 0 & 0 & 0 & 1 \end{bmatrix}$

References

[1] N. M. Bashara, A. B. Buckman and A. C. Hall, Eds., *Proceedings of the Symposium on Recent Developments in Ellipsometry* (North-Holland, Amsterdam, 1969); Surface Sci. **16**, (1969).
[2] R. H. Muller, Surface Sci. **16**, 14 (1969). Also in ref. 1.
[3] R. C. Jones, J. Opt. Soc. Am. **31**, 488 (1941).
[4] W. A. Shurcliff, *Polarized Light* (Harvard University, Cambridge, Massachusetts, 1962).
[5] E. L. O'Neill, *Introduction to Statistical Optics* (Addison-Wesley, Reading, Massachusetts, 1963), Ch. 9.
[6] W. A. Shurcliff and S. S. Ballard, *Polarized Light* (Van Nostrand, Princeton, New Jersey, 1964).
[7] J. W. Simmons and M. J. Guttmann, *States, Waves and Photons* (Addison-Wesley, Reading, Massachusetts, 1970).
[8] D. Clarke and J. F. Grainger, *Polarized Light and Optical Measurement* (Pergamon, New York, 1971).
[9] G. N. Ramachandran and S. Ramaseshan, in *Handbuch der Physik*, S. Flügge, Ed., (Springer-Verlag, Berlin, 1961), Vol. XXV, Ch. 1.
[10] J. M. Bennett and H. E. Bennett, in *Handbook of Optics*, W. G. Driscoll and W. Vaughan, Eds. (McGraw-Hill, New York, 1978), Ch. 10.
[11] R. C. Jones, J. Opt. Soc. Am. **32**, 486 (1942).
[12] J. G. Collins and W. H. Steel, J. Opt. Soc. Am. **52**, 339 (1962).
[13] H. Poincaré, *Théorie Mathématique de la Lumière* (Gauthiers-Villars, Paris, 1892), Vol. II, Ch. 12.
[14] H. G. Jerrard, J. Opt. Soc. Am. **44**, 634 (1954).
[15] D. A. Holmes and D. L. Feucht, J. Opt. Soc. Am. **57**, 466 (1967).
[16] V. H. Rumsey, Proc. IRE **39**, 535 (1951).
[17] G. A. Deschamps, Proc. IRE **39**, 540 (1951).
[18] R. M. A. Azzam and N. M. Bashara, J. Opt. Soc. Am. **62**, 222 (1972).
[19] R. M. A. Azzam and N. M. Bashara, Optics Commun. **5**, 5 (1972).

- [20] B. A. Fuchs and B. V. Shabat, *Functions of a Complex Variable* (Pergamon, New York, 1964), Vol. I, pp. 60ff.
- [21] R. M. A. Azzam and N. M. Bashara, Optics Commun. **5**, 319 (1972).
- [22] C. W. Helstrom, *Statistical Theory of Signal Detection* (Pergamon, New York, 1968), Ch. 2.
- [23] M. Born and E. Wolf, *Principles of Optics* (Pergamon, New York, 1970), Ch. 10.
- [24] E. O. Ammann, in *Progress in Optics*, E. Wolf, Ed., (North-Holland, Amsterdam, 1971), Vol. IX, Ch. 4.
- [25] W. H. Steel, Opt. Acta **11**, 211 (1964).
- [26] G. A. Vanasse, A. T. Stair and D. J. Baker, Eds., *Proceedings of the Aspen International Conference on Fourier Spectroscopy* (AFCRL 71-0019, Special Report 114, 1970).
- [27] A. L. Fymat, Appl. Opt. **11**, 160 (1972).
- [28] O. Bryngdahl, J. Opt. Soc. Am. **62**, 839 (1972).
- [29] S. Kawakami and S. Nishida, Rec. Elec. Commun. Eng. (Tohoko Univ., Japan) **41**, 5, 9 (1972).
- [30] J. C. Cassidy, J. Opt. Soc. Am. **61**, 378 (1971).
- [31] D. A. Holmes, J. Opt. Soc. Am. **54**, 1115 (1964).
- [32] A. Weissberger and B. W. Rossiter, Eds., *Physical Methods of Chemistry* (Wiley, New York, 1972), Vol. I, Part III C.
- [33] A. Fresnel, Ann. Chim. et Phys. **28**, 147 (1825).
- [34] G. A. Deschamps, in *Proceedings of the Symposium on Modern Network Synthesis*, J. Fox, Ed. (Polytechnic Press, Brooklyn, 1952).
- [35] G. Eichmann, J. Opt. Soc. Am. **61**, 1585A (1971).
- [36] H. de Lang, Philips Res. Rep. **8**, 1 (1967).
- [37] R. M. A. Azzam and N. M. Bashara, Appl. Opt. **12**, 62 (1973).
- [38] R. M. A. Azzam and N. M. Bashara, Optics Commun. **4**, 203 (1971).
- [39] W. K. Pratt, *Laser Communication Systems*, (Wiley, New York, 1969).
- [40] R. M. A. Azzam and N. M. Bashara, Appl. Opt. **11**, 2210 (1972).
- [41] R. M. A. Azzam and N. M. Bashara, Optics Commun. **7**, 317 (1973).
- [42] H. C. Hurwitz and R. C. Jones, J. Opt. Soc. Am. **31**, 493 (1941).
- [43] R. C. Jones, J. Opt. Soc. **31**, 500 (1941).
- [44] H. Hsu, M. Richartz and Y. Liang, J. Opt. Soc. Am. **37**, 99 (1947).
- [45] E. F. Dawson and N. O. Young, J. Opt. Soc. Am. **50**, 170 (1960).
- [46] A. S. Marathay, J. Opt. Soc. Am. **55**, 969 (1965).
- [47a] R. M. A. Azzam and N. M. Bashara, Appl. Phys. **1**, 203 (1973).
- [47b] R. M. A. Azzam and N. M. Bashara, Appl. Phys. **2**, 59 (1973).
- [48] A. E. Taylor, *Calculus with Analytic Geometry* (Prentice-Hall, Englewood Cliffs, New Jersey, 1959), Ch. 7.
- [49] E. Malus, Mém. Soc. Arcueil **1**, 113 (1808).
- [50] R. C. Jones, J. Opt. Soc. Am. **38**, 671 (1948).
- [51] R. C. Jones, J. Opt. Soc. Am. **46**, 126 (1956).
- [52] R. M. A. Azzam and N. M. Bashara, J. Opt. Soc. Am. **62**, 1252 (1972).
- [53] E. L. Ince, *Ordinary Differential Equations*, (Dover, New York, 1956).
- [54a] R. M. A. Azzam, B. E. Merrill and N. M. Bashara, Appl. Opt. **12**, 764 (1973).
- [54b] B. E. Merrill, R. M. A. Azzam, and N. M. Bashara, J. Opt. Soc. A. **64**, 731 (1974).
- [55] A. S. Marathay, J. Opt. Soc. Am. **61**, 1363 (1971).
- [56] C. W. Oseen, Trans. Faraday Soc. **29**, 883 (1933).
- [57] G. Parrent and P. Roman, Nuovo Cimento **15**, 370 (1960).
- [58] H. Mueller, J. Opt. Soc. Am. **38**, 661 (1948).

[59] N. G. Parke, Ph.D. dissertation, MIT, 1948 (unpublished); J. Math. Phys. (MIT) **28**, 131 (1949).
[60] M. C. van de Hulst, *Scattering of Light by Small Particles* (Wiley, New York, 1957) p. 44.
[61] M. H. Cohen, Proc. IRE **46**, 172, 183 (1958).
[62] S. Suzuki and A. Tsuchiya, Proc. IRE **46**, 190 (1958).
[63] P. Drude, Ann. Phys. Chem. **36**, 532, 865 (1889).
[64] E. Passaglia, R. R. Stromberg, and J. Kruger, Eds. *Ellipsometry in the Measurement of Surfaces and Thin Films*, Natl. Bur. Std. Misc. Publ. 256 (U.S. Govn't. Printing Office, Washington, 1964).
[65] A. Rothen, Rev. Sci. Instr. **16**, 26 (1945).
[66] R. M. A. Azzam and N. M. Bashara, J. Opt. Soc. Am. **62**, 336 (1972).
[67] R. C. Jones, J. Opt. Soc. Am. **37**, 110 (1947).
[68] F. L. McCrackin, E. Passaglia, R. R. Stromberg and H. L. Steinberg, J. Res. Natl. Bur. Std. **67A**, 363 (1963).
[69] R. M. A. Azzam and N. M. Bashara, Opt. Acta **21**, 497 (1974).
[70] A. B. Winterbottom, J. Sci. Instr. **14**, 203 (1937).
[71] A. B. Winterbottom, in *The Royal Norwegian Sci. Soc. Rept. No. 1* (F. Bruns, Trondheim, 1955).
[72] R. J. King, J. Sci. Instr. **43**, 924 (1966).
[73] R. J. Archer and C. V. Shank, J. Opt. Soc. Am. **57**, 191 (1967).
[74] T. Yolken, R. Waxler, and J. Kruger, J. Opt. Soc. Am. **57**, 283 (1967).
[75] W. G. Oldham, J. Opt. Soc. Am. **57**, 617 (1967).
[76] C. E. Moeller and D. R. Grieser, Appl. Opt. **8**, 206 (1969).
[77] P. H. Smith, Surface Sci. **16**, 34 (1969).
[78] H. G. Jerrard, Surface Sci. **16**, 67 (1969).
[79] F. Lukeš, Surface Sci. **16**, 74 (1969).
[80] M. Ghezzo, Brit. J. Appl. Phys. (J. Phys. D) **2**, 1483 (1969).
[81] D. Moisil, Opt. Commun. **1**, 257 (1969).
[82] F. L. McCrackin, J. Opt. Soc. Am. **60**, 57 (1970).
[83] J. A. Johnson and N. M. Bashara, J. Opt. Soc. Am. **60**, 221 (1970).
[84] W. R. Hunter, D. H. Eaton, and C. T. Sah, Surface Sci. **20**, 355 (1970).
[85] H. G. Liljenvall and A. G. Mathewson, Appl. Opt. **9**, 1489 (1970).
[86] M. J. Dignam and M. Moskovits, Appl. Opt. **9**, 1868 (1970).
[87] G. Forgács, Brit. J. Appl. Phys. (J. Phys. D) **3**, 1513 (1970).
[88] E. Schmidt, J. Opt. Soc. Am. **60**, 490 (1970).
[89] M. R. Steel, Appl. Opt. **10**, 2371 (1971).
[90] R. H. W. Graves, Appl. Opt. **10**, 2679 (1971).
[91] D. E. Aspnes and A. A. Studna, Appl. Opt. **10**, 1024 (1971).
[92] D. E. Aspnes, Appl. Opt. **10**, 2545 (1971).
[93] A. P. Lenham, J. Opt. Soc. Am. **62**, 1175 (1972).
[94] D. E. Aspnes, J. Opt. Soc. Am. **61**, 1077 (1971).
[95] R. M. A. Azzam and N. M. Bashara, J. Opt. Soc. Am. **61**, 600 (1971).
[96] R. M. A. Azzam and N. M. Bashara, J. Opt. Soc. Am. **61**, 773 (1971).
[97] R. M. A. Azzam and N. M. Bashara, J. Opt. Soc. Am. **61**, 1118 (1971).
[98] R. M. A. Azzam and N. M. Bashara, J. Opt. Soc. Am. **61**, 1236 (1971).
[99] R. M. A. Azzam and N. M. Bashara, J. Opt. Soc. Am. **61**, 1380 (1971).
[100] R. M. A. Azzam and N. M. Bashara, J. Opt. Soc. Am. **62**, 700 (1972).
[101] W. R. Hunter, J. Opt. Soc. Am. **63**, 951 (1973).
[102] R. M. A. Azzam, T. L. Bundy, and N. M. Bashara, Opt. Commun. **7**, 110 (1973).
[103] S. R. Rajagopalan and S. Ramaseshan, Proc. Ind. Acad. Sci. **60A**, 297 (1964).

[104] J. Monin and G.-A. Boutry, Nouv. Rev. Optique **4**, 159 (1973).
[105] S. C. Som and C. Chowdhury, J. Opt. Soc. Am. **62**, 10 (1972).
[106] D. Brewster, Phil. Trans. **69**, 133 (1830).
[107] H. M. O'Bryan, J. Opt. Soc. Am. **26**, 122 (1936).
[108] M. Yamamoto, Opt. Commun. **10**, 200 (1974).
[109] R. M. A. Azzam, A.-R. M. Zaghloul, and N. M. Bashara, J. Opt. Soc. Am. **65**, 252 (1975); **65**, 1464 (1975).
[110] S. Roberts, in ref. 64, p. 157.
[111] B. D. Cahan and R. F. Spanier, Surface Sci. **16**, 166 (1969).
[112] R. Greef, Rev. Sci. Instr. **41**, 532 (1970).
[113] J. C. Suits, Rev. Sci. Instr. **42**, 19 (1971).
[114] D. E. Aspnes, Opt. Commun. **8**, 222 (1973).
[115] P. S. Hauge and F. H. Dill, IBM J. Res. Devel. **17**, 472 (1973).
[116] S. N. Jasperson and S. E. Schnatterly, Rev. Sci. Instr. **40**, 761 (1969).
[117] S. N. Jasperson, D. K. Burge and R. C. O'Handley, Surface Sci. **37**, 548 (1973).
[118] D. E. Aspnes, J. Opt. Soc. Am. **64**, 639 (1974).
[119] D. E. Aspnes, J. Opt. Soc. Am. **64**, 812 (1974).
[120] R. M. A. Azzam and N. M. Bashara, J. Opt. Soc. Am. **64**, 1459 (1974).
[121] M. Abramowitz and I. A. Stegun, *Handbook of Mathematical Functions* (Dover, New York, 1965), p. 361.
[122] R. C. O'Handley, J. Opt. Soc. Am. **63**, 523 (1973).
[123] H. F. Hazebroek and A. A. Holscher, J. Phys. E: Sci. Instr. **6**, 822 (1973).
[124] A. B. Buckman and N. M. Bashara, J. Opt. Soc. Am. **58**, 700 (1968).
[125] A. B. Buckman and N. M. Bashara, Phys. Rev. **174**, 719 (1968).
[126] A. B. Buckman, Surface Sci. **16**, 193 (1969).
[127] L. G. Holcomb and N. M. Bashara, J. Opt. Soc. Am. **61**, 608 (1971).
[128] S. Gottesfeld and B. Reichman, Surface Sci. **44**, 377 (1974).
[129] O. S. Heavens, *Optical Properties of Thin Solid Films* (Dover, New York, 1965).
[130] O. S. Heavens, *Thin Film Physics* (Methuen, London, 1970), Ch. 6.
[131] A. Vašiček, *Optics of Thin Films* (North-Holland, Amsterdam, 1960).
[132] F. Abelès, in *Progress in Optics*, Vol. II, E. Wolf, Ed. (North-Holland, Amsterdam, 1963).
[133] O. S. Heavens, in *Physics of Thin Films*, Vol. 2, G. Hass and R. E. Thun, Eds. (Academic Press, New York, 1964).
[134] H. E. Bennett and J. M. Bennett, in *Physics of Thin Films*, Vol. 4, G. Hass and R. E. Thun, Eds. (Academic Press, New York, 1967).
[135] F. Abelès, in *Physics of Thin Films*, Vol. 6, G. Hass and R. E. Thun, Eds. (Academic Press, New York, 1971).
[136] A. V. Sokolov, *Optical Properties of Metals* (American Elsevier, New York, 1967).
[137] J. N. Hodgson, *Optical Absorption and Dispersion in Solids* (Barnes and Noble, New York, 1971).
[138] F. Wooten, *Optical Properties of Solids* (Academic Press, New York, 1972).
[139] F. Abelès, Ed. *Optical Properties and Electronic Structure of Metals and Alloys* (North-Holland, Amsterdam, 1966).
[140] F. Abelès, Ed., *Optical Properties of Solids* (North-Holland, Amsterdam, 1972).
[141] J. M. Stone, *Radiation and Optics* (McGraw-Hill, New York, 1963), Ch. 15.
[142] R. T. Baumel and S. E. Schnatterly, J. Opt. Soc. Am. **61**, 832 (1971).
[143] H. B. Holl, *The Reflection of Electromagnetic Radiation*, Vols. I and II, Report No. RF-TR-63-4, U.S. Army Missile Command, Redstone Arsenal, Alabama, 1963.
[144] G.B. Airy, Phil. Mag. **2**, 20, (1833).

[145] G. Gergely, Ed., *Ellipsometric Tables of the Si–SiO$_2$ System for Mercury and He–Ne Laser Spectral Lines*, (Akadémiai Kiadó, Budapest, 1971).
[146] P. C. S. Hayfield and G. W. T. White, in ref. 64, pp. 157ff.
[147] P. C. S. Hayfield, in ref. 1, pp. 370ff.
[148] P. C. S. Hayfield, in *Advances in Corrosion Science and Technology*, Vol. 2, M. G. Fontana and R. W. Staehle, Eds. (Plenum Press, New York, 1972), Ch. 2.
[149] K. Kinosita, M. Nishibori, M. Yamamoto, and Hideshi Yokota, Optica Acta **17**, 115 (1970).
[150] R. J. Archer, J. Opt. Soc. Am. **52**, 970 (1962).
[151] K. H. Beckman, Ber. Bunsenges. Phys. Chemie **70**, 842 (1966).
[152] R. H. Muller, in *Advances in Electrochemistry and Electrochemical Engineering*, Vol. 9, R. H. Muller, Ed. (Wiley, New York, 1973), Ch. 3.
[153] L. Tronstad, Trans. Faraday Soc. **31**, 1151 (1935).
[154] C. E. Leberknight and B. Lustman, J. Opt. Soc. Am. **29**, 59 (1939).
[155] Ref. 131, p. 315.
[156] D. K. Burge and H. E. Bennett, J. Opt. Soc. Am. **54**, 1428 (1964).
[157] D. W. Peterson and N. M. Bashara, J. Opt. Soc. Am. **55**, 845 (1965).
[158] A. N. Saxena, J. Opt. Soc. Am. **55**, 1061 (1965).
[159] R. J. Archer, *Manual on Ellipsometry* (Gaertner Scientific Corp., Chicago, 1968).
[160] E. C. Rowe and J. Shewchun, J. Opt. Soc. Am. **59**, 1385 (1969).
[161] J. D. E. McIntyre and D. E. Aspnes, Surface Sci. **24**, 417 (1971).
[162] R. C. Smith and M. Hacskaylo, in ref. 64, pp. 83ff.
[163] R. H. Muller, R. F. Steiger, G. A. Somorjai, and J. M. Morabito, in ref. 1, pp. 234ff.
[164] R. M. A. Azzam, M. Elshazly-Zaghloul, and N. M. Bashara, Appl. Opt. **14**, 1652 (1975).
[165] N. J. Harrick, *Internal Reflection Spectroscopy* (Wiley, New York, 1967).
[166] W. N. Hansen, J. Opt. Soc. Am. **63**, 793 (1973).
[167] F. L. McCrackin and J. P. Colson, in ref. 64, pp. 61ff.
[168] W.-K. Paik and J. O'M. Bockris, Surface Sci. **28**, 61 (1971).
[169] W.-K. Paik, in *Electrochemistry*, Vol. 6 of *Physical Chemistry, Series One*, J. O'M. Bockris, Ed. (University Park Press, Baltimore, 1973).
[170] F. Lukeš, W. H. Knausenberger and K. Vedam, in ref. 1, pp. 112ff.
[171] S. S. So and K. Vedam, J. Opt. Soc. Am. **62**, 16 (1972).
[172] W. G. Oldham, in ref. 1, pp. 16ff.
[173] D. G. Shueler, in ref. 1, pp. 104ff.
[174] J. Shewchun and E. C. Rowe, J. Appl. Phys. **41**, 4128 (1971).
[175] M. M. Ibrahim and N. M. Bashara, J. Opt. Soc. Am. **61**, 1622 (1971).
[176] J. R. Rice, in *Numerical Solutions of Nonlinear Problems* (Computer Sci. Center, Univ. of Maryland College Park, 1970), p. 80.
[177] J. Kowalk and M. R. Osborne, *Methods of Unconstrained Optimization Problems* (American Elsevier, New York, 1968).
[178] D. L. Marquardt, J. Indus. Appl. Math. **2**, 431 (1963).
[179] J. A. Johnson and N. M. Bashara, J. Opt. Soc. Am. **61**, 457 (1971).
[180] A. W. Crook, J. Opt. Soc. Am. **38**, 954 (1948).
[181] F. Abelès, in ref. 64, pp. 41ff.
[182] J. Billard, Thesis, University of Paris (1966).
[183] S. Teitler and B. Henvis, J. Opt. Soc. Am. **60**, 830 (1970).
[184] D. W. Berreman and T. J. Scheffer, Phys. Rev. Lett. **25**, 577 (1970).
[185] Ref. 23, pp. 51ff.
[186] D. W. Berreman, J. Opt. Soc. Am. **62**, 502 (1972).
[187] P. Drude, *The Theory of Optics* (Dover, New York, 1965), Ch. VI.

[188] M. Born, *Optik* (Springer, Berlin, 1933).
[189] D. W. Berreman, J. Opt. Soc. Am. **63**, 1374 (1973).
[190] T. P. Sosnowski, Opt. Commun. **4**, 408 (1972).
[191] R. A. W. Graves, J. Opt. Soc. Am. **59**, 1225 (1969).
[192] D. den Engelsen, J. Opt. Soc. Am. **61**, 1460 (1971).
[193] K. B. Blodgett, J. Am. Chem. Soc. **57**, 1007 (1935).
[194] K. B. Blodgett and I. Langmuir, Phys. Rev. **51**, 964 (1937).
[195] H. Schopper, Z. Phys. **132**, 146 (1952).
[196] J. C. Maxwell Garnett, Phil. Trans. **203**, 385 (1904); **205**, 237 (1906).
[197] C. S. Strachan, Proc. Camb. Phil. Soc. **29**, 116 (1933).
[198] D. V. Sivukhin, Sov. Phys. JETP **3**, 269 (1956).
[199] Ref. 159, pp. 18ff.
[200] D. W. Berreman, J. Opt. Soc. Am. **60**, 499 (1970).
[201] S. Y. Yamaguchi, J. Phys. Soc. Japan **15**, 1577 (1960).
[202] S. Yoshida, T. Yamaguchi, and A. Kinbara, J. Opt. Soc. Am. **61**, 62 (1971); **61**, 463 (1971).
[203] T. Yamaguchi, S. Yoshida, and A. Kinbara, J. Opt. Soc. Am. **62**, 634 (1972).
[204] C. A. Fenstermaker and F. L. McCrackin, in ref. 1, pp. 85ff.
[205] R. M. A. Azzam and N. M. Bashara, Phys. Rev. **B5**, 4721 (1972).
[206] S. Silver, *Microwave Antenna Theory and Design* (McGraw-Hill, New York, 1947), p. 161.
[207] I. Ohlídal and F. Lukeš, Opt. Commun. **5**, 323 (1972).
[208] I. Ohlídal and F. Lukeš, Opt. Acta **19**, 817 (1972).
[209] I. Ohlídal, F. Lukeš, and K. Navrátil, Surface Sci. **45**, 91 (1974).
[210] I. Ohlídal and F. Lukeš, Opt. Commun. **7**, 76 (1973).
[211] E. L. Church and J. Zavada, J. Opt. Soc. Am. **64**, 547A (1974).
[212] A. A. Maradudin and D. L. Mills, Phys. Rev. **B11**, 1392 (1975).
[213] V. Celli, A. Marvin, and F. Toigo, Phys. Rev. **B11**, 1779 (1975).
[214] A. Marvin, F. Toigo, and V. Celli, Phys. Rev. **B11**, 2777 (1975).
[215] M. Ruiz-Urbeita and E. M. Sparrow, J. Opt. Soc. Am. **62**, 1188 (1972); **63**, 194 (1973).
[216] R. J. King and M. J. Downs, in ref. 1, pp. 288ff.
[217] K. K. Svitashev, A. I. Semenenko, L. V. Semenenko, and V. K. Sokolov, Opt. Spectrosc. **30**, 288 (1971); 34, 542 (1973).
[218] I. Wilmanns, Surface Sci. **16**, 147 (1969).
[219] S. S. So, W. H. Knausenberger, and K. Vedam, J. Opt. Soc. Am. **61**, 124 (1971).
[220] *Photo Sensitive Devices, Radio Corporation of America Electron Tube Handbook*, (RCA Electron Tube Division–Commercial Engineering, Harrison, New Jersey).
[221] J. R. Zeidler, R. B. Kohles, and N. M. Bashara, Appl. Opt. **13**, 1115 (1974).
[222] J. R. Zeidler, R. B. Kohles, and N. M. Bashara, Appl. Opt. **13**, 1938 (1974).
[223] R. R. Stromberg, L. E. Smith, and F. L. McCrackin, in *Symposium on Optical Studies of Adsorbed Layers at Interfaces*, Faraday Society (University Press, Aberdeen, 1971).
[224] E. F. Roberts and A. Meadows, J. Phys. E: Sci. Instrum. **7**, 379 (1974).
[225] A. B. Winterbottom, in ref. 64, pp. 97ff.
[226] H. Takasaki, Appl. Opt. **5**, 759 (1966).
[227] L. Ord and B. L. Wills, Appl. Opt. **6**, 1673 (1967); ref. 1, p. 155.
[228] H. J. Mathieu, D. E. McClure, and R. H. Muller, Rev. Sci. Instrum. **45**, 798 (1974).
[229] R. H. Muller and H. J. Mathieu, Appl. Opt. **13**, 2222 (1974).
[230] T. Yamaguchi and H. Hasunuma, Sci. Light (Japan) **16**, 64 (1967).
[231] W. Budde, Appl. Opt. **1**, 201 (1962).
[232] Y. J. van der Meulen and N. C. Hien, J. Opt. Soc. Am. **64**, 804 (1974).
[233] D. E. Aspnes and A. A. Studna, Appl. Opt. **14**, 220 (1975).

REFERENCES

[234] D. E. Aspnes, Bull. Am. Phys. Soc. **20**, 348 (1975).
[235] Ref. 129, p. 199.
[236] D. E. Aspnes, Phys. Rev. Lett. **28**, 168 (1972).
[237] D. E. Aspnes and J. E. Rowe, Solid State Commun. **8**, 1145 (1970).
[238] T. E. Faber and N. V. Smith, J. Opt. Soc. Am. **58**, 102 (1968).
[239] Ref. 159, p. 20.
[240] Ref. 159, p. 22.
[241] F. Meyer, E. E. Kluizenaar and D. den Engelsen, J. Opt. Soc. Am. **63**, 529 (1972).
[242] R. T. Jacobsen and M. Kerker, J. Opt. Soc. Am. **57**, 751 (1967).
[243] J. Kruger and W. J. Ambs, J. Opt. Soc. Am. **49**, 1195 (1959).
[244] L. S. Bartell and D. Churchill, J. Phys. Chem. **65**, 2242 (1961).
[245] N. M. Bashara and D. W. Peterson J. Opt. Soc. Am. **56**, 1320 (1966).
[246] M. J. Dignam, B. Rao, M. Moscovitz, and R. W. Stobie, Can. J. Chem. **49**, 1115 (1971).
[247] M. J. Dignam, B. Rao, and R. W. Stobie, Surface Sci. **46**, 308 (1974).
[248] R. J. Archer, in ref. 64, pp. 255ff.
[249] S. Brunauer, P. H. Emmett, and E. Teller, J. Am. Chem. Soc. **60**, 309 (1938).
[250] T. Smith, J. Opt. Soc. Am. **58**, 1069 (1968).
[251] F. Meyer and G. A. Bootsma, in Ref. 1, pp. 221ff.
[252] J. J. Vrakking and F. Meyer, Appl. Phys. Lett. **18**, 226 (1971).
[253] F. Meyer, Phys. Rev. **B9**, 3622 (1974).
[254] R. R. Stromberg, D. J. Tutas, and E. Passaglia, J. Phys. Chem. **69**, 3955 (1965).
[255] R. R. Stromberg, E. Passaglia, and D. J. Tutas, in ref. 64, pp. 281ff.
[256] E. A. DiMarzio and F. L. McCrackin, J. Chem. Phys. **43**, 539 (1965).
[257] R. J. Archer, J. Electrochem. Soc. **104**, 619 (1957).
[258] F. Lukeš, Surface Sci. **30**, 91 (1972).
[259] T. Smith, Surface Sci. **38**, 292 (1973).
[260] N. J. Chou, J. M. Eldridge, R. Hammer, and D. W. Dong, J. Electron. Mater. **2**, 115 (1973).
[261] R. J. Lederich, J. Opt. Soc. Am. **62**, 1524 (1972).
[262] J. Kruger, J. Electrochem. Soc. **108**, 503 (1961).
[263] D. K. Burge, J. M. Bennett, R. L. Peck, and H. E. Bennett, in ref. 1, pp. 303ff.
[264] W.-K. Paik, M. A. Genshaw, and J. O'M. Bockris, J. Phys. Chem. **74**, 4266 (1970).
[265] L. Young and F. G. R. Zobel, J. Electrochem. Soc. **113**, 277 (1966).
[266] L. Young, *Anodic Oxide Films* (Academic Press, New York, 1961).
[267] A. K. N. Reddy, M. A. Genshaw and J. O'M. Bockris, J. Chem. Phys. **48**, 671 (1968).
[268] J. Horkans, B. D. Cahan, and E. Yeager, Surface Sci. **46**, 1 (1974).
[269] C. J. Dell'Oca, J. Electrochem. Soc. **120**, 1225 (1973).
[270] J. Kruger, in ref. 152. Ch. 4.
[271] J. O'M. Bockris, A. K. N. Reddy, and M. G. B. Rao, J. Electrochem. Soc. **113**, 1133 (1966).
[272] J. Kruger, and J. P. Calvert, J. Electrochem. Soc. **114**, 43 (1967).
[273] J. Kruger, Corrosion, **22**, 88 (1966).
[274] C. L. McBee and J. Kruger, Nat. Phys. Sci. **230**, 194 (1971).
[275] M. Novak, A. K. N. Reddy, and H. Wroblowa, J. Electrochem. Soc. **117**, 733 (1970).
[276] A. Rothen, in *Progress in Surface and Membrane Science*, Vol. 8, D. A. Cadonhead, E. F. Danielli and M. D. Rosenberg, Eds., (Academic Press, New York, 1974), and references therein.
[277] L. Vroman and A. C. Adams, in ref. 1, pp. 438ff, in ref. 64, pp. 335ff. See also refs. 276 and 278.
[278] G. Poste and C. Moss, in *Progress in Surface Science*, Vol. 2 (Part 3), S. G. Davison, Editor (North-Holland, Amsterdam, 1972).
[279] M. M. Wintrobe, *Clinical Hematology*, (Lea & Febiger, Philadelphia, 1967), Ch. 5.

[280] R. E. Baier and R. C. Dutton, J. Biomed. Mater. Res. **3**, 69 (1969).
[281] H. J. Woodliff, *Blood and Bone Marrow Cell Culture*, (Lippincott, Philadelphia, 1964).
[282] E. R. Gold and D. B. Peacock, Eds., *Basic Immunology* (Wright, Bristol, 1970).
[283] I. Langmuir and V. M. Schaefer, J. Am. Chem. Soc. **39**, 1406 (1937).
[284] M. F. Shaffer and J. H. Dingle, Proc. Soc. Exp. Biol. Med. **38**, 523 (1938).
[285] E. F. Porter and A. M. Pappenheimer, J. Exp. Med. **69**, 755 (1939).
[286] J. B. Bateman, H. E. Calkins, and L. A. Chambers, J. Immunol. **40**, 511 (1941).
[287] A. Rothen and K. Landsteiner, J. Exp. Med. **76**, 437 (1942).
[288] I. Giaever, J. Immunol. **110**, 1424 (1973); Bull. Am. Phys. Soc. **19**, 564 (1974).
[289] A. Rothen and C. Mathot, *Immunochemistry* **6**, 241 (1969); in ref. 1, pp. 428ff; also in ref. 276.
[290] M. D. Rosenberg, Biophys. J. **1**, 137 (1960).
[291] G. Weissman and R. Claiborne, Eds., *Cell Membranes* (H. P. Publishing Co., Inc., New York, 1975).
[292] G. Poste and L. W. Greenham, Cytobias. **3**, 5 (1971).
[293] J. B. Schroeder and H. D. Dieselman, J. Appl. Phys. **40**, 2559 (1969).
[294] M. M. Ibrahim and N. M. Bashara, Surface Sci. **30**, 632 (1972).
[295] M. M. Ibrahim and N. M. Bashara, Surface Sci. **30**, 680 (1972).
[296] M. M. Ibrahim and N. M. Bashara, Opt. Commun. **4**, 339 (1972).
[297] M. M. Ibrahim and N. M. Bashara, J. Vac. Sci. Technol. **9**, 1259 (1972).
[298] J. R. Adams and N. M. Bashara, Surface Sci. **49**, 441 (1975).
[299] D. K. Brice, Radiation Effects **6**, 77 (1970).
[300] P. Sigmund and J. B. Sanders, in *Proceedings of the International Conference on Applications of Ion Beams to Semiconductor Techniques*, P. Glotin, Ed. (Editions OPHRYS, 1967), p. 215.
[301] H. Yokota, H. Sakata, M. Nishibori, and K. Kinosita, in ref. 1, pp. 265ff.
[302] K. Vedam and S. S. So, Surface Sci. **29**, 379 (1972).
[303] T. Smith, Surface Sci. **45**, 117 (1975).
[304] K. Knorr and J. D. Leslie, Solid State Commun. **12**, 615 (1973).
[305] E. Merzbacher, *Quantum Mechanics* (Wiley, New York, 1961).
[306] N. J. Chou, Y. J. van der Meulen, R. Hammer, and J. Cahill, Appl. Phys. Lett. **24**, 200 (1974).
[307] W. D. Cornish and L. Young, Proc. R. Soc. Lond. **A335**, 39 (1973).
[308] A. B. Buckman, N. H. Hong, and D. Wilson, J. Opt. Soc. Am. **65**, 914 (1975).
[309] J.-F. Reber and R. Steiger, Surface Sci. **49**, 236 (1975).
[310] K. Nakamura and M. Kondo, Jap. J. App. Phys. **11**, 1205 (1972).
[311] W. A. Pliskin, in *Progress in Analytic Chemistry*, Vol. II, E. M. Murt and W. G. Guldner, Eds. (Plenum Press, New York, 1969).
[312] Y. A. Kontsevoi, R. R. Rezvyi, and V. M. Galolobov, Zavodskaya Laboratoriya (USSR) **37**, 184 (1971).
[313] A. R. Hilton and C. E. Jones, J. Electrochem. Soc. **113**, 472 (1966).
[314] R. O. DeNicola, M. A. Saifi, and R. E. Frazee, Appl. Opt. **11**, 2534 (1972).
[315] D. Moisil and G. Moisil, *Teoria Și Practica Elipsometriei* (Editura Technică, Bucharest, 1973) [in Romanian].

*Contents of the International
Conferences on Ellipsometry*

Ellipsometry in the Measurement Of Surfaces and Thin Films

Symposium Proceedings
Washington 1963

Symposium held September 5–6, 1963, at the
National Bureau of Standards, Washington, D.C.

Edited by E. Passaglia, R. R. Stromberg, and J. Kruger

National Bureau of Standards Miscellaneous Publication 256

Issued September 15, 1964

1. Historical Review

Measurements of the thickness of thin films by optical means, from Rayleigh and Drude to Langmuir, and the development of the present ellipsometer. ALEXANDRE ROTHEN 7

2. Theory

Optical study of a thin absorbing film on a metal surface. ANTONÍN VAŠÍČEK 25

Optical properties of inhomogeneous films. FLORIN ABELÈS 41

3. Techniques

Computation

Computational techniques for the use of the exact Drude equations in reflection problems. FRANK L. McCRACKIN and JAMES P. COLSON 61

Factors influencing the experimental sensitivity of the Drude technique. RICHARD C. SMITH and MICHAEL HACSKAYLO ... 83

Instrumentation

Increased scope of ellipsometric studies of surface film formation. A. B. WINTERBOTTOM 97

Electronic polarimeter techniques. JEROME M. WEINGART and ALAN R. JOHNSTON 113

On determining optical constants of metals in the infrared. SHEPARD ROBERTS 119

4. Applications

Oxidation and Corrosion of Metals

Use of ellipsometry for *in situ* studies of the oxidation of metal surfaces immersed in aqueous solutions. JEROME KRUGER ... 131

An assessment of the suitability of the Drude-Tronstad polarized light method for the study of film growth on polycrystalline metals. P. C. S. HAYFIELD and G. W. T. WHITE 157

Studies of thin oxide films on copper crystals with an ellipsometer. J. V. CATHCART and G. F. PETERSEN 201

Optical study of the formation and stability of anodic films on aluminum. M. A. BARRETT 213

Ellipsometry in electrochemical studies. A. K. N. REDDY and J. O'M. BOCKRIS 229

Application of ellipsometry to the study of phenomena on surfaces prepared in ultra-high vacuum. J. F. DETTORRE, T. G. KNORR, and D. A. VAUGHAN 245

Adsorption

Measurement of the physical adsorption of vapors and the chemisorption of oxygen and silicon by the method of ellipsometry. R. J. ARCHER 255

Application of ellipsometry to the study of adsorption from solution. ROBERT R. STROMBERG, ELIO PASSAGLIA, and DANIEL J. TUTAS 281

Determination of thickness and refractive index of thin films as an approach to the study of biological macromolecules. J. B. BATEMAN 297

Other

Blood coagulation studies with the recording ellipsometer. L. VROMAN 335

Ellipsometry for frustrated total reflection. T. D. YOUNG and J. M. FATH 349

CONTENTS OF THE INTERNATIONAL CONFERENCES ON ELLIPSOMETRY

PROCEEDINGS OF THE SYMPOSIUM ON

RECENT DEVELOPMENTS IN ELLIPSOMETRY

ELECTRICAL MATERIALS LABORATORY
UNIVERSITY OF NEBRASKA
LINCOLN, NEBRASKA, U.S.A., AUGUST 7–9, 1968

Guest Editors:

N. M. BASHARA
*Electrical Materials Laboratory – Department of Electrical Engineering
University of Nebraska, Lincoln, Nebraska*

A. B. BUCKMAN
*Electrical Materials Laboratory – Department of Electrical Engineering
University of Nebraska, Lincoln, Nebraska*

A. C. HALL
Mobil Research and Development Corporation, Dallas, Texas

1969

NORTH-HOLLAND PUBLISHING COMPANY – AMSTERDAM

CONTENTS

A. C. Hall, A century of ellipsometry 1
R. H. Muller, Definitions and conventions in ellipsometry 14

ANALYSIS AND COMPUTATION

P. H. Smith, A theoretical and experimental analysis of the ellipsometer . . . 34
H. G. Jerrard, Sources of error in ellipsometry 67
F. Lukeš, The accuracy of the measurement of the ellipsometric parameters Δ and ψ . . 74
C. A. Fenstermaker and F. L. McCrackin, Errors arising from surface roughness in ellipsometric measurement of the refractive index of a surface 85
W. G. Oldham, Numerical techniques for the analysis of lossy films 97
D. G. Schueler, Error analysis of angle of incidence measurements 104

TECHNIQUES AND INSTRUMENTATION

F. LUKEŠ, W. H. KNAUSENBERGER and K. VEDAM, Ellipsometric liquid immersion method for the determination of all the optical parameters of the system: non-absorbing film on an absorbing substrate 112
P. C. S. HAYFIELD, Ellipsometry and multiple reflection between parallel plates . . 126
H. G. JERRARD, A high precision photoelectric ellipsometer 137
I. WILMANNS, A double-modulation photoelectric ellipsometer 147
J. L. ORD, An ellipsometer for following film growth 155
B. D. CAHAN and R. F. SPANIER, A high speed precision automatic ellipsometer . 166
H. P. LAYER, Circuit design for an electronic self-nulling ellipsometer 177

FIELD EFFECTS

A. B. BUCKMAN, Modulated ellipsometry for band structure studies of solids and films 193
W. N. HANSEN, Electromodulation of the optical properties of metals 205
J. A. JOHNSON and D. W. PETERSON, The relative electric fields in a thin film on an opaque substrate 217

SORPTION STUDIES

F. MEYER and G. A. BOOTSMA, Ellipsometric investigation of chemisorption on clean silicon (111) and (100) surfaces 221
R. H. MULLER, R. E. STEIGER, G. A. SOMORJAI and J. M. MORABITO, Gas adsorption studies by ellipsometry in combination with low energy electron diffraction and mass spectrometry 234
J. J. CARROLL and A. J. MELMED, Ellipsometry–LEED study of the adsorption of oxygen on (011) tungsten 251

FILMED SURFACES

H. YOKOTA, H. SAKATA, M. NISHIBORI and K. KINOSITA, Ellipsometric study of polished glass surfaces 265
H. YOKOTA, M. NISHIBORI and K. KINOSITA, Ellipsometric study of a thin transparent film overlaid on a transparent substrate having a surface layer 275
R. J. KING and M. J. DOWNS, Ellipsometry applied to films on dielectric substrates 288
D. K. BURGE, J. M. BENNETT, R. L. PECK and H. E. BENNETT, Growth of surface films on silver 303
B. J. BORNONG, Ellipsometric study of films of polar organic molecules on chromium 321
C. J. DELL'OCA and L. YOUNG, Ellipsometry of non-uniform anodic oxide films . 331

OTHER APPLICATIONS

C. L. MCBEE and J. KRUGER, Ellipsometric-spectroscopy of films formed on metals in solution 340
L. A. WEITZENKAMP, Ellipsometric study of amorphous selenium on vacuum-evaporated gold 353
W. C. MALM, A. J. D. LIU and H. H. SOONPAA, Optical constants of $Bi_8Te_7S_5$. 365
P. C. S. HAYFIELD, Studies by ellipsometry on high pressure gas and liquid reactions 370
J. R. ADAMS and K. K. RAO, Study of thin film annealing by ellipsometry . . 382
D. L. JOHNSON and L. C. TAO, Application of ellipsometry to oxidation studies of binary titanium–aluminum alloys 390
W. PRIMAK, Determination of small dilatations and surface stress by birefringence measurements 398
C. MATHOT and A. ROTHEN, The immunoelectroadsorption method 428
L. VROMAN and A. L. ADAMS, Findings with the recording ellipsometer suggesting rapid exchange of specific plasma proteins at liquid/solid interfaces . . . 438

AUTHOR INDEX 447

SUBJECT INDEX 449

CONTENTS OF THE INTERNATIONAL CONFERENCES ON ELLIPSOMETRY

ELLIPSOMETRY

PROCEEDINGS OF THE THIRD INTERNATIONAL
CONFERENCE ON ELLIPSOMETRY

UNIVERSITY OF NEBRASKA, LINCOLN, NEBRASKA, USA
23–25 SEPTEMBER 1975

Guest Editors:

N.M. BASHARA and R.M.A. AZZAM
University of Nebraska

1976

NORTH-HOLLAND PUBLISHING COMPANY – AMSTERDAM

A.B. Winterbottom, Summary of conditions for precision in applications of ellipsometry to surface and film studies	1
R.M.A. Azzam, A perspective on ellipsometry	6
R.H. Muller, Present status of automatic ellipsometers	19
F. Meyer, Ellipsometric studies of adsorption reactions on clean surfaces	37
M. Malin and K. Vedam, Generalized ellipsometric method for the determination of all the optical constants of the system: optically absorbing film on an absorbing substrate	49
K. Kinosita and M. Yamamoto, Principal angle-of-incidence ellipsometry	64
M.A.B. Gültepe, Some considerations on the determination of optical properties of adsorbed films	76
A.-R.M. Zaghloul, R.M.A. Azzam and N.M. Bashara, Inversion of the nonlinear equations of reflection ellipsometry on film–substrate systems	87
S.S. So, Ellipsometric analyses for an absorbing surface film on an absorbing substrate with or without an intermediate surface layer	97
A. Rothen, Immunologic reactions carried out at a liquid–solid interface	109

S.M. Ma, D.L. Coleman and J.D. Andrade, Ellipsometry studies of albumin films on tantalum oxide and SiO_2	117
R.M.A. Azzam, Use of a light beam to probe the cell surface in vitro	126
A.C. Lowe, Practical limitations to accuracy in a nulling automatic wavelength-scanning ellipsometer	134
P.S. Hauge, Generalized rotating-compensator ellipsometry	148
D.E. Aspnes, A photometric ellipsometer for measuring flux in a general state of polarization	161
D.N. Henty and H.G. Jerrard, A universal ellipsometer	170
O. Hunderi and R. Ryberg, A simple automatic ellipsometer for a wide energy range	182
E.G. Lluesma, C.A. Pela and I. Wilmanns, A new automatic ellipsometer	189
C.C. Matheson, J.G. Wright, R. Gundermann and H. Norris, A high precision, polychromatic, automatic ellipsometer	196
T. Smith, An automated scanning ellipsometer	212
K. Vedam, Characterization of defects in real surfaces by ellipsometry	221
F. Abelès, Surface electromagnetic waves ellipsometry	237
T. Smith, Effect of surface roughness on ellipsometry of aluminum	252
D. den Engelsen, Optical anisotropy in ordered systems of lipids	272
M. Elshazly-Zaghloul, R.M.A. Azzam and N.M. Bashara, Explicit solution for the optical properties of a uniaxial crystal in generalized ellipsometry	281
D.J. De Smet, Generalized ellipsometry and the 4×4 matrix formalism	293
J.R. Adams, Complex refractive index and phosphorus concentration profiles in P_{31}^+ ion implanted silicon by ellipsometry and Auger electron spectroscopy	307
S. Kawabata and K. Ichiji, Optical anisotropy in the plane parallel to the surface of obliquely evaporated gold films	316
D.E. Aspnes, Extending scanning ellipsometric spectra into experimentally inaccessible regions	322
R.W. Stobie, B. Rao and M.J. Dignam, Infrared ellipsometry of adsorbed molecules: CO on evaporated Cu films	334
B.D. Cahan, Implications of three parameter solutions to the three-layer model	354
S. Gottesfeld, M. Babai and B. Reichman, Combined ellipsometric and reflectometric measurements of surface processes on noble metal electrodes	373
J. Kruger and J.R. Ambrose, Qualitative use of ellipsometry to study localized corrosion processes	394
J.L. Ord, An optical study of the deposition and conversion of nickel hydroxide films	413
E.E.J. Roberts and D. Ross, Anomalous optical constants of thin films	425
R.H. Muller and C.G. Smith, Ellipsometry of mass-transport boundary layers	440
G.R. Boyer and B.S. Prade, Atmospheric and wind tunnel air-flow-birefringence measurements	449
E.P. Honig and B.R. de Koning, Ellipsometric investigation of the skeletonization process of Langmuir–Blodgett films	454
T.H. Allen, Ellipsometric measurements on vapor deposited organic films	462
M.A. Hopper, R.A. Clarke and L. Young, Removal of oxide films from silicon surfaces in a high vacuum system: an in situ ellipsometric study	472
J.R. Adams and D.K. Kramer, A study of the oxidation of tantalum nitride by ellipsometry and Auger electron spectroscopy	482
P.C.S. Hayfield, Ellipsometry as an aid in studying metallic corrosion problems	488
Author index	508
Subject index	511
Critique participants	518

ELLIPSOMETRY

PROCEEDINGS OF THE FOURTH INTERNATIONAL
CONFERENCE ON ELLIPSOMETRY

LAWRENCE BERKELEY LABORATORY, UNIVERSITY OF CALIFORNIA
BERKELEY, CALIFORNIA, 20–22 AUGUST 1979

Guest Editors:
R.H. MULLER
Lawrence Berkeley Laboratory
R.M.A. AZZAM
University of New Orleans
D.E. ASPNES
Bell Laboratories

1980

NORTH-HOLLAND PUBLISHING COMPANY – AMSTERDAM

CONTENTS

Preface vii

OPTICAL THEORY

O. Hunderi, Optics of rough surfaces, discontinuous films and heterogeneous materials . 1
F. Abelès and T. Lopez-Rios, Ellipsometry of metallic films and surfaces with nonlocal effects 32
P. Yeh, Optics of anisotropic layered media: a new 4 × 4 matrix algebra 41
T. Inagaki, H. Kuwata and A. Ueda, Dispersion relations and sum rules for the ellipsometric function 54
R.M.A. Azzam, Ellipsometry of transparent films on transparent substrates 67
P.S. Hauge, R.H. Muller and C.G. Smith, Conventions and formulas for using the Mueller–Stokes calculus in ellipsometry 81

INSTRUMENTATION AND TECHNIQUE

P.S. Hauge, Recent developments in instrumentation in ellipsometry 108
N. Umeda and H. Takasaki, New ellipsometry realized by the use of a stabilized two-frequency laser 141
G.R. Boyer, B.F. Lamouroux and B.S. Prade, Fast three-dimensional ellipsometry 149
E. Collett, Determination of the ellipsometric characteristics of optical surfaces using nanosecond laser pulses 156
A.-R.M. Zaghloul and R.M.A. Azzam, Single-element rotating-polarizer ellipsometer: psi meter 168
D.W. Stevens, A perpendicular-incidence microellipsometer 174
M. Yamamoto and O.S. Heavens, A vacuum automatic ellipsometer for principal angle of incidence measurement 202
A. Straaijer, L.J. Hanekamp and G.A. Bootsma, The influence of cell window imperfections on the calibration and measured data of two types of rotating-analyzer ellipsometers 217
A.-R.M. Zaghloul and M. Elshazly-Zaghloul, Reflectance-aided null ellipsometry (RANE) for film–substrate systems 232

SOLID STATE APPLICATIONS

E.T. Arakawa, T. Inagaki and M.W. Williams, Optical properties of metals by spectroscopic ellipsometry 248
J.B. Theeten, Real-time and spectroscopic ellipsometry of film growth: application to multilayer systems in plasma and CVD processing of semiconductors 275
D.E. Aspnes and A.A. Studna, An investigation of ion-bombarded and annealed ⟨111⟩ surfaces of Ge by spectroscopic ellipsometry 294

Q. Kim and Y.S. Park, Ellipsometric investigation of ion-implanted GaAs 307
S.C. Mushakarara and K. Vedam, Characterization of surfaces of laser-annealed samples by ellipsometry 319
G.H. Bu-Abbud and N.M. Bashara, Characterization of fabrication damage in $SrTiO_3$ by internal and external measurements 329
A.B. Buckman and S. Chao, Ellipsometric characterization of the glassy layer at metal/semiconductor interfaces 346
D. Beaglehole and D. Nason, Transition layer on the surface on ice 357

ELECTROCHEMISTRY AND CORROSION APPLICATIONS

J.J. Ritter and J. Kruger, A qualitative ellipsometric-electrochemical approach to the study of film growth under organic coatings 364
R.H. Muller and C.G. Smith, The use of film-formation models for the interpretation of ellipsometer observations 375
W. Paik and Z. Szklarska-Smialowska, Reflectance and ellipsometric study of anodic passive films formed on nickel in sodium hydroxide solution 401
R. Nishimura, K. Kudo and N. Sato, Intensity-following ellipsometry of passive films on iron 413
C.Y. Chao and Z. Szklarska-Smialowska, Ellipsometric study on the film formation of nickel in phosphate solutions 426
M. Yamashita, K. Omura and D. Hirayama, Passivating behavior of copper anodes and its illumination effects in alkaline solutions 443
M. Babai and S. Gottesfeld, Ellipsometric study of the polymeric surface films formed on platinum electrodes by the electrooxidation of phenolic compounds 461
A. Moritani, H. Kubo and J. Nakai, In situ ellipsometric study of successive anodization of monolayers in some layered semiconductors 476

ADSORPTION AND BIOLOGICAL STUDIES

F.H.P.M. Habraken, O.L.J. Gijzeman and G.A. Bootsma, Ellipsometry of clean surfaces, submonolayer and monolayer films 482
J.J. Carroll, T.E. Madey, A.J. Melmed and D.R. Sandstrom, The room temperature adsorption of oxygen, hydrogen and carbon monoxide on ($11\bar{2}0$) ruthenium: an ellipsometry–LEED characterization 508
U. Merkt and P. Wissmann, Ellipsometric response to CO adsorption and temperature change for thin silver films 529
R.B. Davis, R.M.A. Azzam and G. Holtz, Ellipsometric observations of surface adsorption and molecular interactions of native and modified fibrinogen and factor VIII . . 539
P.A. Cuypers, M.P. Janssen, J.M.M. Kop, W.Th. Hermens and H.C. Hemker, Temperature dependent transitions in several phospholipids measured by ellipsometry 555

Papers presented at the conference, but not included in the proceedings 564
List of Conference Participants 565

Author index 569
Subject index 572

JOURNAL DE PHYSIQUE

Tome 44 — Colloque C10, supplément au n° 12 — Décembre 1983

Conférence Internationale sur

Ellipsométrie et autres Méthodes Optiques pour l'Analyse des Surfaces et Films Minces

Ellipsometry and other Optical Methods for Surface and Thin Film Analysis

7-10 juin 1983
Paris (France)

les éditions de physique

Avenue du Hoggar,
Zone Industrielle de Courtabœuf, B.P. 112
91944 Les Ulis Cedex, France

TABLE DES MATIERES

NEW DEVELOPMENTS IN ELLIPSOMETRY AND OTHER OPTICAL TECHNIQUES - PROGRES RECENTS EN ELLIPSOMETRIE ET DANS LES AUTRES TECHNIQUES OPTIQUES.

D.E. ASPNES.- Optical characterization by ellipsometry - a prospective.... C10-3

A.J. SIEVERS and Z. SCHLESINGER.- IR surface plasmon spectroscopy......... C10-13

D. CHANDLER-HOROWITZ and G.A. CANDELA.- On the accuracy of ellipsometric thickness determinations for very thin films........................... C10-23

J.C. CHARMET and P.G. de GENNES.- Ellipsometric formulas for an index profile of small amplitude but arbitrary shape........................... C10-27

S. LOGOTHETIDIS and J. SPYRIDELIS.- Optical constants of layer structures from ellipsometric data.. C10-31

M. YAMAMOTO.- Ellipsometry of thick ($d>\lambda$) transparent films using channelled spectrum.. C10-35

H.W. DINGES.- Determination of the optical properties of GaAs and InP by multiple angle of incidence ellipsometry............................... C10-39

S.M. KELSO and D.E. ASPNES.- Ellipsometric studies of $GaAs_{1-x}P_x$: Fourier analysis of critical point complexes........................... C10-45

E. ZAMFIR and C. OANCEA.- The anisotropy of the PbTe films obliquely deposited in vacuum studied by ellipsometry and photovoltaic effect........ C10-49

S.V. PIHLAJAMÄKI and J.J. KANKARE.- A microprocessor controlled spectroellipsometer for study of electrochemical systems....................... C10-53

R.H. MULLER and J.C. FARMER.- Fast, self-nulling spectroscopic ellipsometer : instrumentation and application................................. C10-57

A. MORITANI, Y. OKUDA and J. NAKAI.- A retardation modulation ellipsometer for studying fast surface transients................................... C10-63

R.M.A. AZZAM.- Ellipsometry of unsupported and embedded thin films....... C10-67

R.M.A. AZZAM.- Theory, implementation and applications of AIDER.......... C10-71

T. SANDSTRÖM.- Differential ellipsometry................................. C10-75

L. STIBLERT and T. SANDSTRÖM.- Geometrical resolution in the comparison ellipsometer... C10-79

M. STENBERG and H. NYGREN.- The use of the isoscope ellipsometer in the study of adsorbed proteins and biospecific binding reactions........... C10-83

R.R. SCHAEFER.- Infrared ellipsometry - A new technique for characterization of dopant parameters in silicon..................................... C10-87

B. YOUS, A. DONNADIEU et S. ROBIN.- Réflectivité hémisphérique et diffuse de couches minces de tungstène préparées par décomposition en phase vapeur... C10-91

I.M. REDA, A. WAGENDRISTEL, H. BANGERT and P. SCHATTSCHNEIDER.- Determination of the growth rates of Au_2Al diffusionally formed in thin step-shaped Au-Al couples... C10-95

H.-E. KORTH.- A computer integrated spectrophotometer for film thickness monitoring... C10-101

R.E. HUMMEL.- Differential reflectometry................................. C10-105

G. HINCELIN.- Accuracy in the determination of thin films optical constants by the ATR method... C10-109

D. RIVIÈRE and G.R. ROGER.- Investigation of a very thin and inhomogeneous film of SiO from surface plasmon optical excitation................... C10-113

M. ABRAHAM, A. MAHMOUDI, A. TADJEDDINE and J.P. ROLLAND.- Determination of the dielectric constant of CdS by ellipsometry, interferometry and by optical excitation of surface plasmons................................ C10-119

M. OLIVIER, J.C. PEUZIN and A. CHENEVAS-PAULE.- Broadband spectroscopy of ultrathin layers by optical guided waves techniques. Application to A 80 Å thick a - Si : H layer... C10-123

K. SASAKI, T. SHIMIZU, O. NONAKA, O. HAMANO and M. SERIZAWA.- Anisotropic refractive indices of cadmium sulfide thin film on a slab-type optical waveguide... C10-127

J.-P. GRUSON, P.-J. GOIRAND, F. COCHET and O. PARRIAUX.- Characterization of dielectric films on semiconductor substrates by leaky-modes measurement... C10-131

OPTICAL CHARACTERIZATION OF INTERFACES (SOLID/LIQUID, LIQUID/VAPOUR, SOLID/SOLID) AND SURFACE LAYERS - CARACTERISATION OPTIQUE DES INTERFACES (SOLIDE/LIQUIDE, LIQUIDE/VAPEUR, SOLIDE/SOLIDE) ET DES REACTIONS DE SURFACE.

D.M. KOLB.- Electroreflectance spectroscopy in the study of metal-electrolyte interfaces.. C10-137

D. BEAGLEHOLE.- Ellipsometry of liquid surfaces........................... C10-147

D. LANGEVIN and J. MEUNIER.- Light scattering and reflectivity of liquid interfaces... C10-155

M. TOMKIEWICZ and W. SIRIPALA.- Characterization of surface states at a semiconductor electrolyte interface by electrolyte electroreflectance spectroscopy... C10-163

M. TOMKIEWICZ and R. GARUTHARA.- Characterization of semiconductor electrolyte interface by modulated photoluminescence....................... C10-167

J.G. GORDON II.- ATR studies of water on emersed electrodes............... C10-171

P. SCHMIDT and W. PLIETH.- Optical absorption of molecules in the electric field of the double layer... C10-175

E.D. PALIK and V.M. BERMUDEZ.- Ellipsometric investigation of the silicon / anodic-oxide interface.. C10-179

J. JOSEPH et A. GAGNAIRE.- Etude "in situ" de la croissance d'oxyde anodique par ellipsométrie.. C10-183

M. FROELICHER, A. HUGOT-LE GOFF, V. JOVANCICEVIC, R. DUPEYRAT and M. MASSON.- Study of surfaces covered by thin films, by SPEC and Raman spectroscopy.. C10-187

T. OHTSUKA, K. AZUMI and N. SATO.- An in-situ reflection-spectroscopic study applied to anodic oxide films on iron, nickel, and titanium...... C10-191

A. GAGNAIRE et J. JOSEPH.- Ellipsomètre à épaisseur variable de film. Application TiO_2/Ti.. C10-195

E.F.I. ROBERTS-SENGIER, M.A. HAMILTON and R. HUNT.- Polyphase substrate ellipsometry - a kinetic investigation................................... C10-199

P. DUMAS, J.P. DUBARRY-BARBE, D. RIVIERE, Y. LEVY and J. CORSET.- Growth of thin alumina film on aluminium at room temperature : a kinetic and spectroscopic study by surface plasmon excitation...................... C10-205

P. PICOZZI and S. SANTUCCI.- Transmittance measurements on discontinuous copper films : a sensitive method for studying the oxidation process... C10-209

W. VISSCHER.- Ellipsometry of nickel-oxides and -hydroxides in alkaline electrolyte.. C10-213

T. HOSHINO, A. MORITANI and J. NAKAI.- Ellipsometric study of nucleation and growth in the anodic film formation on Bi_2Te_3...................... C10-217

J.L. ORD.- Application of ellipsometry to studies of electrostriction and electrochromism in anodic oxide films.................................. C10-221

J.J. RITTER and J. KRUGER.- Further development of the qualitative ellipsometric technique for the study of corrosion processes under organic coatings... C10-225

C. ALIBERT, S. GAILLARD, M. ERMAN and P.M. FRIJLINK.- Electroreflectance and spectroscopic ellipsometry studies of GaAs/GaAlAs heterojunctions.. C10-229

G. CHABRIER, A. MOUSTAIDE, G. NIQUET and P. VERNIER.- Electroreflectance of thin film Al-ZnS-Al sandwiches.. C10-235

J.A. LAHTINEN and T. TUOMI.- Electroreflectance of thin zinc sulfide films grown by atomic layer epitaxy and electron beam evaporation techniques. C10-239

J.W. SCHMIDT and M.R. MOLDOVER.- Ellipsometry of thin films on vapor-liquid interfaces.. C10-243

J. PERRIN and B. DREVILLON.- Growth characterization of a Si:H films by multiple angle of incidence spectroscopic ellipsometry................. C10-247

Y. DEMAY, P. MAUREL and S. GOURRIER.- Interactions of Si (III) surface with H_2, NH_3, SiH_4 multipolar plasmas studied by in situ ellipsometry.. C10-253

A. VAREILLE, A. STRABONI, B. VUILLERMOZ et Ph. GED.- Etude par ellipsométrie en temps réel d'anodisations plasma de silicium monocristallin.... C10-257

M. ERMAN, P. CHAMBON, B. PREVOT and C. SCHWAB.- Spectroscopic ellipsometry and Raman scattering analysis of Be shallow implantation in GaAs... C10-261

Ph. GED and M.H. DEBROUX.- Evidence of structural modification of implanted silicon during furnace annealing.................................... C10-267

G. CELOTTI, R. ROSA, C. SUMMONTE and G. MARTINELLI.- A new ellipsometric programme applied to the characterization of transparent conducting titanium nitride films... C10-273

G.M. JAKUBOWSKA.- Electromigration in d.c. electroluminescent thin films of ZnS.. C10-277

OPTICAL SURFACE EFFECTS (SERS, SURFACE AND VOLUME INHOMOGENEITIES, NON-LOCAL EFFECTS, ETC...) - EFFETS OPTIQUES DE SURFACE (SERS, INHOMOGENEITES DE SURFACE ET DE VOLUME, EFFETS NON LOCAUX, ETC...)

R.K. CHANG.- Surface enhanced Raman scattering : an overview of experiments... C10-283

M.R. PHILPOTT.- Theory of surface enhanced Raman scattering : a prospective view.. C10-295

S. LUNDQVIST and P. APELL.- Non-local electromagnetic effects at metal surfaces.. C10-305

A.G. MAL'SHUKOV.- Raman scattering from a molecule adsorbed at the metal surface. The molecular charge oscillation mechanism..................... C10-315

T. MANIV.- Raman side-bands in the reflectivity from metals due to surface impurities... C10-321

A.V. BOBROV, A.N. GASS, O.I. KAPUSTA and N.M. OMEL'YANOVSKAYA.- Surface enhanced Raman scattering of ethylene adsorbed on vapor deposited silver in UHV. Effects of the silver film thickness and coverage.............. C10-327

J. BILLMANN and A. OTTO.- The role of charge transfer excitations between silver and pyridine or cyanide in surface Raman enhancement............ C10-331

B. PETTINGER and L. MOERL.- Evidence for low density of active sites in surface-enhanced Raman scattering....................................... C10-333

C. PETTENKOFER and A. OTTO.- On the "first layer" enhancement in Raman scattering of pyridine on silver.. C10-337

A. CAMPION, V.M. GRIZZLE, D.R. MULLINS and J.K. BROWN.- Raman spectroscopy of molecules adsorbed on single crystal metal surfaces without enhancement... C10-341

S. GAROFF, D.A. WEITZ, M.S. ALVAREZ and J.C. CHUNG.- Electromagnetically induced changes in intensities, spectra, and temporal behavior of light scattering from molecules on silver-island films....................... C10-345

M. NEVIÈRE and R. REINISCH.- Electromagnetic theory of enhanced nonlinear optical process.. C10-349

T. LÓPEZ-RÍOS, Y. BORENSZTEIN and G. VUYE.- Investigation of electromagnetic fields at a rough Ag surface by differential reflectometry of Cu and Al adsorbates... C10-353

G.A. FARIAS and A.A. MARADUDIN.- Surface plasmons on a randomly rough surface.. C10-357

J. VLIEGER and M.M. WIND.- Optical properties of thin films on rough surfaces.. C10-363

G. RASIGNI, M. RASIGNI, F. VARNIER, J.P. PALMARI and A. LLEBARIA.- Restitution of surface profiles from microdensitometer analysis of electron micrographs of surface replicas... C10-367

A.G. MAL'SHUKOV.- Non-local theory for the far-infrared absorption in small metal particles... C10-371

P. AHLQVIST, P. DE ANDRÉS, R. MONREAL and F. FLORES.- Quantum-size effects on the optical properties of small particles........................... C10-375

D. BEDEAUX and J. VLIEGER.- A statistical theory for the dielectric properties of thin island films, application and comparison with experimental results.. C10-379

M. YAMAMOTO, K. ICHIJI, Y. WADA, K. TAKEUCHI and K. KINOSITA.- Immersion-type transmission ellipsometry of very thin metal films................ C10-383

S. DAMASKINOS, F.E. GIROUARD, A. PINARD and V.V. TRUONG.- Structures and optical behavior of overcoated granular films......................... C10-387

K. KEMPA, F. FORSTMANN, R. KÖTZ and B.E. HAYDEN.- Ellipsometric study of oxygen on a Ag(110) surface around the plasma frequency of silver...... C10-393

F. FORSTMANN, K. KEMPA, G. PIAZZA and D.M. KOLB.- Bulk plasmon resonances in electroreflectance spectra of thin silver layers.................... C10-395

C. WIJERS.- Numerical calculation of the optical properties of linear chain systems.. C10-397

V.M. AGRANOVICH, V.E. KRAVTSOV and T.A. LESKOVA.- Spectroscopy of surface polariton diffraction.. C10-403

OPTICAL STUDIES OF ADSORBED MOLECULES AND POLYMERS - ETUDES OPTIQUES DES MOLECULES ADSORBEES ET DES POLYMERES

B.N.J. PERSSON.- Lateral interactions in small particle systems........... C10-409

R. RYBERG.- Infrared spectroscopy of adsorbed molecules ; some experimental aspects.. C10-421

E. BURSTEIN, A. BROTMAN and P. APELL.- The valence electron excitations and the optical properties of adsorbed atoms and molecules on metal surfaces.. C10-429

D. MÖBIUS.- Light reflection by complex dye monolayers................... C10-441

Ph. AVOURIS, J.E. DEMUTH and N.J. DINARDO.- The nature of the excited states of adsorbates and their decay mechanisms........................... C10-451

F. BOCCUZZI and G. GHIOTTI.- IR study of CO chemisorbed on metallic Cu dispersed in ZnO... C10-455

M. WATANABE and P. WISSMANN.- Ellipsometric studies of oxygen adsorbed on silver films.. C10-459

C. PETTENKOFER, I. POCKRAND and A. OTTO.- Surface enhanced Raman spectra from oxygen on silver.. C10-463

T.M. CHRISTENSEN and J.M. BLAKELY.- Oxidation studies of the Be(0001) surface.. C10-465

L.J. HANEKAMP and A. VAN SILFHOUT.- The behaviour of oxygen on Ni(110) investigated by ellipsometry and AES................................... C10-469

G. QUENTEL and R. KERN.- Ellipsometric study of the cleaved germanium (111) surface as a function of temperature........................... C10-473

Y. BORENSZTEIN, T. LÓPEZ-RÍOS and G. VUYE.- Comparative study of the optical properties of silver clusters and silver continuous overlayers on aluminium substrate by differential reflectometry..................... C10-475

M. ORRIT, J. BERNARD, J.M. TURLET and Ph. KOTTIS.- Monolayer surface Raman and fluorescence spectroscopy of the anthracene crystal............... C10-479

S. GAROFF and C.J. SANDROFF.- Surface-enhanced Raman studies of the liquid/solid interface.. C10-483

E. KOGLIN and J.M. SÉQUARIS.- The short-range component of SERS : results of biopolymers.. C10-487

J.M.M. KOP, J.W. CORSEL, M.P. JANSSEN, P.A. CUYPERS and W.Th. HERMENS.- Ellipsometric measurement of the association of prothrombin with phospholipid monolayers.. C10-491

P.A. CUYPERS, J.W. CORSEL, J.M.M. KOP, M.P. JANSSEN and W.Th. HERMENS.- Ellipsometric measurement of the specific volumes of phospholipids and proteins at solid-water interfaces.................................... C10-495

H. ELWING, A. PALMQVIST and A. ASKENDAL.- Ellipsometric quantification of antigen-antibody precipitates adsorbed on antigen-coated silicon surfaces. A new principle for immunochemical analysis..................... C10-499

J.D. SWALEN and J.F. RABOLT.- Raman spectroscopy in guided optical waves.. C10-501

R. GREEF and P.J. PEARSON.- Ellipsometry in absorbing solutions........... C10-505

Y. BOUIZEM, F. CHAO, M. COSTA and A. TADJEDDINE.- Ellipsometric study of acrylonitrile electropolymerisation on nickel........................... C10-509

C. NGUYEN VAN HUONG and C. HINNEN.- Application of the visible electroreflectance spectroscopy to the characterization of processes occuring on gold electrodes... C10-513

H. NEUGEBAUER, A. NECKEL, G. NAUER, N. BRINDA-KONOPIK, F. GARNIER and G. TOURILLON.- In situ IR-ATR-spectroscopy of electrode surfaces....... C10-517

K. TOTH, G. VUYE, H. VAINER, Gy. RONTO and D. ASLANIAN.- Ellipsometric studies of biological macromolecules................................. C10-521

INDEX AUTEURS.. C10-525

Author Index

Abelès, F. [132] 269; [135] 269; [139] 269; [140] 269; [181] 340
Abramowitz, M. [121] 261
Adams, A.C. [277] 464, 465, 466
Adams, J.R. [298] 476, 479, 480
Airy, G.B. [144] 248
Ambs, W.J. [243] 428, 429
Ammann, E.O. [24] 66
Archer, R.J. [73] 187, 371; [150] 298, 299; [159] 212, 302, 373, 383, 403, 423, 424, 425; [199] 359
Aspnes, D.E. [91] 187, 379; [92] 187; [94] 187; [114] 257, 411, 414, 419; [118] 260; [119] 260; [161] 303; [233] 414; [234] 418; [236] 420; [237] 419
Azzam, R.M.A. [18] 29, 85, 86; [19] 40, 240, 242; [21] 51, 85; [37] 88, 90, 94; [38] 90; [40] 91; [41] 94; [47] 103, 113, 246; [52] 123; [54] 131, 136, 137; [66] 157, 187, 233, 240; [69] 184; [95] 187; [96] 187; [97] 187, 383; [98] 187; [99] 187, 213; [100] 187, 212, 475; [102] 236, 242; [109] 254, 289, 319, 370; [120] 260; [164] 307, 318; [205] 361

Baier, R.E. [280] 466
Baker, D.J. [26] 67
Ballard, S.S. [6] 13, 28
Bartell, L.S. [244] 429, 430
Bashara, N.M. [1] 11, 155, 270, 317; [18] 29, 85, 86; [19] 40, 240, 242; [21] 51, 85; [37] 88, 90, 94; [38] 90; [40] 91; [41] 94; [47] 103, 113, 246; [52] 123; [54] 131, 136, 137; [66] 157, 187, 233, 240; [69] 184; [83] 176, 187; [95] 187; [96] 187; [97] 187, 383; [98] 187; [99] 187, 213; [100] 187, 212, 475; [102] 236, 242; [109] 254, 289, 319, 370; [120] 260; [124] 265; [125] 265, 426, 427, 428; [127] 268, 419, 420; [157] 302; [164] 307, 318; [175] 319, 320, 329, 331, 332; [179] 332; [205] 361; [222] 392, 395; [245] 431, 432; [294] 473, 475, 477; [295] 476, 478; [296] 476; [297] 476; [298] 476, 479, 480
Bateman, J.B. [286] 467
Baumel, R.T. [142] 274, 370
Beckman, K.H. [151] 298
Bennett, H.E. [10] 13, 28, 74, 79, 282, 366, 367, 373; [134] 269; [156] 302, 307, 322, 327, 332; [263] 450, 451
Bennett, J.M. [10] 13, 28, 74, 79, 282, 366, 367, 373; [134] 269; [263] 450, 451
Berreman, D.W. [184] 340; [186] 340, 345, 348, 352; [189] 347; [200] 360
Billard, J. [182] 340

AUTHOR INDEX

Blodgett, K.B. [193] 356, 467; [194] 356, 467
Bockris, J.O'M. [168] 318, 319; [169] 318, 319; [264] 452, 453; [267] 455, 456, 457; [271] 457, 461, 462
Bootsma, G.A. [251] 437, 439, 440
Born, M. [23] 28, 58, 60, 73, 42, 332; [185] 340; [188] 345
Boutry, G.A. [104] 248
Brewster, D. [106] 253
Brice, D.K. [299] 477
Brunauer, S. [249] 434
Bryngdahl, O. [28] 67
Buckman, A.B. [1] 11, 155, 270, 317; [124] 265; [125] 265, 426, 427, 428; [126] 265; [308] 483, 484
Budde, W. [231] 411
Bundy, T.L. [102] 236, 242
Burge, D.K. [117] 260, 262, 415, 416; [156] 302, 307, 322, 327, 332; [263] 450, 451

Cahan, B.D. [111] 257, 411, 475; [268] 456
Cahil, J. [306] 481, 482
Calkins, H.E. [286] 467
Calvert, P. [272] 458, 462, 463
Cassidy, J.C. [30] 67
Celli, V. [213] 363; [214] 363
Chambers, L.A. [286] 467
Chou, N.J. [260] 446, 447; [306] 481, 482
Chowdhury, C. [105] 249
Church, E.L. [211] 363
Churchill, D. [244] 429, 430
Claibone, R. [291] 470
Clarke, D. [8] 13, 28, 257, 366, 367, 373
Cohen, M.H. [61] 154
Collins, J. [12] 22
Colson, J.P. [167] 316, 317, 319, 332
Cornish, W.D. [307] 481, 482, 483
Crook, A.W. [180] 339

Dawson, E.F. [45] 103
De Lang, H. [36] 88, 98
Dell'Oca, C.J. [269] 456, 458, 459, 460
Den Engelsen, D. [192] 356; [241] 424, 426
Deschamps, G.A. [17] 28; [34] 85
Dieselman, H.D. [292] 473
Dignam, M.J. [86] 187, 383; [246] 431, 432; [247] 433
Dill, F.H. [115] 257, 260, 402, 411, 412, 413, 486
DiMarzio, E.A. [256] 442, 443

Dingle, J.H. [284] 467
Downs, M.J. [216] 372, 401
Drude, P. [63] 155, 467; 307; [187] 345
Dutton, R.C. [280] 466

Eaton, D.H. [84] 187
Eichmann, G. [35] 85
Elshazly-Zaghloul, M. [164] 307, 318
Emmett, P.H. [249] 434

Faber, T.E. [238] 422
Fencht, D.L. [15] 28
Fenstermaker, C.A. [204] 361, 362
Forgács, G. [87] 187, 383
Fresnel, A. [33] 84
Fuchs, B.A. [20] 34, 44, 45, 51, 86
Fymat, A.L. [27] 67

Galolobov, V.M. [312] 486
Garnett, J.C. Maxwell [196] 359, 418
Genshaw, M.A. [264] 452, 453; [267] 455, 456, 457; [271] 457, 461, 462
Gergely, G. [145] 289, 298
Ghezzo, M. [80] 187, 383
Giaever, I. [288] 467, 468
Gold, E.R. [282] 466
Gottesfeld, S. [218] 268
Grainger, J.F. [8] 13, 28, 257, 366, 367, 373
Graves, R.H.W. [90] 187, 383; [191] 356.
Greef, R. [112] 257, 411
Greenham, L.W. [292] 471, 472, 473
Grieser, D.R. [76] 187
Guttmann, M.J. [7] 13, 28

Hacskaylo, M. [162] 306
Hall, A.C. [1] 11, 155, 270, 317
Hammer, R. [306] 481, 482
Hansen, W.N. [166] 315
Harrick, N.J. [165] 314
Hasunuma, H. [230] 410
Hauge, P.S. [115] 257, 260, 402, 411, 412, 413, 486
Hayfield, P.C.S. [146] 292, 332, 448; [147] 292; [148] 292, 457
Hazebroek, H.F. [123] 262
Heavens, O.S. [129] 269; [130] 269; [133] 269; [235] 269, 282, 286
Helstrom, C.W. [22] 54

Henvis, B. [183] 340, 348
Hien, N.C. [232] 411, 421, 425
Hilton, A.R. [313] 486
Hodgson, J.N. [137] 269, 421
Holcomb, L.G. [127] 268, 419, 420
Holl, H.B. [143] 282
Holmes, D.A. [15] 28; [31] 74, 371
Holscher, A.A. [123] 262
Hong, N.H. [308] 483, 484
Horkans, J. [268] 456
Hsu, H. [44] 103
Hunter, W.R. [84] 187; [101] 187, 383
Hurwitz, H.C. [42] 103

Ibrahim, M.M. [175] 319, 320, 329, 331, 332; [294] 473, 475, 477; [295] 476, 478; [296] 476; [297] 476
Ince, E.L. [53] 123

Jacobsen, R.T. [242] 426, 427
Jasperson, S.N. [116] 260, 415; [117] 260, 415, 416
Jerrard, H.G. [14] 28; [78] 187
Johnson, J.A. [83] 176; [179] 332
Jones, C.E. [313] 486
Jones, R.C. [3] 13, 15, 28, 69, 212; [11] 22, 28, 98, 106; [42] 103; [43] 103; [50] 119; [51] 120; [67] 157

Kawakami, S. [29] 67
Kerker, M. [242] 426, 427
Kinbara, A. [202] 360; [203] 360
King, R.J. [72] 187; [216] 372, 401
Kinosita, K. [149] 298; [301] 477
Kluizenaar, E.E. [241] 424, 426
Knausenberger, W.H. [219] 375, 402
Knorr, K. [304] 478, 481
Kohles, R.B. [222] 392, 395
Kondo, M. [310] 484, 486
Kontsevoi, Y.A. [312] 486
Kowalk, J. [177] 324, 325
Kruger, J. [64] 155, 317; [74] 187, 371; [243] 428, 429; [262] 449, 450, 451; [270] 457, 459, 460, 463; [272] 458, 462, 463; [273] 458; [274] 460

Landsteiner, K. [287] 467
Langmuir, I. [194] 356, 467; [283] 467
Leberknight, C.E. [154] 302, 307
Lederich, R.J. [261] 449

Lenham, A.P. [93] 187
Leslie, J.D. [304] 478, 481
Liang, Y. [44] 103
Liljenvall, H.G. [85] 187, 383
Lukeš, F. [79] 187, 392; [170] 318, 319; [207] 362; [208] 362; [209] 362; [258] 444, 445; [310] 363
Lustman, B. [154] 302, 307

Malus, E. [49] 111
Maradudin, A.A. [212] 363
Marathay, A.S. [46] 103, 142; [55] 134
Marquardt, D.L. [178] 329
Marvin, A. [213] 363; [214] 363
Mathewson, A.G. [85] 187, 383
Mathieu, H.J. [228] 408, 409; [229] 410
Mathot, C. [289] 468
McBee, C.L. [274] 460
McClure, D.E. [228] 408, 409
McCrackin, F.L. [68] 168, 380, 382; [82] 187; [167] 316, 317, 319, 332; [204] 361, 362; [223] 402; [256] 442, 443
McIntyre, J.D.E. [161] 303
Meadows, A. [224] 403, 406
Merrill, B.E. [54] 131, 136, 137
Merzbacher, E. [305] 481
Meyer, F. [241] 424, 426; [251] 437, 439, 440; [252] 441; [253] 441
Mills, D.L. [212] 363
Moeller, C.E. [176] 187
Moisil, D. [81] 187
Monin, J. [104] 248
Morabito, J.M. [163] 306, 441, 442
Moskovits, M. [86] 187, 383; [246] 431, 432
Moss, G. [278] 464, 466, 469, 470, 471, 472, 474
Mueller, H. [58] 148
Muller, R.H. [2] 11, 270; [152] 298; [163] 306, 441, 442; [229] 410

Nakamura, K. [310] 484, 486
Nishibori, M. [149] 298; [301] 477
Nishida, S. [29] 67
Novak, M. [275] 460, 464, 465

O'Bryan, H.M. [107] 253, 254, 419
O'Handley, R.C. [117] 260, 262, 415, 416; [122] 262
Ohlidal, I. [207] 362; [208] 362; [209] 362; [210] 363

Oldham, W.G. [75] 187, 390; [172] 319, 326, 332
O'Neill, E.L. [5] 13, 28, 60, 142, 148
Ord, J.L. [227] 406, 408
Osborne, M.R. [177] 324, 325
Oseen, C.W. [56] 134

Paik, W.K. [168] 318, 319; [264] 452, 453
Pappenheimer, A.M. [285] 467
Parke, N.G. [59] 148
Parrent, G. [57] 142
Passaglia, E. [64] 155, 317; [68] 168, 380, 382; [254] 442; [255] 442
Peacock, D.B. [282] 466
Peck, R.L. [263] 450, 451
Peterson, D.W. [157] 302, 307; [245] 431, 432
Pliskin, W.A. [311] 485
Poincaré, H. [13] 28, 47
Porter, E.F. [285] 467
Poste, G. [278] 464, 466, 469, 470, 471, 472, 474; [292] 471, 472, 473
Pratt, W.K. [39] 90

Rajagopalan, S.R. [103] 236
Ramachandran, G.N. [9] 13, 28, 73, 117, 367, 370, 373
Ramaseshan, S. [9] 13, 28, 73, 117, 367, 370, 373; [103] 236
Reber, J.-F. [309] 483, 485
Reddy, A.K.N. [267] 455, 456, 457; [271] 457, 461, 462; [275] 460, 464, 465
Reichman, B. [128] 268
Rezvyi, R.R. [312] 486
Rice, J.R. [176] 324, 325
Richartz, M. [44] 103
Roberts, E.F. [224] 403, 406
Roberts, S. [110] 257
Roman, P. [157] 142
Rosenberg, M.D. [290] 469
Rossiter, B.W. [32] 77, 155
Rothen, A. [65] 156, 366; [276] 464, 467, 468, 469; [287] 467; [289] 468
Rowe, E.C. [160] 302, 307; [174] 319; [237] 420
Ruiz-Urbeita, M. [215] 370
Rumsey, V.H. [16] 28, 40

Sah, C.T. [84] 187
Sakata, H. [301] 477

Sanders, J.B. [300] 477
Saxena, A.N. [158] 302, 307, 346
Schaefer, V.M. [283] 467
Schmidt, E. [88] 187
Schnatterly, S.E. [116] 260, 415; [142] 274, 370
Schopper, H. [195] 357
Schroeder, J.B. [292] 473
Semenenko, A.I. [217] 374
Semenenko, L.V. [217] 374
Shabat, B.V. [20] 34, 44, 45, 51, 86
Shaffer, M.F. [184] 340; [284] 467
Shank, C.V. [73] 187, 371; [239] 421; [240] 421; [248] 434, 435, 436; [257] 443, 444
Shewchun, J. [160] 302, 307; [174] 319
Shueler, D.G. [173] 319, 326
Shurcliff, W.A. [4] 13, 28, 75, 76, 98, 366, 367, 369; [6] 13
Sigmund, P. [300] 477
Silver, S. [206] 361
Simmons, J.W. [7] 13, 28
Sivukhin, D.V. [198] 359
Smith, L.E. [223] 402
Smith, N.V. [238] 422
Smith, P.H. [77] 187, 392, 398, 399
Smith, R.C. [162] 306
Smith, T. [250] 436, 439; [259] 445, 446, 447; [303] 478
So, S.S. [171] 319; [219] 375, 402; [302] 478
Sokolov, A.V. [136] 257, 269, 270, 421; [217] 374
Som, R.F. [105] 249
Somorjai, G.A. [163] 306, 441, 442
Sosnowski, T. [190] 355
Spanier, R.F. [111] 257, 411, 475
Sparrow, E.M. [215] 370
Stair, A.T. [26] 67
Steel, M.R. [89] 187, 383
Steel, W.H. [12] 22; [25] 67
Stegun, I.A. [121] 261
Steiger, R. [309] 483, 485
Steiger, R.F. [163] 306, 441, 442
Steinberg, H.L. [68] 168, 380, 382
Stone, J.M. [141] 271
Stone, R.W. [246] 431, 432; [247] 433
Strachan, C.S. [197] 359
Stromberg, R.R. [64] 155, 317; [68] 168, 380, 382; [223] 402; [254] 442; [255] 442
Studna, A.A. [233] 414
Suits, J.C. [113] 257, 411

AUTHOR INDEX

Suzuki, S. [62] 154
Svitashev, K.K. [217] 374

Takasaki, H. [226] 406, 407
Taylor, A.E. [48] 103
Teitler, S. [183] 340, 348
Teller, E. [249] 434
Togio, F. [213] 363; [214] 363
Tronstad, L. [153] 302, 307
Tsuchiya, A. [62] 154
Tutas, D.J. [254] 442, [255] 442

Vanasse, G.A. [26] 67
Van de Hulst, M.C. [60] 149, 155
Van der Meulen, Y.J. [232] 411, 421, 425; [306] 481, 482
Vašiček, A. [131] 269, 270, 274; [155] 302
Vedam, K. [171] 319; [302] 478
Vrakking, J.J. [252] 441
Vroman, L. [277] 464, 465, 466

Waxler, R. [74] 187, 371
Weissberger, A. [32] 77, 155
Weissman, G. [291] 470
White, G.W.T. [146] 292, 332, 448
Wills, B.L. [227] 406, 408

Wilmanns, I. [218] 374
Winterbottom, A.B. [70] 187; [71] 187, 354, 357, 392, 397, 467; [225] 408
Wintrobe, M.M. [279] 465
Wolf, E. [23] 28, 58, 60, 73, 142, 332; [24] 66; [185] 340
Woodliff, H.J. [281] 466, 469
Wooten, F. [138] 269, 369, 418, 421
Wroblowa, H. [275] 460, 464, 465

Yamaguchi, S.Y. [201] 360
Yamaguchi, T. [202] 360; [203] 360; [230] 410
Yamamoto, M. [108] 254; [149] 298
Yeager, E. [268] 456
Yokota, H. [149] 298; [301] 477
Yolken, T. [74] 187, 371
Yoshida, S. [202] 360; [203] 360
Young, L. [265] 453, 454, 457; [266] 454, 457; [307] 481, 482, 483
Young, N.O. [45] 103

Zavada, J. [211] 363
Zaghloul, A.-R.M. [109] 254, 289, 319, 370
Zeidler, J.R. [221] 379; [222] 392, 395
Zobel, F.G.R. [265] 453, 454, 457

Subject Index

Absolute intensity, 16
  phase, 7, 8
Absorption,
  coefficient, 75
  index, 270
  spectrum, 430
Acceptance angle, 368
Acetylene, 441
Achromatic retarder, 372
Active optical systems, 111–112
Adsorption
  of blood proteins on foreign surfaces, 464–466
  of polymer from solution, 442, 443
Adsorption, chemical
  of several gases on silicon, 435–440
Adsorption, physical
  isotherms, 434, 435, 441
  of molecules of known dimensions on mercury, 436–439
  of oxygen, krypton and acetylene on silver, 441, 442
  of water vapor on silicon, 434, 435
  of xenon on silicon, 440
Aerosols, 155, 156
Albumin, bovine serum, 467, 468
Alignment, ellipsometer, 376–379

Aluminum, 419, 481
Ambient
  -film–substrate system, 283, 288, 301, 356–358
  phase (medium), 417–418
Ammonium dihydrogen phosphate (ADP), 406
Amplitude
  of a field component, 2
  of an elliptic vibration, 7, 8
  complex, *see* Complex amplitude
  transfer function, 94–97
Analog-to-digital converters, 413, 414
Analogy, 85, 91
Analysis, Fourier, 2, 257
  of measurements in ellipsometers, 153–268
Analytic functions, 96, 292
Analytical inversion, 315–317
Analyzer, *see* Polarizer
  elliptic, 172
  imperfect, 201, 224
  imperfections, 200–202, 212–213, 221–224, 231, 389
  linear, 163
  polarization, 154
  rotating, 257, 411

SUBJECT INDEX

Angle
  azimuth, see Azimuth
  Brewster, 79, 275, 276, 281, 307, 308, 311, 370, 383
  critical, 276, 280, 312
  ellipticity, 7, 8
  of incidence, 271
  nulling, 166, 168, 204, 207, 215, 217, 226, 389
  polarizing, 276, 293, 297
  principal, 254, 282, 297
  pseudo-Brewster, 282
  of refraction, 271
Angular deviation, see Beam deviation
Anisotropic crystals
  biaxial, 345, 371, 426
  optically active, 371
  uniaxial, 370, 424
Anisotropic films
  biaxial, 357
  uniaxial, 356, 358, 360, 342
Anisotropic media
  differential propagation Jones matrix, 121
  differential propagation 4×4 matrix, 343–347
  eigenpolarizations, 125, 138
  eigenvalues, 138, 139
  eigenwaves, 138, 139, 345
  elliptically birefringent, 128–130
  elliptically dichroic, 128, 130
  evolution of complex amplitude in, 137–141
  evolution of ellipse of polarization in, 122–137
  generalized complex refractive indices, 139
  homogenous, 123
  inhomogenous, 131
  linearly birefringent and/or linearly dichroic, 73, 76, 112, 131
  optically active, 77, 83, 345, 346, 371, 402
  propagation of polarized light in, 119–141
  reflection and transmission of polarized light by, 340–358
  stratified, 340–358
Annealing, 476
Anodic oxidation
  of niobium, 453–455
  of platinum, 455–457
  of silicon nitride films, 456–460

Anodic oxide films, see Anodic oxidation
Anodization, see Anodic oxidation
Anodization stripping, 477
Antibodies, 466, 467
Antigens, 466
Antigen–antibody reactions, 466–469
Anti-reflection coatings, 398, 405
Anti-static agent, 484
Applications of ellipsometry, 417–486
Approximation of ellipsometry equations, 301–315
Arrays of coupling coefficients, 188, 199, 200, 219, 221, 222
Astronomical ellipsometry, 154
Atmospheric oxidation, 443–445
Auger spectroscopy, 441, 446, 482
Automatic ellipsometers, 405–416
  classification, 405, 406
  null, 405–411
  photometric, 411–416
Automation, see Automatic ellipsometers
Azimuth
  angle errors in ellipsometry, 186, 193, 194, 206, 207, 209, 215, 388
  angles of optical components, 159, 160
  evolution in an anisotropic medium, 123, 129, 130
  evolution in a liquid crystal, 135, 136
  invariant, 88, 89, 93
  loci of polarization states of equal, 35, 38, 42, 45, 48, 50
  measurement, 167, 248–250
  of vibration, 7

Babinet–Soleil compensator, 371–372
Band structure, 418, 423, 426
Bandwidth, 2, 213, 398
Barium, 419
Barium stearate, 467
Basis states
  of polarization, 22, 180, 184, 487
  change of, 24, 25
Basis vectors
  arbitrary (generalized), 25
  Cartesian, 22
  circular, 22
Beam deviation, 392–397
Beam splitters, 253, 263
Behenic acid, 436–439
Beilby layer, 418

Beryllium, 419
Bessel functions, 261
Biaxial
  crystals, 345, 371, 426
  films, 357
  substrates, 355
Bilinear transformation
  change of basis states, 44, 181, 184
  circle-to-circle mapping property, 87, 88, 90, 93, 126, 177
  coordinate rotation, 47
  derivative, 91
  determination by generalized ellipsometry, 157, 158
  invariant points, 87
  from the mapping of three points, 86
  network impedance, 91
  and propagation of light in anisotropic media, 125, 126, 128
  ratio of transmission coefficients, 300
  reflection coefficient, 290
Biology, applications of ellipsometry in, 464–473
Birefringence
  cell-window, 205, 233, 374
  circular, 83, 140, 155, 158, 183
  electro-optic, 406
  elliptical, 128–130, 140, 155
  Faraday-cell, 405
  induced, 155
  lens, 374, 401
  linear, 73, 74, 131, 140, 155, 158, 174
  liquid crystal, 131
  natural, 155
  piezo, 415
  streaming, 155
  stress, 233
Birefringent, *see* Birefringence
  networks, 66
  retarder, 73, 99, 370
Blood
  antibodies, 466, 467
  antigens, 466
  banking, 465
  carcinoembryonic antigen, 466
  clotting factors, 465
  coagulation, 465
  growth hormone, 469
  Hageman factor, 466
  immunological reactions, 464, 466–469
  interaction with foreign surfaces, 465–466
  plasma, 464–466
  proteins, 464–469
  serodiagnosis, 467, 469
Bombardment, *see* Ion bombardment
Boundary conditions, 271, 272, 349, 353
Breakdown of passivity, 459, 463
Brewster angle 79, 275, 276, 281, 307, 308, 311, 370, 383
  polarizers, 79, 370
  pseudo, 282
  second, 282
Brunaur–Emmett–Teller theory, 434
Built-up films, 467
Butanol, 433

Calcite, 367, 371, 373
Calcium, 419
Calculator, programmable, 415
Calibration, ellipsometer, 380–385
Caproic acid, 436–439
Carbon dioxide, 428, 429, 448
CARIS, 485
Cartesian
  basis states, 22
  coherency matrix, 63–65
  complex-plane representation of polarization, 28–39
  Jones matrix, 80
  Jones vector, 22, 81
Cascade of optical systems, 70
Cells, biological, 469–474
Cell coat, 469–474
Cell culture, 446, 469
Cell membrane, 470
Cell windows, 187, 192, 195–197, 199, 205–207, 209, 233, 388, 391, 419
Cerium, 419
Chemisorption, *see* Adsorption
Cholesteric liquid crystal, 131
Chopper, 403
Chromium, 429, 430
Circles
  constant amplitude ratio, 31
  equal intensity transmittance, 105
  equal nearness, 115
  equi-azimuth, 35, 38, 45, 48, 50
  equi-ellipticity, 36, 38, 42, 45, 48, 50
  optical system, 242

SUBJECT INDEX

steepest variation of transmittance, 107
Circles, coaxial families of, 38, 45, 105, 107, 115
Circle-to-circle mapping, 34, 45, 48, 88–89, 92, 93, 177, 241
Circular
  coherency matrix, 63–65
  complex plane representation, 39–43
  ideal polarizer, 100, 490
  Jones matrix, 81–84, 181, 488–490
  Jones vector, 22, 82, 83, 180
  partial polarizer, 100, 489
  polarization, 9
  retarder, 488
Classification
  of optical devices, 97–103
  of automatic ellipsometers, 405, 406
Clausius–Claperyon relation, 441
Cleaning, surface, 476
Cleanliness, surface, 421
Coherence, 53
Coherency matrix
  Cartesian, 63–65
  circular, 63–65
  intensity transmittance in terms of, 145
  partially polarized light, 64–65
  relationship to the Stokes vector, 62
  totally polarized light, 64
  transformation by change of basis states, 62
  transformation by coordinate rotation, 61
  transformation by an optical system, 142, 143
  unpolarized light, 64
Coherency-matrix formulation, 141–147
Coherency vector, 63
Collimating lenses, 374
Collimation of light, 400
Collodion, 429
Colloidal suspensions, 156
Compensation, conditions of, 238–245
  fixed-analyzer nulling scheme, 239–240
  fixed-compensator nulling scheme, 240–242
  fixed-polarizer nulling scheme, 242–245
Compensator, *see* Retarder
  Babinet–Soleil, 371, 372
  imperfections, 187, 195, 198, 200, 206–210, 212, 221, 224, 231, 389, 390
Complex amplitude

of an elliptic vibration, 95, 96
evolution in anisotropic media, 137, 138
reflection coefficients, 272, 285
transfer function, 94, 96, 98
transmission coefficients, 272, 285
Complex exponential function, 125–129, 290
Complex number
  representation of time-harmonic scalars, 10
  representation of polarization states, 28, 160
Complex optical density function, 431
Complex plane *of polarization*
  Cartesian, 28–39, 85, 89, 93
  circular, 39–43
  conditions of compensation, 240–245
  contours of equal phase shift, 118
  contours of equal transmittance, 103–105
  domains, 39
  equi-azimuth contours, 33–35, 42, 236
  equi-ellipticity contours, 33, 36–37, 42, 236
  generalized, 43–47, 184
  paths of steepest change in transmittance, 107
  points of simply related polarizations, 40, 41
  polarization transformation by an optical system in the, 85–91, 241–243
  relationship to the Poincaré sphere, 48, 92
  resolution of an arbitrary polarization into two orthogonal states, 113
  symmetry operations, 40, 115
  trajectories for light propagation in anisotropic media, 124–130, 133–136
Complex plane *of the ellipsometric function*
  constant-angle-of-incidence contours (polar curve), 289–293
  constant-thickness contours, 293–298
Complex vector, 10
Computers
  in ellipsometric data reduction, 315
  interfaced with ellipsometers, 408, 413, 414
Components, optical (devices), 66
  ellipsometer, 366–376
  imperfections, 186–189, 195–233, 388–397
Condensed gases, 428–429

Conformal transformation, 90, 96
Contact potential, 436, 446
Contamination films, 418
Convergence
  in numerical inversion, 324
  to the null, 386–388
Coordinate axes
  choice of, 68, 69
  rotation, 19, 47, 79, 150
Copper, 449–451
Correction factors
  four-zone, 210, 227
  two-zone, 208, 209, 227, 392
Correction of errors, 386–403
Correlation
  and coherence, 53
  parameter, 321–324
Corrosion, 452, 457–460, 462, 463
Coupling coefficients
  2×2 arrays, 188, 196–200, 202
  4×4 arrays, 219, 221, 222
  azimuth-angle error, 188, 193, 194
  component imperfection, 188, 219
  tables, 194, 199–200, 202, 221, 222
  for film-thickness changes, 303
  minimization, 211
  modified, 189
Critical angle, 276, 280, 312
Cross-correlation function, 61
Crystals, *see* Anisotropic crystals
  Ag 441, Cu 449, GaAs 444, GaSe 424, Pb 446, Si 423, 434, U 448
Cytopathic effect, 472

Damage, surface
  mechanical, 418, 477, 478
  by high-energy ion bombardment, 473
  by low-energy ion bombardment, 473, 476
  radiation, 473
Dangling bonds, 440
Degenerate anisotropy, 88
Degree of polarization, 59, 63, 220
  change by an optical system, 141, 142
  in terms of the coherency matrix, 63
  at the output of a polarizer, 220
  in terms of the Stokes vector, 59, 220
Delta, *see* Ellipsometric parameters
Depolarization, 141, 148
  ellipsometer component, 229, 230
  Mueller matrix elements that cause, 230
  as a source of error in ellipsometry, 213, 229, 398
Depolarizer, 373
Depolarizing optical systems, 148–152
Desorption, 434
Devices, optical
  classification, 97–103
  Jones matrices, *see* Jones matrix
  Mueller matrices, *see* Mueller matrix
  of an optical system, 66
  reflection-type, 77–79
  transmission-type, 72–77
Derivative
  bilinear transformation, 91
  Jones matrix and Jones vector, 121
  Mueller matrix and Stokes vector, 215–217
Diagonal matrix, 83, 158, 164
  elements of the Hessian matrix, 342
Dichroic, *see* Dichroism
  anisotropic medium, 75, 140, 174
  crystal, 369
  polarizer, 76, 369
  retarder, 76
Dichroism
  apparent, 371
  circular, 140, 155, 158, 183
  elliptical, 128, 130, 140, 155
  induced, 369
  linear, 75, 76, 140, 155, 158, 174
  natural, 155, 369
Dielectric function, 418, 420–422, 428, 429
Differential equation
  complex amplitude, 137
  generalized field vector, 343
  Jones vector, 121
  polarization ellipse, 123
  Ricatti, 123
  wave, 343
Differential propagation matrix, 121, 343–347
Differential reflection spectroscopy, 303
Diffraction
  electron, 441, 446, 482
  grating, 361
  integral, 362
  X-ray, 447, 450
Diffusion, 484, 486
Discontinuous films, 359

SUBJECT INDEX

Dynamic range, 409

Eigenpolarizations
  as basis for classifying optical devices, 79, 97
  circular, 158, 180
  degenerate anisotropy, 88, 102
  elliptic, 158, 184
  fast and slow (retarder) 99, 488
  homogeneous anisotropic medium, 125
  linear, 158, 159
  low- and high-absorption, 100, 127, 129, 130
  optical rotator, 83
  optical system, 87, 97, 98, 487–490
  reflection, 173, 197, 272
  transmitted and extinguished (polarizer), 101, 490
Eigenvalues
  anisotropic medium, 138, 139, 347
  ideal polarizer, 101, 367
  isotropic reflector, 173, 272
  optical system, 97, 98, 487
  partial polarizer, 100, 367
  retarder, 98, 99
Eigenvalue equation
  Differential propagation 4×4 matrix, 347
  Jones matrix, 97, 98
Eigenvalues, measurement of the ratio of (associated with)
  circular eigenpolarizations, 180–184
  elliptic eigenpolarizations, 184–186
  known eigenpolarizations, 158
  linear eigenpolarizations, 159–180
Eigenvectors
  anisotropic medium, 347
  Jones matrix, 98
Eigenwaves, 138, 139, 345
Electric field (vector), 1
  in anodic films, 454–456, 460
  elliptic vibration, 2–8
  in immunoelectroadsorption, 468
  modulation, 265, 426
  monochromatic wave, 2
  quasi-monochromatic wave, 52
  reflection coefficients, 272
Electrical networks, two-port, 91
Electrochemistry, 452–464
Electrodes, 426, 452–453, 455, 458
Electrode/electrolyte interface, 452–453

Electromagnetic field, 1, 2, 341
Electromagnetic waves, 1, 2, 269–363
Electro-modulated ellipsometry, 426–428
Electron diffraction, 441, 446, 482
Electron microscopy, 452, 472
Electro-polishing, 460–462, 464
Electro-optic
  modulators, 406, 407
  self-nulled ellipsometer, 410–411
  birefringence, 406, 410
  coefficient, 406
  effect in anodic films, 481–483
Electro-strictive effect
  in anodic films, 481–483
Ellipse of polarization, 2–8
  differential equation, 123
  evolution in an anisotropic medium, 122–137
  along the helical axis of a cholesteric liquid crystal, 131–137
  representation by a complex number, 28–47
  representation on the Poincaré sphere, 47–52
  short-term and long-term, 53
Ellipsometer
  alignment, 376–379
  analysis of light propagation in an, 160–165, 170–172
  astronomical, 154
  based on azimuth measurement, 248–251
  calibration, 380–385
  without a compensator, 248–255
  construction, 364–366
  correction of errors, 386–403
  dynamic photometric, 257–262, 411–416
  general arrangement, 154
  imperfections, 386–403
  interferometric, 262–265, 406
  multiple-reflection, 252–255
  null, 166–247, 364–411
  nulling schemes, 238–245, 385–388
  optical devices, 366–376
  origin of the term, 156
  photometric, 165, 247, 255–362
  as a polarization analyzer, 154
  polarization-modulated, 260–262, 415–416
  polarizer–compensator–sample–analyzer (PCSA), 159–169, 211–213, 234–246, 366

Ellipsometer (*cont.*)
  PSA, 254, 296
  PSCA, 169–173, 212–213, 246–247, 366
  PSWSW'A, 192–210, 216–233
  precision, 403–405
  return-path (principle-angle), 253–255, 419
  rotating-analyzer, 257–260, 411–414, 420
  self-nulled, 405–411
  static photometric, 255–257
  unified treatment, 175–180

Ellipsometry
  applications, 417–486
  astronomical, 154
  definition, 153
  generalized, 86, 156–157, 233–247, 428
  with imperfect components, 186–233
  instrumentation and techniques, 364–416
  interferometric, 262–265
  inversion of the equations of, 315–332
  linear approximations, 301–315
  modulated, *see* Modulated ellipsometry
  multiple-angle-of-incidence, 317, 320–332
  null, 166–255, 364–411
  of optical systems, 156–186, 233–247
  photometric, 165, 255–262, 411–416
  and polarimetry, 153
  principal-angle (return-path), 253–255, 419, 477
  reflection (surface), 155, 173–174, 233, 251–255, 262–268, 418–486
  scattering, 155, 156
  spectroscopic, 269, 371, 414, 417–433
  total reflection, 314
  transmission, 155–156, 173–174, 274, 287–288, 299–301, 304–308, 311–314

Ellipsometric measurement
  effect of azimuth errors on, 186, 193, 194, 206, 207, 209, 215, 388
  effect of component imperfections on, 186–189, 195–233, 388–397
  of the Jones matrix, 153, 156–157, 233–247
  of the Mueller matrix, 153
  of the polarization transfer function, 233–247
  of the ratio of eigenvalues, 159–186

Ellipsometric parameters (angles) – psi and delta
  ambient–film–substrate system, 287–288, 296, 299
  corrected, 202–208
  errors, 189, 202–210, 211–222
  interface, 274, 277–278, 280, 282
  reflection, 174

Elliptic (or Elliptical)
  analyzer, 172
  birefringence, 128–130, 140, 155
  dichroism, 128, 130, 140, 155
  eigenpolarizations, 158, 184
  ideal polarizer, 101, 172, 200, 490
  partial polarizer, 100, 489
  retarder, 98, 488
  vibration, 2–8

Ellipticity, 7, 8
  angle, 7, 8
  invariant, 88, 89, 93
  sign, 8

Elovich equation, 443
Epitaxial films, 486
Equi-azimuth contours, 35, 38, 45, 48, 50
Equi-ellipticity contours, 36, 38, 42, 45, 48, 50

Error
  azimuth-angle, 186, 187, 193, 194, 206, 207, 209, 215, 388
  component-imperfection, 186–189, 195–233, 388–397
  correction, 386–403
  function, 317–319
  model, 316, 402–403
  random, 329
  sources of, 386–403
  systematic, 329

Extinction coefficient, 75, 270
  of the eigenwaves, 139
  ordinary and extraordinary, 75, 426
  profile, 480
  spectra, 419, 423, 425–427, 430, 432, 477

Extinction ratio, 367–369
Extraordinary refractive index, 73, 174, 368, 426
  extinction coefficient, 75, 426
Eye, 366
Eyepiece, Gauss, 375, 377

Faraday
  cells, 404, 408
  effect, 155, 158, 248, 345
  law, 454

SUBJECT INDEX

modulators, 404, 405
rotation, 346
rotators, 404, 408
Fatigue, 478
Feedback, 407, 409, 410
Fiber optics, 412
Fibrin, 465, 466
Fibrinogen, 465, 466
Film (layer)
    adsorbed on a surface, see Adsorption
    anisotropic, 340–358
    anodic-oxide, 453–460
    anodization, 456–460
    antigen–antibody reactions, 466–469
    biaxial, 357
    condensed-gas, 428–429
    contamination, 418
    contours of constant-thickness, 293–298
    discontinuous, 359–360
    electropolishing, 460–462, 464
    epitaxial, 486
    evaporated, 416, 418–420
    growth kinetics, see kinetics
    inhomogeneous, 340, 442
    initial growth, 304, 312–315
    ionic, 452–453
    Langmuir–Blodgett, 356, 467
    molecular, 356, 436
    microexudate, 470
    optical properties, 428–433, 462
    oxidation, 448–449, 456–460
    oxide, 443–452
    passivation, 457
    phase thickness, 284
    polish, 477
    reflection, see Reflection
    refractive index and extinction coefficient, see Optical properties
    state of aggregation, 418
    spectroscopy, 429–433
    stratified anisotropic, 340–358
    stratified isotropic, 332–340
    submonolayer, 360, 314
    -substrate polarizer, 293, 297, 370
    -substrate retarder, 297
    -substrate system, 283–332
    thickness measurement, 299, 316, 433–486
    thickness period, 284
    transmission, see Transmission
    uniaxial, 356, 358, 360, 482
Filter
    low pass, 415, 416
    Lyot, 66
    polarization, 234–239
    Šolc, 66
    spatial, 374
Free electron theory, 421
Frequency, 12, 13
Fresnel
    coefficients, 272–281
    rhomb, 79, 280, 372
    theory of optical activity, 84
Function
    amplitude transfer, 94–97
    complex amplitude transfer, 94, 96
    complex dielectric, 418, 428
    complex exponential, 125–129, 290
    complex optical density, 431
    complex sensitivity, 303–305
    cross-correlation, 61
    delta sensitivity, 303–305
    ellipsometric, 289–315
    error, 317–319
    nearness, 113–117, 246
    phase transfer, 94–97, 117–119
    polarization transfer, 84–85, 96, 97, 175
    psi sensitivity, 305–306
Fusion, cell, 472

Gallium arsenide, 445
Gallium selenide, 426
Gauss eyepiece, 375, 377
Gaussian distribution, 400, 477
General ellipsometer arrangement, 154
Generalized
    basis states, 43
    basis vectors, 25
    complex plane representation, 43–47, 184
    complex polarization number, 44
    field vector, 343
    Jones matrix, 81
    Jones-matrix formulation, 81
    Jones vector, 25, 81
Generalized ellipsometry, 86, 156–157, 233–247, 428
Generalized Law of Malus, 110–111
Geometric series, 285
Germanium, 423–425
Glan–Foucault prism, 367

Glan–Thompson prism, 78, 368
Glass, 275–277, 281, 282, 308–309, 312–314, 477
Gold, 279–280, 308, 313, 314, 387, 388, 421, 431, 432
Gradient, 107
    operator, 321
Grating, diffraction, 361
    Moiré, 403

Hageman factor, 466
Half-shade devices, 366, 372, 405
Half-wave retarder, 370
Handedness, 8
Helical axis, 131
    structure, 131, 158
Helium, 428
Hermitian matrix, 100
Hessian matrix, 324
Hyperbola, 238

Ideal
    compensator, 167
    component, 189, 199
    ellipsometer, 214
    polarizer (analyzer), 101, 172, 200, 213, 490
    zone relations, 204
Imaging optical systems, 66
Immunoelectroadsorption test, 468–469
Immunological reactions, 464–466, 468–469
Imperfect component, 189, 199
    see Imperfections
Imperfections, component, 186–189, 195–233, 388–397
Imperfection plate, 189–191
Implantation, ion, 476
Incoherent effects, 213–233
Industrial applications, 485
Infrared ellipsometer, 486
    polarizers, 369, 370, 486
Initial conditions, 124
    growth of thin films, 304, 312–315
In situ, 416, 433, 442, 452, 453, 455, 460
Instrumentation of ellipsometry, 364–416
Interface
    ambient/substrate, 306, 418
    gas/solid, 275–282, 433–437, 439–442
    electrode/electrolyte, 452–453
    liquid/solid, 433, 442, 443, 449

    silicon oxide/silicon, 481
    air/liquid, 436–439
Interference in thin films, 283
Interferometers, 61, 157, 263
Interferometric ellipsometers, 263–265
Intensity, 16, 21, 24, 26
Intensity transmittance
    condition for constant, 111, 112
    expressions for the, 107, 109, 144–147, 151
    loci of equal, 103–105
    maximum and minimum, 104, 105
    of optical systems, 103–113, 144–147, 151
    paths of steepest change of, 107
    polarization states of maximum and minimum, 106
Invariant
    azimuth states, 88, 89, 93
    ellipticity states, 88, 89, 93
    intensity, 21, 24, 26
    intensity transmittance, 110
    nearness function, 114, 115
    points of the bilinear transformation, 87
Inversion of the equations of ellipsometry, 315–332
In vitro, 469, 471–472
In vivo, 464, 466
Ion
    adsorption, 452–453
    bombardment, 473–480
    implantation, 476
Ionic conduction, 453–455, 457–458
Iron, 459, 460, 462, 463
Isosteric heat, adsorption, 442
Isotherms, adsorption, 434, 435, 441
Isotropic
    interface, 78, 270
    stratified media, 332–340

Jones matrix
    of active optical systems, 111
    of basic optical devices, 72–79
    and the bilinear transformation, 85
    Cartesian, 80, 488, 490
    of a cascade of optical systems, 70–72
    circular, 81–84, 181, 488–490
    differential propagation, 121, 139–141
    in terms of the eigenpolarizations and eigenvalues, 98, 487
    eigenvalues, 98, 487

eigenvalue equation, 97, 98
eigenvectors, 98
ellipsometer, 165, 170, 171
  of the ellipsometer components, 160
  of elliptic, linear and circular ideal polarizers, 101–102, 440
  of elliptic, linear and circular partial polarizers, 100–101, 489
  of elliptic, linear and circular retarders, 98–100, 488
  generalized, 81
  Hermitian, 100
  imperfect component, 187, 190
  imperfection plate, 191
  intensity transmittance in terms of, 103
  from the mapping of three polarizations, 157
  measurement, 153, 156
  Mueller matrix, relation to, 148, 149, 224, 232
  normalized, 157, 490
  operating on the coherence matrix, 142–145
  operating on the Jones vector, 69
  optical system, 69
  perfect polarizer, 112
  of reflection-type devices, 77–79
  significance of the elements of, 69–70
  singular, 86, 91, 101, 112
  tables, 488–490
  transformation by change of basis, 81, 82
  transformation by coordinate rotation, 79–80
  of transmission-type devices, 72–77
  of transposed optical components, 191–192
  unitary, 99
Jones-matrix formulation, 67–85
  application to ellipsometer analysis, 160–165, 170–172
  Cartesian, 67–80
  circular and generalized, 81–85
  separability of amplitude and phase in, 94–103
  separability of polarization in, 84–85
Jones eigenvectors, 98
Jones vector
  basis, 22
  Cartesian, 22, 81
  circular, 22, 82, 83, 180

  in terms of complex amplitude and complex polarization variables, 95
  differential equation, 121
  generalized, 25, 81
  of a given elliptical state, 26–28
  normalized, 17
  orthogonal and orthonormal, 19, 32, 95
  quasi-monochromatic wave, 53, 142
  of some states of polarization, 17
  TE plane wave, 13
  transformation by change of basis, 24, 25
  transformation by coordinate rotation, 19, 80
  transformation by an optical system, 69, 95, 142

KDP crystals, 410
Kinetics
  anodization, 454, 459
  cell-coat synthesis, 471–474
  chemisorption, 436
  diffusion, 484, 486
  oxidation, 443–452
  passivation, 462–463
  photolysis, 484–485
  polymer adsorption, 442–443
Kramers–Kronig relations, 369, 418, 426
Krypton, 428–429, 441–443

Langmuir–Blodgett films, 356, 467
Langmuir trough, 436
Lanthanum, 419
Lasers
  ellipsometer alignment using, 378, 379
  He-Ne, 375
  as light sources, 375–376
  tunable, 376
  water-vapor, 386
Lateral deviation, 368
Lauric acid, 436–439
Layer, see Film
Lead, 446–447
Lead chloride, 483–485
Least squares, 318–319, 321, 324, 329
Lemniscate of Bernoulli, 236
Lenses, collimating, 374
  focusing, 374, 401
Light, partially polarized
  coherency matrix, 64–65
  correlation between field components, 53

Light, partially polarized (*cont.*)
  cross-correlation function, 61
  degree of polarization, 59, 63, 220
  propagation through depolarizing optical systems, 148–152
  propagation through non-depolarizing optical systems, 141, 144, 146, 147
  representation on the Poincaré sphere, 59, 60
  representation in Stokes subspace, 59, 60
  resolution into totally polarized and unpolarized components, 58, 59, 64, 146
  Stokes vector, 59
Light, polarization-modulated, 90, 91, 248, 260, 415
Light, polarized, *see* Polarization
  propagation in anisotropic media, 119–141
  propagation in liquid crystals, 131–137
  propagation through optical systems, 66–152
  reflection and transmission, 269–363
Light, unpolarized (natural)
  coherency matrix, 64
  lack of correlation between field components, 53
  propagation through depolarizing optical systems, 148
  propagation through non-depolarizing optical systems, 141, 143, 144
  representation on the Poincaré sphere, 54
  representation in Stokes subspace, 59, 60
  resolution into orthogonal components, 145
  Stokes vector, 58
Light waves, polarization of, 1–65
Light sources, 375–376
Linear
  approximations, 301–315
  birefringence, 73, 74, 140, 155, 158, 174
  dichroism, 75, 76, 140, 155, 158, 183
  eigenpolarizations, 158, 159
  electro-optic effect, 406
  ideal polarizer, 76, 367, 490
  partial polarizer, 76, 489
  polarization, 9
  retarder, 73, 79, 98, 99, 488
Lipids, 470
Liquid crystals
  optical model, 131

privileged polarizations, 133
propagation of light in, 132–133
Ricatti equation, 132
Liquid
  helium, 428
  metals, 470
  nitrogen, 423
  –solid interface, 433, 442, 443, 449
Loci
  of equal nearness states, 115
  of equal phase shift states, 118
  of equal transmittance states, 103
  of equi-azimuth states, 35, 38, 42, 45, 48, 50
  of equi-ellipticity states, 36, 38, 42, 45, 48, 50
  of invariant-azimuth states, 88, 89, 93
  of invariant-ellipticity states, 88, 89, 93
Lock-in amplifiers, 403, 415
Logarithmic law of oxidation, 443–445
  spiral, 294, 302
Lorentz force, 1
Lorentz–Lorenz equation, 453
Low-energy electron diffraction, 441, 446, 482

Magnesium, 419
Magnesium fluoride, 308–312
Malus, law of, 110–111
Materials, *see* Optical properties
Matrix
  coherence, *see* Coherency matrix
  differential propagation, 121, 343–345
  differentiation, 121, 215–217
  direct product, 60, 148
  eigenvalue equation, 97, 98, 347
  Hermitian, 100
  Hermitian adjoint, 16
  interface, 334–336
  inverse, 21, 24, 26
  Jones, *see* Jones matrix
  layer, 334, 337, 345–352
  Mueller, *see* Mueller matrix
  rotation, 21, 61
  singular, 101, 112
  transformation, 24, 25, 44
  unitary, 24, 26, 99
Maxwell's equations, 341
Measurement

ellipsometric, *see* Ellipsometric measurement
film-thickness, 299, 316, 433–486
optical properties, 417–433
polarization, 153, 258
Medicine, applications of ellipsometry in, 464–473
Membrane, cell, 470
Mercury, 289, 376, 422
Metals
optical properties, 419–422
oxidation, 445–451
Metal–insulator–metal junctions, 478
Meteorology, 156
Method of swings, 386, 403
Mica, 371
Microelectronics, 486
Microexudate, 470
Microscope, ellipsometer laser, 486
Microscopy, electron, 452, 472
light, 469
Model errors, 316, 402–403
Modulated ellipsometer, polarization-, 260–262, 406, 415–416, 420
Modulated ellipsometry, 265–268
of gold and silver, 426–428
PCSA arrangement, 267–268
PSA arrangement, 266–267
Modulation, polarization, 260–262, 406, 410, 415–416
Modulators
ADP, 406
Faraday, 404
KDP, 410
piezobirefringence, 415
Moiré grating, 403
Molecular films, 356, 436
structure, 155
Monochromatic light, 2
Monochromator, 67, 414, 416
Monolayers, 435, 436, 467
Motors
computer-controlled, 408
servo, 407, 412
stepping, 408
Mueller matrix
first row, 214, 218
imperfect component, 216
intensity transmittance in terms of, 151
Jones matrix, relationship to, 148, 149, 224, 232
measurement, 153
operating on the Stokes vector, 149
of optical elements (table), 492
optical system, 149
train of devices, 150
transformation by coordinate rotation, 150
Mueller-matrix formulation, 148–152
in ellipsometry, 213–233
Multiple-angle-of-incidence ellipsometry, 317, 320–332
Multiple-null measurements, 245–247
Multiple reflections, 74, 252, 283, 371, 431

Nearness function
in the complex plane, 113–116
collective, 246
modified, 115, 116
on the Poincaré sphere, 116, 117
properties, 114
Neon, 429
Neper, 270
Networks, birefringent, 66
two-port, 91
Niobium, 453–455
Nitrogen, 421, 423
Noise, 376, 403
Non-depolarizing optical systems, 141–147
Normalized Jones vector, 17
Jones matrix, 157, 490
Null
convergence, 386–389
determination, 385–386
ellipsometer, 166–247, 364–411
ellipsometry, 166–255, 364–411
symmetry, 386, 400–401
Nulling angles, 166, 168, 204, 207, 215, 217, 226, 389
in conjugate zones, 168
ideal zone relations, 204, 217
in an imperfect ellipsometer, 207, 215, 226, 389
Nulling schemes, 167, 238, 385
fixed-analyzer, 239, 387
fixed-compensator, 167, 215, 240, 385, 387
fixed-polarizer, 242, 388

One-zone ellipsometry, 206, 207, 389

On-line data reduction, 408, 409, 411, 414
Optical
  activity, 77, 83, 345, 346, 371, 402
  components, 66, 366–376
  density function, 431
  devices, see Devices
  matrix, 341–342
  properties, see Optical properties
  rotation, 77, 78
  rotation tensors, 342
  rotator, 77, 79, 248
  rotatory power, 77, 155, 158
  system, see Optical system
Optical properties
  of biaxial crystals, 426–427
  of condensed gases, 428–433
  effect of adsorption on, 439–441
  effect of ion bombardment on, 477
  effect of polishing on, 477–478
  effect of temperature on, 425
  of materials in bulk, 417
  of materials in thin films, 428–433, 462
  of a uniaxial crystal, 424–426
Optical system
  active, 111–112
  amplitude transfer function, 94–97
  analogy with two-port electrical networks, 91
  bilinear transformation, 85
  cascade, 70
  in the coherency-matrix formulation, 141–143
  complex-amplitude transfer function, 94, 96, 98
  depolarizing, 141, 148
  eigenpolarizations, 87, 487
  eigenvalues, 97, 98, 487
  eigenvalue problem, 97
  with elliptic eigenpolarizations, 158, 184
  general classification, 97
  generalized Law of Malus, 110
  imaging, 66
  intensity transmittance, 103, 144, 145, 147, 151
  with isotropic attenuation or amplification, 111
  Jones eigenvectors, 98
  Jones matrix, 69
  with known eigenpolarizations, 158
  loci of equal transmittance states, 103
  loci of invariant-azimuth states, 88, 93
  loci of invariant-ellipticity states, 88, 93
  maximum and minimum transmittances, 105
  measurement of the Jones matrix of, 153, 156
  measurement of the Mueller matrix of, 153
  measurement of the ratio of eigenvalues of, 158–180
  Mueller matrix, 149
  non-depolarizing, 141
  with orthogonal circular eigenpolarizations, 158, 180–184
  with orthogonal linear eigenpolarizations, 158–180
  phase transfer function, 94–97, 117–119
  polarization mapping properties, 85–87, 92
  polarization transfer function, 84, 96, 97
  polarizing, 66
  response to polarization-modulated light, 90
  response to right- and left-circular polarization, 87
  simplified expressions for the intensity transmittance of, 107–109
  states of minimum and maximum transmittance, 106
Ordinary refractive index, 73, 174, 368, 426
  extinction coefficient, 75, 426
Orthogonal Jones vectors, 19, 32, 95
  polarization states, 30, 32
Orthonormal Jones vectors, 19, 95
Oscillator strength, 359
Oxidation
  anodic, 453–460
  atmospheric, 443–445
  in controlled atmospheres, 445–451
  of copper in water, 449–451
  dependence on crystal orientation, 445, 450
  dependence on pressure, 446–447
  dependence on temperature, 447
  kinetics, 443–452
  of metals (Ti, U, Pb, Mo), 445–451
  of semiconductors (Si, Ge, GaSe), 443–445
  of thin films, 448
Oxides, see Oxidation

SUBJECT INDEX 535

Oxygen, 429, 435–437, 439–442

Parameter correlation, 321–324
  space, 321
Parameters, Stokes, 55
Parasitic beams, 397–398
Partially polarized light, see Light
Partial polarizers, 100–101, 489
Passivation (passivity), 457–463
Paths of steepest change, 107
Pentane, 436–439
Perfect polarizer, 112, see Ideal polarizer
Phase
  of an elliptic vibration, 7, 8
  of a field component, 2
  sensitive detection, 403
  shifts upon reflection, 79, 273, 275–282, 287
  transfer function, 94–97, 117–119
  velocity, 12
Phasor, 10–11
Phosphorous, 479–481
Photodetectors, 376, 401
Photodiodes, 376
Photoelasticity, 155
Photolithography, 486
Photolysis, 483–485
Photometric ellipsometers, 165, 247, 255–362
  dynamic, 257–262, 411–416
  polarization-modulated, 260–262, 415–416, 420
  rotating-analyzer, 257–260, 411–414
  static, 255–257
Photometric ellipsometry, 165, 255–262, 411–416
Physical adsorption (or Physisorption), see Adsorption
Pinhole, 374
Plastic, 484
Platinum, 455–457
Poincaré sphere, 47–50
  circle-to-circle mapping by an optical system on the, 92
  relationship to the complex plane, 48, 49, 92
  relationship to the Stokes parameters, 51, 59
  representation of light propagation in anisotropic media and liquid crystals, 127, 128, 134
  representation of parasitic beams in an ellipsometer, 397
  representation of totally, partially and un-polarized light, 54, 55
  resolution of an arbitrary state into two orthogonal states on the, 116–117
Polar curves, 291, 292, 300, 301
Polarimetry, 153, 155, 156
Polarization
  azimuth, 7
  basis states, 22, 180, 184
  circular, 9
  concept of, 1
  degree of, 59, 63, 220
  ellipse, 2–8
  elliptical, 6–8
  ellipticity, 8
  ellipticity angle, 7, 8
  filter, 234–235
  handedness, 8
  linear, 9
  measurement, 153, 258
  modulation, 260–262, 406, 410, 415–416
  monochromatic wave, 2–8
  orthogonal states of, 30
  partial, 53
  quasi-monochromatic wave, 52–65
  representation by a coherency matrix, 60
  representation by complex numbers, 28–47
  representation by Jones vectors, 13–15
  representation on the Poincaré sphere, 47–50
  representation by a Stokes vector, 51, 55–57
  total, 53
  transfer function, 84–85, 96, 97, 175
Polarization modulation, 260–262, 406, 410, 415–416
Polarized light, see Light and Polarization
Polarizer
  acceptance angle, 368
  angular deviation, 368–394
  birefringent, 367
  Brewster-angle reflection, 79, 276
  circular ideal, 490
  circular partial, 100, 489
  depolarization, 223
  dichroic, 369

Polarizer (cont.)
  eigenpolarizations, 100, 367, 489, 490
  eigenvalues, 100, 367
  elliptic ideal, 101, 172, 200, 490
  elliptic partial, 100, 489
  extinction ratio, 367–369
  extinction and transmission axes, 76
  film–substrate reflection, 293, 297, 370
  Glan–Faucault prism, 367
  Glan–Thompson prism, 78, 368
  imperfect, 187, 200, 213, 219, 367
  infrared, 369, 370, 486
  Jones matrices, 100, 101, 489–490
  lateral deviation, 368
  linear ideal, 76, 102, 367, 490
  linear partial, 76, 489
  Mueller matrices, 491, 492
  perfect (ideal), 112
  wire-grid, 369, 486
  Wollaston prism, 263
Polarizing angle(s)
  of an ambient–film–substrate system, 293, 297
  of an interface, 276
Polarizing optical systems, 66
Polaroid, 76, 369
Polishing films, 477–478
Polybutadiene, 486
Polychromatic waves, 2
Polykaryocytes, 472
Polymer, 442–443
p-polarization, 78, 272
Precision of null ellipsometers, 403–405
Principal angle(s)
  ambient–film–substrate system, 297
  interface, 254, 282
Principal-angle ellipsometry, 253–255, 419, 477
Principal frame, 160, 219, 222
  complex refractive indices, 424, 425
Prosthetic implants, 465
Proteins, 464–470
Psi, *see* Ellipsometric parameters

Quarter-wave retarder, 370
Quartz, 371, 384, 415
Quasi-monochromatic waves, 2, 52–65

Radiation damage, 423
Reciprocity, 212

Reflectance, 274, 276, 278, 279, 281, 318
Reflection
  by an ambient–film–substrate system, 283–288
  by an anistropic stratified structure, 340–358
  by an anisotropic substrate, 352–354
  by a biaxial film on an isotropic substrate, 357–358
  by a biaxial substrate, 355–356
  change of polarization by, 79, 154, 273, 287
  devices, 77–79
  eigenpolarizations (p, s), 272
  ellipsometric angles (ψ, Δ), 174, 274
  ellipsometry, 155, 173, 233, 251–255, 262–268, 418–486
  by a finite anisotropic layer, 348–352
  by an interface, 76, 154, 270–283
  by an isotropic stratified structure, 332–340
  Jones matrix, 78
  multiple, 74, 252, 283, 371, 431
  partial polarizer, 79
  phase shifts, 79, 273, 275–282, 287
  of polarized light, 269–363
  polarizers, 79, 276, 293, 297, 370
  retarder, 79, 280, 297
  by rough surfaces, 361–363
  by surfaces covered with discontinuous films, 359–360
  by a uniaxial film on an isotropic substrate, 356–357
  by a uniaxial film on a uniaxial substrate, 358
  by a uniaxial substrate, 354–355
Reflection coefficients
  ambient–film–substrate, 285
  ambient–substrate (Fresnel), 272
  ambient–two film–substrate, 340
  stratified structure, 339
Reflection matrix
  anisotropic layer, 351–352
  isotropic ambient–uniaxial substrate, 354–355
  isotropic reflector, 77
Refraction
  angle of, 271
  change of polarization by, 154, 273
  double, 367

SUBJECT INDEX

Refractive index
  apparent, 361–363
  average, 77, 83
  complex, 75, 270
  effective, 359
  extraordinary, 73, 174, 368
  generalized complex (anisotropic medium), 139
  of an optically active medium, 83, 183
  measurement, 274, 299
  profile, 479
  spectra, 419, 423–427, 430, 431, 477
Resolution
  of an arbitrary polarization into two orthogonal states, 113, 116–117
  of the electric vector into three vibrations, 2–4
  of partially polarized light into totally polarized and unpolarized components 58, 59, 64, 164
  of unpolarized light into orthogonal components, 145
Retardance (retardation), 74, 99, 371
Retarder
  achromatic, 372
  Babinet–Soleil, 371–372
  birefringent (linear), 73, 370
  calcite, 371
  eigenpolarizations (fast and slow), 99
  eigenvalues, 99
  electro-optic, 410
  elliptic, 98, 488
  film–substrate reflection, 297
  Fresnel-rhomb, 79, 280, 372
  half-wave, 370
  imperfect, 187, 212, 232
  Jones matrices, 99, 488
  linearly dichroic, 76
  mica, 371
  Mueller matrix, 492
  multiple reflections, 74, 371
  quarter-wave, 370
  quartz, 371
  retardance (retardation), 74, 99, 371
  tunable, 371–372, 410
Ricatti equation, 123
Rotating-analyzer ellipsometer, 257–260, 411–414
Rotation, *see* Optical rotation
Rotator, optical, 77, 79, 248

Scattering ellipsometry, 155, 156
  matrix, 334, 338–340
Segregation of phosphorous, 481
Self-nulled ellipsometers, 405–411
Semiconductor, 275, 443–445, 485
Sensitivity factors (functions)
  complex reflection and transmission, 303–305
  delta and psi, 305–306
  as functions of angle of incidence, 306–312
Separability of information on the ellipse of polarization, 84–85
Separability of information on the amplitude and phase, 94–97
Silica, 473
Silicon, 277–278, 289, 292, 314, 376, 425, 434–437, 439–440, 485–486
  nitride, 456
  oxide, 444, 456, 458–460, 481–482, 485–486
Silicon oxide–silicon system, 289, 291, 326–332
Silver, 420, 428, 450, 451
  sulfide, 450–451
Simulation, 91
Singular Jones matrix, 101, 112
Slew rate, 405
Snell's law, 271, 286, 348
Sources of error, 386–403
Sources, light, 375–376
Source polarization, 398–400
Spectra
  absorbance, 433
  absorption, 430
  complex dielectric function, 420–422
  complex optical density function, 432
  complex refractive index, 419, 423–427, 430–432
  extinction coefficient, 419, 423, 425–427, 430, 432
  modulated ellipsometry, 427–428
  refractive index, 419, 423–427, 430, 431
Spectral distribution, 2, 66
Spectrometers, 67
Spectroscopic ellipsometry, 269, 371, 414, 417–433
Spectroscopy, differential reflection, 303
Spectrum, *see* Spectra
s-polarization, 78, 272

Sputtering yield, 476
Static photometric ellipsometers, 255–257
Stearic acid, 439
Stereographic projection, 48, 49, 92, 93, 117, 124, 134
Sticking probability, 436–437
Stokes parameters, 55
Stokes vector
  degree of polarizations in terms of, 59, 220
  derivative, 215, 217
  in ellipsometer analysis, 213–233
  measurement, 57, 153
  partially polarized light, 58
  perturbation, 214
  relationship to the coherency matrix, 62
  relationship to the Poincaré sphere, 51, 59
  totally polarized light, 58
  unpolarized light, 58
Stratified structure, 269–358
  anisotropic, 340–358
  isotropic, 332–340
Strontium, 419
Submicroscopic cracks, 478
Submonolayer coverage (on film), 360, 435–437
Substrate, 417
Sulfide, silver, 450–451
Surface
  cleaning, 476
  coverage, 360, 435, 437
  covered with discontinuous films, 359
  damage, *see* Damage
  hydrophilic and hydrophobic, 434, 465
  roughness, 361, 478
  states, 418, 441
  temperature, 476
  tension, 436
  thermodynamics, 441–442
System, *see* Optical system
Systematic errors, *see* Errors
Synthesis, cell-coat, 471–474
  of optical systems, 87

Tapered film, 481
Temperature, effect of
  on anodic oxidation, 455
  on cell-coat synthesis, 472
  on cleaning of silicon surfaces, 476
  on optical properties, 425
  on oxidation kinetics, 447
Temperature, measurement by ellipsometry, 476
Tetraphenylporphine (TPP), 340
Thermodynamics, surface, 441–442
Thin films, *see* Films
Time dependence, choice of, 11, 270
Tissue culture, 446, 469
Totally polarized light, 53, *see* Light
Total reflection, 276, 280–282, 314
Trajectories of the state of polarization, 125–130
Transfer function
  amplitude, 94–97
  complex-amplitude, 94, 96
  modulation, 85
  phase, 94–97, 117–119
  polarization, 84–85, 96, 97, 175
Transformation, *see* Bilinear, Jones matrix, Jones vector, Mueller matrix *and* Stokes vector
Transmission
  by an ambient–film–substrate system, 283–288
  by anisotropic stratified structures, 340–358
  by an anisotropic substrate, 352–354
  change of polarization by, 154, 273, 287
  devices, 72–77
  ellipsometry, 155–156, 173, 288, 299
  by a finite anisotropic layer, 348–352
  by isotropic stratified structures, 332–340
  of polarized light, 269–363
Transmission coefficients
  ambient–film–substrate, 285, 287
  ambient–substrate (Fresnel), 272–273
  ratio, 299–302
Transmission matrix
  anisotropic layer, 351–352
  isotropic ambient–anisotropic substrate, 354–355
Transmittance, *see* Intensity transmittance
  interface, 274
Tunneling junctions, 481
Tunneling resistance, 481
Two-zone
  averages, 208–209, 226, 228, 231–233, 389
  correction factors, 208–209, 227, 392
  residuals, 390

Uniaxial
  crystals, 370, 424
  films, 356, 358, 360, 482
  substrates, 354, 358
Unified treatment of two ellipsometer arrangements, 175–180
Unitary matrix, 24, 26, 99
  transformation, 61
Unpolarized light, see Light
Uranium, 448

Vacuum, 418, 419, 435, 441, 445
VAMFO, 485
Vanadium pentoxide, 427
Vector
  basis, 22
  coherency, 63
  complex, 10
  electromagnetic field, 1, 341
  generalized field, 343
  Jones, see Jones vector
  Stokes, see Stokes vector
Verniers, 364, 366
Viral infection, 472–474
Viruses, 469, 472, 474

Wafer, silicon, 411, 486
Water, oxidation of Cu in, 448–451
Water vapor
  adsorption on Si surfaces, 434–435
  effect on oxidation, 448
  laser, 486
Wave
  absolute phase, 8
  amplitude, 8
  of arbitrary spatial structure, 2
  frequency, 12, 13
  monochromatic, 2
  polarization, see Polarization
  polychromatic, 2
  uniform TE plane, 11–13
Wavefront, 12, 400
Wavelength, 12
Wavevector, 12, 68
Wave velocity, 12, 270
Windows, cell, 187, 192, 195–197, 199, 205–207, 209, 233, 388, 391, 419
Wire-grid polarizer, 369, 486
Wollaston prism polarizer, 263

Xenon, 440
X-ray diffraction, 447, 450

Zone
  average, 208, 210, 226, 389
  conjugate, 168
  definition, 168, 178
  relations, 168, 172, 178, 179, 183, 204

DATE DUE

JUN 1 0 1988			
NOV - 5 '93			
May 16, 1994			

No. 370 Waverly Publishing Co.